AVIATION
WEATHER
HANDBOOK

Other McGraw-Hill Aviation Handbooks of Interest

Bjork • PILOTING BASICS HANDBOOK

Turner • INSTRUMENT FLYING HANDBOOK

AVIATION
WEATHER
HANDBOOK

Terry T. Lankford

McGRAW-HILL

New York   San Francisco   Washington, D.C.   Auckland   Bogotá
Caracas   Lisbon   London   Madrid   Mexico City   Milan
Montreal   New Delhi   San Juan   Singapore
Sydney   Tokyo   Toronto

Library of Congress Cataloging-in-Publication Data

Lankford, Terry T.
   Aviation weather handbook / Terry T. Lankford.
      p.   cm.
   Includes index.
   ISBN 0-07-136103-0
   1. Meteorology in aeronautics—Handbooks, manuals, etc. I. Title.
TL556.L355   2000
629.132'4—dc21   00-062497

McGraw-Hill

A Division of The McGraw·Hill Companies

1 2 3 4 5 6 7 8 9 0 DOC/DOC 0 6 5 4 3 2 1 0

ISBN 0-07-136103-0

The sponsoring editor for this book was Shelley Ingram Carr, the editing supervisor was Stephen M. Smith, and the production supervisor was Sherri Souffrance. It was set in Times Roman following the HB1 design by Paul Scozzari of McGraw-Hill's Hightstown, N.J., Professional Book Group composition unit.

Printed and bound by R. R. Donnelley & Sons Company.

This book is printed on recycled, acid-free paper containing a minimum of 50% recycled de-inked fiber.

CONTENTS

Chapter 9. Pilot Weather Reports

Chapter 10. Satellite Images

Chapter 11. Radar and Convective Analysis Charts

Chapter 12. Air Analysis Charts

Chapter 13. Introduction to Forecasts   13-1

Chapter 14. Weather Advisories   14-1

Chapter 15. Area Forecasts   15-1

Chapter 16. TWEB Route Forecasts   16-1

Chapter 17. Terminal Aerodrome Forecasts   17-1

Chapter 23. Turbulence   23-1

Chapter 24. Icing   24-1

Chapter 25. Thunderstorms   25-1

Chapter 26. Strategies for Weather Systems   26-1

Chapter 27. Risk Assessment and Management   27-1

Appendix A. Abbreviations   A-1

Appendix B. Graphs and Charts   B-1

PREFACE

Throughout the decade of the 1990s, report after report, both government and industry, have recommended improved aviation weather education for pilots, dispatchers, controllers, and forecasters—weather users and providers. On September 27, 1997, Federal Aviation Administration (FAA) Administrator Jane F. Garvey issued the FAA's Aviation Weather Policy statement. It read in part:

> The FAA is committed to improving the quality of aviation weather information and the application of that information by pilots, controllers, and dispatchers. The FAA acknowledges that training is a critical component of this objective, enabling the aviation community to make the best use of weather information to make sound operational decisions and to ensure safety and efficiency.

This policy was echoed by National Weather Service (NWS) Director John J. Kelly, Jr., in a letter written to the National Weather Association (NWA) in October 1998. Director Kelly supported the NWS's commitment to leadership and progress in aviation weather forecasting. He stated:

> Our aviation customers have expressed serious concerns about NWS aviation services.... New aviation products and services and improvements to existing products and services will be designed, validated and implemented.... Finally, I reaffirm the NWS dedication to the aviation program and I look forward to working with you to improve our aviation services.

The FAA, the NWS, and the National Aeronautics and Space Administration (NASA), along with industry groups, such as the Cooperative Program for Operational Meteorology, Education and Training (COMET), have spent millions of dollars on aviation safety programs. Unfortunately, very little has filtered down to the operational user or weather provider. To date, with the noted exceptions of NASA and the Aircraft Owners and Pilots Association's (AOPA) Air Safety Foundation, little has been accomplished. Although AOPA's efforts are commendable, pilot seminars and pamphlets have limited application.

Add to this that during the last decade there has been a revolution in aviation weather observing, reporting, and forecasting. The concurrent communications revolution has made new weather products available for both preflight and inflight use. With continuing flight service station (FSS) and NWS consolidation and the expansion of self-briefing systems, a pilot's ability to decode, translate, and interpret weather information has taken on even greater significance.

Over the years there has been no comprehensive source of weather information for pilots. Most weather texts have been dry and complex, and they often don't relate to the "real world." It's exasperating to read an article or book about aviation weather that espouses the need to understand meteorological terms, then fails to include a definition or explain the application of the concept.

Almost all aviation weather texts, both government and commercial, suffer from poor organization and omission of required subjects. Most, written by meteorologists, contain too much detail in climatology and general meteorological theory that does not relate to flight operations.

As pilots, we want to known how the weather affects our flying activity—the bottom line. This book is written for the pilot, by a pilot, from the pilot's point of view. It does not require the memorization of countless formulas, the use of calculus, or a degree in meteorology. Many complex phenomena are simplified, while still being explained in terms of their operational impact on aviation. (This, undoubtedly, will bring some consternation from the meteorological community. But our goal is operational, not theoretical.)

Weather affects a pilot's flying activity more than any other physical factor. Most pilots agree that weather is the most difficult and least understood subject in the training curriculum. Surveys indicate that many pilots are uneasy with, or even intimidated by, weather. In spite of these facts, or maybe because of them, weather training for pilots typically consists of bare bones, only enough to pass the written test, while weather-related fatal accident statistics remain relatively unchanged.

Air Force pilot training contains a mere 15 h of formal weather instruction, as compared to 50 or 60 h in the past. Navy pilot weather training has been reduced by 25 percent to 30 h, with no refresher. Army aviator weather training consists of about the same number of hours. Most civilian primary ground schools include a mere 9 h of weather instruction. Ironically, regulations perpetuate this trend. Student, private, recreational, and commercial airplane pilots are required only to obtain and use weather reports and forecasts, recognize critical weather situations, and estimate ground and flight visibility, although wind shear and wake turbulence avoidance have been added to the pre-solo requirement. An additional requirement of forecasting weather trends on the basis of information received and personal observations is required for an instrument rating. Only the airline transport pilot certificate requires the applicant to have any serious meteorological knowledge.

A major revision to the practical test standards in 1984 requires that applicants exhibit "knowledge of aviation weather information by obtaining, reading, and analyzing..." reports and forecasts and make "a competent go/no-go decision based on the available weather information." The practical test standards do reference AC 00-6, *Aviation Weather* (basic meteorological theory), and AC 00-45, *Aviation Weather Services* (a basic discussion of weather reports and forecasts), but neither reference relates weather to flight situations or to a go/no-go decision. Most flight tests are given during good weather, and pilots requesting check ride briefings all too often have little idea of the information required or the presentation format of the FSS briefer; this requirement would seem, then, to have little practical consequence.

This situation evolved because, when regulations were originally written, virtually all flying, except military and airline, was visual. The aircraft of the day had neither the performance nor the instruments to take on weather. When weather was encountered, pilots simply landed in the nearest field. Regulations are extremely difficult to change; consider, for example, the aeronautical experience of 40 h required for the private pilot. These hours have been woefully inadequate for years with the constant increase of additional requirements. Many of the people who write regulations and perform flight tests are former military pilots who have had minimal weather training, because after graduation they were under the direct control of older, more experienced pilots and so gained experience in a controlled environment.

Accident reports and commentaries frequently refer to a pilot's poor judgment—namely, the failure to reach a sound decision. Pilot judgment is based on training and experience. Training is knowledge imparted during certification, flight reviews, and seminars, and from literature; experience can best be defined as when the test comes before the lesson. Unfortunately, failure can be fatal. Pilot applicants have only their instructor to prepare them to make competent flight decisions.

Odds are little that judgment training occurred unless situations were actually encountered. General aviation training runs the gamut from the flight school that prohibits its instrument students and instructors from flying in the clouds to the instructor in a Cessna 182 who requested to be vectored into icing conditions to demonstrate the effects of ice on the airplane. (Neither instance seemingly exhibits sound judgment.) The fact remains that the least experienced pilots have minimal weather training. Following certification there is no requirement for additional or refresher weather instruction.

Various government and nongovernment publications—books, articles, pamphlets, and videos—have attempted to provide and improve weather education for pilots. Most books do not relate weather theory and phenomena to actual flight situations; discussions of weather reports, forecasts, and weather charts are limited to decoding, with little, if anything, said about interpretation and applications. Articles and pamphlets typically can address only one issue. The same may be said about videos. The AOPA Air Safety Foundation produced an informative and entertaining Tactical Weather Workshop program. Flight scenarios were based on a turbo, pressurized, radar- and Stormscope-equipped Cessna 210. Unfortunately, many pilots don't have the opportunity to fly this type of equipment. Flying sophisticated, well-equipped aircraft allows the pilot many more options when dealing with the weather. While most of the scenarios used in this book relate to non-turbo, single-engine aircraft, the principles apply to all.

The goal of this book is to provide you, the pilot, with a complete, comprehensive aviation weather resource. Included is information not only to prepare students to pass written and practical examinations but to prepare dispatchers as well as pilots—from student through airline transport, from recreational to biz-jet—to operate safely and efficiently within an ever increasingly complex environment.

Terry T. Lankford

INTRODUCTION

An essential part of flight preparation concerns the weather. No matter how short or simple the mission, regulations place the responsibility for flight planning on the pilot— not on the National Weather Service (NWS) forecaster or on the flight service station (FSS) briefer. To effectively use all available resources, a pilot must understand what weather information is available, how it is distributed, and how it can be applied to a flight situation.

Basic weather theory, weather reports, forecasts, and strategies are interrelated; therefore, it might be necessary for the reader to complete this book to realize the most thorough understanding of this often complex subject.

We've tried to organize subjects using the building block approach. Information from one section will be used to increase understanding of subsequent sections. Elements of one chapter are used as a foundation for the material in later chapters. In this way the reader is not overwhelmed by any section or chapter. Unfortunately, to accomplish this, we occasionally run into the "chicken or the egg" syndrome. We may have to introduce a topic, then provide details in subsequent sections. Because of this, the reader should not be overly concerned if any one element or concept is not completely clear on initial reading.

A pilot must have a basic understanding of weather phenomena and theory. This will provide a sound foundation for the novice and a practical review for the experienced pilot. Most weather is produced by the interaction of moisture, vertical motion, and stability. We'll indicate where information on specific meteorological phenomena can be located. For example, when we discuss frontal systems, we'll indicate that these weather patterns can be located on the surface analysis chart. This is accomplished in Part 1, Basic Weather Theory and Fundamentals.

Technical meteorological concepts and terms are translated into language any pilot can easily understand. On the other hand, such subjects as vorticity, microbursts, and upper-level weather systems will not be omitted because of complexity.

Pilots will be able to recognize atmospheric conditions conducive to icing, turbulence, thunderstorms, and other weather hazards. Student, recreational, private, and commercial pilots will develop abilities to recognize and avoid critical weather situations.

Chapter 1 begins with a description of the atmosphere. Since aircraft (we're not going to count the Space Shuttle) fly within only the lower two layers of the atmosphere, the troposphere and stratosphere, only these layers are discussed, along with their boundary, the tropopause. A detailed discussion of the jet stream describes this phenomenon. Since the influence of the jet stream on weather—both surface and aloft—has such significance, this subject will be revisited in subsequent chapters.

The chapter includes atmospheric properties that affect aviation weather. These include temperature, pressure, moisture, and density. Without moisture, we would not have weather as we know it. Along with moisture, atmospheric density is the key to understanding many weather phenomena; its importance cannot be overemphasized.

Chapter 2 discusses the affects of stability, or the lack thereof, in the atmosphere. This chapter contains a detailed section on lapse rate. Lapse rate is essential to an understanding of vertical motion. And vertical motion is a primary cause of weather. Included are the concepts of the level of free convection and lifted condensation level. Both are important to an understanding of cloud development and thunderstorms.

With an understanding of the atmosphere, its properties, and stability, we move on in Chap. 3 to pressure patterns and the wind. The wind is a mover of atmospheric properties from one location to another. These properties are modified by the terrain over which they pass. Wind and terrain are additional mechanisms that produce and modify weather. This chapter examines local winds, with emphasis on their effect on aviation operations.

The effects and significance of clouds and precipitation are presented in Chap. 4. The effects of terrain on the weather are included. I thought about including a cloud code chart. Since we are restricted to black-and-white photographs, and because significant cloud pictures are included in appropriate sections, I decided a separate cloud code chart would not be necessary. For those who would like an excellent, detailed, color cloud chart, "The Cloud Chart" is available through the National Weather Association, 6704 Wolke Court, Montgomery, AL 36116-2134.

Chap. 5 discusses air masses. Air masses have specific characteristics that affect aviation. These characteristics are modified by underlying terrain. A knowledge of these phenomena explains many weather phenomena and hazards. For example, high pressure usually means good weather, but not always. Moisture trapped under high-pressure inversions may create widespread "zero-zero" conditions that can last for weeks. Strong pressure gradients within high-pressure areas can cause severe turbulence.

Frontal systems, their formation and dissipation, have a major impact on flight operations. Chapter 6 covers these phenomena from their birth, through their evolution, to their dissipation. Most texts cover only classical fronts—predominately those east of the Rockies. Because of mountainous terrain in the west, a frontal surface may have a width of hundreds of miles. This chapter covers these phenomena throughout the United States and North America.

Chapter 7 is devoted to nonfrontal weather systems. An all-too-often common misconception is that in the absence of a front there is not significant weather. This is not the case. Troughs, ridges, and vorticity are major weather producers. A pilot's—even one who flies below 10,000 ft—weather education is woefully lacking without a sound knowledge of upper-level weather systems and their effect on surface weather. A front under a ridge aloft may produce no weather at all. On the other hand, a front supported by an upper trough and the jet stream produces the most severe weather anywhere. Finally, there is the hurricane, possibly the best example of a nonfrontal weather-producing system. We included a section called Putting It All Together. Here we combine the effects of our three-dimensional atmosphere. From this discussion it is clear how all atmospheric phenomena come together to produce our ever-changing weather.

At this point we move on to Part 2, Pilot Weather Resources and Services. Explanations of weather reports and forecasts go beyond decoding and translating, to interpreting and applying information to actual flight situations. Discussions include applying weather information to visual flight rules (VFR) as well as instrument flight rules (IFR) operations, and to high-level as well as low-level flights.

Pilots will learn to obtain weather reports and forecasts through the Federal Aviation Administration (FAA) and other sources and then interpret and apply the information to flight situations. They will understand the principles of forecasting and how to apply forecasts and observations to the flight environment. Pilots will also understand weather collection and distribution, symbols and abbreviations, charts and forecasts, and the way in which terrain affects weather. The student pilot will have access to the knowledge of the airline transport pilot.

Chapter 8 presents a detailed description of surface weather reports including methods and criteria used by the weather observer—manual or automated. Failure to understand the observer's requirements often leads to pilot misunderstanding and unwarranted observation criticism. Limitations on observations are discussed with practical guidance on interpreting and applying information.

Most pilots are unaware of their contributions through pilot weather reports or the impact their reports can have on the system. Inaccurate or overestimated reports of turbulence and icing cause misleading advisories to be issued. Pilots making these reports gain unrealistic confidence in their ability to handle severe conditions. Pilots receiving unwarranted advisories, based on overestimated pilot reports, might conclude all advisories are pessimistic. When severe conditions develop, these pilots can be led down the primrose path to disaster. In Chap. 9, pilot weather report criteria are discussed in detail, along with distribution and application.

Pilots now have access to real-time satellite images. Although satellite interpretation is a science in itself, many flights and all grades of pilots can benefit from these products. But, like all other weather reports and forecasts, satellite images suffer from limitations. These limitations, along with interpretation and application, are discussed in Chap. 10, and they are presented in a manner to allow maximum application to the flight environment.

Beyond written weather reports, a multitude of charts are available. Each chart has a specific purpose and use, as well as limitations. These charts are discussed and analyzed in Chaps. 11 and 12 with respect to their individual significance and application, and are available from the NWS. Weather charts obtained through direct user access terminals (DUATs) and other private vendors are typically not as detailed, but their interpretation and application are similar.

Chapter 13 discusses the methods, criteria, and limitations of the forecaster. Numerous misunderstandings of and misconceptions about forecast products and terminology, potentially catastrophic to the uninformed pilot, are explained. This book tells how the pilot can interpret and apply forecasts to specific flight situations, a skill that will become increasingly important with pilot self-briefing systems.

Approximately one dozen different forecasts—text and graphic—are produced for aviation. Each has criteria, purposes, and limitations; often they overlap and might appear to be inconsistent. Analysis, however, most often reveals that the forecasts are consistent within the scope of each product. Each forecast is analyzed and put into perspective in Chaps. 14 through 18. The pilot can then apply the array of forecasts available, with respect to regulations, aircraft performance, and pilot ability, to efficiently operate within the system and to make sound go/no-go decisions.

Chapters 19 and 20 are devoted to obtaining, updating, and using weather information. Discussions will help the pilot efficiently obtain weather data, through FAA or commercial sources, and apply the information to actual flight situations. Chapter 19 is devoted to applying information to VFR and IFR flight planning situations and the services available from the FAA's FSSs. With DUATs, pilots have the option to obtain weather and Notices to Airmen (NOTAMs) and to file flight plans through a computer terminal. On the surface, this might seem a simple task, but it can be complicated and frustrating to the pilot and has the potential for trouble for the individual who fails to obtain or misinterprets information. Chapter 20 discusses FAA-sponsored DUATs and commercial weather briefing services. Since pilots will have to decode, translate, and interpret NOTAMs, a section on this subject is found at the end of the chapter.

Part 3, Strategies for Interpreting and Flying the Weather, describes how to handle weather hazards. Unfortunately, it's difficult to teach judgment—the ability to evaluate facts and come to a rational, safe decision. However, I believe that judgment, like common sense, is built largely on training and experience. My goal is to share my training and experience. In this way I can, I hope, help pilots avoid the pitfalls of learning by experience—which can best be defined as when "the test comes before the lesson"! In other words, learn from the mistakes of others; you won't live long enough to make them all yourself. The bottom line: Judgment and application of judgment, as well as a knowledge of the aviation weather system and its relation to air traffic control, are essential to a safe, efficient flight.

In Chap. 21, we build on our understanding of atmospheric properties with discussions of altimetry, airspeed, and aircraft performance. These discussions include the effects of the atmosphere on aircraft performance during all phases of flight. We then present strategies to deal with high temperatures and conditions that produce high-density altitude.

Next we discuss low ceilings and visibilities and their effects on aviation operations. This section describes fog, haze and smoke, dust, sand, and precipitation. Over the last decade another aviation weather hazard has been explored—volcanic ash. A discussion of volcanic ash is included in this chapter. Similar to low ceilings and visibilities, terrain obscurement can constitute an aviation weather hazard. Similar to IFR conditions, terrain obscurement typically applies to mountainous regions. Chapter 22 concludes with strategies to deal with the effect of low ceilings and visibilities, for both IFR and VFR operations.

Chapter 23 provides a detailed discussion on turbulence, including its causes and locations and cloud types related to nonconvective turbulence. The significance of rotor and standing lenticular clouds, and Kelvin-Helmholtz (H-K) wind shear clouds, is included. Because of its significance for hazard, we include an artificial form of turbulence—wake turbulence. A change of wind speed with height or horizontal distance causes shear. The effects of wind shear as they relate to turbulence on flight operations are discussed. The chapter ends with strategies to minimize or avoid the effects of turbulence.

Chapter 24 contains a detailed section on the effects of aircraft icing on induction, the airframe, and the carburetor. The final section of this chapter provides strategies to minimize or avoid the icing hazard.

Thunderstorms contain just about every weather hazard, including the potential for high-density altitude. Thunderstorm development, along with its hazards, is presented in Chap. 25. Convective low-level wind shear, with all its perilous implications, is also presented in this chapter. Because of their significance, a discussion of microbursts, related phenomena, and recognition and recovery techniques is included. The final section, Strategies, provides a comprehensive discussion of thunderstorm avoidance procedures.

Chapter 26 contains strategies for dealing with weather systems. Both frontal and nonfrontal weather systems are presented. Pilots need to understand that nonfrontal weather systems—surface and upper-level lows and troughs—present different hazards and require different strategies from frontal systems.

Chapter 27 delves into the often ambiguous subject of risk assessment and management. Risk assessment and management begins with flight planning strategies. These include personal minimums, preparation, evaluating the weather, and consideration of the pilot and passengers. With this accomplished, we can move on to risk evaluation. In the final section of this chapter we apply flight planning strategies and risk assessment, using actual weather scenarios, to risk management techniques.

Appendix A, Abbreviations, contains weather, Notices to Airmen, and international abbreviations. Appendix B, Graphs and Charts, provides temperature conversion and density altitude calculation graphs. Since we refer to cloud types and their characteristics throughout the book, App. B contains a cloud characteristics table. This appendix also provides a reference to NWS area designator maps, area forecast regions, and transcribed weather broadcast (TWEB) route forecast locations.

A glossary is included that serves as a "quick reference" for terms and concepts.

The Aircraft Owners and Pilots Association's (AOPA) Air Safety Foundation publication *Safety Review General Aviation Weather Accidents: An Analysis and Preventive Strategies* (1996) provides a detailed analysis of general aviation accidents. We recommend this publication, which should be part of every pilot's library. We use accident statistics and scenarios throughout the *Aviation Weather Handbook*.

With respect to accident statistics and scenarios, some may say, why emphasize the negative? Our goal, along with AOPA's Air Safety Foundation and the FAA, is accident prevention. People learn through either training or experience. We don't want the learning experience to be one in which "the test comes before the lesson." We hope, through the use of these incidents and scenarios, to help prevent pilots from becoming statistics.

While we're on the subject of accident scenarios, let's concede that hindsight is always 20-20. It's easy to sit back in a favorite armchair and analyze and criticize someone else's performance and decisions. When we review an accident in this book, or any other publication or forum for that matter, let's not judge or attempt to blame. Our goal is prevention through education. Therefore, if the reader perceives any judgment or blame in any of the incidents or scenarios discussed in this book, it is strictly unintentional; that is not our purpose.

From time to time the FAA is accused of having a "tombstone mentality." That is, people have to die before anything changes. This notion certainly has some basis in fact, but, like many generalizations, it is not always true. A good example is the 1994 Roselawn, Indiana, accident. This tragic accident spurred a number of changes in aircraft design, regulations, and weather forecast products. Some would say these changes came 68 human lives too late. On the other hand, aircraft have been operating relatively safely in icing conditions since the 1930s. To quote the U.S. Supreme Court, "Safe does not mean risk free." We should all acknowledge that the FAA must walk a tightrope between safety and overregulation.

Periodically, we will talk about minimums. Minimums are just that—minimums. Minimums do not necessarily mean safe. An operation may be technically legal but not safe! In various chapters we will explore how a number of accidents occurred even though the pilot had "legal minimums."

During the last 10 years approximately 27 percent of all general aviation airplane accidents involved adverse weather conditions. Of these, 30 percent caused fatalities. The biggest causal factor, as in the past, involved continued VFR flight into instrument meteorological conditions (IMC). This has accounted for 56 percent of fatalities for all weather-related accidents. By a wide margin, the most dangerous situation is VFR flight into IMC! Therefore, a major theme of this book will be—to use a 1990s term—risk management.

The following chapters, told with a little humor, I hope, explain how to use the weather system, translate and interpret weather reports and forecasts, and then apply them to a flight situation. Throughout the text we've poked some fun at ourselves and others, especially government agencies like the FAA and NWS. Sometimes we tend to take ourselves a little too seriously. By doing this, we've tried to inject a "reality check." However, incidents are not intended to disparage or malign any individual, group, or organization; the sole purpose is illustration.

Examples of weather reports and forecasts are real and are taken from the weather circuit. None have been created. They reflect current products as they appear on briefing terminals.

Weather report phraseology used is taken from FAA Handbook 7110.10, *Flight Services*. This is the same phraseology pilots can expect in FAA radio communications and broadcasts.

Great effort has been made to ensure that information is current and accurate as of the time of this writing. Unfortunately, especially in aviation and aviation weather, the only thing that doesn't change is change itself.

BASIC WEATHER THEORY AND FUNDAMENTALS

CHAPTER 1
THE REALM OF FLIGHT

The Earth is unique in the solar system. Evolution has provided a thin layer of gas covering the planet. The atmosphere consists of about 78 percent nitrogen, 21 percent oxygen, and lesser amounts of argon, carbon dioxide, and other gases—including water vapor, which is essential to our story. The types and amount of gases have only one implication to our weather discussion: If the gases were in a different ratio, life on Earth, as we know it, would not exist.

Another unique, and essential, property of the Earth consists of its temperature and pressure. The average temperature and pressure at the Earth's surface allow water to exist as a liquid. If this were not true, we would not have weather. (We use the term *average* an awful lot. For our purposes *average* is an arithmetic mean—add up all the temperatures or pressures around the world and divide by the total number of measurements. Actual conditions on the Earth or in the atmosphere are rarely the average, as we shall see.)

The generation of weather requires an energy source. In the solar system the Sun provides virtually all the energy available to the Earth's weather machine. Thus, we may attribute all our weather to the Sun. So, what do we have? A planet where water can exist as a liquid, with a constant, steady source of energy. The result: An atmosphere that sustains life and produces weather.

The Earth's atmosphere has evolved over the last 4.5 billion years. Between 1 billion and 550 million years ago, the oxygen level increased from about 2 percent to the present value of about 21 percent, high enough for land animals to develop. Modern humans have been around for about 10,000—aren't you glad you waited 'til now to study the weather?

STRUCTURE OF THE ATMOSPHERE

The atmosphere can be divided into various layers, or *spheres,* as illustrated in Fig. 1-1. Each sphere has unique properties. Based on temperature, the atmosphere is typically divided into the following:

- Troposphere
- Stratosphere
- Mesosphere
- Thermosphere
- Exosphere

Since most weather occurs within the troposphere, and most flying takes place in the troposphere and stratosphere, we will concern ourselves with these two layers.

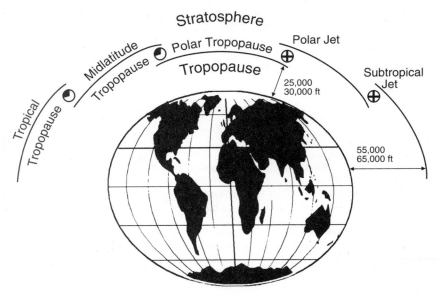

FIGURE 1-1   The atmosphere can be divided into various layers, or spheres, and each has unique properties.

The Troposphere

Beginning at the Earth's surface, the troposphere extends to an average altitude of about 7 mi. Thickness varies from the equator to the poles as shown in Fig. 1-1. The troposphere extends higher over the equator than the poles, and higher in summer than in winter. Why? Warm columns of air are taller than cool columns. Near the equator the Sun heats the surface, which in turn warms the air. Therefore, the troposphere extends to a higher altitude at the equator than at the poles. The same rationale explains why the troposphere extends to a higher altitude in summer: There is more solar radiation during the summer months, and therefore the troposphere is warmer.

The troposphere contains about three-quarters of the atmosphere by weight and almost all the water vapor. Nearly all clouds and weather occur within this layer.

The average thickness of the troposphere is 25,000 to 30,000 ft at the poles and between 55,000 and 65,000 ft at the equator. Average temperature decreases with altitude from 15°C at sea level to about −57°C at 36,000 ft. (More about the average or Standard Atmosphere later in this chapter.) Pressure decreases with altitude from an average of 1013 mb (millibars) at sea level to about 230 mb at 36,000 ft. In the troposphere wind speed tends to increase with height.

The Tropopause

The boundary between the troposphere and the stratosphere is known as the *tropopause*. In this layer temperature remains relatively constant with altitude. Pressure continues to decrease with height.

As shown in Fig. 1-1, the tropopause is not a continuous layer. There are breaks in the tropopause. On average, breaks occur between the polar tropopause and midlatitude tropopause, and the midlatitude tropopause and the tropical tropopause. Where these breaks occur lies the jet stream. In Fig. 1-1 the bullets, on the left, represent the jet stream coming out of the illustration; the arrow feathers, on the right, represent the jet going into the illustration. In the northern hemisphere the jet stream generally blows from west to east. (More about the jet stream later in this chapter.)

The tropopause acts as a lid that resists the exchange of air between the troposphere and the stratosphere. This is why almost all water vapor and weather is found in the troposphere.

The Stratosphere

Above the tropopause is the stratosphere. The average altitude of this layer is from 7 mi to about 22 mi (115,000 ft). There is a slight increase in temperature with height (that is, an isothermal layer—constant temperature with increase in height, or an inversion—increases in temperature with height). Pressure continues to decrease. There is virtually no water vapor or clouds.

Ozone reaches its maximum concentration near the top of the stratosphere. Ozone is important because it absorbs most of the deadly ultraviolet rays from the Sun.

Above the stratosphere are the mesosphere, thermosphere, and exosphere, respectively.

The Jet Stream

The jet stream was virtually unknown until World War II when pilots flying at high altitudes reported turbulence and tremendously strong winds. These winds blew from west to east near the top of the troposphere, in the northern hemisphere. Not until 1946 was the jet stream fully recognized as a meteorological phenomenon.

Sharp horizontal temperature differences cause strong pressure gradients that result in the jet stream. Refer to Fig. 1-2. Note the strong temperature gradients illustrated by *isotherms*—lines of equal temperature (dashed lines) in Fig. 1-2. Temperature changes rapidly with height. In polar regions the atmosphere compensates for extremely cold temperatures in the troposphere with relatively warmer air above. This relatively warmer air above the tropopause extends well up into the stratosphere. In such zones, the slope of constant pressure surfaces increases with height. Since fronts lie in zones of temperature contrast, the jet stream is closely linked to, or associated with, frontal boundaries. When wind speed becomes strong enough, the flow is termed a *jet stream. Isotachs*—lines of equal wind speed—increase to a maximum in the jet core.

> **RELATIVELY** Throughout this book we use the term *relatively* to describe differences in atmospheric characteristics—temperature, pressure, moisture, density, stability. We will continually use terms like *hot* and *cold, warm,* and *cool.* We are not referring to absolute values but to the relationship between values. For example, in Fig. 1-2, −40 is warmer than −70; −55 is colder than −50. The absolute values are all extremely cold, but some are "relatively" warmer than the others. For example, the atmosphere north of the jet is relatively colder at the tropopause than above the tropopause.
>
> Stated another way, we may refer to a parcel of air, or an air mass, as "warm." By this we mean a parcel of air that is warmer than the surrounding air, or an air mass warmer than the surface or warmer than an adjacent air mass.

A jet stream is a narrow, shallow, meandering area of strong winds embedded in breaks in the tropopause (Figs. 1-1 and 1-2). Two such breaks typically occur in the northern hemisphere; the polar jet located around 30° to 60° north latitude at an approximate height of

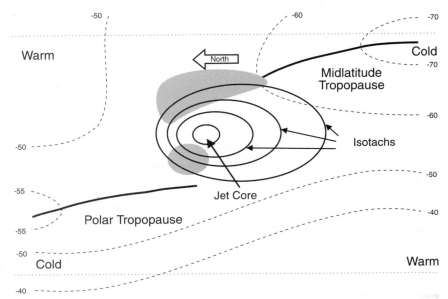

FIGURE 1-2   Sharp horizontal temperature differences cause strong pressure gradients that result in the jet stream.

30,000 ft, associated with the polar front, and the subtropical jet around 20° to 30° north latitude at approximately 39,000 ft. To be classified as a jet stream, winds must be 50 kn or greater, although jet stream winds generally range between 100 and 150 kn. Winds can reach 200 kn along the east coasts of North America and Asia in winter when temperature contrasts are greatest.

A "jet" is most frequently found in segments 1000 to 3000 mi long, 100 to 400 mi wide, and 3000 to 7000 ft deep. The strength of the jet stream increases in winter in mid and high latitudes when temperature contrasts are greatest; the jet stream shifts south with the seasonal migration of the polar front.

The presence of jet streams has significant impact on flight operations. The jet stream can cause a significant head wind component for westbound flights, increasing fuel consumption and requiring additional landings.

Another factor associated with the jet is wind shear turbulence. *Wind shear* is caused by a change in wind speed, either horizontal or vertical, or direction. With a significant wind change over a relatively small distance, severe turbulence can result. Maximum jet stream turbulence tends to occur above the jet core and just below the core on the north side. These areas are illustrated by the gray areas in Fig. 1-2. Notice that the maximum curvature (greatest change in direction) and closest spacing (greatest change in horizontal and vertical distance) of the isotachs surrounding the jet core occur on the north side of the jet. Additional areas of probable turbulence occur where the polar and subtropical jets merge or diverge. With an average depth of 3000 to 7000 ft, a change in altitude of a few thousand feet will often take the aircraft out of the worst turbulence and strongest winds.

ATMOSPHERIC PROPERTIES

For our discussion a *property* is a characteristic trait or peculiarity. The Earth has weather because of its atmosphere, water, and the energy from the Sun. The manifestations of weather result from these properties.

As aviators we're interested in those atmospheric properties that affect our flying activities. These are temperature, pressure, moisture, and density. Two of our most important flight instruments operate on these properties—the altimeter and the airspeed indicator. Therefore, it's appropriate to include discussions of these flight instruments in subsequent chapters. Additionally, atmospheric properties affect aircraft performance. How the atmosphere impacts aircraft performance is also discussed in subsequent chapters.

Temperature

Temperature is a measure of the average speed of molecules. But let's back up for a moment. Heat is the total energy of the motion of molecules; it is a form of energy and has the ability to do work. Temperature can, therefore, be related to the amount of energy available in the atmosphere. And energy drives the weather.

The amount of heat required to raise the temperature of air, or the amount of heat lost when air is cooled, is known as *heat capacity*. The fact that this is called "heat capacity" is not important. What is important? That energy is released or absorbed during the process. How this relates to aviation weather will soon become apparent.

It's important to understand how the Sun heats the Earth. Both dry air and water vapor are transparent to visible light, most of the solar radiation coming from the Sun. So what happens to the solar radiation other than visible light? A small amount is absorbed and heats the atmosphere. Some is reflected back to space by clouds and the atmosphere, and some by the Earth's surface. A sizable portion of the Sun's energy (approximately 33 percent) is absorbed by the Earth's surface. This radiant energy is converted to heat and raises the temperature of land and water surfaces.

During the day the Earth's surface absorbs solar radiation and in turn the atmosphere near the surface is heated by conduction. *Conduction* is the process of heat transfer by contact from one substance to another. At night the Earth radiates heat back into space. This is known as *terrestrial radiation* or *radiational cooling*. This process, in turn, cools the atmosphere in contact with the surface.

The amount of energy absorbed by the Earth depends on several factors. One factor is the angle at which solar radiation strikes the surface. More energy is absorbed at the equator than at the poles; more during summer than during winter; more during midday than during early morning or late afternoon. At night the surface radiates energy back into space and cools.

Different surfaces absorb and reflect different amounts of the Sun's radiation. Dense cloud layers greatly reduce the amount of energy reaching the surface. Snow covered areas reflect most incoming solar radiation. Land areas absorb radiation to a much greater degree than water surfaces, with different land surfaces absorbing different amounts of radiation. For example, plowed fields absorb more radiation than green pastures, and green pastures more than rivers or lakes.

As a result land becomes warmer during the day, while the temperature of water areas remains relatively constant. At night the land surfaces radiate heat back into the atmosphere, while water areas, again, remain relatively unchanged. Additionally, large bodies of water tend to retain heat much longer than land areas. This is why temperatures tend to be more moderate along coastal areas and deep lakes do not freeze in the winter.

Liquid water droplets, ice particles, and water vapor are transparent to incoming solar radiation but are opaque—appear solid—to outgoing terrestrial radiation. At night clouds absorb nearly all terrestrial radiation and reflect it back to the surface. This is why temperatures remain relatively cool on cloudy days and relatively warm during cloudy nights. In dry, desert areas terrestrial radiation cools the land rapidly after sunset, but humid areas remain relatively warm.

These concepts will be important to our discussions of fog, lake effect, and air masses.

The *diurnal temperature range* is the Earth's daily range in temperature. The Earth receives its greatest amount of solar radiation at noon (sun time) and the least heating at sunset and sunrise. But if the greatest heating is at noon, why doesn't maximum temperature occur at noon? Surface air temperature is also affected by the amount of reflected radiation. At noon, the Earth is receiving its greatest heat, but it still is not warm enough to radiate as much energy as it is receiving. Therefore, surface air temperature continues to get warmer. Shortly after noon the Earth gets still warmer, but it begins to receive less solar radiation. Finally, incoming solar radiation exactly balances outgoing terrestrial radiation, and that is the time of maximum temperature. Maximum temperature typically occurs just before midafternoon.

During the night the surface receives no solar radiation but continues to radiate and become cooler. Just after sunrise, solar radiation begins again but at a very low angle. Very little heating occurs while the surface continues to radiate and get cooler. Not long after sunrise, solar radiation again balances terrestrial radiation and begins to warm again. This is the time of minimum temperature.

Maximum and minimum temperatures affect aircraft performance and density altitude. Minimum temperature also plays a part in fog formation and frost hazards.

The international standard for temperature measurement is the Celsius scale. Winds aloft forecasts have always used this scale. Since July 1, 1996, with the adoption of the METAR (Aviation Routine Weather Report) weather code, surface temperatures for aviation in the United States, as well as the rest of the world, are reported in Celsius. Therefore, we will use this system. (Like the airspace reclassification in 1993 and the METAR/TAF codes in 1996, we might as well get used to using Celsius.) To this end the following benchmark temperatures should help us cope with the change from Fahrenheit (F) to Celsius (C):

$-40°C = -40°F$	Eskimos stay in the igloo.
$-20°C = -4°F$	Chilly, even in North Dakota.
$-10°C = 14°F$	Break out the gloves and earmuffs.
$0°C = 32°F$	Water freezes.
$10°C = 50°F$	Need a light jacket.
$20°C = 68°F$	Room temperature.
$30°C = 86°F$	Turn on the air-conditioner.
$40°C = 104°F$	We're having a heat wave.

OK, for those who still need a crutch, App. B, Graphs and Charts, contains a Celsius-Fahrenheit conversion table.

A concept that will be continually used is standard temperature. Standard temperature at sea level is 15°C. This value is nothing more than an average—we previously discussed average—temperature taken from measurements all over the Earth at sea level over a period of many years. Actual temperature for any location is rarely standard. The purpose of this value is to provide a common reference point.

Aircraft performance charts are based on the Standard Atmosphere, more precisely, the International Standard Atmosphere (ISA). The *standard lapse rate*—decrease of temperature with height—is approximately 2°C per 1000 ft. Temperature decreases to a value of

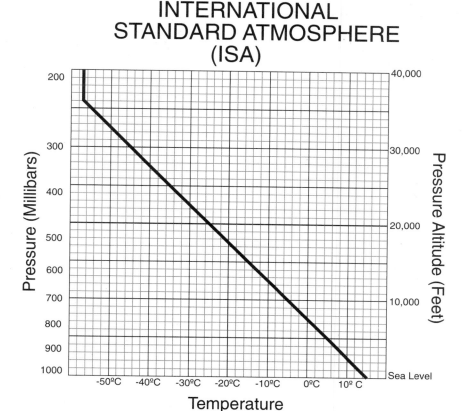

INTERNATIONAL STANDARD ATMOSPHERE (ISA)

FIGURE 1-3 Aircraft performance charts are based on the Standard Atmosphere, more precisely the International Standard Atmosphere (ISA).

$-57°C$ at approximately 36,000 ft. Above 36,000 ft, in the standard atmosphere, temperature remains constant to about 66,000 ft, as illustrated in Fig. 1-3, the International Standard Atmosphere (ISA).

As mentioned, standard conditions rarely occur in the real world and performance charts are based on standard conditions. Accommodation—most significantly a temperature correction—must be made for a nonstandard environment. Manufacturers sometimes provide an ISA conversion with cruise power setting charts for high, low, and standard temperatures or simply note that performance is based on standard conditions. The aircraft doesn't understand any of this and performs based on the prevailing flight environment.

Pressure

Pressure is defined as force per unit area. For example, tires are inflated to a specified pressure. Units are typically described as pounds per square inch (lb/in^2). The atmosphere exerts pressure. At sea level, atmospheric pressure is approximately 14.7 lb/in². (We don't directly experience this pressure because our bodies exert an opposite and equal force. However, we do experience pressure changes with our ears, which must be cleared periodically with changes in pressure or altitude.)

Atmospheric pressure is the sum of all the air molecules above a specific point on the Earth. This can be illustrated by an imaginary column of bricks. There are 10 bricks in the column, each weighing 1 lb. If we weigh the bottom of the stack, the scale reads 10 lb. However, if we weigh the stack at the fifth brick, the scale reads 5 lb; at the first brick, 1 lb. The atmosphere behaves in the same way.

HOW MUCH DOES THE EARTH'S ATMOSPHERE WEIGH?   It is estimated to weigh about 5 million billion tons!

For aviation purposes we commonly relate atmospheric pressure to inches of mercury–altimeter setting, or pressure in millibars. Inches of mercury (in Hg), however, is not a direct expression of force per unit area. The international unit of atmospheric pressure is the hectopascal (hPa), which is equivalent to the millibar (mb). [We like to name units of measurement after famous folks. Blaise Pascal (1623–1662) was a French philosopher and mathematician.]

So why do we measure pressure in inches of mercury? In 1643 an Italian, Evangelista Torricelli (1608–1647), invented an instrument for measuring atmospheric pressure—the mercurial barometer. Mercury was used because it was the heaviest liquid available that remained liquid at normal temperatures. (If water were used, the column would be 30 ft high!) Sea level pressure balances a column of mercury approximately 30 in high.

Like temperature, pressure has a standard. At sea level it is 29.92 in Hg, or 1013.2 hPa (1013.2 mb). More about this when we discuss altimetry. And, guess what? For every level there is a standard temperature and pressure in the—you know what's coming next—standard atmosphere.

In aviation weather, we often refer to a "constant pressure surface." We have already mentioned that the troposphere is higher at the equator than at the poles (Fig. 1-1). Why? Air is warmer at the equator; and, therefore, molecules are farther apart. Columns of warm air are taller than columns of cold air. Let's take the 500-mb constant pressure surface. Since pressure is the sum of air molecules above, the 500-mb surface will, typically, be higher at southern latitudes in the northern hemisphere. Layers in the atmosphere are described in this manner. In aviation we most often refer to the 850, 700, 500, 300, and 200 mb constant pressure surfaces. From Fig. 1-3 we see that these constant pressure surfaces occur at approximately 5000, 10,000, 18,000, 30,000, and 39,000 ft, respectively.

The slope of the pressure surface, caused by horizontal temperature differences, determines approximate wind speed. Typically wind speed at the 500-mb slope is greater than at the 700-mb slope, and the wind speed at the 300-mb slope is greater than that at the 500-mb slope. When pressure surface slope increases with height, wind speed increases with height. This is the general case in the troposphere. Winds are light or calm in areas of little or no horizontal temperature difference. And, in some cases, winds can decrease with height in the troposphere. This tends to occur within large high-pressure areas.

This can be related to our discussion of the jet stream. Typically the highest winds in the atmosphere occur with the jet stream, the area of greatest temperature difference.

The atmosphere, a gas, is compressible. That is, the weight of air molecules above compresses, or pushes together, the molecules below. This is illustrated in Fig. 1-3. Notice that the pressure at sea level is approximately 1000 mb. Pressure decreases with height but not at a constant rate. Half the atmosphere by weight exists below 18,000 ft—the 500-mb level.

We have already discussed how pressure decreases with altitude. Atmospheric pressure also changes with the movement of weather systems. We have all heard the local weather forecaster talking about the movement of high- and low-pressure areas. So we have an atmosphere where pressure decreases with height and where pressure at the surface and aloft will vary with latitude and weather systems.

Moisture

If we can think of heat as the energy that drives the weather machine, then moisture, in the form of water vapor, is the fuel. Without moisture there would be no clouds or precipitation—in other words, no weather as we know it.

Moisture in the atmosphere occurs in the form of ice crystals (a solid), water (a liquid), or water vapor (a gas). Clouds and precipitation are made up of ice crystals or liquid water. *Precipitation* is any form of water (solid or liquid) that falls to the Earth's surface. Water vapor is invisible, suspended in the air. The effects of water vapor are important phenomena in meteorology and aviation weather.

The amount of water vapor in the atmosphere is measured in several ways. The term that is most familiar is *relative humidity*. A term often used in aviation is *dew point*.

Relative humidity is the ratio, expressed as a percentage, of the water vapor present in the air compared to the maximum amount of water vapor the air could hold at its present temperature. So, what does that mean?

Because air is a mixture of gases, it has no inherent capacity to hold moisture. The space available in a particular sample of air determines the amount of water vapor molecules it can hold. Since air molecules in warm air are farther apart, warm air can hold more water vapor than cold air. We will use this principle many times in our discussion of weather producing systems.

In Fig. 1-4 there are three air samples. The air samples have temperatures of 25°, 15°, and 5°, respectively. Notice that each sample contains the same amount of water vapor. However, the warmer samples have capacity to hold additional water in the form of vapor. Because of this the relative humidities of the warmer samples are lower than the cooler samples.

When relative humidity reaches 100 percent, the air is saturated. This means that the air can no longer hold any additional water in the form of vapor. This is illustrated by the 5° sample in Fig. 1-4. Should any more water vapor be added or the air be cooled to a lower temperature, condensation will occur in the form of clouds or precipitation—visible moisture.

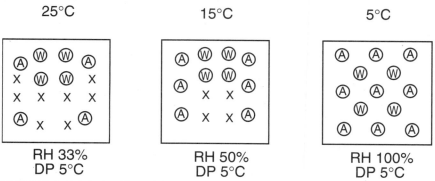

FIGURE 1-4   Relative humidity and dew point. The warmer the air, the greater its capacity to hold water in the form of vapor. RH, relative humidity; DP, dew point; A, air; W, water vapor; X, space available for water vapor.

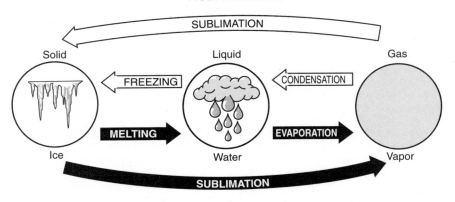

FIGURE 1-5 Change of state. Energy, in the form of heat, is required to change water from a solid to liquid and from liquid to gas; the amount of heat exchanged is called *latent heat*.

Dew point is the temperature to which air must be cooled, water vapor remaining constant, for air to become saturated. The dew point is usually available in METAR (Meteorological Aviation Weather Report) and SPECI (Aviation Selected Special Weather Report) reports, along with temperature, reported in degrees Celsius. In Fig. 1-4 all samples have a dew point of 5°. Therefore, the first sample would have to be cooled by 20° and the second by 10° to become saturated.

Energy, in the form of heat, is required to change water from a solid (ice) to liquid and from liquid to a gas (water vapor). The amount of heat exchanged (absorbed or released) is called *latent heat*. Refer to Fig. 1-5. When ice melts, heat is absorbed. The heat required must be supplied from somewhere. It comes from the surrounding environment. In the process of freezing, heat is released to the surrounding environment.

When water evaporates (changes to water vapor), an even greater amount of heat is absorbed—without any change in temperature. The heat is supplied mostly by the liquid, with a smaller portion coming from the surrounding environment. We can all relate to the latent heat of evaporation. What happens when we jump out of a swimming pool? As the water evaporates, our skin cools. When water vapor *condenses* (changes to liquid), heat is transferred, or released, to the environment. A primary source of energy in the weather machine comes from the heat released from condensation—the latent heat of condensation.

Can water vapor change to ice without going through the liquid stage? Yes. And ice can evaporate, bypassing the liquid stage. This process is called *sublimation*. For example, an aircraft with structural ice will lose the ice once it is out of the icing environment, even though the temperature remains below freezing. In the process of frost formation, water vapor in the atmosphere sublimates directly into ice. In the process of sublimation, a large amount of heat is absorbed or released.

Melting, freezing, evaporation, condensation, and sublimation are important factors in the weather machine. We will constantly refer to this process in subsequent chapters.

In aviation we often use the term *freezing level*. This is actually a misnomer. Water melts at 0°C, but it can exist in the liquid state at temperatures well below freezing! Liquid water, or water vapor, that exist at temperatures below freezing is described as *supercooled*.

Because water can exist at temperatures below freezing as supercooled cloud and water droplets, structural icing can occur.

Another important concept is the heat capacity of water vapor. We have already mentioned that water vapor is transparent to solar radiation, but it is largely opaque to terrestrial radiation. That is, water vapor in the air absorbs terrestrial radiation and converts it to heat, raising the temperature of the air. Water vapor absorbs outgoing terrestrial radiation much as heat is trapped in a greenhouse—the greenhouse effect.

Density

Although temperature and pressure are important atmospheric properties, to understand the weather machine a thorough knowledge of density is second in importance only to an understanding of moisture. As we shall see, density is a key player in most weather phenomena.

Density is the weight of air per unit volume often expressed as pounds per cubic foot (lb/ft^3) or grams per cubic meter (g/m^3). Occasionally, we hear the question: "Which weighs less, a pound of dry air or a pound of moist air?" This is the same type question as: "Which weighs more, a pound of lead or a pound of feathers?" The answer to both is that they weigh the same—a pound is a pound. However, the volume of a pound of dry air compared to a pound of moist air, like a pound of lead compared to a pound of feathers, is different! We're not really interested in these units. But the fact that atmospheric density varies both horizontally and vertically is vitally important. At this point we can begin to put together some of the concepts we've discussed thus far and draw some practical relationships.

As we fly higher in the atmosphere, density decreases. As noted, density is related to weight, and we know that the higher we go, the fewer molecules there are; therefore, the air is less dense.

Atmospheric pressure also affects density. When temperature and humidity—moisture—remain constant, we know that if pressure is higher than standard, density is higher; conversely, if pressure is lower than standard, density is lower than standard.

Although pressure is an important factor in the density equation, temperature is often the most significant. Typically, the higher the temperature, the lower the density of the air. Remembering that density is the weight of molecules, at higher temperatures there are fewer molecules per unit volume, pressure and moisture remaining constant. This can be seen in Fig. 1-4. There are fewer air molecules per unit volume in the warmer samples.

The final factor in air density is moisture. The higher the moisture content of the air, the lower its density. Why? Water molecules weigh less than air molecules. This is not a major factor, however. What we need to remember is that when it's humid, air density is less than when conditions are dry.

The bottom line:

- Density has a direct relationship to pressure—as pressure decreases, density decreases.

- Density has an inverse relationship to temperature—as temperature increases, density decreases.

- Density has an inverse relationship to moisture (water vapor)—as moisture increases, density decreases.

CHAPTER 2
EVER-CHANGING WEATHER

Now we can build on the knowledge of Chap. 1. A major theme of that chapter was the standard atmosphere. But, as we know, in the real world the atmosphere is rarely standard.

A dominant factor in the weather is stability, or lack thereof. Stability is a major element in vertical motion. What happens when air is forced upward? It cools; but at what rate? It's rarely standard. We will discuss and apply the different types of atmospheric stability: absolute and neutral stability, and absolute and conditional instability. These processes occur in the real world and have significant effects on our flying activity.

Vertical motion is a primary weather producer. Clouds, weather, and aviation weather hazards are closely related to vertical motion. We will introduce eight vertical motion mechanisms, any one of which, under the right conditions, is capable of initiating thunderstorms and other hazards. In subsequent chapters we will refine and expand our knowledge of these phenomena. The bottom line: Both stability and vertical motion are important to our understanding of weather phenomena.

LAPSE RATE

The *Glossary of Meteorology* defines *lapse rate* as the decrease of temperature with height. In Chap. 1 we discussed the standard lapse rate in the troposphere as approximately 2°C per 1000 ft, as illustrated in Fig. 1-3. Well, what if the atmosphere isn't standard? [Here is where we apply our discussion of the change of state of water from the previous chapter (recall Fig. 1-5).]

The lapse rate of dry air is 3°C per 1000 ft, which is the rate at which unsaturated air cools as it is forced upward—known as the *dry adiabatic lapse rate.* Conversely, when unsaturated air is forced downward, it heats at the rate of 3°C per 1000 ft. But what about saturated air? The moist, or saturated, lapse rate varies from 1 to 2°C per 1000 ft. To understand how all this works, we will need the process of adiabatic expansion and compression.

From Chap. 1 we know that atmospheric pressure is the weight of air above, which includes water vapor. What happens if air is forced upward? To understand, we need the concept of an air parcel. (I know when I first heard the term, it made absolutely no sense whatsoever.) A *parcel of air* is a small volume of air arbitrarily selected for theoretical study. It retains its composition. That is, it does not mix with the surrounding air. It responds to all meteorological processes. For example, it can expand or be compressed; thus its temperature can change. Its moisture remains constant as long as its temperature allows moisture to remain in the vapor state—dew point remains lower than temperature. As an example, uneven surface heating causes parcels of air near the surface to be heated and rise. (This is illustrated in Fig. 3-8. Each cloud resulted from a parcel of air that had been lifted from the surface.) This process causes thermal turbulence. If the concept of an air parcel is not clear now, it will be by the end of this section.

As a parcel of air is forced upward, pressure decreases and the parcel expands—its volume increases. Expansion pulls the molecules apart. Energy is required to keep the molecules apart. The energy comes from the gas itself in the form of a temperature decrease. If the parcel is returned to its starting point, the reverse process occurs. As pressure increases, temperature increases and volume decreases. This is the *adiabatic process,* defined as the internal changes within a body of gas during expansion or compression when no energy is added or removed.

In the general circulation, many comparatively large areas have gradually ascending or descending air masses. Moving air masses may be forced upward by sloping terrain. Large bodies of warm air often override colder air and are forced aloft by the cold air. In all these processes, the vertically moving air undergoes adiabatic changes.

Changes follow definite universal gas laws. The mathematical computation of the process is complex and cumbersome. However, we can illustrate the process graphically. Several charts have been developed for this purpose. For illustration we will use the adiabatic chart beginning with Fig. 2-1.

Like the standard atmosphere chart in Fig. 1-3, the adiabatic chart shows temperature on the horizontal axis and pressure on the vertical—millibars on the left, pressure altitude on the right. The solid diagonal lines represent dry adiabats. Dashed lines are mixing ratios—a measurement of water vapor content. Dotted lines show the saturated adiabats. We'll define all of these in a moment.

Refer to Fig. 2-1. Let's take a sea level parcel of air with a temperature of 22°C and a dew point of 3°C. If we lift the parcel, it will cool at the rate of 3°C per 1000 ft, along the dry adiabats (diagonal lines). As the parcel rises, the dew point temperature decreases along the mixing ratio (dashed) lines. When a parcel is lifted, the actual amount of water vapor does not change. However, its relative humidity increases and the dew point temperature changes, but at a much slower rate than the decrease in air temperature.

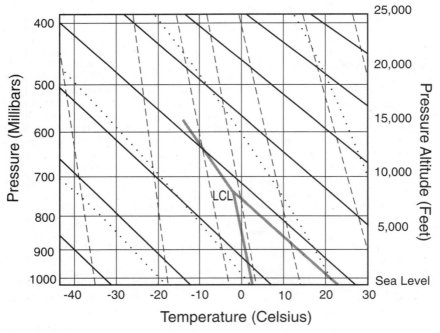

FIGURE 2-1   The adiabatic chart allows us to study the behavior of lifted air parcel.

When unsaturated air rises, why does its temperature-dew point spread decrease and humidity increase? The temperature of a rising air parcel decreases due to expansion. The dew point of unsaturated, rising air remains approximately the same. This is true because the water vapor content of the air remains constant. Dew point temperature reflects the amount of water vapor in the air. The dew point temperature of unsaturated air actually decreases slightly as the air rises to lower pressure, but the change is less than 1°C per 1000 ft. For our discussions we will assume the dew point remains constant as long as the air remains unsaturated.

Since the temperature-dew point spread of rising air decreases, then, by definition, the air's relative humidity must increase. Conversely, the reverse process is true for descending air. The temperature of descending air increases while the dew point remains approximately constant. Thus the temperature-dew point spread of sinking air increases, which means the relative humidity decreases.

Where the dry adiabat and mixing ratio lines intersect, the temperature and dew point are the same—the parcel has reached saturation. In the example this occurs at 8000 ft. Our parcel has cooled to saturation due to expansion as it has moved upward in the atmosphere, and yet no heat has been added or removed from the parcel. This intersection is known as the *lifted condensation level* (LCL). Relative humidity is 100 percent.

This process can be used to estimate the base of convective clouds. When a cloud layer is produced by surface heating, bases can be approximated by applying the following formula: $T - Td \times 4 \times 100$ = cloud base above ground level (AGL), where T = temperature n degrees Celsius, Td = dew point in degrees Celsius, and 4 and 100 are constants. For example, the temperature and dew point on the METAR are 15/05. Temperature minus dew point ($15°C - 5°C$) equals 10°C; apply the formula $10 \times 4 \times 100 = 4000$, and the approximate cloud bases are 4000 ft AGL.

As another example, use this observation for Denver:

METAR KDEN 182350Z ...SCT090 SCT150 BKN250 33/08...

Applying the formula to the Denver METAR, the cloud base works out to be approximately 10,000 ft AGL ($33 - 08 \times 4 \times 100 = 10,000$). This is consistent with the observer's report of 9000 scattered. An article in a popular aviation magazine claimed this procedure could be used for any cloud layer. That's not true. It's important to remember the procedure applies only to convective clouds produced by surface heating—air lifted adiabatically.

YESTERDAY'S TECHNOLOGY TODAY! Pilots, at least as of this writing, can expect to see written test questions on the above procedure. Temperature and dew point are provided in Fahrenheit, and applicants are expected to use the dry adiabatic lapse rate of 5.4°F per 1000 ft. Applying this to our Denver example, the cloud base works out to be approximately 7000 ft AGL. This procedure is more simplified than the preceding method.

Why include it on written tests? Well, the people who write the tests were in the field some 30 years ago. This was a viable method at that time; but, not today.

To continue our discussion, we must employ the final set of lines on the chart, the saturated adiabats. Remember that condensation is a warming process—heat is released. This adds heat to the air and partially offsets adiabatic cooling due to expansion. As we continue to lift the parcel above saturation, it no longer cools at the dry adiabatic rate because it absorbs the heat produced by condensation. Therefore, it cools at a slower rate. Cooling is now along the saturated adiabats (dotted lines).

Vertical motion also affects the moisture content of saturated air. Saturated air that rises must have its water vapor content decrease. This decrease occurs because the lowering temperature means its capacity to hold water vapor decreases. Where does the water vapor go? Water vapor must change from a gas within the air to water, in either a liquid or solid form, depending of the temperature of the air. Visible water condenses in the form of clouds and precipitation.

If we continue to lift the parcel, it cools at the saturated adiabatic lapse rate. In the example, if the parcel is lifted to 15,000 ft, its temperature will be $-13°C$. The temperature and dew point are the same; the relative humidity remains 100 percent. However, moisture has decreased. Where has the moisture gone? Precipitation. Precipitation is the process by which liquid water is removed.

When saturated air sinks and it contains no visible moisture, it will sink at the dry adiabatic rate. As the air descends, its relative humidity will decrease.

If visible water in the form of clouds or precipitation is present, the relative humidity of the descending air will tend to decrease. The decrease will be at the saturated adiabatic rate rather than the dry adiabatic rate. As relative humidity decreases, water particles evaporate. The evaporation increases the air's dew point. The latent heat loss in the air due to evaporation slows the temperature increase. The slower rate of increase of temperature-dew point spread means a slower decrease in relative humidity. In some instances with the presence of liquid droplets or precipitation, it is possible that the relative humidity will remain near the saturation point. The amount of visible water in the form of clouds or precipitation will decrease due to evaporation within the sinking air. Thus precipitation tends to diminish and clouds tend to dissipate.

What happens if we return our parcel to sea level? Refer to Fig. 2-2. We'll assume that our parcel is saturated but contains no visible moisture. The air will warm at the dry adiabatic lapse rate of $3°C$ per 1000 ft. At sea level its temperature will be $30°C$. If we follow the mixing ratio (dashed) lines, its dew point is $-7°C$. The air is hot and dry. This process explains some major meteorological phenomena, for example, the Santa Ana conditions of Southern California and the Chinook that develops along the eastern slopes of the Rockies. In fact, on rare occasions a phenomenon known as a *heat burst* is produced by a thunderstorm. The temperature rapidly, but briefly, jumps to extreme values. Near Kopperl, Texas, in 1966 the temperature was estimated at $60°C$. This was thought to be caused by adiabatic warming in the extreme downdrafts associated with storm. More about this later.

STABILITY

Before we proceed, we need to understand atmospheric stability. *Stability* is defined as the ability of a mass of air to remain in equilibrium—its ability to resist displacement from its initial position. If we move a parcel of air, then remove the lifting mechanism, one of several things will occur: The parcel will tend to return to its original position; the parcel will continue to rise without any additional lifting force; the parcel will initially resist upward displacement to a certain point, where it will then spontaneously continue upward; or the parcel will remain at the level where the external force ceases. These processes are known as the following:

• Absolute stability

• Absolute instability

• Conditional instability

• Neutral stability

To explain how stability is maintained, we will return to the adiabatic chart where we can compare properties of a lifted parcel to the actual or existing lapse rate—which from now on will be referred to as the *lapse rate.*

A parcel is *absolutely stable* when it resists vertical displacement whether it is saturated or unsaturated. This condition is illustrated in Fig. 2-3. The lapse rate is shown as the large, solid black line. Notice that if a parcel is forced upward, it cools at the dry adiabatic rate when unsat-

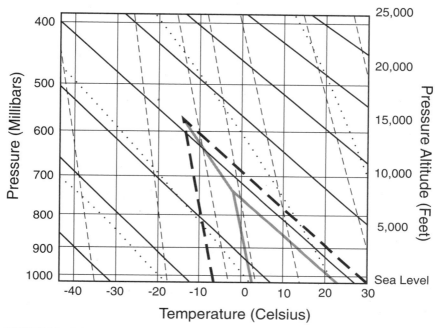

FIGURE 2-2 When a lifted parcel is returned to the surface, its properties are changed.

urated and then at the saturated adiabatic rate. In either case, it remains cooler than the surrounding air. As a matter of fact, as long as the existing lapse rate remains in the gray area of Fig. 2-3, the parcel is absolutely stable. Since the parcel is cooler (denser) than the surrounding air, it wants to sink. Thus, vertical motion is impossible, unless it is caused by an external force.

A stable lapse rate may exhibit the properties of an isothermal layer or inversion (both are illustrated in Fig. 2-3). That portion of the sounding between the 700- and 600-mb levels is isothermal. That is, the lapse rate is constant. The lapse rate between the 600- and 500-mb levels is an inversion — the temperature increases with altitude. An inversion strongly resists lifting and in many cases puts a "cap" or "lid" on weather, the ultimate example of which being the tropopause, which essentially restricts almost all weather to the troposphere.

Air is *absolutely unstable* when vertical displacement of a parcel within the layer is spontaneous, whether it is saturated or unsaturated. This is illustrated in Fig. 2-4. Again, the lapse rate is shown as the large, solid black line. If a parcel is lifted, its temperature is always warmer than the surrounding air. The cooler more dense air surrounding the parcel forces it upward. Vertical motion is spontaneous, and the layer is absolutely unstable. In fact, as long as the existing lapse rate remains in the gray shaded area, a lifted parcel is absolutely unstable.

Let's look at the third case, conditional instablility. *Conditional instability* refers to the structure of a column of air that will produce free convection of a parcel as a result of its becoming saturated when forced upward. *Free convection* means that once saturation occurs, upward movement will continue spontaneously. This is the *level of free convection* (LFC) as shown in Fig. 2-5. A lifted parcel is stable to the point at which saturation occurs. Below the lifted condensation level, the parcel exhibits absolute stability. However, upon saturation upward displacement becomes spontaneous. This occurs when the lapse rate lies within the gray shaded area of Fig. 2-5. The parcel becomes unstable on the condition that

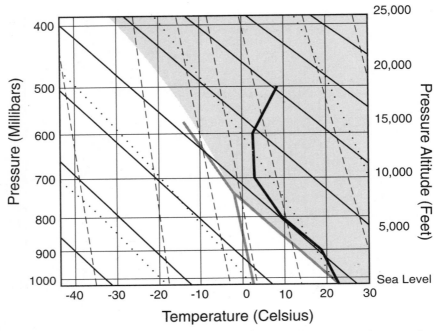

FIGURE 2-3   A parcel is absolutely stable when it resists vertical displacement whether saturated or unsaturated.

it reaches saturation. High moisture content in low levels and dry air aloft favor instability. Conversely, dry air in low levels and high moisture content aloft favor stability.

The concepts of absolute stability and absolute instability are relatively straightforward. Conditional instability is more complex and includes many subclassifications. It depends not only on temperature but also on water vapor distribution.

When a parcel is displaced and remains at rest—even when the displacing force ceases—the layer is *neutrally stable.* For a layer of unsaturated air to be neutrally stable, its lapse rate must be equal to the dry adiabatic rate. For a layer of saturated air to be neutrally stable, its lapse rate must be equal to the moist adiabatic rate.

Table 2-1 contains general characteristics of stable, conditionally unstable, and unstable air. When ceilings exist in stable air, they tend to be low, associated with stratiform clouds and fog. The stable air typically produces poor visibilities. Flight conditions tend to be smooth. Any precipitation is typically light to moderate, and steady. Conditional instability tends to exhibit the characteristics of stable air below the lifted condensation level and unstable air above. Precipitation is steady, and light to moderate in intensity, with embedded areas of heavy showers. Ceilings and visibilities are typically good in unstable air. Flight conditions are turbulence. Cloud types tend to be cumuliform, with showery, heavy precipitation. We will build on these characteristics throughout the book.

VERTICAL MOTION IN THE ATMOSPHERE

In the previous section we discussed stability and what happens when a parcel of air is lifted. A logical question is, what atmospheric processes can lift a parcel? An understand-

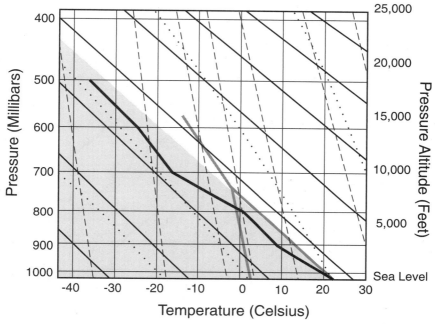

FIGURE 2-4 Air is absolutely unstable when vertical displacement of a parcel within the layer is spontaneous, whether saturated or unsaturated.

ing of vertical motion (both upward and downward) explains many weather phenomena and aviation weather hazards. In this section we will briefly explore the methods employed by nature to move air vertically and the effects of the properties of air on its vertical motion. We will build on this section throughout the book.

Vertical motion can be produced, enhanced, or dampened by one or all of the following:

- Convergence and divergence
- Frontal lift
- Dry line
- Vorticity
- Upslope
- Pressure systems
- Convection
- Warm- and cold-air advection

Convergence refers to an inflow or squeezing of the air. When the air flowing horizontally into an area is greater than the outflow, the air literally piles up. Since the ground prevents the air from going downward, there is only one way left for it to go—up. The bottom line: An area of convergence is an area of rising air. Convergence can occur aloft over dense, cold air, and it is not necessarily confined to a layer bounded by the surface. When moisture is adequate and convergence is great enough, condensation occurs.

Convergence occurs along surface low-pressure troughs and at the center of low-pressure areas. An area of strong winds blowing into an area of lighter winds causes

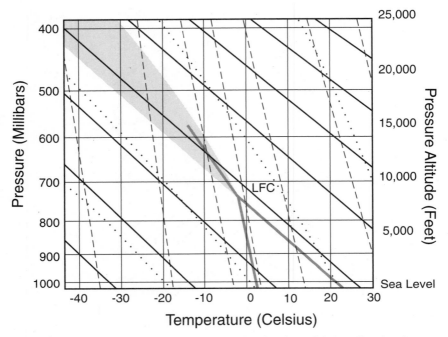

FIGURE 2-5 *Conditional instability* refers to the structure of a column of air that will produce free convection of a parcel as a result of its becoming saturated when forced upward.

TABLE 2-1   Characteristics of Atmospheric Stability

Stable	Conditionally unstable	Unstable
Ceiling poor	Ceiling poor	Ceiling good
Visibility poor	Visibility poor	Visibility good
Smooth	Smooth below/turbulent above	Turbulent
Stratiform	Stratiform/embedded cumuliform	Cumuliform
Precipitation steady	Steady/embedded showers	Precipitation showery

wind speed convergence. When this occurs near the surface, the result is a region of rising air.

Divergence is the opposite of convergence. Downward motion (subsidence) of air causes it to spread out at the Earth's surface. From our knowledge of lapse rate, we can see that divergence is a drying and stabilizing process. Like convergence, divergence may occur in a layer aloft not extending to the ground. Divergence occurs along high-pressure ridges and at the center of high-pressure areas. When air near the surface blows from an area of light winds into an area of stronger winds, wind speed divergence results.

Areas of convergence and divergence can be located on surface analysis and 850- and 700-mb constant pressure charts.

Air masses of different properties (temperature and moisture) do not tend to mix. Temperature, humidity, and wind may change rapidly over short distances. Where there are temperature and moisture differences, there is a difference in density. A zone of rapid change separating two air masses is a *frontal zone*, more commonly referred to as a *front*. In these zones the less dense air is lifted. This causes vertical motion in the atmosphere. The type of weather produced is dependent on the stability of the atmosphere.

A *dry line*, or temperature-dew point front, marks the boundary between moist, warm air from the Gulf of Mexico and dry, hot air from the southwestern United States. Dry lines usually develop in New Mexico, Texas, and Oklahoma during the summer months. Since the moist air from the gulf is less dense than the dry, hot desert air, it is forced aloft. If the air mass is unstable, thunderstorms and tornadoes develop along the boundary.

Areas of frontal lift and dry lines can be located on the surface analysis chart.

Anything that spins has vorticity, which includes the Earth. *Vorticity* is a mathematical term that refers to the tendency of the air to spin; the faster that air spins, the greater its vorticity. A parcel of air that spins counterclockwise (cyclonically) has positive vorticity; a parcel of air that spins clockwise (anticyclonically) has negative vorticity.

Air moving through a ridge, spinning clockwise, gains anticyclonic relative vorticity. Air moving through a trough, spinning counterclockwise, gains cyclonic relative vorticity; therefore, there tends to be downward vertical motion in ridge-to-trough flow and upward vertical motion in trough-to-ridge flow. More about this later.

Orographic is a term used to describe the effects caused by terrain, especially mountains. An orographic effect is *upslope* and *downslope*. As well as through the adiabatic process, the air can also take on the characteristics of the terrain through the process of *conduction*. That is, the air can absorb heat or moisture by direct contact with the surface. Air moving up a slope rises and tends to cool; air moving down a slope sinks and tends to warm. Areas of upslope can be determined from the surface analysis and weather depiction charts—with a knowledge of terrain.

Circulation around low-pressure areas produces upward vertical motion; around high-pressure areas, downward vertical motion. Typically, but not always, low pressure means poor weather and high pressure good weather. Low- and high-pressure areas are depicted on the surface analysis chart.

Upward vertical motion occurs in low-pressure troughs (usually referred to as *troughs*) and surface lows due to convergence. Conversely, downward vertical motion occurs in high-pressure ridges (usually referred to as *ridges*) and surface highs caused by divergence.

Vertical motion also occurs at higher levels. For example, air moving from an upper-level trough to a ridge produces upward vertical motion; air moving from an upper ridge to a trough produces downward vertical motion. Upper-level pressure patterns are depicted on constant pressure charts.

Atmospheric convection is the transport of a property vertically. This was mentioned in our discussion of stability, specifically in relation to the level of free convection. For our purposes, near the surface convection is caused by surface heating. Surface heating, and the resulting convection, is a primary vertical motion producer. In later chapters we will see how this process affects the weather.

Advection is a term used to describe the movement of an atmospheric property from one region to another. Temperature is one property that can be advected.

From the surface to about 10,000 ft, warm-air advection produces upward vertical motion, and cold-air advection, downward vertical motion. As warmer, less dense air moves into an area, it will tend to rise. Warm-air advection causes surface pressures to fall. This results in convergence and upward vertical motion. When cooler, more dense air moves into an area, it tends to sink. Cold-air advection causes surface pressures to rise. This results in divergence and downward vertical motion. Therefore, warm-air advection destabilizes conditions, whereas cold-air advection tends to stabilize the weather at and near the surface.

Above the 500-mb level, the opposite occurs. Cold-air advection destabilizes conditions, and warm-air advection stabilizes the atmosphere. Cold-air advection above the 500-mb

TABLE 2-2   Vertical Motion Producers

Phenomena	Vertical motion	Observed	Forecast
Convergence	Up	Surface 850 mb/700 mb	SIG WX Prog
Divergence	Down	Surface 850 mb/700 mb	SIG WX Prog
Fronts	Up	Surface	SIG WX Prog
Dry line	Up	Surface	
Vorticity	Up/down	500-mb heights/vorticity	500-mb Prog heights/vorticity
Orographic	Up/down	Surface	SIG WX Prog
High pressure (SFC)	Down	Surface	SIG WX Prog
Low pressure (SFC)	Up	Surface	SIG WX Prog
Ridge (SFC)	Down	Surface	SIG WX Prog
Trough (SFC)	Up	Surface	SIG WX Prog
High pressure (aloft)	Down	Constant pressure charts	Constant pressure progs
Low pressure (aloft)	Up	Constant pressure charts	Constant pressure progs
Ridge (aloft)	Down	Constant pressure charts	Constant pressure progs
Trough (aloft)	Up	Constant pressure charts	Constant pressure progs
Convection	Up		
Warm-air advection (SFC)	Up	850-mb/700-mb charts	850-mb/700-mb progs
Cold-air advection (SFC)	Down	850-mb/700-mb charts	850-mb/700-mb progs
Warm-air advection (aloft)	Up	300-mb/200-mb charts	300-mb/200-mb progs
Cold-air advection (aloft)	Down	300-mb/200-mb charts	300-mb/200-mb progs
Thunderstorms	Up	Radar summary	SIG WX Prog/SVR WX outlook

level decreases the lapse rate. This enhances any convective activity that might develop. Conversely, warm-air advection aloft stabilizes the atmosphere by increasing the lapse rate, thus retarding any convection.

Areas of low-level cold- and warm-air advection can be determined from the 850- and 700-mb constant pressure charts. Areas of high-level cold- and warm-air advection can be determined from the 300- and 200-mb constant pressure charts.

A quick reference guide for vertical motion producers is provided in Table 2-2. The table also lists the places where these phenomena can be found on observed and forecast products.

CHAPTER 3
MOTION OF THE AIR

With a sound understanding of the atmosphere, its properties, stability, and vertical motion, we move on to movement of the air. Wind transports atmospheric properties. This is known as *advection*. As the air moves, its characteristics are influenced by the temperature and moisture properties of the surface; the atmosphere can be modified in this way. In essence the atmosphere is a giant heat exchanger, moving warm air up from the equator and cold air down from the poles. As we shall see, this has a tremendous effect on our weather.

We begin with planetary-scale, or global, circulation. This deals with what is usually called the *general circulation*. It consists of the jet stream, subtropical high, polar front, and intertropical convergence zone. We'll discuss the process by which the atmosphere transports warm air from the equator northward and the cold air from poles southward. These winds are caused by pressure differences in the atmosphere.

Synoptic scale (mesoscale) circulation consists of highs and lows, troughs and ridges, surface and aloft. Additional synoptic-scale events are thermal advection, fronts, and terrain effects. Highs and lows, and troughs and ridges are discussed in Chap. 5. Fronts are presented in Chap. 6. Terrain effects are contained in this chapter.

From synoptic-scale motion we move on to *subsynoptic scale (microscale) events*. That is, sea/land breezes, mountain/valley winds, mountain waves, and wind shear. As well as pressure, density plays a role in some of these events. In the case of mountain waves, terrain and atmospheric stability are key elements.

Finally, we touch on nonconvective low-level wind shear, a microscale event—that is, low-level wind shear caused by phenomena other than thunderstorms. (Because of the significance of thunderstorm hazard, including convective low-level wind shear, they are covered separately in Chap. 25.)

Turbulence is a fact of life to the aviator. It ranges from small annoying bumps to severe jolts that can literally tear an aircraft apart. Turbulence can occur on the mesoscale, covering areas the size of several states, or on the microscale, affecting one airport and leaving adjacent fields untouched. With one exception, turbulence is caused by natural phenomena (OK, except in the case of poor pilot technique). Wake turbulence is caused by the aircraft producing lift. Because of the hazard caused by wake turbulence, we decided to include it in our discussion of turbulence in Chap. 23.

On the surface it may seem that pilots are doomed to live with turbulence (that's an oxymoron, isn't it?). This is not always true. There are strategies to eliminate or reduce the effect of this phenomenon. Chapter 23, Turbulence, addresses these issues.

GENERAL CIRCULATION

Three-dimensional motion in the atmosphere results from differential heating of the Earth's surface, which in turn causes pressure differences. In 1735, George Hadley added an important notion to the understanding of the general circulation. He proposed that the

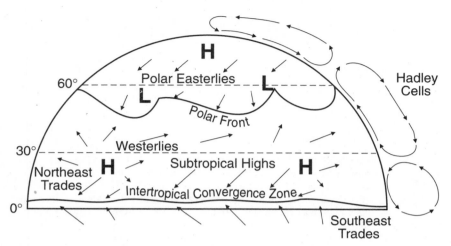

FIGURE 3-1 The general circulation at the surface consists of the polar high, polar front, and subtropical high and intertropical convergence zones.

Earth's rotation produced a deflective force on large-scale wind flows. Hadley was the first to propose a direct thermally driven and zonally symmetric circulation as an explanation for the trade winds. This consisted of the equatorward movement of the trade winds between about 30° latitude and the equator in each hemisphere, with rising wind components near the equator, poleward flow aloft, and finally descending components at about 30° again. The Hadley Cell is a direct thermally driven vertical motion system. There are typically three in each hemisphere, as illustrated in Fig. 3-1. This causes zonal wind patterns with strong easterly and westerly components, which cause the basic motion system of the general circulation.

The converging northeast and southeast trades produce the *intertropical convergence zone* (ITCZ). The ITCZ fluctuates in position and intensity; it can be weak and discontinuous. This can be seen in Fig. 3-2, which is a satellite view of the North Pacific Ocean. The source of the trades are the subtropical highs. Note the location of the high-pressure cells (H) in Fig. 3-2. It should be no surprise that in the vicinity of the ITCZ there is rising air—convergence—and in the area of the subtropical highs, descending air—divergence.

Moving north, about 30° latitude, are the prevailing westerlies. These are also due to the subtropical highs. At the poles is another area of subsidence. This is the region of the polar high—some of the highest atmospheric pressures ever recorded have occurred in these areas. This is also the area of the polar easterlies at about 60° latitude.

HIGH PRESSURE IN POLAR REGIONS   In the late 1980s pressures well above 31.00 in Hg occurred in Alaska. This precipitated emergency regulations because most altimeters can be corrected for pressure only up to 31.00 in Hg.

Between the polar easterlies and the midlatitude westerlies is the polar front. This is another area of global convergence of warm air from the south and cold air from the north. The polar front is more or less continuous around the world as shown in Fig. 3-2. However, where it is weak there may be areas of little or no weather such as the areas over the central Pacific and western continental United States, as seen in Fig. 3-2.

Wind flow at the surface is depicted on surface analysis charts.

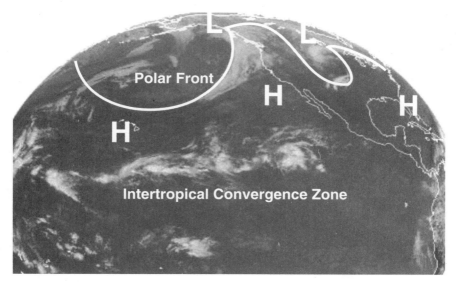

FIGURE 3-2 The components of the general circulation are easily identified on satellite imagery.

Wind is directed by three forces: pressure gradient force, Coriolis force, and frictional force. The effects of these forces are illustrated in Fig. 3-3.

Differences in pressure cause *pressure gradient force*. At least initially, the wind wants to blow from areas of high pressure to those of low pressure. On weather maps, lines of equal pressure are called *isobars*. Isobars typically are nonconcentric circles surrounding centers of high and low pressure. If pressure gradient were the only force acting on the wind, wind would always blow perpendicular to isobars—directly from areas of high pressure to areas of low pressure. This is illustrated in Fig. 3-3, top portion, left. However, this is not the case.

Pressure gradient force determines the strength of the wind. The stronger the pressure gradient force, the stronger the wind. On the surface analysis chart, areas of strong winds are identified by close spacing of isobars; areas of weak winds, by wide spacing of isobars.

Because of the Earth's rotation, there is an apparent force that deflects the wind from a straight path relative to the Earth's surface.

CORIOLIS FORCE This is *Coriolis force*. In 1856 an American, William Ferrel, published the "Essay on the Winds and Currents of the Ocean." Ferrel showed mathematically that winds are affected by the rotation of the Earth. Gaspard de Coriolis, a French mathematician, in 1835 had already developed the theory of an apparent deflection force produced by angular rotation. Although Ferrel applied the theory to the winds, in meteorology, we refer to this effect as a *Coriolis force*.

In the northern hemisphere, the Coriolis force deflection is to the right; in the southern hemisphere, it is to the left. Coriolis force is maximum at the poles and zero at the equator. Coriolis forces always act at a right angle to wind direction and directly proportional to wind speed. The effect of Coriolis is to balance the pressure gradient force, causing wind to blow parallel to the isobars. Note in Fig. 3-3, top portion, right, that the pressure gradient and Coriolis are equal and opposite, causing the wind to blow parallel to the isobars.

EFFECTS OF CORIOLIS AND FRICTION

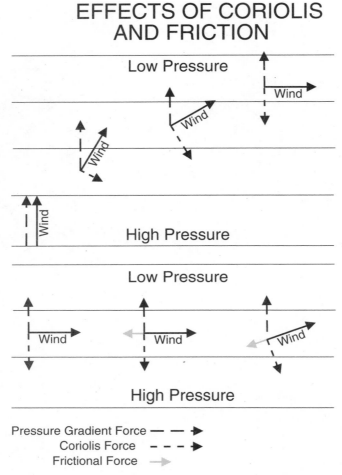

FIGURE 3-3   Wind is directed by three forces: pressure gradient force, Coriolis force, and frictional force.

As we shall see, Coriolis force also plays a role in vorticity and the development of hurricanes.

Frictional force between moving air and the ground slows the wind at and near the surface. This frictional layer is also called the *planetary boundary layer*. The rougher the terrain, the greater the frictional effect. Over oceans, the frictional effect is present to about 1000 ft; over land, to about 2000 ft. Frictional force always acts opposite to wind direction. This is illustrated in Fig. 3-3, lower portion. As frictional force slows the wind, Coriolis force decreases. However, friction has no effect on pressure gradient force. At and near the surface, the three forces eventually reach equilibrium (Fig. 3-3, lower portion, right). Pressure gradient balances the combined effect of frictional and Coriolis forces. Due to the three forces, the wind blows across isobars at an angle out of higher-pressure zones toward, and into, lower-pressure zones.

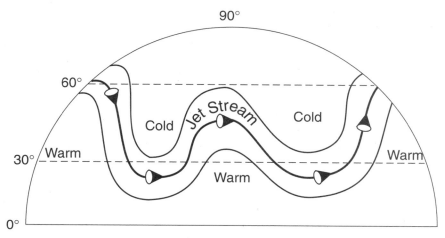

FIGURE 3-4 Planetary, or Rossby, upper-level waves have lengths of 2000 to 4000 mi, with normally three to seven circling the globe.

As a result, at the surface in the northern hemisphere, wind blows out of high-pressure areas toward low pressure in a clockwise direction and into low-pressure areas in a counterclockwise direction. Directions are reversed in the southern hemisphere. This general flow can be seen in Fig. 3-1.

We briefly touched on the transport of air at upper levels with the discussion of Hadley Cells. To complete the circulation pattern of major wind systems requires an exchange of air at upper levels. This is required to compensate for, or balance, the movement of air at the bottom of the troposphere.

Strong winds blow in the upper troposphere and lower stratosphere. When these winds become strong enough, they are called a *jet stream,* as discussed in Chap. 1. Refer to Fig. 3-4. These winds, which generally occur between 30° and 60° latitude, flow more nearly along parallels of latitude than those at the surface.

These waves apparently result from the tendency of winds in large-scale motion systems to retain a constant vorticity about the Earth's rotation. The resulting flow, known as *planetary,* or *Rossby, waves* (planetary scale), have lengths of 2000 to 4000 mi. There are normally three to seven circling the globe. These global, or long-wave, ridges and troughs extend for thousands of miles. Long waves move generally eastward at up to 15 kn, but they can remain stationary for days or even retreat. Their length, amplitude, and position are influenced by differential heating at the surface and mountain barriers, such as the Rockies.

ICE AGE One theory holds that the Ice Age was, at least in part, caused by the deflection of the jet stream due to the creation of the Himalayan mountain range.

Only pressure gradient and Coriolis forces act on upper-level winds. Without frictional force in play, these winds blow parallel to contours as depicted on upper-level constant pressure charts. The solid lines in Fig. 3-4 represent lines of constant pressure—contours— in the upper atmosphere.

Upper-level wind flow can be found on constant pressure charts.

Because the Earth's axis is tilted about 23°, the Sun's maximum heat strikes different latitudes during the year. This causes the seasons. In the northern hemisphere the general circulation, described in Fig. 3-1, moves south during the winter and north during the summer.

In the summer the eastern Pacific high blocks weather systems approaching the northwest Pacific Coast, forcing them on a more northerly track into Canada. The Bermuda high brings a moist, warm southerly flow to the southeastern United States.

In the winter the highs migrate southward. This allows storm systems to move from the Pacific into the Pacific Northwest, sometimes reaching into southern California. Storms continue eastward through the central and southern plains, and along the eastern seaboard. As the polar high moves south, it reaches well into the United States several times a year. Occasionally, the polar air can reach as far south as Texas and Florida, bringing subzero temperatures.

The jet stream also moves south and at a lower altitude than in summer. In fact, the jet stream is strongest in spring in the western Pacific and eastern United States because temperature differences are greatest during this period.

The location and strength of the jet stream can be found on the 300- and 200-mb constant pressure charts.

WINDS

Winds, especially near the surface, are a major factor in nonfatal, noninjurious aviation accidents. The most troublesome phase is landing, followed by takeoffs. Crosswinds and gustiness account for the largest percentage of problems. Of these, many occur due to loss of control. However, long landings, downwind or crosswind landings, delayed go-around, and overshoots account for a significant number.

Winds aloft, accompanied by poor planning, account for a number of fuel exhaustion accidents every year. All are preventable.

Mountains, valleys, and water surfaces affect small-scale (subsynoptic) wind systems, which are superimposed on the larger-scale (synoptic) circulation. These include sea/land breezes, mountain/valley winds, upslope/downslope winds, and mountain waves. Of these, downslope winds and mountain waves are often the most significant to aviation operations. Wind direction and speed is crucial for determining crosswind components. Variable wind direction can make takeoff and landing difficult, even at relatively slow speeds.

Gusts are rapid fluctuations in speed. *Gustiness* is a measure of turbulence. The greater the difference between sustained speed and gusts, the greater the turbulence. When a sudden increase of at least 15 kn, sustained at 20 kn or more for at least 1 min occurs, a squall is reported. Usually associated with thunderstorm activity, the term *squall* implies severe low-level wind shear as well as severe turbulence.

Wind direction, speed, and character (gusts or squall) must be considered when determining crosswind components, or the advisability of landing at a particular airport.

Winds aloft forecasts provide the pilot with two valuable pieces of information: wind direction and speed, plus temperature. Both significantly affect aircraft operation and performance. Failure to properly consider and apply either can be potentially hazardous.

In spite of its limitations, winds aloft forecasts can never be ignored. Pilots are required by regulations to consider "fuel requirements" and are prohibited from beginning a flight either VFR or IFR "unless (considering wind and forecast weather conditions)," the aircraft will have enough fuel to fly to its destination, or an alternate if required, and still have appropriate fuel reserves. Fuel reserve minimums required by

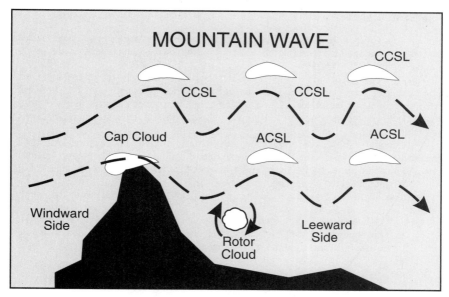

FIGURE 3-5 Major mountain waves occur east of the Cascade Range, the Sierra Nevada, the Rocky Mountains, and the Appalachian Mountains.

regulations do not necessarily equate to "safe," and they in no way relieve the pilot from keeping careful track of ground speed and revising the flight plan accordingly.

When reserves are marginal, good operating practice dictates the careful tracking of position and ground speed. *Marginal* is not necessarily synonymous with *legal*; in sparsely populated areas, a fuel reserve of 30 min with clear weather reported and forecast might be sufficient. But with marginal weather or thunderstorms, and the nearest suitable alternate 35 min away, a 30-min reserve doesn't make any sense. Legal fuel reserves might not be satisfactory with a busted forecast.

The venturi effect of mountains and mountain passes accelerates winds over ridges and through passes. This is illustrated in Fig. 3-5. The dashed lines in the figure are streamlines. *Streamlines* show the flow of the winds. On the windward side, the winds are accelerated due to the compressional effect of the mountain range. On the leeward side, wind speed decreases. Stronger-than-forecast winds should be expected in these areas, especially within 5000 ft of terrain.

CASE STUDY On a flight from Van Nuys to Fresno, California, the winds aloft were forecast out of the northwest at 20 kn. At 8500 ft in a Piper Arrow with a true airspeed of 130 kn, ground speed was only 90 kn! The venturi effects of the mountains had doubled the wind speed.

Pilots must continually keep track of ground speed. Weather updates must be obtained far enough in advance to be acted upon, before fuel runs low. Hoping a stronger-than-forecast head wind will abate is folly.

Surface wind conditions can be obtained from METAR reports. Forecast surface winds are available in TAFs and the area forecast. Winds aloft forecasts are available in both tabulated or chart form.

Sea/Land Breezes

Recall from Chap. 1 that land surfaces warm and cool more rapidly than water surfaces. Therefore, typically land is warmer than the sea during the day and cooler than the sea during the night.

During the day the land heats, but the water remains at relatively the same temperature. The land heats the air near the surface through the process of conduction. The air warms, becomes less dense, and rises. The cooler, more dense air over the water moves into the relatively lower pressure area. Since the wind blows from the sea to the land, it is called a *sea breeze*.

With enough moisture and lift, clouds develop at the lifted condensation level (LCL) over the land. This is particularly true in the southeast United States with its abundant moisture and unstable air. Activity tends to be widespread, with some areas receiving torrential downpours and adjacent areas remaining dry.

Sea breezes, since they occur during the day, can often be seen on visual satellite imagery once clouds develop.

At night this circulation is reversed. The land cools more rapidly than the sea. The wind blows from the cool land toward the warmer water, creating a land breeze. Again, if moisture, lift, and instability are right, thunderstorms and rain showers develop.

Mountain/Valley Winds

In the daytime, mountain slopes become warm, heating the adjacent layer of air. This layer is warmer than the air at the same altitude farther from the slope. The resulting density difference creates a convective current in which the air over the valley sinks, forcing the warmer air up the mountains as a *valley wind*—again so named because the wind blows from the valley toward, or up, the mountain.

At night, the layer of air near the mountain slope cools more rapidly than the air over the valley. There is greater air density near the slope than at the same levels some distance horizontally from the slope. The cool air flows down the slope as a mountain wind. The mountain wind—sometimes called a *gravity wind*—often continues to flow down the more gentle slopes of canyons or valleys, and in such cases it takes on the name *drainage wind*. It can become quite strong over some terrain when atmospheric conditions are favorable, and in extreme cases it becomes hazardous when flowing through canyons.

Upslope/Downslope Winds

Winds blowing upslope cause the air to cool adiabatically. The result is increased humidity, and clouds and precipitation—with sufficient moisture. The type of clouds—for example, stratus and fog, or cumulonimbus—will depend on stability. Upslope allows larger supercooled water droplets to be suspended and carried to a higher altitude, enhancing icing potential and severity. Soaring pilots use upslope winds to gain altitude and prolong flight.

> **CASE STUDY**   There are islands in the Hawaiian chain that have tropical forests on half the island and deserts on the other half. The terrain squeezes out the moisture on the windward side of the island—upslope, leaving other parts of the island in a virtual permanent drought.

Upslope promotes rain shadows, with semiarid conditions, downwind of mountain ranges in the western United States. Another effect of upslope is fog, discussed in Chap. 22. In slightly unstable conditions, upslope will produce stratocumulus clouds. In a moist, unstable atmosphere, upslope may be all that's needed to trigger thunderstorms.

Any wind blowing down an incline, where the incline causes the wind, is a *katabatic wind*. Thus the mountain wind is a type of katabatic wind. If the wind is warm, it is called a *foehn* (pronounced "fān") *wind*; if cold, it may be a *fall wind*—such as the *bora*. (The bora is the cold northeast wind on the Dalmatian Coast of Croatia and Bosnia/Herzegovina in winter caused by cold air from Russia crossing the mountains and descending over the relatively warm coast to the Adriatic.) Or a fall wind may be a gravity wind, such as a mountain wind.

Following its ascent over a mountain barrier, a foehn wind is a katabatic wind. During its movement upslope, it cools at the dry, then moist adiabatic lapse rate, losing its moisture as precipitation. On descent, it warms through the adiabatic process; the air will be warmer than at the same altitude during ascent. Recall our discussion in Chap. 2 and Fig. 2-2.

Strong downslope flows are sometimes given local names, such as the *Chinook* (Native American name for "snow eater") that develops along the eastern slopes of the Rockies. It may raise temperatures by 10°C to 20°C in 15 min and can exceed 85 kn. The Chinook often reaches hundreds of miles into the high plains.

The *Santa Ana* of southern California is another foehn wind—named for strong winds that blow through Santa Ana Canyon. High pressure over the Great Basin and lower pressure off the southern California coast cause the Santa Ana. Pressure gradients determine their severity. The stronger the pressure gradient force, the stronger the wind. Like the Columbia River Gorge in Oregon, the winds are accelerated through the passes and canyons. The result is strong gusty winds, severe turbulence, wind shear, and crystal clear skies—except in local areas of blowing dust and sand. The Santa Ana can last from a day or two to a week at a time. These winds will continue as long as the pressure gradient is there to support them.

A *foehn wall* is the steep leeward boundary of flat, cumuliform clouds formed on the peaks and upper windward sides of mountains during foehn conditions—such as the Santa Ana and Chinook. This cloud formation should alert pilots to strong winds, possible turbulence, and wind shear. The cap cloud in Fig. 3-5 is similar to the foehn wall cloud. When sufficient moisture is present, the cloud forms on the windward side and evaporates on the leeward side of the range.

Mountain Waves

When winds aloft blow in excess of about 40 kn, approximately perpendicular to a mountain range, and speed increases with height in a stable atmosphere, a *mountain wave,* or *standing wave,* can develop. Refer to Fig. 3-5. Turbulence can become severe to extreme. Updrafts and downdrafts occasionally reach 3000 ft/min and can exist to the tropopause or slightly higher. Downdrafts may dip to the surface on the leeward side of the mountains. Large waves may form to the lee of mountains and may extend hundreds of miles downstream. Complete overturning may occur under the wave crests at lower levels. Altimeter errors might exceed 1000 ft. Mountain wave activity typically can be seen on visible, and sometimes infrared satellite imagery. With lack of adequate moisture, waves occasionally occur in clear air.

Major mountain waves occur east of the Cascade Range in Washington and Oregon, the Sierra Nevada in California and Nevada, the Rocky Mountains, and the Appalachian Mountains. Smaller waves can develop over any hill or mountain.

These waves are termed *standing* because the crests and troughs remain stationary while the wind undulates rapidly through them. With sufficient moisture, clouds form. And, although the clouds appear to be stationary, condensation continually occurs at the leading edge of the cloud and evaporation at the trailing edge. The clouds are continuously forming and dissipating.

FIGURE 3-6   Lenticular clouds appear smooth and remain stationary to the observer.

Cap clouds hug the tops of the mountains and appear to flow down the leeside. They are indicators of strong downdrafts on the leeside of mountain range.

Specific names are given to mountain wave clouds: *standing lenticular stratocumulus* (SCSL), a low cloud; *standing lenticular altocumulus* (ACSL), a middle cloud; and *standing lenticular cirrocumulus* (CCSL), a high cloud. Lenticular clouds appear smooth and remain stationary to the observer; they might develop as horizontal bands produced by long ridges as in Fig. 3-6 or circular and stacked from isolated peaks as in Fig. 3-7. A cap cloud may develop over the mountain crest. Although these clouds imply turbulence, turbulence will not always be present.

Lenticular clouds represent visual identification of the wave crests. They may extend to 40,000 ft. Depending on the vertical distribution of moisture in the atmosphere, these clouds may be stacked, as in Fig. 3-7. The distance from the ridge to the first wave crest depends on wind speed, lapse rate, and the shape of the ridge.

Rotor, bell-shaped, clouds that often appear as tubular lines of cumulus or fractocumulus clouds parallel to the ridge line underneath the lenticulars, always imply severe or greater turbulence. Depending on the strength of the flow and atmospheric conditions, less developed lines may form downstream beneath the crests of subsequent waves. The base of the rotors is about the same height of the ridge and may extend vertically 3000 to 5000 ft. Violent updrafts are common in the vicinity of rotor clouds.

CASE STUDY   One such situation occurred at Reno, Nevada, where surface winds were reported gusting to 73 kn. The pilot of a corporate jet reportedly abandoned the approach when all the bottles in the cabin's liquor cabinet broke. (A new definition for severe turbulence?)

Moderate to severe turbulence may be encountered on the windward (updraft) and leeward (downdraft) side of the lenticulars. However, the most dangerous situation occurs when the smooth laminar flow through the wave breaks down, which can result in severe

FIGURE 3-7 Lenticular clouds develop as horizontal bands produced by long ridges, or circular and stacked from isolated peaks.

turbulence throughout the vertical depth of the wave. When this occurs, the highest lenticular clouds, and to a lesser extent the lower wave clouds, will have a jagged, irregular edge.

CASE STUDIES We encountered a mountain wave in California's Owens Valley, east of the Sierra Nevada, while flying a Cessna 150. With cruise power and attitude, the airplane rode the wave, at the rate of 500 ft/min, from 8500 to 13,500 ft, and back down again. The ride was absolutely smooth! Once below 8500 ft, the turbulence was again moderate.

Notwithstanding the previous example, a Navy T-39 trainer was flying a low-level, high-speed navigational training route in mountainous terrain when it encountered severe turbulence. Gust acceleration loads were so high that aircraft design limits were exceeded, resulting in separation of the tail. All aboard were killed. Mountain waves should never be taken lightly. In addition to the T-39 crash, mountain waves were identified in the crash of a C-118 and in extensive damage to a B-52. While this type of turbulence is obviously critical to traditional low fliers like helicopters, all aircraft are susceptible.

Visual satellite images often confirm the presence of wave clouds. An example is included in Chap. 10, Satellite Imagery.

CHAPTER 4
SIGNPOSTS IN THE SKY

Pilots can determine quite a lot about the state of the atmosphere from clouds, specifically cloud types. We have already discussed some cloud types and their significance on aviation operations. As mentioned in the introduction, App. B contains a cloud characteristics table for easy reference.

Cloud types are divided in four categories: low, middle, and high and those with vertical development. However, for aviation purposes, other than some specific types, clouds fall into two major divisions: stratiform and cumuliform. We will discuss the significance of each. Before leaving this subject, we will discuss effects of terrain on clouds and precipitation (for example, lake effect).

CLOUD TYPES

Clouds form through a cooling process. The process initiates and then must sustain condensation or sublimation. Cooling processes are adiabatic or diabatic. We have previously discussed the adiabatic process. *Diabatic* involves the exchange of heat with an external source, or *nonadiabatic*.

The adiabatic process cools the air by raising it through the processes of convection, convergence, or orographic lifting. The diabatic process produces a loss of heat. The loss may occur through terrestrial radiation resulting in fog or low clouds. Conduction through contact with a cold surface may result in dew, frost, or fog. The process may be associated with the movement of air across a cold surface—advection. Finally, the process may occur through mixing with colder air. If the mixture has a temperature below its dew point, clouds or fog may form.

There are a number of methods of cloud classification. Clouds may be classified according to appearance, how they are formed, or the height of their bases. It was not until 1803 that Luke Howard, an Englishman, first classified cloud forms. He divided clouds into three main categories using Latin names:

- *Cirrus,* meaning curly
- *Stratus,* meaning spread out
- *Cumulus,* meaning heaped up

Two affixes may be added:

- *Alto,* meaning high
- *Nimbo,* meaning rain

Today, meteorologists divide clouds into four main groups:

- *Low clouds,* bases near the surface to about 6500 ft
- *Middle clouds,* bases from 6500 feet to 20,000 ft
- *High clouds,* bases at or above 20,000 ft
- *Clouds with vertical development,* bases near the surface, tops of cirrus

Sometimes clouds are classified as one of two general types: stratiform and cumuliform. *Stratiform* are clouds with extensive horizontal development, associated with a stable air mass. Stratiform clouds consist of small water droplets. The following cloud types are classified as stratiform:

- Stratus
- Stratocumulus
- Nimbostratus
- Altostratus
- Cirrostratus

Cumuliform are clouds that are characterized by vertical development in the form of rising mounds, domes, or towers, associated with an unstable air mass. Because of upward-moving currents, cumuliform clouds can support large water droplets. In the case of cumulonimbus, updrafts can support hail. The following cloud types are classified as cumuliform:

- Altocumulus
- Cirrocumulus
- Cumulus
- Cumulonimbus

Table 4-1 shows the general differences between stratiform and cumuliform clouds. Notice that there is a correlation between Tables 4-1 and 2-1, Characteristics of Atmospheric Stability.

Although not one of the two previously mentioned general classifications, a third generic cloud type may be used. It describes the entire group of high clouds—cirroform. *Cirroform* is often used during pilot weather briefings and text forecasts to translate one or all of the high cloud types: cirrus, cirrostratus, or cirrocumulus.

Figure 4-1 illustrates four cloud types and two cloud groups. Low stratus is in the valleys topping a surface based inversion. This indicates a stable air mass at lower levels. Notice the poor visibility in the valley. Moderate winds over the mountains have produced

TABLE 4-1   Characteristics of Cloud Types

Stratiform	Cumuliform
Stable lapse rate	Unstable lapse rate
Poor visibility	Good visibility
Smooth	Turbulent
Widespread cloud mass	More localized cloud mass
Steady precipitation	Showery precipitation
Rime icing	Clear icing

FIGURE 4-1 Cloud types reveal stable air in the valleys and at high levels. Stratocumulus over the mountains indicates some instability and turbulence.

stratocumulus. The presence of stratocumulus indicates some instability, with good visibility and turbulence, over the mountains. High-level moisture has produced cirrus and cirrostratus—stable air at high levels.

Low Clouds

Stratus indicates a stable air mass. Stratus may lower, becoming fog, after sunset due to the absence of surface heating and decreased winds. During the morning, fog may lift, due to surface heating, to become a stratus layers. Ceilings and visibilities are typically poor—recall Fig. 4-1. Flight through stratus clouds is smooth. Precipitation, when it occurs, is usually light, often appearing in the form of drizzle.

Stratocumulus represent a moist layer with some convection. Stratocumulus may form from the spreading out of cumulus, which indicates decreasing convection. Stratocumulus can develop from stratus with winds of moderate to strong intensity—as shown over the mountains in Fig. 4-1. Updrafts and turbulence develop below and within the layer.

Fractostratus and *fractocumulus* (*scud*—shreds of small detached clouds moving rapidly below a solid deck of higher clouds) are normally associated with bad weather. Figure 4-2 shows a layer of scud. Pilots who fly low in poor weather conditions associated with these clouds, are known as "scud runners." This practice has given rise to the aviation axiom "There are old pilots and there are bold pilots, but there are no old, bold pilots." The life expectancy of scud runners is rather poor.

CASE STUDY An unfortunate pilot attempted to negotiate Oakland's east bay hills under these conditions. He took his two grandchildren with him. As John Hyde, an excellent pilot and ex-Army Aviator frequently laments, "Cowardice is the better part of valor."

FIGURE 4-2   Scud—shreds of small detached clouds moving rapidly below a solid deck of higher clouds—
are normally associated with bad weather.

Nimbostratus are low clouds, usually uniform, and dark gray in color. Nimbostratus
usually evolve from altostratus that have thickened and lowered, sometimes with a ragged
appearance. This is the ordinary rain cloud that produces light to moderate, steady precipitation. Flight in, and in the vicinity of, nimbostratus is usually smooth. The presence of
nimbostratus indicates a stable air mass.

Middle Clouds

Middle clouds fall into two general types: altostratus and altocumulus.

Altostratus indicate a stable atmosphere at midlevels. Some altostratus are thin and semitransparent, while others are thick enough to hide the sun or moon. Figure 4-3 shows a thickening band of altostratus. The Sun backlights the cloud in the upper center of the photograph.
As in this case, these altostratus indicate the approach of a warm front. Ceilings are high and
visibilities good below the cloud because no precipitation is falling. These clouds can produce precipitation in the form of rain or snow, even heavy snow at times.

Altocumulus indicate vertical motion and instability at midlevels. Altocumulus may be
thin, mostly semitransparent. Some altocumulus are thick, developed, and may be associated with other cloud forms. In Chap. 3 we discussed altocumulus standing lenticular, associated with mountain waves. Figure 4-4 shows a think band of altocumulus below a band
of cirrus. This cloud often signals the approach of a cold front.

Altocumulus castellanus (ACC), a midlevel cloud, indicates moisture and instability at
that level. ACC might indicate thunderstorm development. Showers falling from these
clouds can evaporate before reaching the surface as illustrated in Fig. 4-5. This phenomena
is known as *virga*. Virga is rain that evaporates before reaching the surface. Evaporative

FIGURE 4-3   A thickening band of altostratus, which indicates stable air at midlevels.

FIGURE 4-4   Thickening and lowering altocumulus may signal the approach of a cold front.

FIGURE 4-5   Virga is rain that evaporates before reaching the ground.

FIGURE 4-6   Altocumulus castallanus indicate instability aloft; a stratus layer attests to a stable layer at the surface.

cooling turbulence develops in the vicinity of virga. Precipitation evaporates and cools the air, causing downdrafts and wind shear. A pilot penetrating these areas will encounter wind shear turbulence, which can be severe. Turbulence can be avoided by circumnavigation of these areas.

Figure 4-6 shows altocumulus castallanus that have developed above a stratus layer. The lower layer of the atmosphere is stable, as indicated by the stratus and haze. The upper layers are unstable, as testified by the presence of ACC. This could be hazardous to pilots flying below the stratus who suddenly encounter heavy rain and turbulence. A group of hot-air balloons were downed in California's Napa Valley under similar circumstances.

Altocumulus floccus is a cloud with a cumuliform or rounded appearance; the lower portion is ragged and often accompanied by virga. Altocumulus floccus may evolve from altocumulus castellanus. Like altocumulus castellanus, they indicate moisture and instability at midlevels in the atmosphere.

High Clouds

High clouds are known as *cirrus, cirrostratus,* and *cirrocumulus.* Cirrus clouds are composed entirely of ice crystals. Usually the air is so cold that they present no serious icing hazard.

Cirrus often consist of filaments, commonly known as *mares' tails.* Other cirrus are associated with cumulonimbus clouds. A thickening cirrus layer may indicate the approach of a front. Figure 4-7 shows a band of cirrus over a stratus deck.

At times, ice crystals or snowflakes can fall from cirrus clouds. As they fall into dry air, they sublimate—change directly from a solid to a gas. These dangling white streamers are called *fall streaks.*

FIGURE 4-7 Cirrus often consist of filaments, commonly known as *mares' tails.*

Cirrostratus are sheets or layers of cirrus. Sun or moon halos may occur, as illustrated in Fig. 4-8. When cirrostratus appear within a few hours after cirrus in midlatitudes, there is a good probability of an approaching front. Thick bands of cirrostratus often mark the location of the jet stream. Jet stream cirrus can be seen on satellite imagery and are often observed in the area of an upper-level ridge. The poleward edge of the cirrus shield ends in a line parallel to the jet axis.

Cirrocumulus indicate vertical motion at high levels and might indicate high-altitude turbulence. Figure 4-9 shows a bank of cirrocumulus clouds. As discussed in Chap. 3, cirrocumulus standing lenticular often develop in a mountain wave. Transverse lines are often observed in jet stream cirrus. These cloud patterns appear as small-scale lines at an angle almost perpendicular to the jet. These lines look somewhat like waves, but they are much more irregular than mountain waves. Cirrus streaks are also associated with the jet stream. However, cirrus streaks are parallel to the jet core.

Cirrus, of itself, has no significance to low-level flights. Thick, extensive shields of cirrus may prevent or lessen the effect of thermal turbulence. As previously mentioned, cirrus is often associated with the jet stream and high altitude turbulence. Cirrus that forms as transverse lines or cloud trails, and cirrus streaks indicates moderate or greater turbulence. These clouds might be reported as cirrocumulus. Jet stream cirrus, transverse lines, and cirrus streaks are easily identified from satellite imagery.

A *contrail,* or condensation trail, is a cloudlike streamer frequently observed to form behind aircraft flying in clear, cold, moist air. Contrails typically form in the upper troposphere. Figure 4-10 shows an aircraft contrail over West Virginia. Contrails can form from the addition of water vapor produced by engine combustion. They may also form in air that is almost fully saturated; aerodynamic forces around the propeller tips or wings can cool the air and induce condensation.

FIGURE 4-8 When cirrostratus appear within a few hours after cirrus, there is a good probability of an approaching front.

In a very rare occurrence, an aggregate of ice crystals will form in the stratosphere. These are known as mother-of-pearl clouds.

Clouds with Vertical Development

Clouds with vertical development are cumulus and cumulonimbus (CB). Some cumulus indicate fair weather and are seemingly flattened, with little vertical development, and there is no significant weather with turbulence below the bases and no turbulence on top. Other cumulus contain considerable vertical development, generally towering. This type precedes the development of cumulonimbus and thunderstorms. Surface weather observations may contain TCU (towering cumulus) to describe this cloud. What's the difference? Towering cumulus resemble a cauliflower, but the tops of the cumulus have not yet reached the cirrus level. Cumulonimbus have a least part of the upper cloud smooth, fibrous, and almost flattened, often in the form of an anvil or plume. Figure 4-11 shows towering cumulus developing during the summer over California's Sierra Nevada mountains. Note the bubbling, cauliflower appearance of the tops.

FIGURE 4-9   Cirrocumulus indicate vertical motion at high levels and might indicate high-altitude turbulence.

FIGURE 4-10   Contrails typically form in the upper troposphere.

FIGURE 4-11   Towering cumulus are growing cumulus that resemble a cauliflower, but the tops of the cumulus have not yet reached the cirrus level.

With cumulus clouds, expect usually good ceilings and visibilities, except in the vicinity of precipitation. These clouds imply turbulent flying conditions. Precipitation is showery and heavy at times.

Cumulonimbus clouds exhibit great vertical development with tops composed, at least in part, of ice crystals. Tops no longer contain the well-defined cauliflower shape of towering cumulus. Cumulonimbus may develop a clearly fibrous (cirroform) top, often anvil shaped. Regardless of vertical development, a cloud is classified as cumulonimbus only when all or part of the top is transformed or is in the process of transformation into a cirrus mass. Any cumulonimbus cloud should be considered a thunderstorm with all the ominous implications.

Cumulonimbus mammatus (CBMAM), from the Latin word meaning udder or breast, result from severe updrafts and downdrafts. They are characterized by lobes that protrude from the bottom of the cloud as illustrated in Fig. 4-12. They indicate probable severe or greater turbulence, and often appear just before or at the beginning of a squall. Avoid these areas.

Heavy, showery precipitation falls from cumulonimbus clouds. (Showers are characterized by the suddenness with which they start and stop and rapid changes in their intensity.) Outside the areas of precipitation the air is usually clear and relatively smooth in the unstable air mass. In the vicinity of the showers, expect severe turbulence and strong up and down drafts. Figure 4-13 illustrates these conditions. Notice how visibility is good outside of the rain showers. However, the mountains are partly to completely obscured on the right side of the photograph. Even in the absence of clouds, precipitation is so heavy that VFR flight is not possible (not to mention the turbulence and wind shear).

Associated with convective activity, three additional cloud types that pilots should be familiar with may appear in the remarks of surface weather observations. A *shelf cloud,* layered and resembling shelves, can appear under a thunderstorm. A *wall cloud,* usually on the

FIGURE 4-12 Cumulonimbus mamatus indicate probable severe or greater turbulence and often appear just before or at the beginning of a squall.

FIGURE 4-13   Heavy showers, even in the absence of clouds, may preclude VFR flight (especially so due to the turbulence and wind shear).

southwest edge of the thunderstorm, has a lowered base and indicates the storm might be severe. A *roll cloud* appears as a detached, dense, horizontal cloud at the lower front part of the main cloud. All three indicate thunderstorms with a potential for severe weather.

Occasionally a severe thunderstorm will break through the tropospause into the stratosphere. This is one of the few instances when clouds appear in this layer of the atmosphere. The abbreviations TCU and CB may appear in body of METAR and SPECI reports and CB on terminal aerodrome forecasts to indicate the presence of these clouds. Manual surface observation may contain other significant cloud types, as discussed in this section.

PRECIPITATION

We've used the term *precipitation* throughout the book thus far. To ensure that we all have the same understanding of this phenomena: *Precipitation* is any or all of the forms of water particles, whether liquid or solid, that fall from the atmosphere and reach the ground. Precipitation is a major classification, but it does not include clouds, fog, dew, or frost. Precipitation is distinguished from cloud, virga, and fall streaks in that the water particles must reach the ground. Precipitation includes the following:

• Drizzle
• Rain
• Freezing drizzle and freezing rain
• Snow grains

- Snow
- Snow pellets
- Ice crystals
- Ice pellets
- Hail

Drizzle is made up of very small, numerous, and uniformly dispersed water drops that may appear to float. Unlike fog droplets, drizzle falls to the ground.

Drizzle usually falls from stratus clouds and indicates a relatively shallow cloud layer. It usually takes a cloud thickness of 4000 ft to produce precipitation. Drizzle restricts visibility to a greater degree than rain because it falls in stable air often accompanied by fog, haze, and smoke.

Rain is precipitation in the form of liquid water drops. Rain drops have diameters greater than 0.5 mm, which distinguish them from drizzle drops that have diameters less than 0.5 mm. (Right, we're all going out to measure the stuff!) From an observational point of view, other than size, drizzle appears to float or fall very slowly, while rain has a definite vertical speed.

We have already discussed how rain reduces forward visibility, but visibility to the side and downward remains good. In moderate rain forward visibility can be reduced to less than VFR. If the rain is heavy, it can obscure terrain. Flight through rain tends to be turbulent. Avoiding areas of rain will not only improve forward visibility but result in a smoother ride.

Freezing rain and freezing drizzle are caused by liquid precipitation falling from warmer air into air that is at or below freezing. Droplets freeze upon impact. Structural icing can be expected while flying through freezing precipitation. Freezing precipitation is probably the most dangerous of all icing conditions. It can build hazardous amounts of ice in a few minutes and is extremely difficult to remove. Freezing rain can flow back along the aircraft, covering the static port with the resultant loss of accurate pitot-static instruments—airspeed, altimeter, and vertical speed.

Snow grains are small, white, opaque grains of ice, the solid equivalent of drizzle. Since snow grains are already frozen, they typically do not present an icing hazard.

Snow is composed of white or translucent ice crystals, chiefly in complex branched hexagonal form and often integrated into snowflakes. Snow can fall about 1000 ft below the freezing level before melting. Snow can often begin with temperatures of 2°C; it's even possible to see snowflakes at temperatures around 10°C. This occurs only when the air is very dry. As snow falls into above-freezing air, it begins to melt. The water evaporates and cools the air. Evaporation cools the snow, which retards melting. Water vapor is added to the air, which increases the dew point. Finally, the air cools and becomes saturated at 0°C.

Dry snow does not lead to the formation of aircraft structural ice. However, wet snow—snow that contains a great deal of liquid water—produces structural icing.

Snow flurry is a popular term for snow showers, particularly of a very light and brief nature. On the other hand, a *blizzard* is used to describe a severe weather condition characterized by low temperatures, strong winds, and large amounts of snow fall.

BLIZZARD   The term *blizzard* is thought to have originated in Virginia, but it is now applied to similar occurrences in other countries. In North America blizzards occur with the northwesterly winds in the rear of low-pressure areas in winter. In popular usage the term often refers to any heavy snowstorm accompanied by strong winds.

Ceiling and visibility can be good in light snow. Heavy snow, however, can reduce ceiling and visibility to zero and produce white-out effects.

Snow pellets—small, white, opaque grains of ice—form when ice crystals fall through supercooled droplets and the surface temperature is at or slightly below freezing. Falling from cumuliform clouds, snow pellets are more prone to cause structural icing than snow grains. Snow pellets are also known as *soft hail* or *graupel.*

Ice crystals might first appear as suspended forms, after which they fall from a cloud or clear air. They frequently occur in polar regions in stable air and only at very low temperatures. Ice crystals are not assigned an intensity.

CASE STUDY   At the Lovelock, Nevada, FSS on a very cold, clear day, ice crystals were sublimating to the right of the clear air. It was a beautiful sight, but it was of absolutely no significance to aviation or anything else.

Ice pellets, formerly called *sleet,* are grains of ice consisting of frozen raindrops, or largely melted and refrozen snowflakes. They fall as continuous or intermittent precipitation. Ice pellet showers are pellets of snow encased in a thin layer of ice formed from the freezing of droplets intercepted by the pellets, or water resulting from the partial melting of the pellets. Ice pellets do not bring about the formation of structural ice, except when mixed with supercooled water. Frequently, ice pellets or ice pellet showers indicate areas of freezing rain above.

Hail is precipitation in the form of balls or irregular lumps of ice, always produced by convective clouds, nearly always cumulonimbus. An individual ball is called a *hailstone.* Thunderstorms that are characterized by strong updrafts, large liquid water content, large cloud-drop size, and great vertical height are favorable for hail formation. The violent updrafts keep hailstones suspended for several up and down cycles. Each cycle adds a layer to the hailstone until it can no longer be suspended in the cloud.

Often hail size is reported in generic terms. The following are hail size categories in inches:

0.25	Pea
0.50	Marble
0.75	Dime
1.00	Quarter
1.25	Half dollar
1.50	Walnut
1.75	Golf ball
2.00	Egg (hen)
2.50	Tennis ball
2.75	Baseball
4.00	Grapefruit
4.50	Softball

Hail can cause severe damage to objects on the ground as well as aircraft. Blunted leading edges, cracked windscreens, and frayed nerves are a common result of a hail encounter. Like most thunderstorm hazards, avoidance is the only solution.

CASE STUDY   On a flight from Lancaster's Fox Field to Van Nuys in California, we flew the Cessna 172 into a big, black ugly cloud. Thunderstorms were not forecast. To this day I'm not sure if it was small hail that we encountered. The echo within the cabin was deafening. It sounded like the windscreen was about to implode, and I was sure it would, at the very least, have to be replaced. We survived and the windscreen was OK, but I don't want to fly into anything like that again. The solution: Don't fly into big, black, ugly clouds!

We have already discussed how terrain affects the weather. The land and water surfaces underlying the atmosphere significantly affect cloud and precipitation development. As moist air moves upslope and cools adiabatically, condensation produces clouds and precipitation. Showers occur if the air is unstable; when the air is stable, precipitation will be more widespread and steady. As the air moves over the crest and downslope, it heats adiabatically; precipitation ceases and clouds dissipate.

One effect of terrain is *rain shadows*. These occur predominately in the West. Pacific storms shed most of their moisture over the Cascades of Washington and Oregon and the Sierra Nevada of California. This results in semiarid climates east of these ranges. After frontal passage in the western United States there is typically enough wind and moisture to produce mountain obscurement and trap low clouds where the wind flows up the mountains. This may occur for a day or two after the front passes. Figure 4-14 illustrates this condition. Air flowing up from California's Central Valley has caused clouds to develop along the west slopes of the Sierra Nevada. Because the air is slightly unstable and the wind moderate, stratocumulus have developed. If the air were stable and the winds light, a condition more like the upslope of the western plains would occur.

Another phenomenon is lake effect. Often in winter, cold air moves over relatively warm lakes. The warm water adds both heat and water vapor to the air. The added heat makes the air unstable, resulting in showers to the lee of the lakes. Since it's winter, snow showers develop downwind. These snow showers can be heavy and produce severe aircraft icing. This often occurs in the Great Salt Lake area of Utah and the Great Lakes. In November 1996 a severe lake effect caused heavy snow and the closing of Cleveland's Hopkins International Airport for days. During the period several aircraft slid off the runway.

FIGURE 4-14   Air flowing up from California's Central Valley has caused stratocumulus to develop along the west slopes of the Sierra Nevada.

CASE STUDY   During the first week of January 1988, lake effect dumped 70 in of snow on the Tug Hill region of New York State. Areas only a few miles north and south experienced less than 6 in of snow!

Type and intensity of precipitation are found in the body of surface weather reports, and they are forecast in terminal aerodrome forecasts and the area forecast. Manual surface observation remarks may include significant types of precipitation, such as WET SNOW.

CHAPTER 5
FAMILIES OF THE AIR

In Chap. 1 we discussed the structure of the atmosphere and in Chap. 3 general circulation. As we have done in our earlier discussions, we will build on the knowledge gained in those chapters and this chapter to better understand weather theory and phenomena and its application to flight planning and inflight decision making.

This chapter begins with a discussion of air masses, their source regions, and properties. From here we move on to air mass modification; then to air mass influence on aviation weather.

Within our discussion of air masses we will present the effects of low- and high-pressure areas. Pressure patterns indicate areas at the surface under the influence of high or low pressure, troughs, or ridges. The terms *high* and *low,* like *hot* and *cold,* are relative. A *high* is simply an area completely surrounded by lower pressure. Conversely, a *low* is an area surrounded by higher pressure. Convergence occurs in areas of low pressure, divergence in areas of high pressure. Surface convergence and divergence affect only the lower 10,000 ft of the atmosphere, but they can be major factors in the weather machine—vertical motion. In general, areas of low pressure are associated with poor weather, and high pressure with good weather. Of course, there are no absolutes, and we will find a number of exceptions to this general rule.

CASE STUDY "A disastrous thunderstorm accident close to Bowling Green, Kentucky, in 1943 that involved an American (Airlines) DC-3 started a chain of events that eventually led to the first systematic research into thunderstorm behavior. The plane crashed onto the ground either near or under a severe thunderstorm. Buell [C. E. Buell, chief meteorologist, American Airlines, 1939–1946] initiated a letter to the Civil Aeronautics Board pointing out the appalling dearth of understanding of what actually occurs inside a thunderstorm, as evidenced by the accident investigation. He recommended a massive research effort be organized to probe into thunderstorms and document their internal structure." (Peter E. Kraght, *Airline Weather Services, 1931–1981*, Peters Books, 1986)

Thunderstorms were and are a major hazard to aviation operations. It has been said that sooner or later, a pilot must contend with thunderstorms. This would seem to be an oxymoron. In this section we will define and dissect this phenomenon.

As aircraft instruments and navigational system capabilities improved, pilots began taking on more and more weather. Hazards of fog and low clouds were solved with the instrument landing system (and increased fuel reserves). Icing, to a large extent, was overcome in the 1930s, and, for the most part, turbulence and thunderstorms were mastered with high cruise altitudes and radar. With these hazards virtually solved, major air carrier accidents most often fall into the categories of mechanical failure, pilot error, and wind shear caused by microbursts. (Typically major air carrier accidents are caused by a combination of the previous causal factors. Any one of the factors would not have resulted in an accident, but in combination they have proven disastrous. More about this when we discuss risk assessment and management.)

AIR MASSES

An *air mass* is a widespread body of air with homogeneous—that is, similar—properties. Air masses take on the properties of the region where they originate. These areas are known as *source regions*. Air masses have similar temperature and moisture content through both their horizontal and vertical cross sections. As air masses migrate, they undergo modifications to their temperature, moisture, and lapse rate.

To better understand air mass source regions, it might be helpful to review Fig. 3-1. As discussed in Chap. 3, at the pole is the polar high—stable. Between the polar easterlies and the midlatitude westerlies is an area of global convergence that accounts for the polar front—unstable. At around 30° latitude are the subtropical highs, an area of surface divergence—stable. In the vicinity of the equator is the intertropical convergence zone, another area of surface convergence—unstable.

When air masses develop and remain relatively stationary in their source regions, they take on the characteristics (temperature, moisture, and stability) of that region. Air mass source regions are arctic, polar, and tropical, which reflects their temperature. To indicate moisture content, air masses are classified as continental (dry) or maritime (moist). The following air mass source regions affect North America:

- Polar ice cap
- Plains of northern Canada
- North Pacific Ocean
- North Atlantic Ocean
- Gulf of Mexico
- Southwestern United States
- Northern Mexico

Air stagnating over northern continental regions forms continental polar or continental arctic air masses, depending on temperature. (Remember from Chap. 1 that temperatures are relative. *Cold* means colder than the underlying surface; *warm* means warmer than the underlying surface.) These air masses are characterized by cold, dry, air, reflecting their source region, and are very stable. Maritime polar air masses form over northern oceanic areas and are generally not as cold as continental polar air masses, especially in winter. Why? As discussed in Chap. 1, water surfaces tend to retain heat, while land areas cool rapidly. Maritime polar air masses tend to contain more moisture than their continental counterparts, and they are conditionally unstable. Maritime tropical air masses form over warm oceanic areas in lower latitudes, and they are very moist and usually unstable. Arid continental regions produce continental tropical air masses that are hot, dry, and unstable.

Refer to Fig. 5-1. Continental polar, and sometimes arctic, air masses develop over Canada and Alaska. They push down through Canada into the United States, sometimes reaching as far south as California and the Gulf of Mexico—cold, dry, and stable. The North Pacific and North Atlantic oceans spawn maritime polar air masses—cold, moist, and conditionally unstable. Since the general flow, over midlatitudes, is from west to east, the United States is rarely affected by maritime polar air masses that develop over the North Atlantic. The west coast, Gulf coast, and Atlantic coast are affected by maritime tropical air masses—warm, moist, and unstable. Finally, from northern Mexico and the desert southwest come continental tropical air masses—warm, dry, and unstable.

Dust devils, or *dust whirls* as they are known in the international METAR code, form in continental tropical air masses. These microscale whirlwinds form over hot, dry land in fair weather and light winds as the result of intense local daytime convection and surface friction. They have diameters of 10 to 50 ft and extend from the surface to several thousand

AIR MASS SOURCE REGIONS

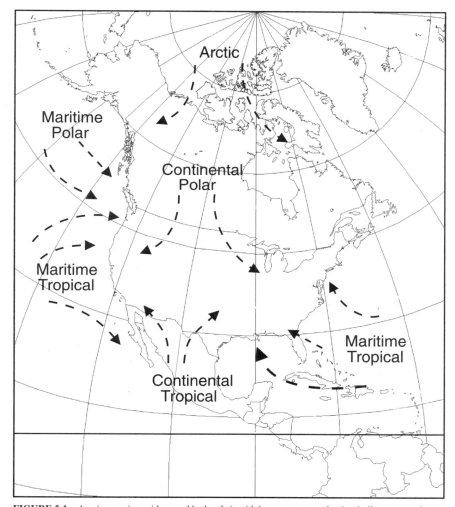

FIGURE 5-1   An air mass is a widespread body of air with homogeneous—that is, similar—properties.

feet. Wind speed within the rotations vary from 25 to more than 75 kn. Dust devils are capable of substantial damage, but the majority are small.

CASE STUDY   While doing pattern work with a student at Lancaster's Fox Field in California's Mojave Desert, we flew into a dust devil. The encounter was equivalent to light-to-moderate turbulence; it shook the Cessna 150, and the low pressure in the vortex caused both windows to pop open! Dust devils most likely form in the same manner as fair-weather waterspouts (discussed later in this chapter), but they lack the energy released from latent heat.

TABLE 5-1 Characteristics of Air Masses

Warm—cooled from below	Cold—warmed from below
Stable lapse rate	Unstable lapse rate
Poor visibility	Good visibility
Smooth	Turbulent
Stratiform clouds	Cumuliform clouds
Steady precipitation	Showery precipitation
Rime icing	Clear icing

When an air mass moves over an area of different properties, the air mass takes on some or all of the surface properties. Horizontal changes in temperature, moisture, and lapse rate gradually occur. For example, let's take a maritime air mass that moves over the continent. In winter the air mass is cooled from below. Its lapse rate becomes more stable. In the summer the air mass is warmed from below, resulting in greater instability. These effects play an important part in our discussion of fronts in Chap. 6. When an air mass moves over a colder surface, the term *cold* is added to the type. For example, when our maritime air mass moves over the continent, it is now termed a *maritime polar cold air mass*. In the summer, if it moves over warmer land, it is called a *maritime polar warm air mass*. Table 5-1 compares the general characteristics of air masses. Like Tables 2-1 and 4-1, there are many similarities.

Figure 5-2 shows a cross section of a low- and a high-pressure area. The central pressure in the low is 800 mb, and in the high 1000 mb. Constant pressure surfaces (850, 700, and 500 mb) are at different heights within each air mass. A sage aviation axiom states: "When flying from high to low, look out below." From Fig. 5-2 we can easily see that without correcting the altimeter, the aircraft would indeed be flying at a lower true altitude as it approached the low-pressure center.

Convergence

Low-Pressure Areas and Troughs

moist T-storm
↑ clouds ↑ humidity

A *low* is an area completely surrounded by higher pressure. A *trough* consists of an elongated area of low pressure, almost always associated with cyclonic wind flow (counterclockwise in the northern hemisphere). In common practice the terms *cyclone* and *low* are used interchangeably. Both terms refer to a closed circulation; a cyclone or low is not a trough.

Convergence—upward vertical motion—occurs in lows and troughs. Convergence destabilizes the atmosphere, which increases relative humidity and clouds. Convergence in itself does not necessarily produce poor weather. Other factors, such as moisture and stability, must be considered. For example, a surface low or trough may only mix—stir up— the lower atmosphere. Without moisture and instability this would produce good visibility and possibly low-level turbulence. However, a surface low or trough with a moist, unstable air mass may be all that's needed to trigger thunderstorms.

Surface convergence occurs along curved isobars surrounding a low or trough. Maximum convergence takes place at the center of the low or along the trough line.

Troughs are not fronts, although fronts normally lie in troughs. A front, as we shall see in the next chapter, is the boundary between air masses of different temperatures, whereas a trough is simply a line of low pressure. Both phenomena produce upward vertical motion.

A *thermal low* is an area of low atmospheric pressure caused by high temperatures produced by intensive heating of the Earth's surface. Thermal lows are common to the conti-

SURFACE PRESSURE

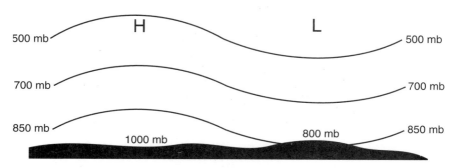

FIGURE 5-2   It can easily be seen that without correcting the altimeter, the aircraft would indeed be flying at a lower true altitude as it approached the low-pressure center.

nental subtropics in summer. They remain stationary over the area. Air masses exhibiting cyclonic circulation are generally weak and diffuse. They are not associated with fronts.

Thermal lows develop over the desert southwest in late spring through early fall. At times they can extend to the arid regions east of the Cascade Range in Washington and Oregon and the Snake River Valley of Idaho. Their primary effect is to lower pressures inland, increasing coastal marine stratus layers. They provide thermal convection and weak convergence, which can trigger the development of air mass thunderstorms.

divergence

High-Pressure Areas and Ridges

H *stabilizes ↓humidity ↓clouds*

A *high* is an area completely surrounded by lower pressure. A *ridge,* the opposite of a trough, is an elongated area of high pressure, almost always associated with anticyclonic wind flow (clockwise in the northern hemisphere). Like *cyclone, anticyclone* refers to a closed circulation system and is used interchangeably with *high.*

Divergence—downward vertical motion—occurs in highs and ridges. Divergence stabilizes the atmosphere, which decreases relative humidity and clouds. Divergence does not necessarily produce good weather.

There is a misconception that high pressure always means good flying weather. Although good weather often occurs, there are exceptions. Strong pressure gradients at the edge of high cells can cause vigorous winds and severe turbulence. Near the center of a high, or with weak gradients, moisture and pollutants can be trapped at lower levels, causing reduced visibilities and even producing zero-zero conditions in fog for days or even weeks.

A *thermal high* results from the cooling of air by a cold underlying surface. For this to occur, the air mass must remain relatively stationary over the cold surface. This is a factor in the development of continental polar and arctic air masses.

better weather but not always

THUNDERSTORMS

A *thunderstorm* is a local storm produced by cumulonimbus clouds. The storm itself may be a single cumulonimbus cloud or cell, a cluster of cells, or a line of cumulonimbus clouds that in some cases may extend for several hundred miles. Thunderstorms, as the name implies, are always accompanied by lightning and thunder. These storms usually produce strong, gusty winds, heavy rain, and sometimes hail. Individual cells last for a short period, seldom over two hours. *Lightning* is the visible electrical discharge produced by thunderstorms; *thunder* is the sound produced by rapidly expanding gases along the channel of a lightning discharge.

For manual weather observation, a thunderstorm is reported when thunder is heard or when overhead lightning or hail is observed. The Automated Lightning Detection and Reporting System (ALDARS), which acquires lightning information from the National Lightning Detection Network, allows automated surface observations to report the occurrence of a thunderstorm. As the definition of a thunderstorm in the previous paragraphs indicates, the presence of a cumulonimbus cloud implies a thunderstorm. Therefore, any report of thunder, lightning, or cumulonimbus clouds means a thunderstorm—with all of its ominous implications.   *Heat pump*

FACT A thunderstorm is like a giant heat pump. It transports warm air from the surface into the cold upper atmosphere. If thunderstorms did not occur, it is estimated that the Earth's average temperature would increase by more than 10°C! This would melt ice caps, placing many coastal regions under water.

When surface winds gust to 50 kn or more, or hail three-quarters inch or greater, accompanies a thunderstorm, the storm is classified as severe. Before the adoption of the international METAR code, severe thunderstorms were reported in surface observations. In the METAR code a severe thunderstorm is implied with winds of 50 kn or more or with hail three-quarters inch or greater. Surface observations contain the intensity of precipitation produced by thunderstorms. It's important to remember that although precipitation and thunderstorms are always associated, they are two separate phenomena. Thunderstorms may be reported with no precipitation or some intensity of precipitation—light, moderate, or heavy. Some pilots mistranslate reports as a light thundershower or a light thunderstorm and rain showers; neither is correct. Light rain or rain showers may accompany a thunderstorm. However, there is no such thing as a light thunderstorm!

When a thunderstorm exists completely within nonconvective clouds and precipitation, it is termed *embedded*. Embedded thunderstorms are especially dangerous because they can be impossible to detect without storm detection equipment—radar or lightning detectors.

Three conditions are necessary for thunderstorm formation:

- An initial lifting mechanism
- Sufficient water vapor
- Unstable air

warm air rises & then cooled

If any one of the three elements is missing, thunderstorms will not develop.

All thunderstorm cells progress through three stages called the *life cycle:*

- Cumulus
- Mature
- Dissipating

The thunderstorm life cycle is illustrated in Fig. 5-3.

The first ingredient required for thunderstorm development is an initial lifting mechanism. In Chap. 2, we discussed vertical motion in the atmosphere. Any of those phenomena can produce the required lifting. Most often, lifting is produced by a frontal surface,

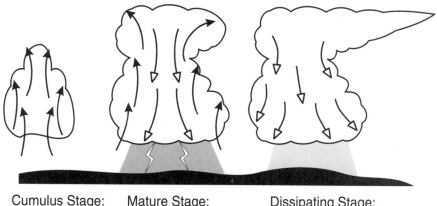

Cumulus Stage:
All upward vertical
currents.

Mature Stage:
Both updrafts and
downdrafts; rain
begins.

Dissipating Stage:
All weak downdrafts;
rain ends.

FIGURE 5-3   All thunderstorms progress through the three stages of their life cycle.

sloping terrain, convergence, or surface heating. However, upslope and low-level warm-air advection may be all that's required to trigger thunderstorms.

This lifting creates an initial updraft. At the lifted condensation level, adiabatic cooling in the updraft results in condensation and the beginning of cloud development. This is the beginning of the cumulus stage.

The cumulus stage gives birth to the thunderstorm. Although most cumulus clouds do not become thunderstorms, the initial stage is always a cumulus cloud or collection of cumulus clouds. Figure 4-11 shows towering cumulus, which represents the cumulus stage of the thunderstorm. The main feature of the cumulus or building stage is the predominant updrafts that may extend from the surface to several thousand feet above the visible cloud top. During the early period of this stage, cloud droplets are very small, but ultimately they grow into raindrops as the cloud builds upward. Note in Fig. 4-11 that the clouds are in various development levels of the cumulus stage.

If we think of the thunderstorm as an engine: The initial lifting creating the updraft is the "starter." "Ignition" is the initial condensation. "Fuel" is the water vapor.

Without sufficient water vapor, the initial lifting may not be sufficient to sustain continued vertical movement and a thunderstorm will not form. If the atmosphere is stable, vertical motion ceases and stratiform clouds form. However, if the air is unstable—the third requirement for thunderstorm development—cumulus clouds form.

Condensation releases latent heat, which partially offsets the adiabatic cooling in the saturated updraft, increasing buoyancy. That is, the rising air is warmer than the surrounding air. The increased buoyancy drives the updraft still faster, drawing more water vapor into the cloud. The updraft becomes self-sustaining in the unstable air, and a thunderstorm develops. Cold-air advection aloft destabilizes the atmosphere and enhances thunderstorm development.

As the cloud towers upward, the updraft cools adiabatically until in the upper levels of the storm it becomes colder than the surrounding air. This difference in temperature, together with the increasing weight of water drops and ice particles, retards the upward motion, ultimately turning it to a downdraft or outflow. The change of flow may directly reverse the updraft, or it may arch outward as an outflow, allowing the updraft to continue unabated. This is the mature stage of the thunderstorm.

In the mature stage drops are ejected from the updraft, or they become so large that the updraft can no longer support them and they begin to fall. The mature stage begins when rain first falls from the bottom of the cloud. Updrafts continue to gain strength in the early mature stage, which may exceed 80 ft/s.

Figure 5-4 shows the mature stage. This occurs roughly 10 to 15 min after the cloud has built upward above the freezing level. As the raindrops fall, they pull air downward. This is a major factor in the formation of downdrafts in the mature stage. The air dragged downward by the falling rain cools the surrounding air, which accelerates downward motion. Throughout the mature stage, downdrafts continue to develop and coexist with updrafts. Maximum updrafts tend to be found in the upper two-thirds of the storm. Updrafts as strong as 6000 ft/min and downdrafts of up to 2000 ft/min can develop.

The mature state represents the most intense period of the thunderstorm. Lightning activity is greatest, and hail, if present, most often occurs in the mature stage. The essential difference between a nonsevere and severe thunderstorm is the mechanism of the mature stage, which will be discussed later in this section.

Back to our engine analogy: "Energy" to drive the storm is the latent heat released by condensation. With this energy the engine, or storm, now becomes self-sustaining—it no longer depends on the starter or initial lifting. "Combustion products" are clouds and precipitation.

The dissipating stage begins when water vapor is cut off and downdrafts predominate. The dissipating stage is characterized by weak downdrafts throughout the cloud. Figure 5-5 shows the dissipating stage. Condensation gradually decreases, and when all the water has fallen from the cloud or evaporated, the dissipating stage is complete. At the surface all signs of the thunderstorm disappear, and any clouds that remain are stratiform.

A final look at our engine: "Exhaust" is the outflow or downdraft. The "throttle" is the cutting off of fuel—water vapor. Individual storms vary, particularly in the structure of the exhaust and throttle, a determining factor as to the severity of the storm.

Structure while in the mature stage has an important bearing on the duration and potential severity of a storm and leads to two general thunderstorms classification categories: limited-state thunderstorms—generally referred to as *air mass thunderstorms*—and steady-state thunderstorms.

Limited-State Thunderstorms

Thunderstorm cells that progress rapidly through the mature stage are usually *limited-state thunderstorms*. The mature stage is self-destructive. When the updraft is too weak to support the rain drops, precipitation falls through the updraft. Falling precipitation induces frictional drag, which retards the updraft, finally reversing it to a downdraft. The downdraft and precipitation cool the lower portion of the storm cloud and the surface below. This cuts off the inflow of water vapor and dissipates the storm. This self-destructive cell usually lasts from 20 min to about 1.5 h, and it rarely produces extreme turbulence or large hail. However, even a limited-state thunderstorm is capable of producing severe icing and turbulence and hail that can cause structural damage.

In the summer season air mass thunderstorms that occur west of the Rockies are most frequent over the higher mountains, deserts, and plateaus. Moisture can be brought in from as far away as the Gulf of California or even the Gulf of Mexico. The anticyclonic flow of high pressure over the Gulf of Mexico pushes moist, unstable air through Arizona and New Mexico, into Utah and Colorado. This moist air usually spreads north, and can even reach as far as Idaho at least two or three times during the season.

FIGURE 5-4   The mature stage is characterized by both updrafts and downdrafts.

In the western United States weak fronts, and most importantly the passage of upper-level troughs, help to set off summer thunderstorms. When moist tropical air is lifted by a cold front, severe thunderstorms can develop. Although infrequent, lines or clusters of thunderstorms can occur. They most often develop along California's coastal mountains, the Sierra Nevada, the Cascade, west slopes of the Wasatch Mountains in Utah, the mountains east of Burley and Idaho Falls in Idaho, and the mountains of northeast Nevada. Mountains provide a primary lifting mechanism for the development of these thunderstorms.

FIGURE 5-5   The dissipating stage is characterized by weak downdrafts throughout the cloud.

Steady-State Thunderstorms

Supercell and *squall line thunderstorms* are the main types of severe, *steady-state thunderstorms.* The supercell thunderstorm, with its enormous updrafts and downdrafts, is able to maintain itself as a single entity for hours. This type of storm produces tornadoes and large hail. These storms can grow to over 60,000 ft.

When both updrafts and downdrafts coexist in the mature stage and are about equally balanced—not significantly affecting the other—steady-state thunderstorms result. The most obvious consequence is that the mature stage continues in this "steady state" and intensifies, becoming severe with extreme turbulence and large hail. The life of an individual cell may be considerably longer than that of a limited-state cell. The *severe storm complex,* often consisting of many dissipating and developing cells, may last as long as 24 h and may move as far as 1000 mi. The thunderstorm continues in this steady state until it is affected by some outside influence or until the mechanics of the thunderstorm cell itself change and it becomes self-destructive.

Dissipating groups of thunderstorms can produce outflow boundaries. An *outflow boundary* is a surface boundary left by the horizontal spreading of thunderstorm-cooled air. The boundary is often the lifting mechanism needed to generate new thunderstorms. Outflow boundaries are often depicted on the surface analysis chart.

When thunderstorms form lines, they are often referred to as *squall lines.* Squall lines can develop along a cold front, but more often they appear a hundred or so miles ahead of the front. Squall lines are caused by air aloft flowing over the cold front and developing into waves, much like mountain waves. Where the wave crests—a lifting mechanism—thunderstorms develop in the moist, unstable air.

Another cause of squall lines occurs in the central plains, especially during the spring. Behind the cold front, cold, dry continental polar air pushes in from the north. Ahead of the front, warm, dry continental tropical air moves in from the southwest. To the east warm, moist maritime tropical air advances from the gulf. Where the continental tropical and maritime tropical air masses meet, a dry line forms, which produces a squall line. Once the thunderstorms develop, the outflow of cold air along the ground initiates lifting, which generates new, and possibly more severe, storms—an outflow boundary.

When conditions are right, individual thunderstorms can grow into large, organized convective weather systems. These are known as *mesoscale convective complexes* (MCCs). MCCs are large, covering an area the size of several states. The following are characteristics of MCCs:

- Light winds aloft
- High moisture content at low levels
- Low-level warm-air advection
- Late afternoon development
- Persistence all night
- Widespread areas of rain or rain showers
- IFR conditions over wide areas
- Heavy rain showers and thunderstorms
- Thunderstorms usually circumnavigable with storm detection equipment
- Area tending to remain stationary

Individual thunderstorms within the MCCs combine to generate a long-lasting, slow-moving weather system. The thunderstorms feed on themselves, producing widespread precipitation. MCCs can produce widespread severe weather, including high winds, hail, and tornadoes.

Thickness is the vertical depth of a layer in the atmosphere between two pressure surfaces. The NWS produces a 1000- to 500-Millibar Thickness Chart. MCCs, unlike air mass thunderstorms, tend to move parallel to the thickness lines on the 1000- to 500-Millibar Thickness Chart.

Figure 5-6 illustrates a mesoscale convective complex. The satellite image shows the cloud band produced by the MCC covering half the state of Texas. The area of precipitation is restricted to southern Texas (see the radar chart at the bottom of Fig. 5-6). The rain showers and thunderstorms cover tens of thousands of square miles, and precipitation tops range up to 47,000 ft. However, as is typical in the case of MCCs, the convective activity is not organized in lines, like the cells in northeast Texas, Kansas, and Missouri. Nor are severe thunderstorms or tornadoes reported or forecast.

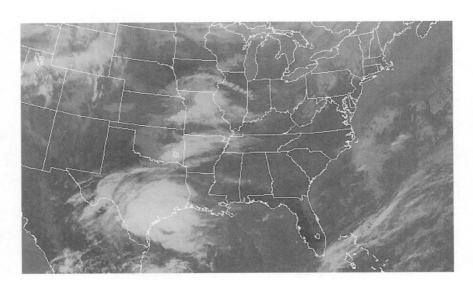

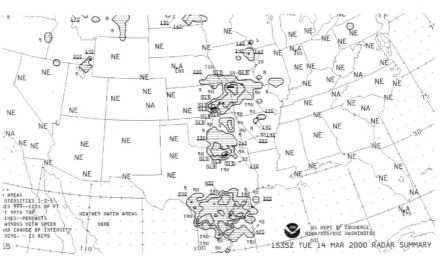

FIGURE 5-6   Individual thunderstorms can grow into a large, organized convective weather systems known as a *mesoscale convective complexes* (MCCs).

Large areas of the Midwest are plagued by night thunderstorms. Nighttime variations in the large-scale wind systems create convergence in low levels. These thunderstorms show a peak occurrence between midnight and 4 a.m. and are not the remnants of evening thunderstorms left over from daytime convection.

Microbursts   *most severe wind shear*

As aircraft instruments and navigational system capabilities improved, pilots began taking on more and more weather. Hazards of fog and low clouds were solved with the instrument landing system (and increased fuel reserves). Icing was overcome in the 1930s, and, for the most part, turbulence and thunderstorms were mastered with high cruise altitudes and radar. With these hazards solved, major air carrier accidents have most often fallen into the categories of mechanical failure, pilot error, and wind shear.

The hazards of wind shear can be even more critical to general aviation. *Wind shear,* a rapid change in wind direction or speed, has always been around. Convective activity produces *severe shear,* which is defined as a rapid change in wind direction or velocity causing airspeed changes greater than 15 kn or vertical speed changes greater than 500 ft/min. The microburst produces the most severe wind shear threat.

Rain-cooled air within a thunderstorm produces a concentrated rain or virga shaft less than one-half mile in diameter, which forms a downdraft. The downdraft, or downburst, has a very sharp edge and forms a ring vortex upon contact with the ground, whereupon it spreads out causing gust fronts that are particularly hazardous to aircraft during takeoff, approach, and landing. Reaching the ground, the burst continues as an expanding outflow.

A *microburst* consists of a small-scale, severe storm downburst less than 2.5 mi across. This flow can be 180° from the prevailing wind, with an average peak intensity of about 45 kn. Microburst winds intensify for about 5 min after ground contact and typically dissipate about 10 to 20 min later. Microburst wind speed differences of almost 100 kn have been measured. On August 1, 1983, at Andrews Air Force Base, indicated differences near 200 kn were observed. Some microburst events are beyond the coping capability of any aircraft and pilot. Although normally midafternoon, midsummer events, microbursts can occur any time, in any season.

CHAPTER 6
AIR MASSES IN CONFLICT

In previous chapters we discussed how air of different properties does not tend to mix. In the last chapter we learned about air masses, their source regions, and how they are modified. Now we can move on to see what happens when air masses collide. Because of the dynamic nature of weather, our discussion begins with generalizations—with "classic" examples of weather systems. Then we can move on to specific examples of weather that affects North America.

When an air mass moves from its source region, it comes in contact with other air masses. In the zone of contact between air masses, temperature and moisture (density), and wind may change rapidly over short distances. This zone of relatively rapid change is called a *frontal zone*—usually referred to as a *front.* Temperature is the most important density factor; therefore, fronts almost invariably separate air masses of different temperatures. Other factors also distinguish a front, such as a pressure trough, change in wind direction, moisture differences, and cloud and precipitation forms.

Norwegian meteorologist Vilhelm Bjerknes, and his son Jakob, developed the polar front theory at the beginning of the twentieth century. World War I had begun, and it was popular to use the language of the conflict. Thus, weather was described using words like *fronts, advances,* and *retreats.* The weather did resemble a war between air masses.

Fronts fall into one of four categories:

- Cold front
- Warm front
- Occluded front
- Stationary front

The Earth's atmosphere is a giant heat exchanger, moving cold air down from the arctic and warm air up from the tropics. Typically in the northern hemisphere the cold air pushes down from the northwest and then lifts and replaces the warm air. The boundary where this action takes place is called a *cold front.* However, to accomplish this, at some point, warm tropical air must replace the colder air. This typically takes place ahead of the cold front as the warm air, moving from the south, overrides and replaces the colder, retreating air; this boundary is known as a *warm front.*

Cold fronts move faster than warm fronts. Sometimes a cold front overtakes a warm front and an occlusion occurs. This front is called an *occluded front.* When frontal speed decreases to 5 kn or less, the front is labeled *stationary.* Fronts produce vertical motion from the surface to about the middle troposphere—about 20,000 ft. This action, the polar front, is more or less continuous around the world at middle latitudes, as illustrated in Figs. 3-1 and 3-2. From the satellite picture in Fig. 3-2, it's easy to see that some portions of the polar front are more active than others. An active weather system is approaching the Pacific coast, with little, if any, activity in the central Pacific area. Benign weather is also occurring over most of the western states, with another active system in the northern plains states.

Fronts have sloping boundaries, which are much shallower than those illustrated in this chapter. An average slope is between 1 in 50 and 1 in 300. That is, for every mile a front extends above the ground, the slope extends 50 to 300 mi downstream. Frontal boundaries lose their identity above 20,000 ft.

The width of a frontal zone depends on wind and temperature differences between the two air masses. The greater the temperature difference—other things being equal—the narrower the front; the stronger the wind component along and behind the front, the narrower the front. For example, a front moving 35 to 45 kn with a temperature contrast of 20°C or more may have a frontal boundary of 1 mi or less at the surface. A front with only a few degrees difference in temperature in a weak circulation may be as much as 50 mi wide. The movement and effects of wide fronts, with low temperature differences, and weak circulation, are difficult to forecast.

Frontal intensity is based on frontal speed, which is determined by the temperature gradient in the cold sector, the region of colder air at a frontal zone. The movement of fronts is affected by many factors, such as temperature, moisture, stability, terrain, and upper-level systems. A front with waves indicates weak low-pressure centers or portions of the front moving at different speeds. A front with waves should be watched. The weak low-pressure areas can intensify and cause significant weather. The closer the front is to the jet stream, the steeper the slope, and typically the stronger the front.

Fronts run the spectrum from a complete lack of weather, to benign clouds that can be conquered by the novice instrument pilot, to fronts that spawn lines of severe thunderstorms that no pilot or aircraft can negotiate. Each front—for that matter any weather system—must be evaluated separately and then a flight decision made based on the latest weather reports and forecast and the pilot's and aircraft's capabilities and limitations.

Often the exact location, and sometimes even the presence, of fronts is a matter of judgment. Additionally, fronts do not necessarily reach the surface; they might be found within layers aloft. This is especially true in the western United States and the Appalachians where mountain ranges break up fronts. Therefore, there might be differences between the charted position of fronts and their location as described in forecasts or plotted on charts.

The location of surface fronts can be obtained from the surface analysis chart or weather depiction chart. The surface analysis chart also provides front type, intensity, and intensity trend. A narrative description of the location of fronts is provided in the synopsis portion of the area forecasts and the *transcribed weather broadcast* (TWEB) route forecasts synopsis. Additional information on the location and movement of fronts, when associated with severe weather, is contained in convective SIGMETs, severe weather watch bulletins, and the convective outlook.

THE LIFE CYCLE OF A FRONT

Frontal formation is called *frontogensis*. There are several mechanisms for this process. One is the interaction of a polar air mass with the polar easterlies and a tropical air mass with the subtropical westerlies. Between these air masses is a convergence zone. As previously mentioned, occasionally fronts develop when a new wave forms on the westward end of the cold front due to varying frontal speed or terrain.

As illustrated in Fig. 3-1, between the subtropical westerlies and polar easterlies lies the more or less permanent polar front. This is the area where most of the midlatitude cyclones—low-pressure areas—develop.

parullel no wx

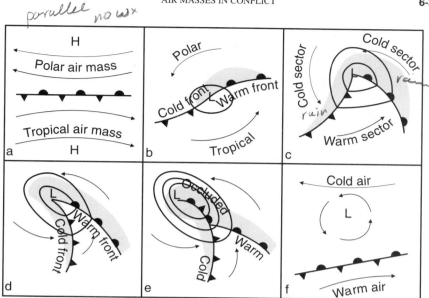

FIGURE 6-1   Life cycle of a front. Between the subtropical westerlies and the polar easterlies lies the more or less permanent polar front.

Refer to Fig. 6-1. Figure 6-1*a* shows a segment of the polar front that is stationary. (On black-and-white weather maps, a stationary front is shown as a line with alternating triangles and semicircles. On color weather maps, it is drawn as an alternating red and blue line.) When winds are parallel to the polar front, there is little if any weather along the front. This is illustrated in the satellite image depicting the portion of the polar front over the intermountain region of western United States, Fig. 3-2. Note the almost total lack of clouds.

This type of flow, however, sets up a cyclonic wind shear. Under the right conditions, a kink or wave forms along the front. Winds are no longer parallel, and frontogensis begins, as illustrated in Fig. 6-1*b*. (The gray areas in Fig. 6-1 represent typical areas of precipitation.) The region of lowest pressure is at the boundary of the two air masses—the front. In Fig. 3-2 frontogensis appears to be developing in the southern plains states.

As the cold air mass of the polar air begins to push southward under the warmer air of the tropical air mass, a cold front develops. (On a black-and-white weather map a cold front is shown as a line with triangles pointing in the direction of frontal movement. On color weather maps it is drawn as a blue line.) This is illustrated in Fig. 3-2. The cold front is on a north-south line off the Washington, Oregon, northern California coast. There is a distinct boundary between the cold and warm air, and the cloud band is relatively narrow.

As the cold air retreats on the north side of the low, the warm subtropical air mass moves in and replaces the colder air, as illustrated in Fig. 6-1*c*. This generates whirlpools between adjacent currents, and waves form along the polar front. As the advancing warm air mass overrides and replaces the retreating cool air, the boundary that forms is called a *warm front*. (On black-and-white weather maps a warm front is shown as a line with semicircles pointing in the direction of frontal movement. On color weather maps it is drawn as a red line.) That portion in Fig. 3-2 north of the polar front in Canada represents this area. Note the relatively large cloud band. The front is weak in Montana, with the cloud band some distance from the frontal boundary.

[handwritten margin notes: "rain behind", "cold", "warm ahead"]

Note in Fig. 6-1c that we refer to the area of warm air as the *warm sector,* the cool air ahead of the warm front and cold air behind the cold front as the *cool* or *cold sector.*

Directed by the upper winds, the wave generally moves to the east or northeast. The central pressure lowers as cyclonic flow becomes stronger. Precipitation forms in a wide band ahead of the warm front and along a narrow band behind the cold front.

The faster-moving cold front constantly moves closer to the warm front. Eventually, the cold front overruns the warm front, and the system becomes occluded—an occluded front. (On black-and-white weather maps an occluded front is shown as a line with alternating triangles and semicircles on the same side indicating the direction of movement of the front. On color weather maps it is drawn as a purple line.) At this point the storm is usually more intense with widespread clouds and precipitation, as illustrated in Figs. 6-1d and 6-1e. The area off the Canadian coast, westward to the low-pressure center, in Fig. 3-2 represents the area of occlusion and the occluded front.

As temperatures and pressures equalize across a front, the front will dissipate. This is called *frontolysis.* Without the energy of the rising warm moist air, the old storm system dies out and gradually dissipates, as illustrated in Fig. 6-1f.

The development or strengthening of a cyclone is called *cyclongenesis.* Over the United States this tends to occur over the Great Basin (the area between the Rockies and Sierra Nevada mountains, consisting of southeastern Oregon, southern Idaho, western Utah, and Nevada), the eastern slopes of the Rockies, the Gulf of Mexico, and the Atlantic Ocean. Some of these waves develop into huge storms; others dissipate within a day. The difference is the upper-level flow, which will be discussed in the next chapter.

Weather associated with any front depends on both the slope of the front and the properties within each of the conflicting air masses. Vertical motion, usually upward, occurs above the frontal surface due to the air flowing along the slope. Thus when a front forces conditionally unstable air upward, showers or thunderstorms form above the frontal zone. With abrupt lifting along a steeply sloping front, significant weather forms in a line near the surface position of the front. A more gradually sloping front may spread the unstable weather for many miles over the frontal surface. Thunderstorms may be embedded in extensive areas of stratified clouds. Moist, stable air overriding a frontal surface produces stratified clouds and steady precipitation.

COLD FRONTS

When a cold air mass replaces a warm air mass at or near the surface, it is by definition a cold front. Since cold air is more dense than warm air, the cold air flows under and replaces the warm air. This forces the warm air upward—a lifting mechanism. A cold front moves at about the speed of the component of the wind perpendicular to the front just above the friction layer between the surface and free air. (Recall the frictional layer is from the surface to about 1000 ft over oceans and 2000 ft over land.) The area ahead of the cold front is called the *warm sector* and behind the front the *cold sector,* as illustrated in Fig. 6-1.

Figure 6-2 shows a cold front with slightly unstable continental polar cold air underrunning moist, unstable, maritime tropical warm air. Abrupt lifting at the surface frontal position releases the instability into a line of thunderstorms. This is the area where squall lines can develop. Note the relatively narrow area of weather associated with the cold front, as illustrated in Fig. 6-2.

Towering cumulus may also develop ahead of the front due to surface heating in the warm air along with wave motion generated by the front—recall our discussion of squall lines from the last chapter. Towering cumulus clouds, technically cumulus congestus, may

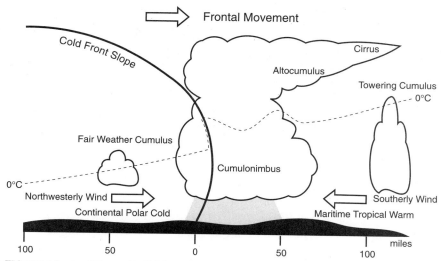

FIGURE 6-2 A cold front with slightly unstable cold air underrunning warm, moist, unstable air produces abrupt lifting and instability.

develop ahead of the front. These clouds often develop into thunderstorms. Towering cumulus should be treated with the same respect as cumulonimbus. Towering cumulus are illustrated in Fig. 4-11, the cumulus stage of a developing thunderstorm.

The approach of the front is often signaled first by cirrus clouds that thicken into cirro-stratus. Next appears a thickening band of altocumulus, similar to those shown in Fig. 4-4. These are quickly followed by cumulonimbus clouds, thunderstorms, heavy rain, and low ceilings and visibilities.

Surface winds are typically out of the southern quadrant in the warm sector and out of the northwest quadrant in the cold sector. Precipitation is generally heavy and showery, associated with an unstable air mass. The freezing level lowers with the approach of the front, drops dramatically during frontal passage, and continues to lower in the cold air behind the front.

Active cold fronts are characterized by the following:

- Fast moving
- Good visibility before and after frontal passage
- Turbulent in the frontal zone
- Relatively narrow band of icing in the frontal zone
- Sharp temperature changes over relatively small distances
- Relatively large changes in moisture content
- Large shifts in wind direction
- Strong pressure gradients
- Relatively small areas of heavy, showery precipitation

Typically the front passes rapidly with clearing skies and cool temperatures. Behind the front, fair weather cumulus develop in the slightly unstable cold air. Fair weather cumulus are illustrated in Fig. 6-3. These clouds signify turbulence below and smooth air above.

FIGURE 6-3 Fair weather cumulus develop in the slightly unstable cold air behind the front.

Pilots beware, even fair weather cumulus, above the freezing level, have a potential for serious structural icing.

Figure 6-4 shows a cold front underrunning warm, moist, stable air. Stable stratified clouds form above the front. Note the stable cloud types: cirrostratus, altostratus, and nimbostratus. This succession of clouds would also announce the approach of the front. The cold air is stable except where surface heating has created a very shallow convective layer, producing stratocumulus clouds behind the front.

Surface winds, again, are typically out of the southern quadrant in the warm sector and out of the northwest quadrant in the cold sector. Precipitation is generally light to moderate and steady, associated with a stable air mass. The freezing level lowers with the approach of the front, drops dramatically during frontal passage, and continues to lower in the cold air behind the front.

Weak or dry cold fronts are characterized by the lack of cloud cover and no precipitation. Both cold and warm air masses are dry and stable. The warm air must reach significant heights before condensation occurs. Clouds will typically be a considerable distance from the surface position of the front and at high levels, the height depending on the moisture content of the warm air.

At the surface, cold frontal passage is characterized by a temperature decrease, a wind shift with often gusty winds—associated turbulence and increasing pressure. With low-level moisture, weak cold fronts are characterized by stable clouds and low tops.

CASE STUDY In California on a flight from Van Nuys to San Luis Obispo, we were able to top the clouds of a weak, dissipating cold front at 6500 ft. With this front, cloud tops were actually lower than those produced by deep coastal stratus layers, which can at times exceed 8000 ft.

Cold fronts are characterized by the following:

- Primarily cumuliform clouds, but stratiform clouds predominant in stable air
- Precipitation generally in narrow bands, showery, and heavy at times

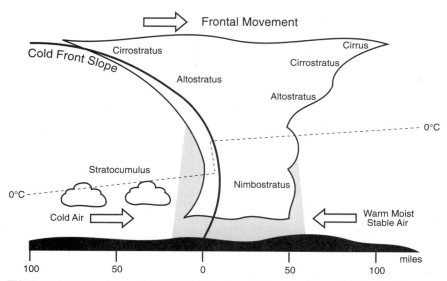

FIGURE 6-4   Cold fronts associated with stable air produce stratiform clouds over relatively short distances.

- Icing above the freezing level
- Wind shift from southwesterly to northwesterly with frontal passage
- Front moving southeasterly at about 20 to 25 kn
- Temperature decreasing with frontal passage
- Pressure falling as front approaches, then rising after frontal passage

WARM FRONTS

Cyclonic flow around the low in the cool sector causes the cool air to retreat. As the cool air retreats, warm air from the warm sector overrides and replaces the cool air. This boundary is called a *warm front.* Winds in the warm sector must exceed the speed of the warm front, which averages about 10 kn. The slope of a warm front is considerably less than that of a cold front because of density differences between the air masses. That is, typically the air masses along a warm front have less density difference than along a cold front boundary.

Figure 6-5 shows a warm front with warm moist, unstable air overriding cool stable air. Lifting along the shallow front, note that the distance as shown at the bottom of Fig. 6-4 is more gradual than it is along the cold front. Unstable showers and thunderstorms are spread out above the frontal surface rather than in the line as in Fig. 6-2. These convective clouds may be embedded in a thick deck of stratiform clouds. Embedded thunderstorms are a very serious hazard to aircraft without storm detection equipment. Stratus fractus may form in the precipitation due to evaporation of water from the relatively warm rain and subsequent condensation in the cool air.

The approach of a warm front is often signaled by the appearance of cirrostratus, altocumulus, and altostratus clouds. These lower to nimbostratus as the front nears. Tower

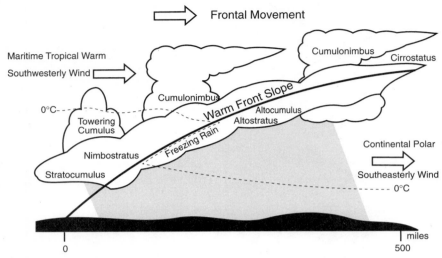

FIGURE 6-5   With a warm front with unstable air, unstable showers and thunderstorms are spread out above the frontal surface rather than forming a line as they would with a cold front, and they often are embedded.

cumulus and cumulonimbus are often hidden by the stratiform layers. Pilots flying on top will typically be able to see the cumulus buildups.

Surface winds are typically out of the southern quadrant in the warm sector and out of the southeastern quadrant in the cool sector. Precipitation is generally light to moderate, steady, and widespread, with embedded areas of heavy, showery precipitation associated with the unstable air mass. The freezing level rises with the approach of the front and continues to rise along the frontal slope—in an area of frontal inversion, as shown in Fig. 6-5. This is another significant hazard associated with a warm front. Rain falling from above-freezing temperatures forms large, supercooled water droplets in the freezing air below. This area of freezing rain can be the most significant icing hazard, and it can affect both VFR as well as IFR flights.

After frontal passage aloft the freezing level continues to rise in the warm air behind the front. Since the slope is shallow, pilots can expect to penetrate the frontal boundary some distance from the location of the front on the ground. With warm front passage, pressure and temperature rise, and skies begin to clear.

Figure 6-6 shows a warm front with warm moist, stable air overriding cool stable air. Here again, lifting along the shallow front is gradual. Cloud types are widespread and stratiform, associated with the stable air. The VFR pilot will be faced with widespread areas of precipitation, and low ceilings and visibilities; the IFR pilot should fare much better, with icing the most significant hazard.

Surface winds are typically out of the southern quadrant in the warm sector and out of the southeastern quadrant in the cool sector. Precipitation is generally light to moderate and steady, associated with the stable air mass. The freezing level rises with the approach of the front, and it continues to rise along the slope of the front. Large, supercooled water droplets produce freezing rain or freezing drizzle. After frontal passage aloft, the freezing level continues to rise in the warm air behind the front.

Active warm fronts are characterized by the following:

• Slow moving
• Poor visibility before frontal passage, usually good after frontal passage

Frontal Movement

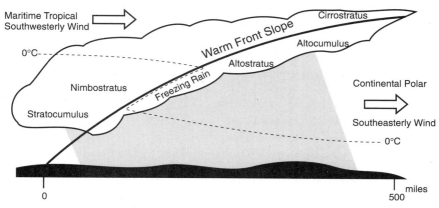

FIGURE 6-6 A warm front with warm moist, stable air overriding cool stable air produces stratiform clouds.

- Smooth, except in areas of embedded cumulonimbus
- Widespread areas of icing and freezing rain
- Large temperature changes over relatively large distances
- Relatively large changes in moisture content
- Moderate shifts in wind direction
- Moderate pressure gradients
- Relatively large areas of steady or showery, moderate to heavy precipitation

At the surface, the passage of a warm front is marked by an abrupt wind shift from southeast to southwest and an increase in pressure. This is usually associated with a clearing trend.

Weak warm fronts are characterized by the lack of cloud cover and no precipitation. If clouds form, they are high and a great distance from the front, as was mentioned in our discussion of the life cycle of a front and illustrated in Fig. 3-2. High clouds with a dry warm front will clear several hours before frontal passage at the surface.

Warm fronts are relatively rare and usually weak in the western third of the United States. This is due to the fact that a moist, southerly flow is typically not found in this region. This does not mean, however, that warm fronts will never exist in the region.

CASE STUDY   I had taken my Turbo Cessna 150 (All right. It wasn't turbocharged. That's an attempt at a little humor.) for its annual inspection from Lovelock to Winnemucca in Nevada. A warm front moving through the area produced thunderstorms and small hail. I had planned to fly back with the mail pilot in a Twin Commanche. It wasn't to be. I had to come back by dog, a Greyhound to be exact—yes, I mean the bus! Object lesson: My experienced airmail pilot knew a light twin without storm detection equipment had no business flying in an area of thunderstorms, possibly embedded.

East of the continental divide is where warm fronts come into their own. Here is where we can expect the classical warm fronts as illustrated in Figs. 6-5 and 6-6. The moist, southerly flow provides the fuel for activity along both the cold front and warm front. In this region a warm front produces extensive areas of low ceilings, visibilities, freezing precipitation, and thunderstorms.

Warm fronts are characterized by the following:

- Primarily stratiform clouds, but cumuliform clouds embedded in unstable air
- Precipitation generally widespread, steady and light to moderate, except with cumuliform clouds that produce showery, heavy precipitation
- Icing above the freezing level, with areas of freezing rain or drizzle, especially during the winter season
- Wind shift from southeasterly to southwesterly with frontal passage
- Front moving northeasterly at about 10 kn
- Temperature increasing with frontal passage
- Pressure falling as the front approaches, then rising after surface frontal passage

OCCLUDED FRONTS

We have mentioned that the speed of a cold front is greater than that of a warm front. So, what happens when a cold front catches up with a warm front? An occlusion or occluded front develops when a cold front overtakes a warm front. Because of density differences, one of the fronts is forced aloft. An occluded front may be either a warm front occlusion or a cold front occlusion. The difference depends on the density of the air ahead of the warm front and behind the cold front.

(Please note that a warm front occlusion does not in itself indicate unstable air; nor does a cold front occlusion indicate stable conditions. Warm front occlusion may be unstable or stable, just as cold front occlusion may be unstable or stable.)

Figure 6-7 shows a warm front occlusion. Here cool air is overriding cold air forcing the cold front aloft. The cool air behind the cold front is less dense than the cold air ahead of the warm front; thus the cold front is forced up the warm front slope. In this example, the cool air is stable, cold air stable, and warm air unstable. This type of occlusion normally forms in the western part of the North American continent. In the United States this occurs when a cold front, backed up by cool maritime polar air from the North Pacific, moves over the continent and catches up with a warm front that is preceded by colder continental polar air. Cloudiness has features of both unstable cold and warm fronts. Maximum convective cloudiness is along the cold front aloft; stratified clouds with possible embedded thunderstorms develop above the warm front. Stratocumulus form in the stable cool air, and stratus fractus in the cold air with warm rain falling from above.

Surface winds are typically out of the northwest in the cool, cold front sector, southerly in the warm air mass above the warm front, and out of the southeastern quadrant in the cold air ahead of the warm front. Precipitation is generally light to moderate and steady, with embedded areas of heavy, showery precipitation associated with the unstable air mass. The freezing level is typical of a warm and cold front with unstable air.

Figure 6-8 shows a cold front occlusion. The cold air of the cold front replaces the cool air at the surface forcing the warm front aloft. Here the dense cold air forces the warm front slope up the cold front surface. Cold front occlusions are most common on the eastern two-thirds of the continent. In the eastern United States, the cold air on both sides comes from continental polar air. The cold air is moving south from its source region. The cool air ahead of the warm front has been over warmer terrain for a longer time; thus it has been modified

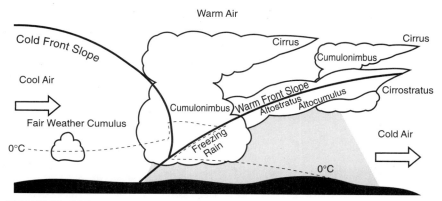

FIGURE 6-7 With a warm front occlusion (unstable occluded front) cool air overrides cold air forcing the cold front aloft.

and is relatively warmer than in the cold sector behind the cold front. The cold air is moderately stable, cool air stable, and warm air stable. Cloud types are stratiform associated with stable air, and a combination of those contained in Figs. 6-4 and 6-6.

Surface winds and freezing level characteristics are similar to those of a cold front occlusion. Strong occluded fronts are characterized by weather associated with both strong cold and strong warm fronts.

Like warm fronts, classic occluded fronts typically occur in the eastern two-thirds of the United States.

STATIONARY FRONTS

As the name implies, a *stationary front* has little or no movement—5 kn or less. Both cold and warm fronts can slow, showing little movement, and develop into stationary fronts. Slope is generally shallow, although it may be steep if density change across the front is large or wind distribution favorable. Winds are parallel but opposite in direction, as illustrated in Fig. 6-1. Weather associated with a stationary front depends on its moisture, stability, and circulation. Clouds, if any, are stratiform, with little, or no, precipitation. When warm, moist air rides up and over the cold air, widespread stratiform clouds and light precipitation occurs. However, with moist, unstable air aloft, even the minimal low-level lifting associated with a stationary front can result in thunderstorms.

When warm, stable air from the west replaces cooler air to the east, the weather will be similar to that of a warm front with stable air. If the front begins to move, a warm front develops. Should colder air from the west replace warmer air to the east, weather will be similar to that of a cold front. If the air is unstable, cumulus clouds and thunderstorms may occur. Convective activity tends to be scattered and not as intense as that of an active cold front. Should the front begin to move, a cold front develops. As mentioned, waves in stationary fronts can develop, spawning new, intense weather systems.

Active stationary fronts are characterized by the following:

- Little movement, less than 5 kn
- Poor visibility
- Little, if any, turbulence

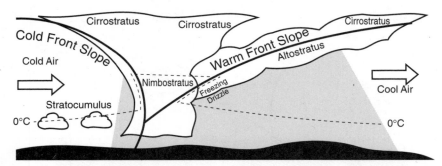

FIGURE 6-8   With a cold front occlusion (stable occluded front), cold air from the cold front replaces the cool air at the surface forcing the warm front aloft.

- Weak temperature changes over relatively large distances
- Relatively small changes in moisture content
- Small shifts in wind direction
- Weak pressure gradients
- Relatively large areas of light, steady precipitation

Weak stationary fronts are characterized by the lack of cloud cover and no precipitation. Pilots flying through a weak stationary front may experience frontal passage only by an area of light turbulence and a small wind change.

Stationary fronts are characterized by the following:

- Primarily stratiform clouds, but cumuliform clouds possibly developing with unstable warm air
- Precipitation, if occurring, generally widespread, steady, and light, except in cumuliform clouds
- Freezing rain or drizzle during the winter season
- Winds generally light and parallel to the front

FRONTAL WEATHER PATTERNS

Most of the western United States does not have four distinct seasons like much of the midwest and east. Even though many parts of the country experience the "four seasons," for aviation purposes weather patterns can be divided into winter weather patterns and summer weather patterns. Summer weather transitions into the winter season in November and December; winter weather changes to summer patterns during May and June. Occasionally, weather from one seasonal pattern overlaps into the other. This is especially true during the transition months.

In the winter, the Gulf of Alaska region produces the most frequent cyclonic activity in the northern hemisphere. Low pressure is centered in the vicinity of the Aleutian Islands, and high pressure persists over the Great Basin of Nevada and Utah. The flow of air over the Pacific Northwest is primarily from the south and southwest. The favored storm track curves from the central Pacific northeastward toward Vancouver Island. These tracks frequently shift southward over Oregon and California, and occasionally to southern California. Unlike the central and eastern United States, in the winter season the source of moisture is brought in with the cold air.

Fronts approaching from the northwest that penetrate southeastward are usually associated with colder, more unstable air masses and produce cumuliform clouds with showers and considerable clearing behind the fronts. These storms are accompanied by strong southerly winds ahead of the front shifting to westerly with passage.

With winter storms, freezing levels typically range from several thousand feet in Washington and Oregon west of the Cascade Range, to near the surface east of the Cascade. The temperature in the cold sector of the front may be higher than at the surface. This results in freezing rain or drizzle east of the Cascade and occasionally as far south as the interior mountains of northern California. Freezing levels in California typically range from 5000 to 8000 ft in the north to 12,000 ft in the south.

Frontal systems south of the surface and upper-level low-pressure centers crossing the Pacific generally weaken as they move on shore.

Storm systems that approach from the west—the central Pacific—tend to be weak. These systems often appear on the surface analysis chart as cold fronts with waves or stationary fronts. These systems might extend well into the Pacific. They bring poor, but seldom severe, weather with prolonged precipitation for several days. There is little chance for clearing between successive systems.

Weather systems approaching the Pacific coast from the southwest are likely to be associated with air masses of relatively higher temperature and moisture content—storms with the "pineapple connection" or the "pineapple express" (from the latitude of Hawaii). These storms are accompanied by substantial cloud layers and bands of precipitation. If the air is stable, widespread areas of steady, heavy precipitation occur; with unstable air, heavy rain, rainshowers, and thunderstorms result. Freezing levels are usually high, often above 8000 to 12,000 ft even into northern California. They are often associated with, or only depicted, as upper-level lows. Poor weather can persist for days.

Because of abundant low-level moisture and an upslope flow, the mountains normally take 24 to 36 h longer than the coastal sections or valleys to clear following frontal passage.

Pacific fronts enter the intermountain region—Idaho, Montana, Wyoming, Nevada, Utah, Colorado, Arizona, and New Mexico—after a substantial overland journey. They tend to produce heavy precipitation, in the form or rain or snow, west of the mountains and light or no precipitation east of the ranges. Freezing levels are typically at or near the surface. These types of cold fronts frequently weaken over the southern portions of Utah and Nevada. They may produce little or no precipitation in these areas. They are often accompanied by gusty, shifting surface winds and blowing dust or sand.

Polar fronts, common in winter months, enter the intermountain region from the north. They are usually frigid and fast moving. Under the influence of these fronts, blizzard conditions are not uncommon throughout the area.

Arctic fronts, descending out of Canada, moving southward or westward, may result in cold arctic air pushing across the mountain barriers all the way to the coast. This, normally, occurs only a few times over a period of several years. Snow and freezing rain may accompany the front, or skies may be clear if the air masses contain little moisture.

Surface frontal positions become difficult to locate as irregular mountain barriers are crossed. The frontal weather becomes diffuse as the air masses ahead of and behind the fronts are forced to ascend the windward slopes, intensifying the weather. On the leeward side, descending air warms and activity diminishes.

Between winter storms, strong high pressure often develops over the area, especially the Great Basin. High pressure often results in the formation of fog in the valleys, Santa Ana winds in southern California, and the Chinook of the Rockies.

Polar fronts still influence the weather, especially east of the continental divide. Here, moisture that drives the weather comes from a moist south to southeasterly flow from the Gulf of Mexico and the Atlantic Ocean. This moisture can, at times, move all the way into the desert southwest.

East of the Rockies cold fronts take on their more classic depictions. Weather deteriorates with the approach of the front, then clears rapidly after frontal passage. Moisture that fuels the front comes from the maritime tropical air of the Gulf of Mexico and the Atlantic Ocean. Fronts are more severe in late spring and early summer because those are the seasons of greatest temperature difference, producing severe thunderstorms and tornadoes.

Occasionally a strong storm system develops near the province of Alberta in Canada. These storms form on the lee side of the Canadian Rockies. The storm moves rapidly east and southeast into the Great Lakes and then into the northeast United States. Severe weather and blizzards may accompany the system, colloquially referred to as the "Alberta Clipper." However, typically the Clipper brings strong winds and cold air, often not producing much snow. However, the strong winds can cause lake effect snow south and east of the Great Lakes.

Arctic and polar air can, occasionally, reach the Gulf of Mexico or the Atlantic seaboard of the southeast United States. Freezing temperature may reach southern Texas and Florida. As the cold air moves through Texas, it may lower temperatures by as much as 10° in a few hours. In this region the cold wind is known as a "Texas norther," or "blue norther," especially when accompanied by snow.

We have discussed how waves can form and storms intensify. One such storm is the "northeaster" or "nor'easter." This storm usually develops in the lower midlatitudes (30° to 40° north latitude), within 100 mi east or west of the Atlantic coast. An intense low-pressure system progresses generally northward to northeastward along the east coast of the United States, and it typically attains maximum intensity near New England and the Maritime Provinces of eastern Canada. These storms nearly always bring heavy precipitation in the form of rain or snow and frequently winds of 30 to 50 kn—occasionally 80 kn—out of the northeast; hence the name. They may occur at any time of year, but they are most frequent and violent between September and April.

As a nor'easter approaches, rain, freezing rain, and then snow occur until drier air from the west moves in behind the storm. These storms can also produce heavy lake effect snows near the Great Lakes.

In the summer season large changes occur in the circulatory pattern. The Aleutian low weakens and moves northward as does the main storm track. The center of the Pacific high moves northwestward, and a ridge extends northeastward over the Pacific Northwest. The Great Basin high disappears, and a thermal low forms over the desert southwest with its trough extending northwestward toward the northern California coast, and at times all the way to eastern Oregon and Washington.

Summertime fronts are usually weak, approaching the west coast from the northwest. They frequently cause only drizzle along the northern coast and scattered showers or thunderstorms over the higher mountains. With their moisture dissipated in Washington and Oregon, they tend to be dry through the intermountain region. But their energy regenerates over the Rockies with the warm, moist air from the Gulf of Mexico and the Atlantic, which can produce severe weather in the midwest and east.

In the midwest and east, the Bermuda High brings moist, warm air to the region. This results in hot, humid conditions with haze reducing visibilities, often to less than 3 mi and tops to above 10,000 ft. Fronts move through, often accompanied by thunderstorms. After frontal passage, a cool, dry north to northwesterly flow brings relief from the hot, humid conditions until the south to southeasterly flow returns.

CHAPTER 7
WHERE'S THE FRONT?

Considerable misunderstanding arises because many aviation weather texts fail to adequately describe and explain nonfrontal weather-producing systems. A pilot presented with such a situation, all too often, will ask the briefer, "Where's the front?" or, "When will the front pass?" only to be told, "There is no front." Recall that weather occurs at all altitudes within the troposphere and sometimes into the lower stratosphere; the surface analysis chart often cannot solely explain the weather, even that occurring at or near the surface. Upper-level troughs, as we shall see, play a key role in the evolution of weather and weather systems.

Upper-level weather systems tend to modify and direct surface weather. They can intensify or stabilize conditions at the surface, cause thunderstorms to occur, and enhance or retard the intensity of frontal zones. Upper-level weather systems can cause severe conditions at the surface, or they can dampen or cancel out the vertical motion required to produce weather. The point that not all weather is caused by frontal systems is one theme of this chapter. In fact, nonfrontal weather producing systems have considerable influence on surface conditions, with the hurricane the ultimate example.

CASE STUDIES We had remained overnight in Las Vegas, Nevada, because of the weather. The next day was also blustery. The route to Van Nuys, California, was plagued with low clouds, mountain obscuration, and turbulence. I went into the, then, Las Vegas Flight Service Station for a briefing. After providing the briefer with necessary background information for the flight, this little old codger replied, "Well, you aren't going today!" This individual's technique was so obnoxious that without even thinking, I replied, "Oh yes I am!" And I hadn't even looked at the weather.

As is often the case in the winter, an upper-level low was over the area. There were no associated fronts, and this kind of system tends to bring poor weather for days. The briefer was correct in part; we were not going to fly a direct course to Van Nuys. As well as low ceilings and visibilities, the route was dominated with scattered rain showers and isolated thunderstorms. However, a careful check of the weather showed that a route from Las Vegas to Needles, California, then to Daggatt and Palmdale was feasible. This route has the lowest terrain and plenty of alternate airfields. By circumnavigating the rain showers, the flight was mostly smooth and without incident.

The preceding incident illustrates an upper-level weather system. A careful check of the weather indicated it was a "Go." Based on the weather, the aircraft's equipment, and a complete weather briefing to make the flight was a rational decision. But when the weather is unflyable, don't go!

We have already mentioned that there are four to seven major upper waves in each hemisphere circling the globe. Embedded within this flow are short wave troughs and ridges. In the summer it's common to find long wave troughs near the west and east coasts of the United States; in winter one is commonly in the central portion of the country. Troughs allow the transfer of colder air down from the north and warmer air up from the south. The strength of a trough is determined by its wind speed, temperature difference across the trough, and speed of movement. With strong winds and large temperature differences, troughs produce more violent weather, such as squall lines and severe thunderstorms, typically in the spring and summer.

When a short wave trough moves through a long wave trough, upward vertical motion is amplified, even sometimes producing upward vertical motion as it moves through a ridge. Short waves can be strong vertical motion producers. Most surface lows and frontal systems are associated with upper-level short wave troughs. Because short wave troughs move faster than associated surface frontal systems, a front can become stationary and left behind only to become active again as the next short wave approaches. In addition to upper troughs and ridges, upper-level closed lows play an important part in the weather picture, along with upper-level highs.

Like surface low- and high-pressure areas, lows and highs develop aloft. These pressure systems have considerable effect on surface weather, as well as that aloft. Cutoff lows are powerful vertical motion producers. They can cause severe surface weather, as well as enhance the development of thunderstorms, and produce severe clear air turbulence.

Highs aloft produce their own type of weather patterns. They can block weather systems, causing droughts, and produce hellish-like heat waves; the classic case is the Dust Bowl of the 1930s.

As a general rule, locations north of a jet stream associated with a surface front are likely to be cold and stormy; locations south of this boundary tend to be warm and dry. Because of its influence on weather at the surface, as well as aloft, we will revisit the jet stream in this chapter on more than one occasion.

Anything that spins has vorticity. In meteorology the term *vorticity* refers to the tendency of the air to spin. Vorticity is a powerful vertical motion producer. Like other atmospheric properties, it can be advected by the upper winds. Vorticity explains strengthening or weakening weather systems, the development of low-pressure areas, and the existence of clear sky associated with weather systems such as upper-level lows and hurricanes.

Upper-level convergence and divergence occur when air piles up or spreads out over a region. Convergence produces downward vertical motion, and divergence upward vertical motion. Upper-level convergence and divergence are one reason for intensifying surface high-pressure areas and deepening surface lows. Convergence and divergence are also associated with the jet stream. An area of upper-level divergence may be all that's needed to trigger thunderstorms and severe weather.

The hurricane is probably the best example of a nonfrontal weather producing system. These systems also produce just about every kind of aviation weather hazard, covering thousands of square miles. Although these storms develop in the tropics, their devastation can reach well into the midlatitudes of the United States. With the exception of the Pacific Northwest coast, their weather can affect all coastal areas, and their remnants can affect areas hundreds miles inland.

We discussed the development of frontal systems from a surface prospective in Chap. 6 and the effects of upper-level weather systems in this chapter. In the next section of this chapter, we will put all of our knowledge of surface and upper-level weather systems together. The discussion contains a three-dimensional view of the weather systems.

UPPER-LEVEL TROUGHS AND RIDGES

Forecasters love troughs. They must, they have even given them a nickname—TROF. (I don't think that's really true. TROF is the official NWS abbreviation for *trough,* but it makes good copy.) Forecasters use this term endlessly to explain all types of weather. They hide behind the term when a forecast goes sour, often blaming a busted forecast on a too strong, too weak, too fast, or too slow trough. Why do they love them so? Let's find out.

(I attended flight watch school in 1986. It was a warm day and after lunch I asked the instructor, Hershel Knolls, what the difference was between a long wave and a short wave. Hershel's eyes lit up! He proceeded to demonstrate with his arm extended, "This is a long wave." Then pulling his arm in, "This a short wave." He continued, turning at 90° angles: "This is a north wave, this is an east wave, this is a south wave, and this is a west wave." Well, I felt pretty silly, but during that afternoon I did learn the difference. Now you will have the advantage of my vast, although somewhat humiliating, experience.)

Global, or *long, Rossby waves,* with their associated ridges and troughs, extend for thousands of miles. A typical wave will have a wavelength of from 50° to 120° of longitude. There is no temperature advection; *isotherms*—lines of equal temperature—are parallel to the contours. Long waves move generally eastward at up to 15 kn, but they can remain stationary for days or even retreat. Air within the waves, however, moves at a much greater speed.

Figure 7-1 illustrates the flow through one of these long waves. The figure is divided into two sections. The top portion shows the wave looking down from above. The lower section shows a profile, or horizontal, view of the air as it moves through the wave. The arrows in Fig. 7-1 show the direction of air flow.

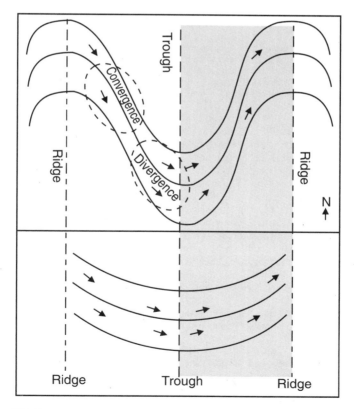

FIGURE 7-1 Upper-level waves produce downward vertical motion in the ridge-to-trough flow and upward vertical motion in the trough-to-ridge flow.

In the northern hemisphere, flow through a long wave is generally from west to east, as illustrated in Fig. 7-1. The air flows parallel to the contours represented by the solid, curved lines in the top portion of the figure. As previously mentioned, air descends in the ridge-to-trough flow and rises in the trough-to-ridge flow. In other words, the long wave produces both downward and upward vertical motion. The gray, shaded area in Fig. 7-1 shows the area of upward vertical motion.

We can better visualize vertical flow within a wave by discussing the lower portion of Fig. 7-1. The solid, curved lines represent constant pressure surfaces. In the ridge-to-trough flow, it can be easily seen that the air is descending as it moves from the ridge to the trough. The flow changes to upward vertical motion in the trough-to-ridge portion of the wave.

Refer back to Fig. 3-2. We can see why some parts of the polar front produce weather and others do not. The upper-ridge axis is on a more or less north-south line through the western Pacific and over the western Rockies. A trough exists on, again, a more or less north-south line through the central Pacific and the eastern Rockies. Note the general absence of clouds in the ridge-to-trough flow over the western Pacific and considerable clouds in the trough-to-ridge flow in the eastern Pacific, western Canada, and Pacific Northwest.

Moisture and clouds sometimes spill over the top or move through the ridge. When this occurs, weather types sometimes refer to it as a *dirty ridge*. I'm not sure if a "dirty ridge" refers to the unsettled character of the weather or the fact that forecasting these conditions is difficult.

CASE STUDY   Several years ago we had just such a situation along the west coast. Forecasters commented that rain fell from every cloud. A frustrated prognosticator wrote, in an internal NWS product, that they had finally determined the difference between a trough and a ridge. "Cold rain falls from a trough and warm rain from a ridge." This would seem to prove the proposition that weather forecasting is as much an art as a science.

Short waves, embedded in the overall flow, tend to pass through the long wave pattern at speeds of 20 to 40 kn. Most surface lows and frontal systems are associated with upper-level short wave troughs. Short waves have wavelengths of from 1° to 40° of longitude. Short waves usually produce thermal advection—isotherms cross contours at some angle. Typically, cold-air advection exists to the west of the trough axis, and warm-air advection to the east. Most surface lows and frontal systems are associated with upper-level short wave troughs. Upper troughs, as we shall see, are a key to the evolution of weather systems.

How do short waves develop? Let's use the following analogy. Air, like water, behaves like a fluid. Let's think of the upper portion of Fig. 7-1 as a river. Water, like air, flows swiftly through the large meanders. Similar to a raging river, the flow comes against one bank and then the other, and it may be disrupted by debris in the river bed. To atmospheric flow, major mountain ranges, like debris in the river bed, are obstructions to the flow. This causes eddies to develop in the flow that move downstream at about the same speed as the overall flow. These eddies correspond to short waves in the atmosphere.

FACT   Some have recently proposed that the Ice Age was, at least in part, due to a change in the jet stream caused by the development of the Himalayan mountain range.

Refer to Fig. 7-2. Again the upper portion of the illustration is a vertical view, the lower portion a horizontal view, and the gray, shaded area represents upward vertical motion. As a short wave trough moves through a long wave ridge-to-trough flow, an area of upward vertical motion can be created. Notice in the lower portion of Fig. 7-2 that in the short wave trough to short wave ridge flow, there is upward vertical movement. Thus, as the short wave trough moves through a ridge, clouds and precipitation can occur. The extent

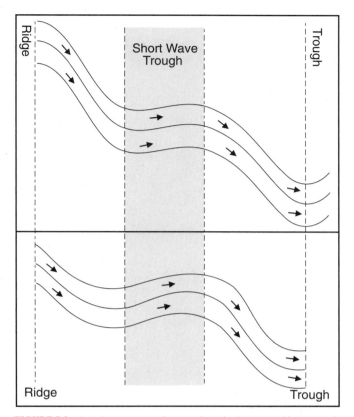

FIGURE 7-2 As a short wave trough moves through a long wave ridge-to-trough flow, an area of upward vertical motion can be created.

of clouds and precipitation, if any, will depend on the relative strengths of the short and long waves.

Conversely, as a short wave trough moves through a long wave trough-to-ridge flow, upward vertical motion is enhanced. This is illustrated in Fig. 7-3. Typically, a short wave ridge follows the short wave trough. The end result, as illustrated in Fig. 7-3, is enhanced vertical motion in the gray, shaded areas of the illustration and even an area of reduced vertical motion in the short wave ridge-to-trough flow. Again, the amount of clouds and precipitation depends on the relative strengths of the troughs.

FACT Forecasters occasionally use the term *flat ridge*. According to Dick Williams, AWC forecaster: "Flat ridge—sounds sort of contradictory, doesn't it? We usually use the term when a trof is crossing over the top of a ridge, reducing the amplitude of the ridge and consequently reducing the 'ridge effects' such as subsidence or downward vertical motion. It's a means of transitioning from high amplitude or meridional flow to zonal flow patterns. A little jargony as terms go, but standard stuff...."

The location of upper-level troughs and ridges can be found on constant pressure charts. Upper-level troughs and ridges transport or advect atmospheric properties aloft. Cold-

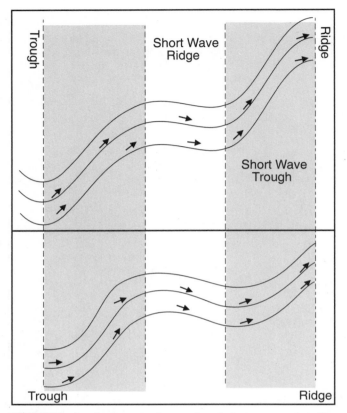

FIGURE 7-3 As a short wave trough moves through a long wave trough-to-ridge flow, upward vertical motion is enhanced.

air advection destabilizes the atmosphere at the 500-mb level. Conversely, warm-air advection at this level stabilizes the atmosphere. This is opposite to the effects of cold- and warm-air advection near the surface. Why? As cold air aloft is advected into an area, any rising air will be warmer than surrounding air. Therefore, cold-air advection aloft enhances the development of weatherlike thunderstorms—by promoting vertical development. Warm-air advection at this level stabilizes the atmosphere, strengthens high-pressure ridges, and diminishes low-pressure troughs.

How does warm or cold-air advection occur? Recall that isotherms are lines of equal temperature. For warm- or cold-air advection to occur, wind must blow across isotherms. This is the mechanism for the transportation of cold or warm air.

Isotherms and areas of cold- or warm-air advection are found on constant pressure charts.

With the approach of an upper-level trough, a period of 8 to 12 h of poor weather can be expected. The surface front will precede the trough, usually bringing IFR weather. However, without a front, the upper trough or low might bring only marginal VFR conditions with localized areas of IFR. Under these conditions, VFR flight might be possible, except in mountainous areas that remain obscured in clouds and precipitation.

On the other hand, the absence of an upper-level trough will tend to weaken and slow a front's progress. A ridge aloft even with a surface front will not tend to produce thunderstorms or severe weather because the ridge prevents the vertical development required. It is not all that unusual to have cloud tops below 10,000 ft with weak fronts.

There are two general rules applied to the 500-mb chart. At latitudes within the United States, weather systems will tend to produce precipitation north of the 564 contour line and snow north of the 540 contour line. These are sometimes referred to as the *rain* and *snow lines.*

UPPER-LEVEL LOWS AND HIGHS

Like low- and high-pressure areas at the surface, upper-level lows and highs also occur. Their definition is the same. An upper-level low is an area completely surrounded by higher pressure, and an upper-level high is an area surrounded by lower pressure. Recall that unlike the surface chart, upper-level charts depict weather patterns through the height of constant pressure surfaces. However, contours can be analyzed the same as isobars on the surface analysis chart.

At and above the 500-mb level, when an area of low pressure is completely surrounded by a contour, it is called a *closed low* or *cutoff low.* These cutoff lows, sometimes also referred to as *cold lows aloft* or *upper-level lows,* are an important winter weather feature. Under these conditions low pressure is reflected from the surface to the tropopause. When the area of low pressure is vertical through the atmosphere, these storms tend to be powerful and erratic. Storms resulting from cutoff lows produce precipitation and low ceilings and visibilities over widespread areas. Forecasting the formation of cutoff lows, and their movement, is difficult.

How do cutoff lows form? Typically an intense high-pressure ridge is present over the eastern Pacific, with strong northerly or northwesterly winds aloft along the coast of the Pacific Northwest, with its associated jet stream. The jet typically works around to the south side of the trough, and finally to the east side. After formation, the lows tend to move southward for the first 12 to 24 h, in response to the strong northerly jet on the west side of the low. Cutoff lows tend to be slow moving and erratic. Under their influence, weather can remain poor for several days or more. When the low develops over the Great Basin of Nevada and Utah, it is often called an "Ely Low." When these lows move over the north central Rockies and plains, widespread blizzard conditions and heavy snow result. During summer months, closed lows aloft support the development of thunderstorms, once surface lifting begins.

The following area forecasts synopsis illustrates the impact of upper-level weather systems:

STRONG UPPER LOW OVER SOUTHWESTERN NEW MEXICO WILL MOVE TO
NORTH CENTRAL TEXAS BY 22Z.

The 0900Z surface analysis chart showed weak surface high pressure over the western United States. However, the 1000Z weather depiction chart discloses extensive areas of IFR and marginal VFR, with rain and snow occurring throughout New Mexico, Texas, and Oklahoma. The weather closed airports for days, and it was blamed for the deaths of dozens of people. The 500-mb analysis reveals the culprit. A deep upper-level low along the southern Arizona-New Mexico border, and associated downstream trough, produced devastating surface conditions.

CASE STUDY During one episode an upper low drifted in the vicinity of Red Bluff, California, for 5 days. Pilots would call day after day wanting to know when the weather would clear. After a while, the common response became, "The low is forecast to move east out of the area tomorrow, but that's what they said yesterday."

Clear air turbulence (CAT) is common in the winter season. The jet stream, lows aloft, and sharp troughs can cause severe CAT. Sharp troughs can produce severe wind shear turbulence as low as 8000 to 10,000 ft.

CAT is implied by the jet stream, lows aloft, and sharp troughs, which can be found on constant pressure charts. Forecasts for CAT are contained in SIGMETs and the AIRMET Bulletin.

Upper-level troughs and lows tend to form bands of weather. The weather deteriorates as a band moves through, then improves, only to deteriorate with the next band. The area forecast cannot, and does not, take this into account. Under these conditions, a pilot must be careful not to get suckered by a temporary improvement.

An upper-level high typically brings warm, dry weather. It tends to block approaching weather systems in the normal west-to-east flow, and it is sometimes known as a *blocking high*. Heat waves can develop when the subsidence from an upper-level high lingers over an area. Blocks form most often over the eastern Pacific in winter and spring.

An upper-level high over a surface low splits the flow. One branch moves poleward, the other toward the equator. Storms that approach the block from the west tend to weaken and either go over the ridge to the north or pass through the trough to the south.

A specific pattern associated with upper-level highs is called an *omega block*. Upper-level flow resembles the Greek uppercase letter omega (Ω), with an upper-level high in the center of the pattern. This is a strong blocking high that can remain stationary for days or weeks. Short waves, and their associated weather, ride up over the high. When this occurs, a drought can result. If the pattern continues for more than one season, severe droughts, such as the Dust Bowl of the 1930s, develop. Note, however, that the effects of the Dust Bowl were aggravated by other factors. Climatologically, major droughts occur in the United States roughly every 22 years, mainly in the midwestern states. However, droughts do occur in other parts of the country.

Flying weather in areas dominated by upper-level highs is similar to conditions under the influence of a stable air mass, and typically poor visibilities may be expected. High temperatures result in high-density altitude. Hot, dry conditions are ideal for brush and forest fires with their resulting smoke layers. If the wind picks up, pilots will often have to contend with widespread areas of blowing dust and sand.

At times upper-level highs develop over surface low-pressure areas. Thermal lows and hurricanes exhibit warm-core lows. This results in a relatively shallow low. Convergence occurs below the 500-mb level, due to thermal heating in the case of a thermal low and the release of latent heat in the case of a hurricane. Above the 500-mb level, divergence occurs. A net outflow occurs at the 500-mb level.

Lifting mechanisms have a cumulative effect. Upper-level troughs parallel to and behind a front intensify the storm. These fronts tend to be fast moving. These lows occur when cold air at the base of the trough is cut off from the cold air to the north. This closed circulation can lead to a circular jet stream. The weather pattern remains moist and unstable even though the surface front has passed. These systems tend to bring extended periods of weather, sometimes severe, in the form of bands one after the other.

Figure 7-4 illustrates distribution of constant pressure surfaces when high pressure aloft forms over low pressure near the surface and vice versa. This situation occurs with temperature advection, either at the surface or aloft. Essentially, when a low aloft forms over high surface pressure, there is relatively warm air near the surface and cold air aloft. When a high aloft forms over low surface pressure, there is relatively cold air near the surface and warm air aloft.

Upper-level lows and highs can be found on constant pressure charts. The location of upper-level lows and highs can be found on constant pressure charts.

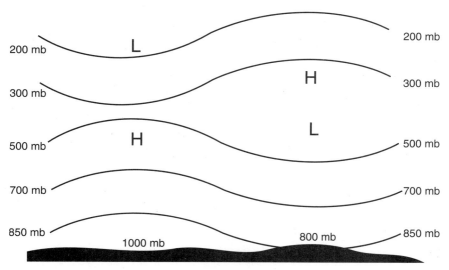

FIGURE 7-4   Pressure patterns. Temperature advection produces low pressure aloft over high pressure near the surface and vice versa.

THE JET STREAM REVISITED

We included a section on the jet stream in Chap. 1. Because of its influence, this player has also been mentioned in sections on turbulence and thunderstorms. We'll talk again about the jet stream in our discussion of convergence and divergence. But now we will expand on its influence, primarily upon surface weather systems. We'll see just how this significant weather phenomenon affects the overall weather picture.

The reason for the existence of the jet stream is the temperature contrast between the higher latitudes and the tropics. Meteorologists use complex computer models to predict the formation, location, and movement of the jet. Due to the complexity of the atmosphere, these predictions do not always come true. However, errors of timing are more prevalent than errors of occurrence. In other words, the forecast will almost always come to pass, but the specific time of occurrence may be off.

Recall that the jet stream is not a continuous band of high winds. The jet stream is composed of segments, sometimes called *jet streaks* or *jetlets*. The *polar jet* marks the boundary between polar air and warmer midlatitude air. The *subtropical jet* separates warm midlatitude air from warmer tropical air. At times there may be a third jet stream: The *arctic jet,* north of the polar jet, occurs between the boundary of extremely cold, arctic air and cold, polar air.

Small, sudden changes in the jet stream often coincide with the development of surface storm systems. These changes also affect the upper wind pattern both upstream and downstream from the occurrence. This has the effect of strengthening or weakening surface weather systems; we'll see how in our discussion of jet stream convergence and divergence.

Recall that as a general rule, locations north of a jet stream associated with a surface front are likely to be cold and stormy; locations south of this boundary tend to be warm and dry. A jet embedded in a long wave can remain relatively stationary for weeks; this usually brings long periods of bad weather to the north of its location and good weather to the south.

The movement of surface high- and low-pressure areas and fronts is related to the movement of the jet stream. Low-pressure areas tend to move with the jet stream flow. As the wave with the jet passes, a ridge builds aloft, usually bringing high pressure and good weather. However, as high pressure builds, surface pressure gradients are often steep, causing strong, sometimes destructive, surface winds.

The jet stream will typically have the following effect on surface weather systems. When the jet stream is parallel to a surface front, the front tends to be slow moving and relatively weak. However, if the jet stream is perpendicular to a surface front, the front is apt to be fast moving and strong.

Often interesting and obtuse terms are used to describe weather systems. For example, when a jet stream behind a trough increases the trough's strength, it is referred to as *digging*. Therefore, a trough "digging southward" describes a trough that is increasing in strength as it moves.

The location of the jet stream can most easily be found on the 300- and 200-mb constant pressure charts. Typically the polar jet appears on the 300-mb chart and the subtropical jet on the 200-mb chart. This is because the polar jet is usually at a lower altitude; hence, it appears on the lower-altitude pressure level chart. Which chart will best reflect the jet varies with season and latitude. In the winter look for the jet on the lower-altitude chart (300 mb); in the summer on the higher-altitude chart (200 mb). The location of the jet may also be revealed on satellite imagery, evidenced in the presence of jet stream cirrus, jet streaks, and transverse lines or cloud trails. The forecast location, speed, and altitude of the jet stream appears on the high-altitude significant weather prognosis (PROG) chart.

VORTICITY

Any nonmeteorologist pilot who wishes to better understand atmospheric phenomena will require a basic knowledge of vorticity. Although some pilot-meteorologists feel this subject is far too technical for pilots, I disagree. At the very least, pilots, especially those using direct user access terminals (DUATs), will come across this term in synopses, convective SIGMETs, and convective outlooks. An understanding of vorticity will help relate the fact that not all weather occurrences can be attributed to pressure and frontal systems alone, as displayed on weather charts.

Anything that spins has vorticity, which includes the Earth. *Vorticity* is a mathematical term that refers to the tendency of the air to spin; the faster air spins, the greater its vorticity. Refer to Fig. 7-5. The spin can occur in any direction, but our concern is with air spinning vertically within the overall horizontal flow. A parcel of air that spins counterclockwise—cyclonically—has *positive vorticity*; a parcel that spins clockwise—anticyclonically—has *negative vorticity*.

How does vorticity occur? Divergence aloft must be compensated by convergence near the surface. In order for the total mass of air to remain constant, the column is elongated vertically. So, what do we have? Upward vertical motion, independent of other vertical motion mechanisms. As the column stretches, spin—vorticity—increases, becoming more positive. Conversely, convergence aloft produces low-level divergence. The column flattens, and its spin decreases, and its vorticity decreases—becoming less positive. The result: downward vertical motion.

The Earth's vorticity is always positive in the northern hemisphere because the Earth spins counterclockwise about its axis. An observer, standing on the north pole, will have maximum vertical spin, one revolution per day. An observer's vertical spin will decrease moving toward the equator, becoming zero at the equator, like the Coriolis force, which is maximum at the poles and zero at the equator. Vorticity is directly related to the Coriolis

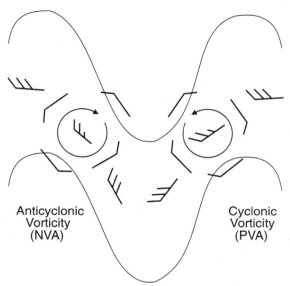

FIGURE 7-5   Areas of vorticity move through the overall flow.

force. The rate, or value, of the vorticity produced by the Earth's rotation is, not surprisingly, known as the *Earth's vorticity*.

The atmosphere is almost always in motion, and therefore it generally will have its own vorticity relative to the Earth—that is, its *relative vorticity*. The sum of the Earth's vorticity plus relative vorticity equals absolute vorticity. The value of absolute vorticity at midlatitudes almost always remains positive because of the Earth's rotation.

Air moving through a ridge, spinning clockwise, gains anticyclonic relative vorticity. Air moving through a trough, spinning counterclockwise, gains cyclonic relative vorticity. Therefore, there tends to be downward vertical motion in ridge-to-trough flow and upward vertical motion in trough-to-ridge flow.

Vorticity is advected just as are other atmospheric properties. As illustrated in Fig. 7-5, areas of vorticity flow with the wind through the overall pattern. Their vorticity values remain unchanged. Therefore, these regions are known as areas of *positive* or *negative vorticity advection*. Pilots can expect to hear the terms *positive vorticity advection* (PVA) and *negative vorticity advection* (NVA) (referring to relative vorticity). Positive vorticity advection indicates the following:

- A trough or low moving into an area
- A ridge or high moving out of an area
- Upward vertical motion probably occurring
- Increasing cloud cover and precipitation

Negative vorticity advection indicates the following:

- A ridge or high moving into an area
- A low or trough moving out of an area

- Downward vertical motion probably occurring
- Decreasing cloud cover

As regions of PVA are swept along within the overall flow, they represent microsystems that can rotate around synoptic-scale highs and lows. These areas may be referred to as "vort maxes" or "vort lobes." If an area of PVA moves over a stationary surface front, a wave can form and a storm develop. An area of PVA might be all that's required (a lifting mechanism) to trigger thunderstorms when moisture and instability are available. On the other hand, an NVA area might retard or prevent thunderstorm development.

Vorticity explains the development of low-pressure areas east of mountain barriers, particularly in eastern Colorado. These lows are most predominant east of the Rockies because of the high mountain elevations, sometimes referred to as *lee-side lows.*

How do lee-side lows form? Across the Rockies, there is typically a westerly flow. Air forced up the west slopes of the mountains is trapped below the tropopause. To compensate for the column's resulting compression, there is a decrease in absolute vorticity. This imparts on the flow an anticyclonic track to the southeast. As the air moves downslope, the column of air expands, and absolute vorticity increases. The air now begins a cyclonic track to the northeast. This creates a low-pressure area on the lee side of the mountain barrier.

We have mentioned that closed upper lows produce bands of weather. Vorticity explains these phenomena. In the areas of weather there is PVA; in the relatively clear areas between the bands, NVA occurs. These areas of NVA are sometimes referred to as the *dry slots.*

Absolute vorticity is analyzed at the 500-mb level. However, vorticity is not directly reflected on constant pressure charts. Even when the air is rotating within the overall system, balloon observation can measure only the mean, or average, wind speed. Refer to Fig. 7-5. Note that the wind speed before and after the areas of NVA and PVA is 30 kn. (The wind arrow shows the direction of the wind; the wind barbs indicate speed. Each barb represents 10 kn, and each half barb is 5 kn.) There is an anticyclonic flow embedded in the ridge-to-trough flow of 10 kn. An area of embedded cyclonic flow of 10 kn has occurred in the trough-to-ridge flow. A weather balloon, or aircraft, flying through the area would see only the overall wind component of 25 kn (represented by the two full barbs and half barb on the wind arrows). The NWS produces composite height/vorticity charts. This chart series, with forecasts to 60 h, depict areas and magnitudes of both PVA and NVA. Although normally not directly available to the pilots, 500-mb composite height/vorticity charts are routinely transmitted. High values of absolute vorticity (greater than 16) have strong cyclonic rotation, indicating strong upward vertical motion, and low values (less than 6) have anticyclonic rotation, indicating strong downward vertical motion.

CONVERGENCE AND DIVERGENCE

Recall that surface convergence and divergence occur between the surface and about 10,000 ft. At the 500-mb level, halfway through the atmosphere, is a level of nondivergence. The air does not tend to converge or diverge at this level. Upper-level convergence and divergence occur above the 500-mb level.

Convergence in the upper atmosphere is the piling up of air over a region. Since air cannot be created or destroyed, it has to go somewhere. The only place for it to go is down. Therefore, convergence aloft is a downward vertical motion producer. Divergence is the spreading out of air over a region. This leaves a void or vacuum. Air is sucked up from below. Therefore, divergence aloft is an upward vertical motion producer.

Upper-level convergence and divergence can be seen on the 300- and 200-mb constant pressure charts.

Recall that at the surface, winds flow out of high-pressure areas. In the upper atmosphere the wind converges over a surface high. If upper-level convergence exceeds surface divergence, the high pressure will build and surface pressure will increase. Conversely, at the surface winds flow into low-pressure areas. Above the low, in the upper atmosphere the winds diverge. If upper-level divergence exceeds surface convergence, the storm system intensifies, and surface pressure decreases.

What causes upper-level convergence and divergence? These phenomena may result from changes in wind direction and speed. Areas of convergence, also called *confluence,* occur in areas where the contours move closer together. The result is similar to the funneling effect of terrain. Wind speed increases. Areas of divergence, also called *difluence,* occur in areas where the contours move farther apart. In these areas wind speed decreases. Areas of convergence and divergence are illustrated in Fig. 7-1.

Areas of convergence and divergence are also associated with the jet stream. Refer to Fig. 7-6. As air enters the jet maximum, it increases rapidly in speed over a relatively short distance due to an increase in pressure gradient forces. The increase in pressure gradient forces temporarily exceeds the Coriolis forces. The Coriolis forces cannot increase fast enough to balance the pressure gradient forces. Thus the air swings slightly to the north across the contours. As the air leaves the jet max, pressure gradient forces decrease. Coriolis forces temporarily exceed pressure gradient forces, and the air swings slightly to the south, across the contours. This causes convergence (regions of upward-moving air from the surface through the jet stream level) in the northwest and southeast quadrants of the jet and divergence (regions of downward-moving air) in the southwest and northeast quadrants. The southwest quadrant is usually the more active because of greater temperature differences.

At the surface, deepening storms develop below the strongest divergence in the northeast quadrant. In the spring, summer, and early fall, one of these quadrants may be all that's required to trigger severe thunderstorms and tornadoes. These are mechanisms that forecasters look for in the development of the convective outlooks and other severe weather products. Beneath the area of strongest convergence, the northwest quadrant, anticyclones build.

As mentioned, divergence aloft develops when contours diverge or move apart as seen on the 300-mb chart. This causes surface convergence and increased cyclonic vorticity. Surface lows can develop in this way. A perfect example occurred one afternoon with scattered thunderstorms forecast for northern California, northern Nevada, and east-

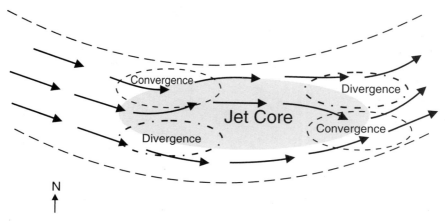

FIGURE 7-6   Areas of convergence and divergence are also associated with the jet stream.

ern Oregon. No weather systems were depicted on either the surface analysis or 500-mb chart. However, thunderstorms did occur, right along a line of difluence. The difluence caused just enough surface convergence, along with a moist, unstable air mass, to trigger thunderstorms.

HURRICANES

Hurricanes produce just about every kind of nasty weather extending over thousands of square miles. Figure 7-7 is a satellite photo of hurricane Andrew on August 24, 1992.

> **FACT** Like great baseball players, the World Meteorological Organization retires the names of great hurricanes, including Andrew—the 44th western hemisphere storm name to be so honored (?). Hurricane Andrew just missed category 5 status as it blasted into southern Florida with a central pressure of 27.23 in.

FIGURE 7-7 Hurricanes produce just about every kind of nasty weather extending over thousands of square miles. The extent of hurricane Andrew can be seen in this satellite photograph taken on August 24, 1992.

Before we proceed, it might be interesting to see how hurricanes get their names. It appears that the practice of giving Atlantic hurricanes women's names began with writer George Stewart in his book *Storm,* published in 1941. A character in the book was a Weather Bureau meteorologist who used this method as he tracked these storms.

During World War II Army and Navy weather types tracked storms over the Pacific as part of the war effort. Using women's names to communicate storm information was short, quick, and less confusing than the identification system that had been used in the past. This practice continued after the war and was applied to Atlantic hurricanes.

Names are alphabetically selected in advance and applied to successive seasonal tropical storms, starting with A and proceeding through the alphabet. Names selected are short, easily pronounced, quickly recognized, and easy to remember. A similar procedure is used for Pacific storms. The only region where tropical cyclones are not named is the north Indian Ocean.

In the 1980s the practice of using only women's names was changed. Tropical storms are still named alphabetically, with A the first storm of the season; however, male (e.g., Andrew) and female names are used alternately.

Tropical cyclone is a general term applied to any low-pressure area that originates over tropical oceans. (*Cyclone* comes from the Greek *kyklon,* which refers to the coil of a snake. *Hurricane* does not have a clear origin, but it seems to have been used by the natives of the West Indies or Central America based on the word for "great wind," hurricane.)

FACT   What happens if a tropical storm or hurricane moves from the Atlantic into the eastern Pacific? The storm may be the same, but the name changes. In 1988 hurricane Joan underwent a "sex change" as it move through Nicaragua, ending up as tropical storm Omar in the Pacific.

Tropical storms typically develop during the mid- to late-summer season. Principal hurricane months are August, September, and October; most occur in September. Ninety-five percent of the intense hurricanes in the Atlantic occur during this period. However, early season hurricanes can develop during May, June, or July. Hurricanes that affect North America evolve in the warm tropical waters of the Atlantic, Caribbean, and Gulf of Mexico, and off the west coast of Mexico.

Tropical cyclones are classified according to their intensity based on average 1-min wind speeds. Wind gusts in these storms may be as much as 50 percent greater than the average 1-min wind speeds. Tropical cyclones are internationally classified as follows:

- *Tropical depression:* Highest sustained winds up to 34 kn
- *Tropical storm:* Highest sustained winds 35 to 64 kn
- *Hurricane:* Highest sustained winds 65 kn or more

Hurricanes are further classified by intensity using the Saffir-Simpson Scale of Hurricane Intensity, as shown in Table 7-1.

Tropical cyclones develop under optimum sea surface temperature and weather systems that produce low-level convergence and cyclonic wind shear. They favor tropical, easterly waves, troughs aloft, and areas of converging northeast and southeast trade winds along the intertropical convergence zone.

The low-level convergence associated with these systems, by itself, will not support the development of a tropical cyclone. The system must also have horizontal outflow—divergence—at high troposphere levels. Recall the earlier discussion of high pressure over low pressure. At high levels there is anticyclonic flow. This combination creates a "chimney" in which air is forced upward, causing clouds and precipitation. Condensation releases large quantities of latent heat, which raises the temperature of the system and accelerates upward motion. The increased temperature lowers surface pressure—recall our discussion

TABLE 7-1 Hurricane Intensity

Category	Central pressure, mb	Storm surge, ft	Mean wind, kn
1. Weak	>980	4–5	64–83
2. Moderate	965–969	6–8	84–96
3. Strong	945–964	9–12	97–113
4. Very strong	920–944	13–18	114–135
5. Devastating	<920	>18	>135

of surface warm-air advection—which increase low-level convergence. This draws more moisture-laden air into the system. When these chain-reaction events continue, a huge vortex is generated, which may culminate in hurricane-force winds.

Tropical cyclones usually originate between 5° and 20° latitude. Tropical cyclones are unlikely within 5° of the equator because the Coriolis force is so small that cyclonic circulation cannot develop. Winds flow directly into an equatorial low, and it rapidly fills.

Tropical cyclones in the northern hemisphere usually move in a direction between west and northwest while in low latitudes. As storms move toward midlatitude, they come under the influence of the prevailing westerlies. Thus a storm may move very erratically, reverse course, or even circle. As the prevailing westerlies become dominant, storms recurve toward the north, then to the northeast, and finally to the east-northeast as they reach well into the midlatitudes.

If a storm tracks along a coastline or over the open sea, it gives up, slowly unleashing its devastation far from tropical regions. However, if the storm moves inland, it weakens due to surface friction and loss of its moisture source. Like the process of air mass modification, as storms curve toward the north or east, they begin to lose their tropical characteristics and acquire characteristics more like low-pressure areas in the midlatitudes. Strength weakens as cooler air flows into the storm.

While developing, the cyclone has the characteristics of a circular area of broken to overcast, multilayered clouds. Numerous rain showers and thunderstorms are embedded in these clouds. Coverage ranges from scattered to almost solid. Lightning is relatively scarce in hurricanes. This is due to the fact that updrafts are relatively weak, compared to those in continental thunderstorms.

As cyclonic flow increases, thunderstorms and rain showers form into broken or solid bands paralleling the wind flow that spirals into the center of the storm. These spiral rain bands frequently are seen on radar and satellite images. In both Figs. 7-8 and Fig. 7-9, these cloud bands can easily be seen. All the hazardous weather effects associated with thunderstorms, including tornadoes, are present. Between the bands, weather is less severe, but like upper-level lows, the situation can "sucker" an unwary pilot.

The "eye" usually forms in the tropical storm stage and continues through the hurricane stage. Near the top of the thunderstorms, the air is relatively dry. Losing its moisture, the air begins to flow outward away from the center, in a diverging anticyclonic flow. This flow extends several hundred miles from the eye. The air begins to sink and warm as it reaches the limit of the storm. This results in clear skies outside the storm.

Surrounding the eye is a wall of clouds that may extend above 50,000 ft. This *wall cloud* contains torrential rains and the strongest winds in the storm. As a result of eye wall thunderstorms, the air warms from the release of latent heat. This initiates a downward motion in the eye, which helps account for the absence of weather at the storm's center. In the eye, skies are cloud free, flight conditions smooth, and winds comparatively light. The average diameter of the eye is between 15 and 20 mi, although it is sometimes smaller or larger.

Figure 7-8 shows two Pacific storms—tropical storm Eugene at 140° and hurricane

FIGURE 7-8 Since their energy comes from the ocean, hurricanes typically dissipate rapidly over land.

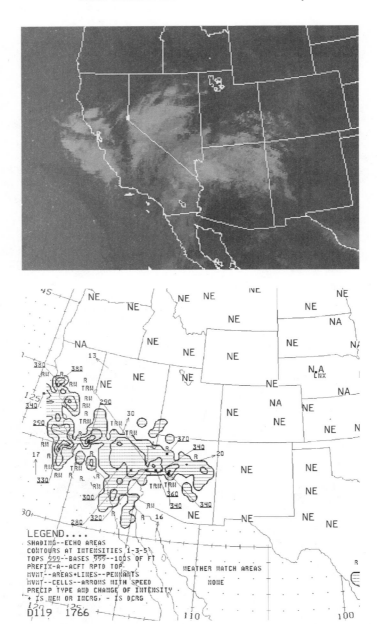

FIGURE 7-9   The moist, unstable remnants of Pacific hurricanes can be carried north and inland to affect central California and the southern part of the intermountain region.

Dora at 125° W longitude. Since their energy comes from the ocean, they dissipate rapidly over land. Occasionally, eastern Pacific hurricanes reach Hawaii and southern California. The moist, unstable remnants of these storms can be carried north and inland to affect central California and the southern part of the intermountain region. In Fig. 7-9 moisture resulting in thunderstorms from the remnants of tropical storm Hillary can be seen as far north as central California and as far east as southern Nevada and Arizona. Analysis of these southerly disturbances was first conducted in the 1930s. Because they sometimes approached from the southeast, they were called "Sonoran storms," after the Mexican state of that name.

The position of tropical cyclones can be found on surface and constant pressure charts. Their forecast position appears on low- and high-altitude significant weather prognosis charts. Additionally, the NWS issues tropical storm and hurricane alerts. These alerts contain the plotted position of the storms, their intensity, forecast movement, and intensity trend.

PUTTING IT ALL TOGETHER

A theme of this chapter has been how upper-level weather systems modify and direct surface weather. We've seen how they intensify or stabilize conditions at the surface, trigger thunderstorms, and enhance or retard the intensity of frontal boundaries. At this point we have enough knowledge to put our three-dimensional atmosphere together. The following discussion will help us understand the development and decay of weather systems at the surface.

In Chap. 6 we discussed frontogenesis and frontolysis strictly from the perspective of events occurring at the surface. Now we are ready to include the influence of upper-level systems.

Let's begin by superimposing a long wave trough over the stationary front in Fig. 6-1, which is illustrated in Fig. 7-10. Colder air is located on the north, polar side of the illustration, and warmer air on the south, tropical side. Winds aloft are strong, which results in a strong wind shear between the surface and upper levels. Now let's move a short wave trough through the area. This causes instability as warmer air rises and colder air subsides—vertical motion. Cyclonic circulation is enhanced by horizontal and vertical air motion.

As a result of the short wave, the flow aloft develops into an area of converging air behind the short wave. An area of diverging air forms ahead of the short wave. At the surface, pressures change and wind speed increases. If there is no surface front, this mechanism can start the chain of events that brings air masses of different properties together. Presto, magic a front is born.

As the converging surface air begins to spin, cold air flows southward and warm air northward—cyclonic flow. The stationary front now develops into distinct cold and warm fronts. Warm- and cold-air advection occurs below the 500-mb level.

Above the 500-mb level cold-air advection occurs behind the cold front, bringing cold air into the trough. The cold-air advection makes the air more dense and lowers the height of the air column between the surface and the 500-mb level. Pressure in the trough lowers, and the trough deepens. Ahead of the trough, warm-air advection occurs, which has the effect of raising the height of the 500-mb surface increasing pressure, and the ridge strengthens. The result of the differential temperature advection is to intensify the strength of the developing cyclone. As condensation in the ascending air releases latent heat, the system strengthens even more. The result is the development of a midlatitude cyclone. Typically the most intense situation occurs when the upper-level trough is one-quarter

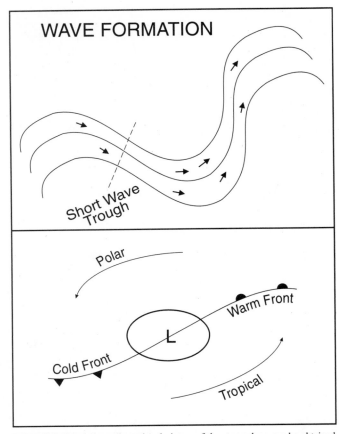

FIGURE 7-10 A three-dimensional picture of the atmosphere can be obtained by superimposing upper-level flow over surface conditions.

wavelength behind the surface cold front. Under these conditions all the vertical motion producers become cumulative. The resulting weather patterns are those described in the life cycle of a front in Chap. 6.

Eventually the supply of warm surface air is cut off. A pool of cold air, which has broken off from the main flow, often lies directly above the surface low. Temperature advection ceases, and the surface system gradually weakens. The upper low itself may persist for days as a cutoff low, bringing clouds and precipitation to a large area with no surface fronts.

PILOT WEATHER RESOURCES AND SERVICES

CHAPTER 8

SURFACE OBSERVATIONS (METAR)

According to Mark Twain, "If you don't like the weather in New England, just wait a few minutes." The same can be said about the format of aviation weather reports. In less than a hundred years there have been evolution and revolution in the way we report the environment. Various evolutions have occurred over the years, such as the change from wind direction arrows to reporting wind in degrees (1960s), and the change from sky cover symbols to contractions (1970s). Revolutions have also taken place, such as the transmission of weather reports electronically (teletype and radio) and the introduction of automated observations. (An *automated report* indicates the observation was derived without human intervention. The terms *manual, augmentation,* and *backup* refer to the fact that a person has overall responsibility for the observation even though some or all of the elements of the report are derived from automated equipment.) How has automation affected the quality of observations? More about this throughout the chapter.

The latest revolution occurred on July 1, 1996, when the Meteorological Aviation Routine (METAR) replaced the North American Surface Aviation (SA) code in the United States. Mexico converted in 1995, and Canada, along with the United States, in 1996. Ostensibly, METAR has standardized surface aviation reports throughout the world. However, with METAR individual member countries are allowed to change certain items in the report. For example, rather than metric, the United States continues to use English units of measurement, except for temperatures.

Another revolution took place in the early 1990s. Prior to this time, when most pilots were asked about coded weather reports, they'd say: "Why do I have to read the weather? When I call flight service, they read and explain the reports." Pilots have been objecting to coded weather information since its inception in the early 1930s. With *direct user access terminals* (DUATs) and other commercially available weather systems, pilots now have to read weather on their own. Even though some systems provide plain-language translations, on occasion they can be confusing and misleading. A pilot's knowledge of weather takes on greater significance with DUAT, and National Weather Service (NWS), and flight service station (FSS) consolidation, with further FSS consolidation already being considered. Whether you're a beginner or an experienced hand, every pilot needs the ability to decode, translate, and interpret weather information.

Most weather reports are history by the time they are transmitted and available. In the teletype era—prior to the late 1970s—reports were often an hour and a half old. Beginning in 1978 the Federal Aviation Administration (FAA) introduced a computer system known as *leased service A* (LSAS). With this and other computerized systems, most reports are available within 20 min of observation. Automated observations received through DUATs or FSSs are accessible within a few minutes of observation time, and they are essentially real-time when obtained via direct phone or radio.

The accuracy and validity of observations, and therefore their usefulness, depend on many factors: who or what is taking the observation; what is the extent of the observer's training and experience; what type of equipment, if any, is available? Automated observations have advantages and disadvantages over manual reports. Observational technique between automated and manual observations differ significantly in four major areas: visibility, weather and obstructions to vision, sky condition, and remarks. It is not a case of whether automated or manual observations are necessarily better, but they are different. How can reports be supplemented, confirmed, or refuted? Now more than ever, pilots need to use all available sources to determine weather conditions. These include adjacent weather reports, *pilot weather reports* (PIREPs), radar reports, satellite imagery, and forecasts.

THE COMPLETE PICTURE   Many have touted the necessity for pilots to get the "big picture." This refers to obtaining a complete weather synopsis—that is, the position and movement of weather producing systems and those that pose a hazard to flight operations. This is important, but it is only one element needed for an informed weather decision. I prefer the term *complete picture*. As well as a thorough knowledge of the synopsis, pilots must obtain and understand all the information available from current weather, discussed in this chapter, and Chap. 9, to forecast weather presented in subsequent chapters. As we shall see, each report, chart, or product provides a clue to the complete picture. Each must be translated and interpreted with a knowledge of its scope and limitations. Then, with a knowledge of the complete picture, we can apply the information to a specific flight. As observed by the U.S. Supreme Court: "Safe does not mean risk free." With the complete picture, a knowledge of our aircraft and its equipment, and ourselves and our passengers, we're ready to assess and manage risk.

The National Weather Service certifies weather observers and automated reporting equipment. The most accurate, valid, and detailed observations come from NWS, military, FSS, tower, and other observers, in that order. Why? NWS and military observers are professionals located at major airports with the latest equipment. At FSS and tower locations, weather observations are a secondary duty, and, generally, equipment is not as sophisticated. As for the others, they are contract observers, retired air traffic controllers, local airport personnel, or even the fire department, often with little or no equipment and the minimum of training and experience. This is not to say that many of them don't provide quality observations, but in many cases they simply don't have the equipment, training, or experience. The FAA, NWS, and Department of Defense (DOD) are replacing manual observations with automated weather observing systems, although many locations are augmented or backed up by human observers.

Attempts at automated weather observations have existed for more than 40 years. Automated surface observations began with the Automatic Meteorological Observing Station (AMOS). Unfortunately, this system was capable of reporting only temperature, dew point, wind direction and speed, and pressure. Occasionally, observers manually entered data to augment the report.

The Automatic Observing Station (AUTOB), a refinement of AMOS, added sky condition, visibility, and precipitation reporting. AUTOB, however, was limited to cloud amount and height measurements of 6000 ft above ground level (AGL), and three cloud layers. Visibility values are reported in whole miles, to a maximum of 7.

In the mid-1980s the FAA published requirements for the Automated Weather Observing System (AWOS). AWOS was originally operationally classified into four levels. These levels consisted of AWOS-A, which reports altimeter setting only; AWOS-1 (the equivalent of AMOS), which reports temperature, dew point, wind, and altimeter setting; AWOS-2, which adds visibility information; and AWOS-3, which reports sky conditions and ceiling along with the other elements. AWOS is a commercially available off-the-shelf system, designed primarily for small airports with published instrument approaches but without weather observations.

The Automated Surface Observing System (ASOS) has been developed to satisfy the requirements of both the FAA and NWS. In addition to the elements available through AWOS, ASOS will report present weather and precipitation, and it will include coded remarks containing climatological data. ASOS will replace the "interim" AWOS system at locations where the government has weather reporting responsibility.

Today's science strives to reach the accuracy and reliability, and reduce some of the limitations, of the human observer. Manual surface observations have been decreasing on a regular basis over the last two decades, with others part-timed. An additional objective is to release air traffic control personnel for their primary duty—separating aircraft.

Approximately 1500 FAA, NWS, DOD, and locally operated automated weather reporting locations will be operational by the turn of the century. The automated observing system routinely and automatically provides computer-generated voice directly to aircraft in the vicinity of airports, using FAA VHF ground-to-air radio. In addition, the same information is available through a dial-in telephone, and most of the data are also provided on the national weather data network.

An additional enhancement is being commissioned. The ASOS-ATIS (Automatic Terminal Information Service) interface switch allows ASOS current weather observations to be appended to the ATIS broadcast, thereby providing real-time weather at part-time tower locations. Upon closing in the evening, the controller will have the ability to add overnight ATIS information to the ASOS automated voice message. Pilots will get normal ATIS when the tower is open, and when it is closed, pilots will receive 1-min weather observations along with the closing ATIS information on the same frequency. This approach allows the pilot to utilize the same frequency 24 h a day for airport information.

To one degree or another, most pilots are dealing with automated observations on a continuous basis. Figure 8-1 shows a typical AWOS/ASOS installation. Like manual observations, automated observations have the potential to provide an extra degree of safety, or to lure the unsuspecting pilot into danger.

The new technology has its faultfinders. FSS weather briefers, NWS forecasters, weather observers, and pilots are among the critics. There is no question that automated observations, like anything new, have had "teething" problems. The two most controversial elements are visibility and sky condition. Critics cite these elements as the least accurate, and they are certainly the least understood. Additionally, accommodation must be made for the lack of certain weather sensors. But we'll find in the final analysis most criticism is due to misunderstanding, a reluctance to change, or personal prejudice.

With the preceding as a background, we can move on to the analysis of a weather observation. METAR reports appear in the following sequence:

- Type of report
- Station identifier
- Date and time of report
- Report modifier, if required
- Wind
- Visibility
- Weather and obstructions to visibility
- Sky condition
- Temperature and dew point
- Altimeter setting
- Remarks

FIGURE 8-1 Automated observations have the potential to provide an extra degree of safety or to lure the unsuspecting pilot to disaster.

Missing, or not reported, elements are omitted. The letter *M* means minus or less than; the letter *P* means plus or more than. Any remarks follow the abbreviation RMK.

TYPE OF REPORT

There are two types of reports: METAR, a routine observation, and SPECI, a special weather report.

METAR observations are reported each hour. Normally, elements are observed between 45 and 55 min past the hour. Whenever a significant change occurs, a SPECI is generated. A complex criterion determines the requirement for specials. Generally they're required when the weather improves to, or deteriorates below, visual flight rules (VFR) or approach and landing instrument flight rules (IFR) minimums. Specials are also required for the following:

• Beginning, ending, or change in intensity of freezing precipitation

• Beginning or ending of thunderstorm activity

• Tornadoes, funnel clouds, or waterspouts

• Wind shifts

• Volcanic eruptions

From this list it's apparent that weather can change considerably without the requirement for a SPECI.

SPECIs are not available for all locations; however, these reports will carry the remark NOSPECI. In such cases significant changes can occur without expeditious notification.

STATION IDENTIFIER

METAR uses standard four-letter International Civil Aviation Organization (ICAO) location identifiers (LOCIDs). For the continental United States this consists of three letters prefixed with the letter *K*. For example, Newark, New Jersey, is KEWR and Philadelphia, Pennsylvania, KPHL. Prior to the United States' accepting the international standard, LOCIDs for weather reports were alphanumeric—that is, they were made up of letters and numbers (O45, Vacaville, California). Now, with more and more weather reports transmitted over FAA and NWS communication systems, all locations will eventually use letters only (KVCB, Vacaville, California). That's why we've seen so many changes to LOCIDs recently, and why they're making even less sense than they used to!

Elsewhere in the world, the first one or two letters of the ICAO identifier indicate the region, state, or country of the station. For example, Pacific locations such as Alaska, Hawaii, and the Marianas Islands start with the letter *P* to indicate the region, followed by the letter *A, H,* or *G,* which represents the state within the region. Canadian station LOCIDs begin with the letter *C,* Mexican and western Caribbean the letter *M,* and eastern Caribbean the letter *T.* A complete worldwide listing is contained in ICAO Document 7910, *Location Indicators.* Along with a catalog of ICAO publications, this document is available from ICAO headquarters: International Civil Aviation Organization (Attention: Document Sales Unit), 1000 Sherbrooke Street West, Suite 400, Montreal, Quebec H3A 2R2 Canada.

LOCIDs for the United States, Canada, and Mexico are contained in FAA Handbook 7350.5, *Location Identifiers,* available for sale from the Superintendent of Documents. Pilots also have access to LOCIDs from aeronautical charts and through FAA flight service stations, direct user access terminals, and the *Airport/Facility Directory*—the green book.

DATE/TIME GROUP

Following the observation of the last element, usually atmospheric pressure, the official time of observation is recorded. This time is transmitted as a six-digit date/time group appended with a *Z* to denote Coordinated Universal Time (UTC), at times—pardon the pun—referred to as ZULU or *Z.*

FACT   It seems that an advisory committee of the International Telecommunications Union in 1970 was tasked with replacing the international time standard of Greenwich Mean Time (GMT). The question became whether to use English or French word order for Coordinated Universal Time—sound familiar? So, instead of CUT or TUC, UTC was adopted and became effective in the late 1980s.

The first two digits represent the day of the month, the last four digits the time of observation—for example, METAR KOAK 142355Z.... This METAR for Oakland, California, was recorded on the 14th day of the month, at 2355 UTC.

REPORT MODIFIER

Two report modifiers may appear after the date/time group. AUTO indicates the report comes from an automated weather observation station without augmentation or backup. The absence of AUTO discloses that the report was produced manually or that an automated report has manual augmentation or backup. When AUTO appears in the body of the report, the type of sensor equipment is contained in the remarks.

Augmented means someone is physically at the site monitoring the equipment. Augmentation requires the observer to manually add data on thunderstorms, tornadic activity, hail, virga, volcanic ash, and tower visibility, which the automated equipment is not capable of reporting. Backup involves correcting nonrepresentative, erroneous, or missing data, such as wind, visibility, ceiling, temperature and dew point, or altimeter setting. It does not, however, necessarily mean that an observer is maintaining a continuous weather watch. The observer may not be changing, correcting, or adding information to the observation. However, the same can be true of manual reports in which the weather observation has a lower priority than air traffic control duties.

COR means the METAR/SPECI report was originally transmitted with an error, which has now been corrected. If a report is a correction, the time entered on the corrected report will be the same as in the original observation. The only way to identify the correction is to compare it with the original report.

Report modifiers appear in METAR/SPECI reports as follows:

METAR KFAT 150055Z AUTO...

SPECI KFAT 150115Z COR....

WIND

Crucial for determining the crosswind component, especially in areas with large magnetic variation, wind direction reported on METAR/SPECI is always reported in relation to true north, given as the direction from which the wind is blowing, and to the nearest 10°. [The only time pilots can expect to receive wind direction in relation to magnetic north is from a control tower, an FSS providing a local airport advisory (LAA), an ATIS recording, or an AWOS/ASOS radio broadcast.]

Wind is reported as a 2-min prevailing wind direction and speed using a five-digit group, six if the speed exceeds 99 kn. The first three digits indicate wind direction. The next two or three digits indicate speed. The units of measurement follow (KT, knots; KMH, kilometers per hour; MPS, meters per second). In the United States knots will continue to be used (34017KT, wind 340° at 17 kn).

Gusts (G) refers to rapid fluctuations in speed that vary by 10 kn or more. Therefore, the report ...18G24KT... describes an average speed of 18 kn with fluctuations between 14 and 24 kn. *Gustiness* is a measure of turbulence. The greater the difference between sustained speed and gusts, the greater the mechanical turbulence. Typically pilots can expect light to moderate turbulence with sustained winds or gusts in the 25- to 30-kn range, and moderate or greater turbulence above 30 kn. Terrain and obstructions—such as buildings or hangars—increase turbulence intensity.

A calm wind is reported as 00000KT. If wind direction varies by 60° or more with a speed greater than 6 kn, a variable group separated by a *V* follows the prevailing group: ...34017KT 310V010..., wind variable between 310° and 010°. A variable wind, especially with speeds of 10 kn or more, may indicate a rapidly changing crosswind. Pilots need to exercise additional caution during takeoff and landing with these conditions. If the wind is 6 kn or less and varying in direction, it may be reported as variable VRB without an assigned direction. For example, ...VRB04KT..., wind variable at 4. VRB may also be used in special cases at higher speeds, such as a wind shift with the passage of a thunderstorm over the station.

Wind direction, speed, and character (gusts) must be considered when determining the crosswind component or the advisability of landing at a particular airport. Since runways are identified by their magnetic orientation, METAR wind direction must be converted to magnetic to determine the crosswind component. This is especially significant in areas with large magnetic variation.

Crosswind or tailwind components can be determined using a flight computer or a graph, as illustrated in Fig. 8-2, a crosswind component chart.

Figure 8-2 allows the pilot to determine crosswind, headwind, and tailwind components. The vertical and horizontal axes represent wind speed in knots; the arc crosswind angle in degrees. Three elements are required: runway orientation (runway number), magnetic wind direction, and wind speed.

Let's say we're planning a departure from runway 25 (runway magnetic orientation 250°). METAR reports surface wind from 290° at 17 kn (29017KT). Local magnetic variation is 12° east. Subtract easterly variation: 290 − 12 ≈ 280. (It doesn't matter if it is a left or right crosswind; we're looking for the difference between runway heading and wind direction.) In this case, the difference is 30° (280 − 250 = 30). Enter the crosswind component chart on the 30° radial line, and proceed toward the origin of the graph—toward zero. Stop at the point where the 30° radial line intercepts the wind speed arc of 17 kn. Moving vertically down to the horizontal axis read 9 kn, the crosswind component for this runway. By moving horizontally, to the vertical axis, the headwind component is approximately 14 kn. Should the angular difference between runway and wind be more than 90° but less than 180°, we would have a crosswind and tailwind component.

Most airplanes have a published "maximum demonstrated crosswind component." This is not an absolute limit but it should be considered along with the pilot's training and experience. All *airplane flight manuals* (AFMs)—also known as *pilot's operating handbooks* (POHs)—contain takeoff and landing performance charts for various headwind, runway, and temperature conditions. Some AFMs also contain performance charts for takeoff with a tailwind.

Favorable conditions for turbulence exist just before, during, and after storm passage, especially when winds blow perpendicular to mountain ridges. For example, consider the following report for Mammoth Lakes, California (KMMH), that occurred just after storm passage with winds perpendicular to the rugged California Sierra Nevada mountains:

METAR KMMH 031545Z 23045G90KT...

The KMMH runway is 09-27, and the magnetic variation is 15° east. To convert wind direction from true to magnetic, subtract easterly variation (east is least). Therefore, the KMMH

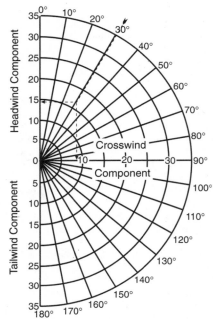

FIGURE 8-2 Crosswind component chart. Most airplane manuals specify a maximum demonstrated crosswind component. Although it is not an absolute value, it represents a speed for which the airplane has been tested.

wind is blowing from 215° magnetic (230 − 015 = 215). At an angle of 55° to the runway (270 − 215 = 55), this results in a 35-kn crosswind component for the sustained speed, and 70 kn for the gusts! This airport would not be suitable for landing. For one thing, the highway patrol would close the roads, and no one could pick you up after the, umm, arrival.

One use of the remarks element in METAR/SPECI is to amplify information in the body of the report. Wind shift WSHFT or peak wind PK WND may appear in remarks of manual and ASOS observations.

Peak wind appears when speed exceeds 25 kn. The direction, speed, and time of occurrence are reported: ...PK WND 3560/40..., peak wind from 350° at 60 kn occurred at 40 min past the hour. Peak wind might substantially exceed the value in the body of the observation.

Wind shift describes a change in direction of 45° or more that takes place in less than 15 min, with sustained speeds of 10 kn or more: ...WSHFT 55..., wind shift occurred at 55 min past the hour. A wind shift of relatively light winds might indicate only a local change; in coastal areas the shift often signals the advance or retreat of stratus or fog. In the midwest the shift might precede the formation or dissipation of upslope fog. Wind shift is usually a good indicator of frontal passage. To indicate frontal passage, FROPA may be included in remarks of manual or augmented stations. In southern California a shift often indicates the advance or retreat of a Santa Ana condition—a dry, often strong, foehn-type wind over and through the mountains and passes.

Should either the wind direction or speed sensor be out of service or unreliable, the wind group is omitted. These sensors consist of a wind vane and anomometer, or a combination known as an *aerovane*. (An aerovane is shown atop the instrument mast in Fig. 8-1.) Each consists of moving parts, which can fail. The next generation of sensors will contain no moving parts at all and will therefore be more reliable.

As automated weather observation systems have become more readily available, they have become a more significant weather resource. This is especially true for uncontrolled fields. By monitoring the automated broadcast, as well as sky conditions, visibility, and altimeter settings, pilots receive surface wind information. From this they can often determine a favored runway. By checking the *Airport/Facility Directory* and published aeronautical charts, pilots can establish traffic direction for a particular runway. This allows a more effective determination of runways in use and traffic patterns.

For airports without weather reporting service, a nearby observation may provide general weather conditions, such as wind direction and speed. A word of caution: In mountainous and hill terrain, wind can change significantly over a very short distance.

VISIBILITY

Visibility is a measure of the transparency of the atmosphere. During the day visibility represents the distance at which predominant objects can be seen; at night, visibility is the distance that unfocused lights of moderate intensity are visible.

CASE STUDY   One NWS observer was quite perplexed that a particular tower always reported increased visibility after sunset. it turned out that the criteria had changed. Pilots should note that daytime values do not necessarily represent the distance that other aircraft can be seen. At night, especially under an overcast, unlighted objects might not be seen at all, and there could be no natural horizon. Aviation has three distinct types of visibility: surface, slant range, and air to air.

Surface visibility represents horizontal visibility occurring at the surface—prevailing visibility or its automated equivalent. Reported surface visibility comes into play when a pilot plans to take off or land or enter a traffic pattern VFR. Surface visibility must be at least the minimum specified for the class of airspace. If surface visibility is not reported, flight visibility is used.

Slant range, or *air-to-ground, visibility* is the distance a pilot can see objects on the ground from an aircraft in flight. Slant range visibility is often reduced by phenomena aloft or on the surface. It may be greater or less than surface visibility, depending upon the intensity of the phenomena aloft or the depth of the surface-based restriction.

Air-to-air, or *inflight, visibility* is visibility aloft. Regulations require specific minimum inflight visibility for VFR operations—to "see and avoid" other aircraft. Visibilities aloft are most often reduced by rain, snow, dust, smoke, and haze.

Visibility from manual observations is reported as prevailing, in statute miles (SM): For example, ...1/2SM..., one-half statute mile; ...7SM..., seven statute miles; or ...15SM..., one five statute miles. *Prevailing visibility* is the greatest visibility equaled or exceeded throughout at least half the horizon circle, which need not be continuous. Figure 8-3 illustrates a reported prevailing visibility of 4 mi. Because one sector has a visibility of only 2, which is operationally significant, remarks contain ...VIS N 2..., visibility north 2. Sectors might also exist with visibility greater than prevailing. These values might or might not appear in remarks. This accounts for some apparent inconsistencies between reported values and those observed by the pilot. In Fig. 8-3 a pilot approaching the airport in the sector where the visibility is 6 mi may question the report. The observation, however, remains consistent within the definition of prevailing visibility.

Pilots are required to be trained in estimating visibility while in flight. Refer to Fig. 8-3. Assuming the observation was taken at an airport in controlled airspace, to legally operate VFR where the sector visibility has been reported as 2 mi, a pilot must have flight visibility of 3 mi. If this is not possible, the pilot must obtain an IFR or request a special VFR

PREVAILING VISIBILITY

FIGURE 8-3 Prevailing visibility represents the greatest visibility equaled or exceeded throughout at least half the horizon circle, which need not be continuous.

METAR: 4SM...RMK VIS S 2

clearance, or depart the area. To land under the provisions of IFR, visibility must not be less than that prescribed for the approach.

CASE STUDY At one southern California airport, it appears local pilots estimate prevailing visibility in the following manner: "1/4 mi to the north, 1/4 mi to the east, 1/4 mi to the south, and 1/4 mi to the west—great! Prevailing visibility is 1 mi; it's VFR." This is not only incorrect but extremely dangerous.

Automated systems determine visibility from a *scatter meter* device, as illustrated in Fig. 8-4. The visibility sensor indirectly derives a value of visibility corresponding to what the human eye would see. The visibility sensor projects light in a cone-shaped beam, sampling only a small segment of the atmosphere—an area about the size of a basketball. The receiver measures only the light scattered forward. The sensor measures the return every 30 s. A computer algorithm—mathematical formula—evaluates sensor readings for the preceding 10 min to provide a representative value. Reported visibility is the average 1-min value for the preceding 10 min.

One misunderstanding is that automated machines extrapolate a prevailing visibility based on sensor data. This is not true. Automated visibility is not prevailing visibility and may be considerably different. The existence of fog banks and visibility in different sectors will not, normally, be reported. Automated visibility may not be representative of surrounding conditions. Regulations specifically take these variables into consideration for IFR, where a suitable alternate may be required. VFR pilots must use the same caution, and though not specifically required by regulations, they must plan for an acceptable VFR alternate during reduced visibility or when conditions are forecast to change.

During rapidly decreasing conditions, it takes between 3 and 9 min for the algorithm to generate a SPECI. During rapidly increasing conditions, the algorithm takes between 6 and 10 min to catch up with actual conditions. This feature adds a margin of safety and buffers rapid fluctuation in visibility.

Siting the visibility sensor is critical. If the sensor is located in areas favorable for the development of fog, blowing dust, or near water, it may report conditions not representative of the entire airport. Airports covering a large area or near lakes or rivers may require multiple sensors to provide a representative observation. To this end, some airports will be equipped with more than one visibility sensor. Site-specific visibility, which is lower than the visibility shown in the body of the report, will appear in remarks (VIS 2 RY11, visibility two, at runway one one).

Automated visibility values are reported from less than one-quarter (M1/4SM) to a maximum of 10 statute miles. Proponents of automated systems point out that they are more consistent, objective, standardized, continuous, and representative. This is certainly

FIGURE 8-4 A scatter meter device indirectly measures how far the human eye can see.

true for IFR operations since the sensors are normally located at the approach end of the instrument runway and are typically more reliable in conditions less than VFR.

CASE STUDY Overheard from a Stockton, California, approach controller: "Visibility minus one-quarter." Can visibility be less than zero? No. The controller must have meant "less than one-quarter."

Pilots and controllers have criticized automated observations because of the perceived frequency that visibilities are too high. Some pilots routinely cut the visibility in half when light precipitation is falling in the area. In the presence of rain, snow, drizzle, and fog, considerable errors have been reported. Automated systems tend to be overly sensitive to ground fog. Under these conditions the human eye is affected by bright *back-scattered light,* which sharply reduces visibility. This is comparable to the headlights of a car shining into fog or snow. The brightly reflected light may blind the observer and limit perceived visibility. Yet the lights of an approaching vehicle seem to penetrate the fog; an observer can see the approaching light source further into the fog. This is caused by light reflected back toward the observer—forward scattered. Under these conditions, research shows that visibility differences between forward- and back-scattered light are on the order of 2 to 1. However, there are no guarantees. For example, in low-visibility conditions at Fresno and Bakersfield in California's San Joaquin Valley, tower visibility has been reported greater than that reported by ASOS.

If conditions are bright enough to use sunglasses, expect the automated systems to report visibility about twice what the human eye perceives. At night, human observers report visibility using forward-scattered light, the same principle used by automated sensors. Therefore, manual and automated visibilities tend to be more consistent.

Manual and automated visibilities may be considerably different, especially during the day. The existence of fog banks and visibility in different sectors may not be reported. Reported visibility, manual or automated, may not be representative of surrounding conditions. Reports from surrounding locations, PIREPs, and satellite images can help fill in the gaps. The bottom line: With manual or automated reports close to VFR or IFR minimums, pilots must exercise additional caution, carry extra reserves of fuel, and have a solid alternate.

Both manual and automated observations face physical limitations, such as site location, rapidly changing conditions, and contrast.

Rapidly increasing or decreasing visibility during the time observation, with average visibility less than 3 mi, is reported as variable in remarks (...VIS 1 1/2V2, visibility variable between one and one-half and two). *Variable visibility* implies conditions rapidly changing at the airport.

METAR KLOL 021555Z ...35SM SKC...RMK VIS S 1/4

Could this observation be valid? Yes, indeed. A fog bank was south of the airport. A little while later, as the fog continued to move over the airport, ...3/4SM BR SKC.../VIS NE 35 was reported.

Conditions can be quite variable and change rapidly in areas affected by stratus and fog. The following METAR/SPECI reports for Crescent City, California, illustrate this point.

METAR KCEC 061755Z ...3SM...RMK VIS E35 S-N 3/4

SPECI KCEC 061805Z ...3SM...RMK VIS N-E 35 S-NW 3/4

SPECI KCEC 061810Z ...25SM...RMK VIS S-NW 1

Considering visibility alone, conditions remained technically VFR. However, a VFR pilot should consider approaching or departing the airport from the east, where remarks indicate good visibility. Regulations require VFR fuel reserves; they don't require VFR alternates, but good judgment does.

At certain tower controlled airports, weather observers report, augment, or back up observations. These facilities are typically not collocated with the tower. Tower controllers report tower-level visibility when it is less than 4 mi. Because visibility can differ substantially over short distances, a complicated formula determines whether *tower* (TWR) or *surface* (SFC) visibility prevails. The remarks portion of the report contains the other value (...1SM...RMK TWR VIS 0; ...1 1/2SM...RMK SFC VIS 2). These remarks alert pilots to variable visibility over the airport. It is not unheard of for the tower to report visibility less than 3 mi or even 1 mi, thus preventing VFR or special VFR operations.

Runway visual range (RVR) measures the horizontal distance at which a pilot can see high-intensity runway lights while looking down the runway—not slant range. A transmissometer transmitter projects a beam of light toward the receiver. A photoelectric cell measures the amount of light reaching the receiver. This value is electronically converted into visibility and displayed at appropriate locations (tower, FSS, weather office, or a combination of locations). RVR applies to instrument approach minimums found on IFR approach and landing and departure procedure charts. Where available, RVR values appear in METAR/SPECI reports whenever prevailing visibility is 1 mi or less, or RVR is 6000 ft or less.

Runway visual range appears after visibility using the following format: the letter *R* followed by the runway number, a solidus (/), and the RVR in feet. For example, ...R29/2400FT..., runway 29 visual range 2400 ft. The following will be added as required:

• V, variability (R32R/1600V2400FT, runway 32 right visual range variable between 1600 and 2400 ft)

• M, less than (R22L/M1600FT, runway 22 left visual range less than 1600 ft)

• P, more than (R36/P6000FT, runway 36 visual range more than 6000 ft)

If the RVR varies by one or more reportable value, the lowest and highest values are reported, separated by the letter *V*. When the RVR is below the minimum value reported by the system, the letter *M* will prefix the value; when above the maximum reported by the system, the letter *P* will prefix the value. Automated stations may report up to four RVR values. When RVR should be reported but it is missing, RVRNO appears in remarks.

A report indicating landing minimums does not necessarily mean that reported visibility exists at the *decision height* (DH) or *minimum descent altitude* (MDA). The RVR reflects the fact that visibility can vary substantially from the normal point of observation; that is the observation point, compared to the runway touchdown zone.

ATMOSPHERIC PHENOMENA

Atmospheric phenomena—precipitation and obstructions to vision—follow the visibility group in METAR, using standard ICAO weather abbreviations, which are contained in Tables 8-2 through 8-4. This can be the most significant portion of the report.

In METAR, precipitation and obstructions to vision may contain some or all of the following elements:

• Intensity
• Proximity
• Descriptor
• Precipitation
• Obstructions to vision

Intensity describes the rate of precipitation, including that associated with thunderstorms or showers. Intensity is entered for all types of precipitation, except ice crystals and hail. Intensity levels may also be shown with obstructions such as blowing dust, sand, or snow. As shown in Table 8-2, there are three levels of intensity: (−) light, (no symbol) moderate, and (+) heavy.

TABLE 8-1   Intensity of Precipitation

Precipitation	Light	Moderate	Heavy
Rain, freezing rain, or ice pellets	0.10 in/h	0.18–0.30 in/h	0.30 in/h
Rain or freezing rain	Scattered, individual drops, easily seen	Drops not identifiable; spray over hard surface	Falls in sheets; heavy spray over hard surface
Ice pellets	Scattered pellets	Slow accumulation; VIS <7 SM	Rapid accumulation; VIS <3 SM
Snow or drizzle	VIS >1/2 SM	VIS >1/4 SM; ≤1/2 SM	VIS ≤1/4 SM

Intensity is based on rate of fall measured in inches per hour (in/h), observed accumulation on the surface, and visibility. These values are summarized in Table 8-1.

The *proximity* modifies the location of a weather event in relation to the airport. Proximity applies only to weather phenomena occurring in the vicinity of the airport. The *vicinity* is defined as precipitation not occurring at the point of observation but within 10 statute miles; or an obstruction to vision between 5 and 10 statute miles. When showery precipitation occurs in the vicinity of the station, the intensity and type are not entered.

The *descriptor* adds additional detail to certain types of precipitation and obstructions to vision. Although *thunderstorms* (TS) and *showers* (SH) are often used with precipitation and may be preceded with an intensity symbol, the intensity applies to the precipitation and not the descriptor.

Intensity, proximity, and descriptor are contained in Table 8-2.

Thunderstorms have a tremendous impact on aviation operations. Thunderstorms contain every aviation hazard, form in lines or clusters, and can even regenerate.

A thunderstorm is reported when thunder is heard or overhead lightning or hail are observed. Note that a thunderstorm may be reported without precipitation, unless precipitation is occurring at the point of observation. Thunderstorms and associated weather observed away from the station appear in remarks (RMK CB NE, cumulonimbus northeast). A report of CBs implies thunderstorm, sometimes translated by FSS briefers as "thunderstorm clouds...." Convective activity reported at or near the station is a clue to possible *low-level wind shear* (LLWS), or *microbursts*. Convective LLWS and microbursts—which produce the most severe wind shear threats—are discussed in Chap. 24.

CASE STUDY   At the Ontario, California, airport the FSS was responsible for weather observations. However, because of the FSS's poor observation point, the tower reported visibility. One night while working the midshift I thought I heard thunder. I asked the tower controller if he had heard thunder or had seen lightning. "Oh yeah," the controller said. "I've been watching it for the last couple of hours." Well, so much for the system. The FAA has taken steps, however, to ensure that tower controllers immediately report this type of activity.

When surface winds gust to 50 knots or more, or hail 3/4 in or greater accompany a thunderstorm, the storm is classified as severe. The three-quarter-inch hail criterion was

TABLE 8-2 Intensity, Proximity, and Descriptor

Intensity		Descriptor	
–	Light	TS	Thunderstorm
No symbol	Moderate	SH	Showers
+	Heavy	FZ	Freezing
		DR	Low drifting
Proximity		BL	Blowing
VC	Vicinity	BC	Patches—*banc*
		MI	Shallow—*mince*
		PR	Partial

established in 1954, and the wind criterion was lowered to 50 kn in 1970 for aviation purposes. Since 1996 there has no longer been a weather code to alert pilots to a severe thunderstorm, but the clues appear in the wind and weather phenomena groups, and remarks. There is a proposal to increase severe thunderstorm criteria to winds of 52 kn (60 mi/h) and hail to 1 in for public use. This would substantially change the number of warnings issued, with the goal to reduce overwarning. If it is adopted, it will be implemented after 2000.

It's important to remember that *intensity* refers to the precipitation, not the descriptor. For example, "+TSRAGR" is a thunderstorm with heavy rain and hail. Since only one descriptor may be used, TS and SH will never appear in the same group. However, FSS controllers may translate TSRA as "thunderstorm, rain showers" because thunderstorms imply showery precipitation.

The weather phenomenon −TSRA, sometimes translated as a *light thundershower* or, worse, a *light thunderstorm and rain showers,* is neither. The correct translation: Thunderstorm accompanied by light rain or light rain showers. There is no such thing as a "light thunderstorm"!

Falling from stratiform clouds—stratus, altostratus, nimbostratus, or stratocumulus—precipitation is usually steady, can be widespread, and usually not heavy in intensity. In contrast, *showery precipitation* (SH) falls from unstable air. Showers fall from cumuliform clouds—cumulus or cumulonimbus—usually briefly and sporadically, and they may be heavy in intensity.

Freezing precipitation (FZ) *in the form of rain* (FZRA) or *drizzle* (FZDZ) is caused by liquid precipitation falling from warmer air into air that is at or below freezing. Droplets freeze upon impact, producing structural icing.

Low drifting (DR) describes snow, sand, or dust raised to a height of less than 6 ft above the surface. In addition to snow, sand, or dust, *blowing* (BL) may be applied to spray. The descriptor blowing describes a condition in which the phenomena are raised to a height of 6 ft or more above the surface. Additionally, when applied to sand, dust, or spray, blowing implies that horizontal visibility is reduced to less than 7 statute miles; applied to snow, blowing identifies snow lifted by the wind to more than 6 ft in such quantities that visibility is restricted at and above that level.

The descriptors *patches* (BC), *shallow* (MI), and *partial* (PR) are only coded with fog. *Patchy fog* (BCFG) and *shallow fog* (MIFG) indicate radiation fog extending to 6 ft or more above the ground; visibility in the fog area is less than 5/8 statute mile and over other parts of the airfield greater than or equal to 5/8 statute mile. Patchy fog randomly covers part of the airport, whereas shallow fog is more extensive and organized, but it does not cover the entire airfield. In both cases visibility is not restricted by fog above the layer. *Partial fog* (PRFG) must meet the same height and visibility criteria as patches and shallow. However, partial fog must cover a substantial part of the station. The bottom line: All three terms describe what used to be called *ground fog*—radiation fog that reduces horizontal visibility at the surface, with little vertical extent, with the coverage of fog around the airport going from *random* (BC), to *limited* (MI), to *substantial* (PR).

Precipitation

Precipitation is any form of water particles, liquid or solid, that fall from the sky and reach the ground. Precipitation does not include clouds, fog, dew, frost, or virga—rain that evaporates before reaching the ground. Table 8-3 contains precipitation that may appear in the body of METAR.

Rain (RA) is liquid precipitation. *Drizzle* (DZ) is liquid precipitation consisting of very small, but numerous and uniformly dispersed drops. As inferred from Table 8-1, drizzle restricts visibility to a greater degree than rain because drizzle drops are smaller and fall in stable air, often accompanied by fog, haze, and smoke.

Hail (GR) is reported in the body of METAR when hailstones are one-quarter inch or greater. Hailstone size appears in remarks (...GR...RMK GR 1/2..., hailstones 1/2 inch in diameter). Hailstones less than one-quarter inch are reported in the body of the report as *small hail* (GS), with no remarks indicating size. GS is also used to report snow pellets.

Snow grains (SG) are the solid equivalent of drizzle. *Snow* (SN) is composed of white or translucent ice crystals, chiefly in complex branched hexagonal form and often integrated into snowflakes. Wet snow—snow that contains a great deal of liquid water—produces structural icing. "WET SN" may appear in remarks.

Accumulations of snow on airport surfaces present different hazards. *Blowing snow* (BLSN) reduces visibility and creates a different type of hazard, especially to the VFR pilot. Blowing snow occurs when strong winds blow over freshly fallen snow. Visibility can be reduced to near zero, but it is typically restricted to within a few hundred feet of the surface. Visibility improves rapidly when the wind subsides. *Drifting snow* (DRSN) is raised by the wind, remains close to the surface, and does not significantly reduce visibility. Snow drifts, which can be inferred by reports of drifting snow, present an airport surface condition hazard; which typically is advertised in a *Notice to Airmen* (NOTAM).

Ice crystals (IC) might appear suspended, and they may fall from a cloud or clear air. *Ice pellets* (PL), formerly *sleet,* are grains of ice consisting of frozen raindrops, or largely melted and refrozen snowflakes. They fall as continuous or intermittent precipitation.

FACT   On November 5, 1998, the previous international abbreviation for ice pellets was changed to "PL." It seems this was required because in certain combinations with other weather contractions, it resulted in offensive language. It's nice to know our METAR and TAF reports are politically correct.

TABLE 8-3   Precipitation

RA	Rain	GS	Small hail/snow pellets—*grésil*	SG	Snow grains
DZ	Drizzle			IC	Ice crystals
GR	Hail (≤1/4 in) —*gréle*	SN	Snow	UP	Precipitation (auto obs)
		PL	Ice pellets		

Automated stations report RA (liquid precipitation that does not freeze) or SN (frozen precipitation other than hail), and intensity. Automated sites may use UP (unknown precipitation) to report precipitation when the precipitation discriminator cannot identify it. For example, this SPECI from Rock Hill, South Carolina:

SPECI KUZA 191942Z AUTO 36007KT 3SM UP BR BKN004 OVC009
02/01 A2976 RMK AO2 RAE11UPB35SNB11E35 P0010 TSNO

We'll use this report again, later in the chapter.

Automated sites are being equipped with freezing precipitation sensors. This will allow automated stations to report the occurrence of *freezing rain* (FZRA) and *freezing drizzle* (FZDZ).

The Automated Lightning Detection and Reporting System (ALDARS), which acquires lightning information from the National Lightning Detection Network, will allow AWOS/ASOS to report the occurrence of a thunderstorm. ALDARS is operational at numerous AWOS sites and is expected to become operational with all of the FAA's commissioned ASOSs. Other forms of precipitation only appear at an augmented site.

A criticism of automated observations is that they are not derived from specific precipitation and weather sensors and therefore they might not alert a pilot to potentially hazardous conditions. Recall the "complete picture"? Automated reports can and should be supplemented with radar products. The system is now virtually complete and covers most of the country. Radar can determine the existence of rain, thunderstorms, tornadoes, snow, and hail. Radar, along with satellite imagery, is better at determining the extent of phenomena than either a manual or automated observation.

To alleviate some of the concerns with automated observations, a government and industry team developed service standards for surface observations. Service standards have been established at four levels, D through A.

Service level D is a completely automated site in which the ASOS observation constitutes the entire observation; there is no augmentation or backup. This service is referred to as *stand-alone D site*. The FAA has determined that ASOS performance appears overwhelmingly satisfactory at locations where there has never been a human observer. Conversely, where a level D site is planned but currently has a human observer, numerous problems and concerns have been reported—many from the observers who would lose their jobs.

Service level C provides augmentation and backup. Service is provided under the provisions of the Limited Aviation Weather Reporting Station (LAWRS) program. Augmentation and backup is provided by tower, FSS, NWS, or Non-Federal Observation Program observers. During the hours that the observing facility is closed, the site reverts to Service level D. In addition to ASOS data, the following elements are reported:

• Thunderstorms

• Tornadoes

• Hail

• Virga

• Volcanic ash

• Tower visibility

• Operationally significant remarks

Service level B is provided at airports that serve as small hubs, or special or remote airports that qualify for additional weather services. In addition to ASOS and level C data, the following elements are reported:

- RVR
- Freezing precipitation
- Ice pellets
- Snow depth and snow increasing rapidly
- Thunderstorm and lightning location
- Significant weather not occurring at the station

Service level A is provided at major hubs or high-volume airports. In addition to ASOS, level C, and level B data, the following elements are reported:

- Sector visibility
- Variable sky conditions
- Cloud layers above 12,000 ft
- Cloud types
- Widespread dust, sand, and other obscurations
- Volcanic eruptions

Obstructions to Vision

Obstructions to vision, typically caused by fog, haze, dust, and smoke, are reported with visibilities less than 7 mi. (Automated sites may report obstructions to vision with visibility greater than 6 mi.) When these phenomena exist with visibilities 7 mi or greater, a remark might describe the condition (...RMK HZ ALQDS, haze all quadrants). FSS briefers sometimes use the term *unrestricted* to describe visibilities of 7 mi or greater. This sometimes causes confusion. A pilot told visibility unrestricted might respond, "What about the haze?" Thus, the phrase *visibility unrestricted* does not imply that smoke, haze, dust, or even fog are not present, just that visibility is 7 mi or greater.

METAR also includes five additional weather phenomena. These phenomena, along with obstructions to vision, are listed in Table 8-4.

(You've probably noticed that a number of atmospheric phenomena abbreviations don't make sense in English. That's because they were derived from French words—shown in italics in the preceding tables. Probably the same thought process that gave us UTC!)

Fog (FG) is reported only when the visibility is less than five-eighths of a mile. With visibility five-eighths or greater, *mist* (BR) designates this phenomenon. [In METAR, *mist* (BR) refers to an obstruction to vision, not precipitation, its generic definition.]

FACT OR FICTION   Allegedly the distinction between fog and mist came about because the English didn't consider it fog unless it could be "cut with a knife." Why five-eighths of a mile? That's equivalent to 1000 m. (Source: FAA)

TABLE 8-4   Obstructions to Vision

FG	Fog (VIS <5/8 SM)	SA	Sand	PY	Spray
BR	Mist (VIS ≥5/8 SM)	DS	Duststorm	SQ	Squall
	—*brume*	SS	Sandstorm	FC	Funnel cloud
HZ	Haze	VA	Volcanic ash	+FC	Tornado
FU	Smoke—*fumée*	PO	Dust/sand whirls	+FC	Waterspout
DU	Dust				

Haze (HZ) is caused by the suspension of extremely small, dry particles invisible to the eye but sufficiently numerous to reduce visibility. Haze, combined with *smoke* (FU), often describes conditions in metropolitan areas. Large anticyclones—high-pressure cells—can dominate the southeast United States, trapping haze and pollutants, especially in industrial areas. Above the haze layer, visibilities are unrestricted and temperatures are cool, resulting in a much more comfortable flight.

Dust (DU) and *blowing dust* (BLDU), a combination of fine dust or sand particles suspended in the air, can be raised to above 16,000 ft by the wind. *Sand* (SA) and *blowing sand* (BLSA), made up of particles larger than dust, usually remains within a few hundred feet of the surface.

The descriptor *drifting* (DR) may be used with dust or sand. This indicates that the *dust* (DRDU) or *sand* (DRSA) has been raised by the wind to less than 6 ft above the ground.

Duststorm (DS) and *sandstorm* (SS) report basically the same phenomena. A duststorm is an unusual condition characterized by strong winds and covering extensive areas. In contrast to a duststorm, a sandstorm causes grains of sand to be blown into the air. In METAR, duststorms or sandstorms are reported with visibilities equal to or greater than 5/16 statute mile and equal to or less than 5/8 statute mile. When visibility is less than 5/16 statute mile with either phenomenon, it is reported as *heavy* or *well developed* (+DS, +SS).

Volcanic ash (VA) consists of fine particles of rock powder, blown out from a volcano. The particles remain suspended in the atmosphere for long periods, extend well into the flight levels, and may drift thousands of miles.

Dust/sand whirls (PO), commonly know as *dust devils,* form in dry, hot regions. They have diameters of 10 to 50 ft and extend from the surface to several thousand feet.

Spray (PY) consists of water droplets blown by the wind from wave crests and carried up a short distance in the air from the surface of a large body of water. Spray may be modified with the descriptor *blowing* (BLPY). Spray is reported in the body of METAR when it reduces visibility to less than 7 mi.

A *squall* (SQ) is a sudden increase in wind of at least 16 kn, sustained at 22 kn or more, for at least 1 min. Usually associated with thunderstorm activity, *squall* implies severe low-level wind shear as well as severe turbulence.

Funnel cloud (FC), *tornado* (+FC), and *waterspout* (+FC) describe tornadic activity—a small, violently spinning column of air, potentially the most destructive of all weather systems. Funnel cloud reports a tornado that has not touched the ground; tornado indicates a funnel cloud that has "touched down;" waterspout represents a tornado over water in contact with the water surface.

Automated stations report FG, FZFG, BR, HZ, and SQ. Other obstructions to vision and weather phenomena will appear only at an augmented site.

SKY CONDITION

The METAR sky condition report consists of the amount of sky cover, height in feet, and under certain conditions cloud type. Heights range from the surface upward, to a maximum of 12,000 ft for automated stations. The next generation of laser CHI will report clouds to 25,000 ft. Cloud heights are reported as three digits, in hundreds of feet AGL, following sky condition (SCT030, 3000 scattered). In METAR, ceiling is not designated. For aviation purposes, the ceiling is the *lowest broken* (BKN) or *overcast* (OVC) *layer,* or *vertical visibility* (VV) into a complete obscuration. (Like many specialties, aviation weather has its own language. For communication and understanding to take place, each party must understand and use the same definitions. Take the pilot that called UNICOM and asked, "What's the ceiling?" After a pregnant pause the operator replied, "I think it's oak.") A manual or

TABLE 8-5 Sky Cover

CLR	Clear below 12,000	Automated reports
SKC	Clear	No clouds
FEW	Few	<1/8 to 2/8 coverage
SCT	Scattered	3/8 to 4/8 coverage
BKN	Broken	5/8 to 7/8 coverage
OVC	Overcast	8/8 coverage
VV	Indefinite ceiling	8/8 coverage

ceiling (handwritten annotation pointing to BKN, OVC, VV)

augmented report will include *towering cumulus* (TCU) or *cumulonimbus* (CB) in the sky condition element of the report. For example: ...BKN035TCU..., ceiling 3500 broken tower cumulus; or ...OVC020CB..., ceiling two thousand overcast cumulonimbus.

The amount of sky cover is reported in eighths—sometimes referred to as *octas*—using the abbreviations contained in Table 8-5.

In manual observations *sky cover* refers to clouds or obscuring phenomena as seen by an observer on the ground from horizon to horizon. Automated observations use a laser ceilometer *cloud height indicator* (CHI) to determine sky cover, a distinct advantage of these systems. METAR reports sky cover as the summation of layers based on specific criteria. Conditions aloft, as seen by a pilot, can differ substantially. The summation principle with terms like *obscuring phenomena* and *obscuration* can be complex and misunderstood.

In Fig. 8-5 the observer sees 2/8 cloud cover at 1000 ft AGL and reports a few clouds at 1000 (FEW010). Another 3/8 cloud cover is observed at 1800 ft AGL. According to the summation principle, a total of 5/8 (2/8 + 3/8 = 5/8) sky cover exists; ceiling 1800 broken (BKN018) is reported. The observer sees the remaining sky covered by cloud at 3000 ft AGL and reports 3000 overcast (OVC030). The observer, unable to determine the extent of higher layers, reports them as continuous. This principle has led many pilots to mistakenly question the accuracy of observations.

Automated stations determine sky cover and height from a laser CHI. Similar to a rotating beam ceilometer, cloud elements reflect the laser. Like visibility, a computer algorithm processes the last 30 min of CHI data. The computer then generates values of sky cover and cloud height for the observation. To be more responsive of the most recent conditions, the algorithm "double-weighs" the last 10 min of data. Up to three layers are reported.

At the transition between scattered and broken, human observers often report too much cloud cover. This is known as the *packing effect,* a condition in which an observer does not detect the opening in the cloud deck toward the horizon. Pilots also tend to overestimate the amount of cloud cover. ASOS is not biased by these limitations.

In rapidly changing conditions, the automated system algorithm tends to lag slightly behind actual conditions. If a sudden overcast layer develops, ASOS will take 2 min to report a scattered layer; within 10 min the system will report broken conditions.

FACT? Sky cover is always an estimate from a manual station. As an FSS briefer for some 25 years, I have on occasion briefed doom and gloom only to find a bright, beautiful day. It would seem that certain tower controllers use the following criteria: They consider the roof of the tower cab as opaque; therefore, one cloud is scattered, two is broken, and three is overcast with breaks!

On rare occasions, ASOS may report a dense moisture layer as clouds before the layer becomes totally visible to the eye. This may occur with an approaching cold front when the sensitive laser detects the large-scale lifting of prefrontal moisture. There have been cases when ASOS has reported a layer 20 min before an observer reported it. ASOS will only report conditions that pass directly over the sensor. During light wind conditions, observers

SUMMATION OF SKY COVER

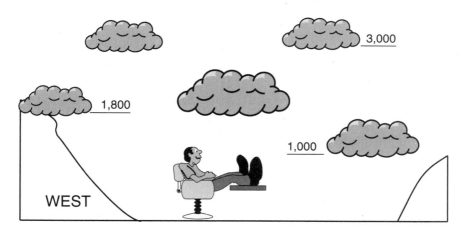

METAR: FEW010 BKN018 OVC030
...RMK MTNS OBSCD W

FIGURE 8-5 The summation principle has led many pilots to mistakenly question the accuracy of observations.

have reported up to three-eighths sky cover when ASOS reported CLR. Ironically, manual observations suffer from the same limitations. Sky cover might not be representative of surrounding conditions, especially at night, or during low visibility, when the observer cannot see or evaluate the whole sky. Refer to Fig. 8-5: If the observer were unable to see the scattered layer at 1000 ft, a ceiling of 1800 ft would be reported. If the CHI or other device went through a hole in the 1800-ft layer, the observer might report a ceiling of 3000 overcast. Such errors can, and do, occur because of limitations on equipment and the observer.

Notwithstanding the previous observation, care must be exercised in climbing or descending VFR through a broken deck. Although it might be possible to safely negotiate the layer, several factors must be considered. Can appropriate distance from clouds as specified in regulations be maintained? Is the weather improving or deteriorating? We don't want to get caught on top or between layers. Is the area congested with other VFR traffic or aircraft operating IFR? The criteria in the regulations are minimums; they do not necessarily equate to "safe." What alternatives are available if needed? Positive answers are required before an attempt.

For manual observation, sky cover and cloud height may be only an estimate, a more or less educated guess by the observer, based on the observer's training and experience. Like visibility, cloud cover and heights should always be viewed with caution, especially at night or close to minimums.

FACT The FAA has conducted "blind" comparisons between manual and automated observations. At many of these locations, ASOS was installed, but not commissioned. A number of the observers complained when they were told they would not be able to use the laser CHI for cloud heights, which they had been previously using to supplement their observations.

At certain automated locations, additional sensors are used to obtain more representative reports. In such cases, remarks will identify site-specific sky conditions that differ and are lower than those reported in the body. For example, ...RMK CIG 020 RY11..., ceiling two thousand at runway one one.

Variable describes a situation in which the amount of sky cover or height changes during the period of observation, normally the 15-min period preceding the time of observation. For example, ...SCT015...RMK SCT V BKN..., scattered layer variable to broken. If more than one cloud layer is being reported, the variable layer height appears in remarks (...SCT015 SCT025...RMK SCT015 V BKN). Cloud height variability is shown as follows: ...BKN010...RMK CIG 008V012, ceiling 1000 broken...ceiling variable between 800 and 1200. A variable ceiling below 3000 ft must be reported; a variable ceiling above 3000 ft, only if it is considered operationally significant. Variability alerts pilots, briefers, and forecasters to rapidly changing conditions over the airport.

How about the following report?

METAR KMWS 131758Z 18005KT 1SM BR OVC///...

When a cloud layer develops below the point of observation, as sometimes happens at Mt. Wilson, 6000 ft above the Los Angles Basin, the layer is encoded "OVC///" (an overcast layer with tops below the point of observation).

A partial obscuration indicates that between one-eighth and seven-eighths of the sky is hidden by surface-based obscuring phenomena. Precipitation—including snow, haze, smoke, and fog—usually causes this condition. Automated systems will normally not report this condition.

Refer to Fig. 8-6. In the example, half, or four-eighths, of the sky is hidden by fog. The observer sees another one-eighth cloud cover at 3000 ft. In METAR a partial obscuration is indicated as FEW, SCT, or BKN on the surface (FEW000, between one-eighth and two-eighths of the sky obscured). In Fig. 8-6 the observer reports visibility reduced by fog (FG), four-eighths of the sky obscured, with a ceiling of 3000 ft broken. The observer uses the summation principle of cloud cover (4/8 + 1/8 = 5/8). Five-eighths is reported as a broken layer. The remark reveals that it is an obscuration (...RMK FG SCT000), not a cloud layer. What else could it be? Well, technically a layer with a base of less than 50 ft would be reported as "000." This is very unlikely. If the observer did indeed mean to report a layer with a base of less than 50 ft, the remark ...FG SCT000... would not appear.

Figure 8-7 is a photograph illustrating the conditions described in Fig. 8-6. A fog layer obscures the sky from the horizon to an elevation of about 45°. Above this point a cloud layer is visible. At the top of what is referred to as the *celestial dome* the sky is clear.

Why such a complicated procedure? The international METAR code has no provision to report a partial obscuration. The FAA and NWS concurred—probably for the first time—and proposed that a partial obscuration be eliminated from United States' reporting procedures. But, there are other players in the game—namely, the Department of Defense. Within DOD is the Department of the Navy and the Marine Corps. Well, the Marines just couldn't get along without a partial obscuration. Semper fi! (As of this writing, the FAA and NWS are still attempting to eliminate partial obscurations from reporting criteria.)

A partial obscuration may be reported without cloud layers, for example, ...3SM HZ FU SCT000...RMK SCT000. These remarks indicate that between three- and four-eighths of the sky is hidden by haze and smoke. A partial obscuration in itself must not be confused with a ceiling. However, with a large amount of sky hidden, cloud coverage amounts might not be representative, and slant range visibility can be less than reported surface visibility.

Normally, a pilot can expect ground contact while flying in areas with a reported partial obscuration. This is why these conditions are not considered ceilings. Penetrating a partial obscuration VFR requires the appropriate horizontal visibility for the class of air-

PARTIAL OBSCURATION

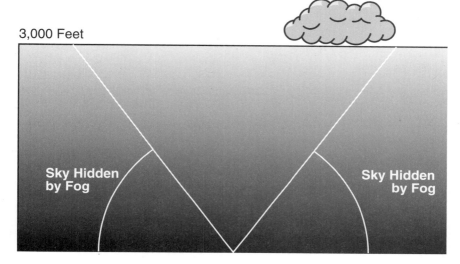

METAR: FG SCT000 BKN030
...RMK FG SCT000

FIGURE 8-6 A partial obscuration indicates that between one-eighth and seven-eighths of the sky is hidden by a surface-based obscuring phenomenon.

space. A partial obscuration with visibilities less than basic VFR can often be safely nego-tiated under the provisions of special VFR.

> **CASE STUDY**   After being briefed that his destination was reporting visibility 2 in mist and haze, clear below 12,000, mist and haze obscuring between six- and seven-eighths of the sky (2SM BR HZ BKN000...RMK BKN000 BR HZ), the Beechcraft Baron pilot emphatically demanded to know the ceiling. The pilot stated this information was required to determine IFR minimums. There was no ceiling. IFR minimums, in this case, would be based on visibility alone. The pilot could expect to maintain ground contact throughout the approach, sighting the airport at about 2 mi.

When the sky is completely hidden by a surface-based obscuring phenomenon, an *indef-inite ceiling* (VV) is reported. An indefinite ceiling is the vertical visibility upward into a surface-based obscuring phenomenon that completely conceals the sky—that is, the dis-tance at which a pilot can expect ground contact when looking straight down on descent, or the point at which the ground disappears on climbout. This condition is most often caused by fog, but snow, smoke, or even heavy rain can also cause this condition (...+RA VV000..., heavy rain, indefinite ceiling zero).

The accuracy of an indefinite ceiling depends on the observer and available equipment. Whether the value is determined by a ceilometer or it is just a guess by the observer, it is reported as indefinite. In Fig. 8-8 the observer has either measured or estimated the verti-cal visibility as 200 ft. The sky is completely obscured. The observer, unable to determine cloud layers above, reports an indefinite ceiling 200: ...VV002....

FIGURE 8-7 The sky from the horizon to about 45° is obscured; above, a cloud layer is visible; at the top, the sky can be seen.

For automated observations, total obscurations are based on visibility, temperature, and the computer cloud algorithm. Fog or precipitation may cause the CHI to report a false layer. During evaluations, automated systems reported more obscurations than the human observers. The overreporting of obscurations can create a false impression of crying wolf.

Indefinite ceilings are most often associated with IFR conditions. Assume that a destination is reporting an indefinite ceiling 200 and that 200 ft is the *decision height* (DH) for the *instrument landing system* (ILS) approach. Assuming the observation is accurate, at DH the pilot should be able to look straight down and see the *approach lighting system* (ALS). However, the pilot would not necessarily be able to see the runway due to increased slant range distance. This is illustrated in Fig. 8-8. In fact, slant range visibility could be less than vertical visibility! That's why approach lights are considered part of the ILS and minimums increase when they're out of service.

INDEFINITE CEILING

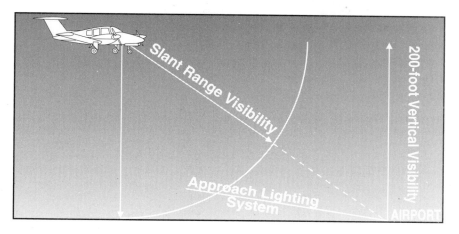

METAR: VV002

FIGURE 8-8 An indefinite ceiling is the vertical visibility into a surface-based obscuring phenomenon.

Would the conditions in the previous paragraph preclude a legal landing? Not necessarily, as long as three landing requirements are met: The aircraft must remain in a position from which a normal descent to landing can be made; the flight visibility must not be less than that prescribed for the approach; and the runway environment (approach lights, threshold, runway, etc.) remains distinctly visible. Any requirement not met or lost after DH requires an immediate missed approach.

How about VFR with an indefinite ceiling of 1000 ft? Even though ceiling and surface visibility might technically meet legal minimums, flight visibility might be substantially less. Now consider the possibility that the observer might not have ceiling height–finding equipment and that the reported ceiling might be only a guesstimate.

At Redding, California, the NWS observer made the following report:

SPECI KRDD 051650Z ...3SM FU HZ VV030....

The reported vertical visibility is 3000 ft, and the sky is completely hidden by smoke and haze. A pilot flying in this area can expect to maintain ground contact within only about 3000 ft of the surface. This situation could be extremely dangerous for the VFR pilot. Should the pilot climb above 3000, the pilot would be in IFR conditions without ground contact and most probably without a natural horizon. Pilots attempting to operate in similar conditions have lost aircraft control with fatal results.

Another typical situation occurs with snow. Consider the following: ...3SM -SN VV015....The visibility is 3 mi in light snow, indefinite ceiling 1500. Is it VFR? Technically, yes. A pilot could expect ground contact at 1500 ft, but slant range visibility would be considerably less. Additionally, should the pilot climb above 1500 ft AGL, the pilot may not be in clouds but he or she will have no visible contact with the ground and no natural horizon! This is a good example of a situation that may be legal but is not safe.

CASE STUDY IFR pilots must also exercise caution operating close to minimums. The VOR-A approach to Ukiah, California, has a *minimum descent altitude* (MDA) of 3400 ft (2784 AGL). Ukiah was reporting SCT020 BKN050.... Sure enough, the scattered layer was at the missed approach point. During the miss the aircraft broke into the clear and was able to land. This was pure luck. Never count on this happening.

TEMPERATURE, DEW POINT, AND ALTIMETER SETTING

Temperature and dew point are reported using two digits in whole degrees Celsius, separated by a solidus (/). Temperature below zero will be prefixed with the letter *M*. For example: ...20/15..., temperature 20°C and dew point 15°C; or ...08/M03..., temperature 8°C and dew point −3°C. Dew point can never be higher than temperature. A reported dew point greater than temperature results from equipment malfunction or transposition of numbers upon transmission. When the dew point is missing, temperature alone will be reported: ...M05/..., temperature minus 5°C, dew point missing. When the temperature is missing, both elements are omitted.

The altimeter, in inches of mercury, follows temperature and dew point in a four-digit group prefixed with the letter *A.*—for example, ...A2992..., altimeter two nine nine two. An extremely cold high-pressure area developed over Alaska in 1989 causing the altimeter setting to rise well above 31.00 in, the upper limit of most aircraft instruments. Because of this situation, the FAA instituted special emergency rules. These are contained in the *Aeronautical Information Manual.*

In countries that report altimeter setting in hectopascals (millibars), the group starts with the letter *Q* rounded down to the nearest whole unit—for example, 1016.6 is reported as ...Q1016....

REMARKS

The absence of operationally significant remarks is another controversial element of automated observations. The existence of weather and cloud types are paramount. The implementation of service-level standards, however, has mostly elevated this concern from all but level D sites.

Remarks are divided into automated, manual, and plain language, and additive and automated maintenance data. *Automated remarks* indicate the type of automated station. *Manual remarks* amplify information already reported, describe conditions observed but not occurring at the station, or they contain information considered operationally significant. *Plain-language remarks* contain other significant data. *Additive data* are used by the NWS and consist of climatological information, usually in numerical code groups. *Automated maintenance data* report sensor outages and maintenance requirements. Remarks use standard aviation weather abbreviations, as contained in App. A, and they follow the altimeter setting, separated by the abbreviation RMK. Remarks can be the most important part of the report. Here again, the NWS, FSSs, and military observers, because of their training and experience, tend to do a better job. However, even they can get carried away.

This observation came from an NWS observer at Denver: METAR KDEN...RMK DSIPTG GUSTNADO N (dissipating gustnado north—a glorified dust devil). Well, *gustnado* is a local term used to describe a funnel cloud that develops along the gust front of a thunderstorm—it is not a tornado that would warrant a special. It is believed that the gustnado receives its initial rotation from the shift in wind directions across the gust front. Cold, dense air behind the gust front lifting the warm air ahead imparts a rotating motion in the wind shear zone.

Remarks can be ambiguous, as this report taken by an FSS illustrates: KDEF SA...BKN000 OVC040...RMK BR BKN000 TOPS 015. This observation would seem to indicate a 4000-ft ceiling with tops at 1500 ft, which is impossible. Could the observer have meant 400 overcast? Upon checking, 7/8 of the sky was obscured by mist, and the observer could see cloud cover at an estimated 4000 ft. The remarks should have read: ...RMK BR BKN000 TOPS OBSCN 015..., tops of the obscuration 1500 ft.

The only way to clarify a report is to check with the observer. For individuals using DUATs, this will be all but impossible. An FSS will usually be able to check through its telecommunications system. But don't even ask unless some very serious—meaning emergency—operational requirement exists.

Automated, Manual, and Plain Language

When used in remarks, the distance of phenomena from the station is reported as follows:

- No modifier: within 5 mi of the station
- Vicinity (VC): between 5 and 10 mi of the station
- Distant (DSNT): beyond 10 mi but less than 30 mi from the station
- Distant (DSNT) for cloud types: beyond 10 mi

Remarks appear in the following order.

Volcanic eruptions when first observed appear in the following sequence:

- Latitude and longitude, or approximate direction and distance from the station
- Date and time of eruption
- Size, description, height, and movement of ash cloud
- Any other pertinent information in plain language

For example, ...MT. ST. HELENS VOLCANO 70 MILES NE ERUPTED 181505 LARGE ASH CLOUD EXTENDING TO APPROX 30000 FEET MOVING SE....

Remarks of manual and augmented observations contain details on tornadoes, funnel clouds, and waterspouts. Data include the status of the phenomenon (beginning, progress, or end), location, and movement. For example, ...RMK TORNADO B13 DSNT NE, tornado began at one three minutes past the hour to the distant northeast.

Automated station type is shown as either AO1 or AO2. AO1 means the station does not have a precipitation type discriminator. AO2 indicates the station is equipped with a precipitation discriminator.

Next in order are:

- Peak wind
- Wind shift
- Tower or surface visibility
- Variable visibility
- Sector visibility
- Visibility at a second site

When *lightning* (LTG) is seen, it will appear in remarks. The frequency of occurrence and type of lightning, as shown below, and location will be included:

- OCNL, occasional: less than 1 flash per minute
- FRQ, frequent: between 1 and 6 flashes per minute
- CONS, continuous: more than 6 flashes per minute
- CG, cloud to ground
- IC, in cloud
- CC, cloud to cloud
- CA, cloud to air

For example, ...RMK FRQ LTGCC VC SE..., frequent lightning, cloud to cloud, in the vicinity of the station, southeast.

When precipitation or a thunderstorm begins or ends, remarks indicate the type of precipitation along with times of occurrence. The purpose of these remarks is climatological, but they do alert pilots, briefers, and forecasters to weather, often significant, occurring at the stations.

Recall the SPECI from Rock Hill, South Carolina:

SPECI KUZA 191942Z AUTO 36007KT 3SM UP BR BKN004 OVC009
02/01 A2976 RMK AO2 RAE11UPB35SNB11E35 P0010 TSNO

In this example, rain ended at 11 (RAE11...); unknown precipitation began at 35 (...UPB35...); and snow began at 11 and ended at 35 (...SNB11E35).

In addition to beginning and ending times for thunderstorms, their location and movement are included in remarks. For example, ...TS VC NE MOV SE..., thunderstorm in the vicinity northeast, moving southeast.

As previously mentioned, hailstone size appears in remarks with GR in the body of the report.

Recall from Fig. 4-5 that remarks will report the presence of *virga*, precipitation falling from a cloud that evaporates before reaching the surface. At times, ice crystals or snowflakes fall from cirrus clouds. As they fall into dry air, they *sublimate*—change directly from a solid to a gas. These dangling white streamers are known as *fall streaks*.

Next in order are the following:

- Variable ceiling height
- Obscurations
- Variable sky conditions
- Significant cloud types
- Sky conditions at a second site

Convective cloud types are reported, along with their direction from the station and movement if known. We have already touched on towering cumulus and cumulonimbus, which may appear in the body of the report. Additional cloud types are as follows:

- CBMAM, cumulonimbus mammatus
- ACC, altocumulus castellanus
- SCSL, standing lenticular stratocumulus
- ACSL, standing lenticular altocumulus

- CCSL, standing lenticular cirrocumulus
- Rotor clouds

Pressure falling rapidly (PRESFR) and *pressure rising rapidly* (PRESRR) signify the approach or passage of a frontal system. Pressure rising rapidly accompanied by a wind shift might be reported as *frontal passage* (FROPA) or *apparent frontal passage* (APRNT FROPA).

Designated stations report sea-level pressure in remarks. This three-number code follows the abbreviation SLP. The code contains the last three digits of the sea-level pressure to the nearest tenth of a hectopascal (hPa). Because average pressure is 1013.2 hPa, the code is prefixed with a 9 or 10, whichever brings it closest to 1000.0. For example, ...RMK SLP102..., sea-level pressure 1010.2 hPa. A word of caution: Sea-level pressure is not the altimeter setting in hectopascals (millibars).

Like SLP, designated stations report snow depth increases by 1 in or more in the preceding hour. The remark contains the abbreviations ...SNINCR... in the past hour, a solidus (/), and the total snow depth on the ground at the time of the report. For example, ...RMK SNINCR 2/10..., snow increase of 2 in during the past hour, total depth on the ground 10 in.

Other operationally significant information is reported at this point. Below are some examples.

Figure 8-5 contains an excellent example of an operationally significant remark. The observer has reported: ...RMK MTNS OBSCD W, mountains obscured west. If an overcast ceiling of 1500 ft exists, VFR flight below the clouds could be conducted. However, due to mountain obscurement, VFR flight into or out of the valley might not be possible. Pilots accustomed to flying over flatlands need to exercise extra caution evaluating conditions in mountainous areas. Studying terrain is as important as checking the weather.

A sky condition example for Ukiah, California, was presented earlier in which a scattered cloud layer existed at the missed approach point. If the observation had carried the remark ...RMK MTNS OBSCD E-W, mountains obscured east through west, a pilot reviewing the approach and terrain could deduce that the scattered layer existed at the missed approach point.

This example also illustrates how observers indicate phenomena coverage. The observer starts at true north and works clockwise. If the remark read ...MTNS OBSCD E S-N, it would be translated as mountains obscured east and south through north. It's important to understand this convention to correctly interpret weather reporting in relation to the reference station.

CASE STUDY I was over Palmdale on a flight to Van Nuys, California, in a Cessna 150 that was not equipped for IFR. Coastal stratus obscured the mountains with Van Nuys reporting OVC030. Because of my experience and familiarity with the area, I knew where there was often a hole. On this occasion the hole was there and I proceeded. I had plenty of fuel to return to the desert. Another pilot attempting to depart the Los Angeles Basin, under similar conditions, was not so fortunate; the pilot wound through the mountains and finally into a blind canyon, and the Cessna 182 came to rest on a 45° slope. Fortunately, the only casualty was the airplane—it was totaled.

Breaks in the overcast (BINOVC) or the synonym *higher clouds visible* (HIR CLDS VSB) means from more than 7/8 to less than 8/8 of the sky is covered. Such a report should never imply VFR flight through a layer is probable or even possible. However, it might be the first indication of a layer beginning to dissipate.

Additive and Automated Maintenance Data

Additive data groups are reported only at designated stations. Groups consists of alphanumeric codes. Since this information does not directly relate to aviation operations, it will not be discussed.

The maintenance data groups are reported only from automated stations. These consist of sensor status indicators and a *maintenance indicator sign* ($).

Sensor status indicators advise users that specific equipment is not available. These consist of the following:

RVRNO	Runway visual range information not available
PWINO	Present weather identifier not available
PNO	Precipitation amount not available
FZRANO	Freezing rain information indicator not available
TSNO	Lightning information not available
VISNO	Visibility sensor information not available
CHINO	Cloud height indicator information not available

When visibility or cloud height indicators at secondary sites are not available, their locations will be included. For example, ...RMK VISNO RY06..., visibility information sensor at runway six not available.

In 1997, at the direction of Congress, the FAA conducted an ASOS operational assessment. The report concluded that overall ASOS performs as designed. ASOS required addition and augmenting 16 percent of the time for "nonrepresentative data." Cloud height, cloud coverage, and visibility accounted for the majority of edits. "At no point did the team consider any of the concerns significant enough to curtail the development of future ASOSs, nor did any of the concerns impact safety, efficiency of operation, or airport capacity...." ASOS availability was excellent. Sensor reliability exceeded 99% percent:

> The ASOS reporting of changing cloud heights and sky coverage was the area in which most of the inconsistencies (whether perceived or real) were found....ASOS lagged behind the observer in reporting lowering ceilings and led the observer in reporting rising ceilings.

> **CASE STUDY**   During a survey conducted at the Oshkosh Fly-in in 1997, 568 pilots were asked this question: How often do the conditions disseminated by AWOS/ASOS match those which you experience in flight? Forty-five percent responded "always" to "often"; another 19 percent responded "occasionally." Only 5 percent responded "never" or "infrequently." The remaining 35 percent indicated "don't use" or did not respond. This would seem to refute the notion that automated observations are less accurate than human reports. In fact, the survey could be interpreted that automated reports are as accurate or more accurate than human observations.

I must point out that human observers average visibility, sky condition, and cloud height, typically over a period of 10 to 30 min, just as ASOS does. The addition of reporting locations and availability of weather reports are undeniable advantages of automated systems. For example, a pilot might cancel a flight rather than "take a look" with IFR or marginal VFR reported. Consider the observation at Mammoth that winds are gusting to 90 kn. A pilot in flight could divert before encountering severe conditions, rather than possibly arriving without enough fuel for a suitable alternate, despite best efforts to follow regulations.

Weather observations are useful only to the extent that a pilot understands them. This comprehension requires a knowledge of the methods of the observer and the limitations due to equipment, training, experience, and time of day. Today's pilot must not only be able to read and translate reports but also interpret their meaning and significance, then apply them to a flight situation (VFR or IFR) and the capability of the aircraft, terrain, and alternates that might be required.

The need to evaluate all available sources cannot be overemphasized—adjacent weather reports, PIREPs, radar reports, satellite imagery, and forecasts. The complete picture and a knowledge of weather can help with evaluation and identification of erroneous observations. For example, on December 1, 1995, Fresno, California, reported a tornado. The synopsis forecast high pressure, radar showed no precipitation, and the satellite indicated no cloud cover. So what happened? The observer, augmenting the report, was practicing entering supplemental data—which was inadvertently transmitted. Other manual and sensor errors can be detected in this way. Whether a manual or automated observation, pilots must remember that reported conditions may not be representative of the surrounding area.

As an old aviation axiom laments: "Aviation weather reports may not be accurate, but they're official." The question is not whether manual or automated reports are better or worse. They are different, with both having advantages and limitations that must be understood.

As far as the codes go, I think the following letter to the editor of *Flying* magazine, which appeared in the April 1997 issue, says it all:

> I am an active general aviation pilot and FSS specialist. I have learned that in aviation the only thing that doesn't change is change itself. Anyone that has a problem with change doesn't belong in Air Traffic Control or aviation. I am tired of pilots and controllers whining about the new MRTAR/TAF codes. We have some serious problems to resolve in our industry, but MRTAR/TAF codes are not one of them.

The response: "Thanks, Terry. Do you know the difference between a jet engine and a pilot (or controller)? The engine finally stops whining.—Ed."

CHAPTER 9
PILOT WEATHER REPORTS

Ever complain about forecasts? Think they were prepared in a sterile environment? Conclude that pilots have no influence on their preparation? If we wish to participate, pilots can and do influence forecasts. According to the National Weather Service, pilot reports (PIREPs) are the most important ingredient for the AIRMET Bulletin, SIGMETs, Center Weather Advisories (CWAs), and winds and temperatures aloft forecast amendments. Furthermore, certain phenomena can be observed only by the pilot. In fact, over the last decade the manual observational network has dwindled and been replaced with automated systems so that the need for accurate pilot reports is greater than ever and cannot be overstated.

Satellite reports, plus upper-air and radar reports, supplement surface observations, but satellites observe only cloud tops, and upper-air observations are infrequent and widely spaced. Radar typically provides only precipitation information. Thus an urgent need exists for information on weather conditions at flight altitudes, along routes between weather reporting stations—especially in mountainous areas—and at airports without weather reporting service. In many cases the pilot is the best and only source of information on the following weather phenomena:

- Cloud layers
- Cloud tops
- Haze, smoke, dust, and sand tops
- Inflight visibility
- Slant range visibility
- Winds and temperatures aloft
- Turbulence
- Icing
- Low-level wind shear

Recognizing the importance of PIREPs, the FAA has directed air traffic controllers to actively solicit PIREPs, especially during marginal or IFR conditions and during periods of hazardous weather. Pilots operating IFR are required to "report... any unforecast weather conditions encountered...." These reports are not only of value to other pilots but to controllers, briefers, and forecasters alike.

PIREPs can be provided to any air traffic facility (center, tower, FSS). Unfortunately, PIREPs reported to center and tower controllers are not always disseminated beyond the sector or control position. To ensure the widest distribution, it's best to report directly to flight service, preferably Flight Watch—Enroute Flight Advisory Service (EFAS). More about Flight Watch is in Chap. 19.

PIREPs are transmitted under the *location identifier* (LOCID) for the *surface report* (METAR) nearest the occurrence using the file type UA (SAC UA, Sacramento pilot report). Reports may be appended, however, to major hub locations to ensure greatest prominence and widest distribution.

PIREP FORMAT

Unfortunately, neither the FAA nor the NWS has taken a strong stand on bringing conformity to the format of PIREPs. Thus PIREPs have appeared in confusing, misleading, and nonstandard formats, which has led to considerable misunderstanding and, in fact, has impacted the safety of our ATC system. Correctly formatted, PIREPs will eliminate any confusion and increase the usefulness of this valuable product.

PIREPs are entered using the standard format illustrated in Fig. 9-1. There is no need to memorize the form because FAA and/or NWS specialists encode the report. However, an understanding of the format will guide the pilot in identifying the information needed, and it will assist him or her in decoding and interpreting reports. Standard abbreviations contained in App. A are used.

Report Type: Urgent (UUA) or Routine (UA)

An *urgent pilot report* (UUA) communicates a hazard, or potential hazard, to flight operations. A UUA message receives special handling and immediate distribution. Urgent PIREPs report the following:

- Tornadoes, waterspouts, or funnel clouds
- Hail
- Severe or extreme turbulence
- Severe icing
- Low-level wind shear when reported with an airspeed change of 10 kn or more
- Volcanic ash
- Any other phenomenon considered hazardous

A *routine pilot report* (UA) receives routine distribution.

Location (/OV)

The location where the phenomena was observed is reported in relation to a three- or four-letter airport or radio *navigational aid* (NAVAID) *identifier.* For example, a report might specify an airport (/OV HAF, Half Moon Bay airport), a fix [/OV LAX or /OV LAX060010, Los Angeles VOR or the Los Angeles 060 radial at 10 nautical miles (nm)], or a location between fixes (SEA-BTG, Seattle VOR and the Battleground VOR; SLC245080-JNC210040, Salt Lake City 245 radial at 80 nm and the Grand Junction 210 radial at 40 nm).

Normally, if a PIREP contains conditions at an airport or specific geographical location such as a mountain pass, the code for that airport or location appears in the remarks:

DEN UUA /OV DEN301021 .../TP MAN/RM OG 1V5 WND 50-80G100+
BLOWING 3/4 GRAVEL

This report contains conditions observed on the Denver 301 radial at 21 nm. The remarks (/RM) indicate the report refers to conditions on the *ground* (OG) at Boulder Municipal Airport (1V5). It seems that the wind is gusting to more than 100 kn and blowing 3/4-in gravel around. This illustrates the importance of such reports.

PIREP FORM

Pilot Weather Report   ➤= *Space Symbol*

3-Letter SA Identifier

1. **UA**➤ ____ **UUA**➤ ____

____ ____ ➤   *Routine*   *Urgent*
                     Report   *Report*

2. **/OV** ➤	Location:
3. **/TM** ➤	Time:
4. **/FL**	Altitude/Flight Level:
5. **/TP** ➤	Aircraft Type:

Items 1 through 5 are mandatory for all PIREPs

6. **/SK** ➤	Sky Cover:
7. **/WX** ➤	Flight Visibility and Weather:
8. **/TA** ➤	Temperature *(Celsius):*
9. **/WV** ➤	Wind:
10. **/TB** ➤	Turbulence:
11. **/IC** ➤	Icing:
12. **/RM** ➤	Remarks:

FAA FORM 7110-2 (1-85) Supersedes Previous Edition

FIGURE 9-1 PIREPs are encoded using the PIREP format. Pilots need not memorize the format, but it does indicate information needed.

Decoding reports can be difficult without a copy of FAA Handbook 7350.5, *Location Identifiers*. DUATs have a decode function, current sectional or world aeronautical charts contain NAVAID identifiers, and sectional charts depict airport identifiers along with other airport data. If necessary, a pilot can always call an FSS.

Time (/TM)

Also included is the time the phenomenon was reported, referenced to *Coordinated Universal Time* (UTC).

Flight Level (/FL)

The *flight level* (/FL), or altitude, in hundreds of feet MSL, that the phenomenon was encountered either during climb (DURGC) or during descent (DURGD) may appear in remarks.

Type of Aircraft (/TP)

A PIREP would note the type of aircraft the pilot was flying, using *standard international* (ICAO) *aircraft type designators*. From time to time, this element will contain /TP PUP (*pickup truck*), /TP CAR, /TP FBO (*airport fixed base operator*), or, as reported on the Denver PIREP, /TP MAN.

Sky Cover (/SK)

Standard *sky cover* abbreviations (SKC, FEW, SCT, BKN, OVC) are used followed by cloud bases in hundreds of feet—like METAR, TAF, and aviation forecasts. It's important to remember that PIREP bases and tops are always reported in reference to *mean sea level* (MSL). Cloud tops are indicated by the word "TOP" followed by the height—similar to aviation forecasts. Additional layers are separated by a solidus (/), and clear above is indicated by SKC. For example:

...SK FEW-SCTUNKN-TOP030/BKN060-TOP100/SKC....

Decoded: A few to scattered (1/8 to 4/8 coverage) clouds with tops at 3000 ft MSL; a broken layer (5/8 to 7/8 coverage) bases 6000 ft MSL, tops 10,000 ft MSL; clear above.

Recall that PIREP bases and tops are always MSL. Conditions in Fig. 8-5 were reported as ...FEW010 BKN018 OVC030.... If the field elevation had been 2000 ft, a PIREP describing these conditions might have read: ...FEWUNKN-TOP035/SCTUNKN-TOP043/SCTUNKN-TOP055... (tops of a few clouds 3500, scattered 4300 and 5500 ft, respectively). Given the summation principle, the observation and pilot report are perfectly consistent.

A pilot report can be one of the most accurate means of determining cloud height, assuming the pilot actually penetrates the clouds or climbs or descends through a scattered or broken layer. Otherwise, it's just a pilot's estimate. As pilots, we should always report cloud bases and tops to the tower or FSS, especially when they are different from reported. But keep in mind that we reference cloud bases to a pressure altimeter set to read MSL, while a ground observer reports clouds AGL. (I recently gave cloud bases to a tower controller. The controller asked if they were AGL or MSL. Hum?)

Weather (/WX)

Weather encountered and flight visibility are reported in this element.

Flight visibility is reported to the nearest whole statute mile, and it is encoded with the suffix SM (FV01SM, FV05SM, etc.). Unrestricted visibility appears as FV99SM. However, FV99SM can be ambiguous. Does it mean visibility 99 mi or greater, or 7 mi or greater? Pilots should report specific values to 98 mi to eliminate any misunderstanding.

Weather, when reported, follows visibility using standard ICAO abbreviations contained in Tables 8-2 through 8-4. Like METAR, should hail be reported, its size appears in remarks in 1/4-in increments (GR1/2, hail 1/2 in).

> **CASE STUDY**   A pilot approaching Lovelock, Nevada (elevation 3900 ft), skeptical of a reported visibility of 2 miles in blowing sand, reported flight visibility 20 mi. Upon landing, however, the pilot concurred, stating the tops of the blowing sand were at 200 ft AGL. This report would appear: ...FL075.../WX FV20SM BLSA000-TOP041....

Air Temperature in Celsius (/TA)

Air temperature is reported in degrees Celsius, with negative values prefixed with letter *M*—as it would be in METAR.

Wind (/WV)

Wind direction and speed is encoded using three digits to indicate direction and two or three digits to indicate speed (/WV 36020KT, observed wind 360° at 20 kn). In spite of the FAA's flight service handbook, which states "wind direction (magnetic)...," direction should be reported in relation to true north, which is consistent with other reports and forecasts. Speed is in *knots* (KT).

Turbulence (/TB)

The intensity (NEG, LGT, MOD, SEV, EXTRM) and altitude (when different from /FL) of turbulence appears in this element. *Clear air turbulence* (CAT) and CHOP should be added when appropriate. (Both terms will be defined shortly.) When turbulence has been forecast but reports indicate smooth, /TB NEG is entered. Therefore, /TB NEG is interpreted as smooth, rather than turbulence that bounces the aircraft down.

Icing (/IC)

The intensity (NEG, TRACE, LGT, MOD, SEV), type (CLR, RIME, MX), and altitude, when different from /FL, are entered in this element. Temperature should also be included with icing reports. Like turbulence, when icing is forecast but not encountered, NEG is entered; a NEG encounter is as important and useful as one reporting the phenomenon!

Remarks (/RM)

This element reports *low-level wind shear* (LLWS), convective activity, surface conditions at airports, or other information to expand or clarify the report.

The following remarks have appeared in some PIREPs:

/RM SMOKE OVER NWS BUILDING DRIFTING EAST-SOMEONE THINKING

/RM VFR NOT RECOMMENDED-THREE AIRCRAFT COULD NOT MAINTAIN
VFR DUE TO ICING IN CLOUDS.

Something odd; oh well, I guess it is difficult to maintain VFR in the clouds when you're
icing up.

/RM HAD TO CLIMB TO FL200 TO REMAIN VFR—NOW LEAVING FREQ TO
CONTACT ZOA.

Because this pilot is already 2000 ft into class A airspace, leaving the frequency to contact
ZOA (Oakland Center) seems like a fairly good idea.

Cloud bases and tops, temperatures, and even winds can be measured. Flight visibility
and weather are direct observations. Intensities of turbulence and icing, however, are some
of the most misunderstood quantities in aviation. That's because they're subjective and
usually based on the pilot's training and experience.

> **CASE STUDY**   A rather shaky voice called flight service to report moderate to severe turbu-
> lence. The specialist asked the novice pilot, "Did you actually lose control of the aircraft?" The
> pilot replied, "Well, no." The specialist then asked, "Would it be okay if we called it light to
> moderate turbulence?" The pilot agreed.

TURBULENCE

The intensity of turbulence is, to some degree, affected by aircraft type and flight configuration. United States Air Force studies have shown the following to generally increase the
effects of turbulence:

- Decreased weight
- Decreased air density
- Decreased wing sweep angle
- Increased wing area
- Increased airspeed

Classifications for the intensity of turbulence can be found in the *Aeronautical Information
Manual* (AIM) and *Aviation Weather Services*; however, I prefer the following.

Light

A turbulent condition during which your coffee is sloshed around but doesn't spill, unless
the cup's too full. Unsecured objects remain at rest; passengers in the back seat are rocked
to sleep.

Moderate

A turbulent condition during which even half-filled cups of coffee spill. Unsecured
objects move about; passengers in the back seat are awakened by a definite strain
against their seat belts.

Severe

A turbulent condition during which the coffee cup you left on the instrument panel whizzes by the passengers in the back seat. The aircraft might be momentarily out of control, but you don't let on. Passengers not using their seat belts are peeling themselves off the cabin ceiling.

Extreme

Usually associated with rotor clouds in a strong mountain wave or a severe thunderstorm, extreme turbulence is a rarely encountered condition where the aircraft might be impossible to control. The turbulence can cause structural damage. Your passengers are becoming concerned by the beads of sweat on your brow, your white knuckles, and the new frequency and new transponder code you have just selected—121.5 and 7700.

The following reports illustrate mountain wave activity:

RNO UUA /OV FMG330025/TM 2345/FL105/TP BE35/TB SEV/RM TMPRY
LOST CONTROL...PILOT CUT ARM IN TURBC...RTNG TO RNO.

A SIGMET was in effect for severe turbulence, Reno surface winds were out of the west gusting to 27 kn, and winds across the Sierra Nevada mountains were gusting to near 50 kn. The pilot had the clues but elected to go. From the PIREP, it would appear the pilot regretted the decision.

Another pilot caught in a mountain wave reported the following:

RNO UUA /OV FMG270012.../TP C404 /TB EXTREME 130-110 MOD SEV
110-090 /RM EXPERIENCING STRUCTURAL DAMAGE.

The airlines are not immune to mountain waves:

DEN UUA /OV DEN313047/TM 0158/FL350/TP L101/TB SEV/RM SEV
MTN WAVE PLUS AND MINUS 6000 FPM.

This Lockheed Tristar (L101) at 35,000 ft over the Rockies experienced severe turbulence and 6000-feet-per-minute updrafts and downdrafts. This illustrates the severity of mountain waves and the fact that they can extend to the stratosphere.

Turbulence encountered in clear air not associated with cumuliform clouds, usually above 15,000 ft and associated with wind shear, should be reported as *clear air turbulence* (CAT). Slight, rapid, and somewhat rhythmic bumpiness without appreciable changes in altitude or attitude defines CHOP. Since CHOP does not cause appreciable changes in altitude or attitude, it would not be classified as severe.

In addition to intensity, the duration of turbulence should be reported.

- *Briefly.* Turbulence encountered for an extremely short period, usually only one or two jolts (/TB LGT /RM 2 MOD JOLTS could be reported /TB LGT BRFLY MOD).
- *Occasional.* Less than 1/3 of the time.
- *Intermittent.* Between 1/3 and 2/3 of the time.
- *Continuous.* More than 2/3 of the time.

Chapter 8 discusses how to use METAR to determine likely areas for turbulence. Recent and accurate PIREPs can verify or refute its presence. If reports or forecasts indicate turbulence, a pilot can minimize the hazard when encountered—strategies are discussed in Chap. 23. Turbulence imposes gust loads that appear to be almost instantaneous. Gust loads increase with the speed of the aircraft and gust velocity.

ICING

As with turbulence, there is a tendency, especially with new or low-time pilots, to overestimate icing intensity. A recently rated instrument pilot, after experiencing his second encounter with icing in a Cessna 172, reported severe icing. The encounter lasted about 30 min, and the pilot was unable to maintain altitude and was forced to descend. This description, however, is only of moderate intensity. Icing intensity has been classified for reporting purposes in the AIM. Perhaps personal definitions are more descriptive.

Trace

Ice becomes perceptible, and the rate of accumulation is slightly greater than the rate of sublimation. It is not hazardous even though ice protection equipment is not utilized unless it is encountered for more than 1 h. Your spouse admires how pretty it looks on the wing; ATC has just instructed you to climb. You advise them icing is probable and request descent. The controller calmly replies that in that case, "you can declare an emergency or land." Shortly, you're handed off to the next controller. You inquire about a lower altitude and the controller responds, "Is that Terry up there?" (a friend at Los Angeles Center). A lower altitude is approved in about 15 min.

Light

The rate of accumulation can create a problem if the flight continues for more than 1 hour. Occasional use of ice protection equipment removes or prevents accumulation. Ice should not present a problem if the protection equipment is used. Your student hasn't noticed the ice yet; your pilot friend in the back seat is hoping he has enough life insurance; you're negotiating with ATC for a lower altitude, which they can approve in 15 mi. This will take only about 8 min, but each minute seems like 10.

Moderate

The rate of accumulation, even for short periods, becomes potentially hazardous, and the use of ice protection equipment or flight course diversion becomes necessary. On his second encounter with ice, a friend and his passengers, in an aircraft without ice protection equipment, survived moderate icing only because the terrain was lower than the freezing level.

Severe

The rate of accumulation is such that ice protection equipment fails to reduce or control the hazard. Immediate diversion is necessary. This is a situation in which the person in the left seat very rapidly ceases being the pilot and becomes a passenger; the wing is an ice cube.

FACT   Until 1968, the maximum intensity reported for icing was "heavy." Certain pilots still insist on using this term. However, it is a misnomer because all ice is heavy! It's time to move on.

The type of icing has also been classified for reporting purposes.

Rime Ice

Rime ice is milky, opaque, and granular, normally forming when small supercooled water droplets instantaneously freeze upon impact with the aircraft. It is most frequently encountered in stratiform clouds at temperatures between 0°C and −20°C.

Clear Ice

Clear ice is glossy and formed when large supercooled water droplets flow over the aircraft's surface after impact and freeze into a smooth sheet of solid ice. It is most frequently encountered in cumuliform clouds or freezing precipitation. Brief, but severe accumulations occur at temperatures between 0°C and −10°C, with reduced intensities at lower temperatures, and in cumulonimbus clouds down to as low as −25°C.

Rime Ice and Clear Ice (Mixed Ice)

Mixed ice is a hard, rough, irregular, whitish conglomerate formed when supercooled water droplets vary in size or are mixed with snow, ice pellets, or small hail. Deposits become blunt with rough bulges building out against the airflow.

A popular notion in some aviation circles is that a pilot's mere mention of ice will receive emergencylike handling. Icing might be an emergency, but remember the controller's job is to separate aircraft within a finite amount of airspace. ATC might have to assign a higher altitude, but ATC cannot, and should not, be expected to fly the aircraft or assume the responsibility of the pilot in command. To paraphrase: An accurate report of actual icing conditions is worth a thousand forecasts.

No one has any business flying in these conditions:

    FAT UUA /OV CZQ090030.../FL160-240 /TP F18/IC SEV CLR
    BFL UUA /OV PMD330040.../FL100 /TP C402/IC SEV RIME/RM PUP 1
    INCH CANT SEE THRU WINDSHIELD

REMARKS

Remarks amplify information or describe conditions not already reported:

    SAC UUA /OV SAC/TM 1753/FL030/TP C172/TB MOD-SEV/RM LTGCG

The pilot observed lightning cloud to ground. (Pilots should use the same frequency and type of descriptions for lightning as described in Chap. 8 on METAR remarks.)

    BFL UA /OV WJF-BFL/TM 1845/FL045/TP UH60/SK OVC050/TB LGT OCNL MOD/RM
    THRU TSP PASS UNDER CLDS. OK FOR HELIO NOT SO HOT FOR FIXED WING

SNA UA /OV SNA-GMN/TM 2218/FL125/TP PA28/TA 00/WV 33020KT
/TB MOD/RM LIKE AN E TICKET AT DISNEYLAND

An "E ticket" was for the big rides.

DEN UA /OV DEN240060/TM 1715/FL350/TP DC8/TB MOT CAT/RM MOD
CAT AT 373 NEG BLO 367 (ZDV)

This report was filed by Denver Center's Weather Service Unit (ZDV). It amplified the turbulence portion indicating that moderate CAT was encountered at 37,300 ft and it was smooth below 36,700 ft.

Remarks also describe *low-level wind shear,* which is shear that occurs within 2000 ft of the surface. Because of its significance, pilot reports of wind shear are extremely important. Wind shear PIREPs should include location, altitude, and airspeed changes:

RNO UA/ OV RNO.../TP DC9/RM LLWS 001-SFC +30 TO 40 KTS

The pilot experienced a 30- to 40-kn increase in airspeed between 100 ft and the surface.

STS UUA/.../FL030/TP C500/TM DURGD RY02 LLWS RESULTING IN 80
KTS CHG AIRSPEED.

Studies have shown that the great majority of nonconvective LLWS reports have in fact been triggered by low-level turbulence. Low-level turbulence will be discussed in Chap. 23. Many pilots lump any turbulence below 2000 ft AGL into the category of wind shear, probably because of FAA and media emphasis on this phenomenon. Both can be severe. So what's the difference? In a wind shear event, a pilot can expect a sudden change in airspeed (plus or minus, but not both)—as illustrated by the preceding examples. Low-level turbulence is characterized by fluctuations in airspeed (plus and minus).

Interesting remarks abound:

/RM STRONGEST TURBC I HAVE EXPERIENCE IN 15 YEARS...
(reported by a King Air)

/RM 2 PASSENGERS INJURED...HAD TO TURN BACK...
(reported by a Navy P3)

/RM IFR NOT RECOMMENDED DUE TO STRONG HEAD WINDS AND 2000 FT
PER MIN UP AND DOWN DRAFTS WILL NEVER DO IT AGAIN...
(reported by a Cessna 182)

/RM SOME REAL GOOD JOLTS PUT KNOT ON HEAD...
(aircraft type missing)

/RM ONE LARGE JOLT, STEW FELL DOWN (SHE IS OK) LOTS OF DRINKS
SPILLED...
(reported by a Fairchild 27)

/RM UNA TO CONTROL HELICOPTER. RETURNED. NURSES KISSED THE GROUND.

/RM WIND O/G AT FCH [Fresno Chandler Airport] IS 300-330 DEGREES AT 40 KTS AND
ALL THE C150'S AND C152'S ARE INTMTLY FLYING ON THEIR CHAINS.

/RM SEV LLWS AFTER 3 APCHS UNABLE TO LAND...
(reported by a Lear Jet)

/RM LOTS OF BAD TURBC. THIS ISN'T THE SMARTEST THING I'VE
EVER DONE. HUGHES COPTER.

/RM ROUGHER THAN A COBB...
(a good old standby, but not very useful)

/OV AVX .../TP FBO/SK SKC/WX SKC/TB NEG/RM LET'S GO FLY...
(AVX is Catalina Island's Airport in the Sky. If you're in the Los Angeles area, try to get out
there; you may want to order a Buffalo Burger at the restaurant.)

LAX UA /OV SXC213186/TM 1911/FL070/TP VYGR/SK BKNUNKN-
TOP040/TB LGT /RM VOYAGER 1...
(This is an actual report filed by the *Voyager* on their record-setting around-the-world flight.)

A strong Santa Ana wind in southern California was responsible for the conditions
described in the following PIREPs at Ontario, California (ONT). Surface wind was gusting
to 45 kn:

ONT UUA /OV ONT/TM 1445/FL050/TP B727/TB MOD OCNL SEV 050-SFC /RM CRCLG
FAP LNDG ONT

ONT UUA /OV ONT/TM 1450/FL020/TP B727/RM UNABLE TO LAND DUE TO X-WINDS

FSS and NWS specialists sometimes tend to editorialize on PIREPs, usually around the
time of championship sporting events. Although unauthorized and unprofessional, pilots
can expect to see these, usually humorous, reports from time to time. Others contain comments of a personal, social, or political nature. For example:

HWD UA /OV OAK110007/TM 1600/FL060/TP BE33/SK SKC/WX FV99SM/TB NEG/RM
DURC HWD NBND SEVL H LYRS AT FL015/FL042/FL060. SLANT VSBY 15-30SM.
PTCHY ST OFSHR-THRU GOLDEN GATE OVR CITY OF OAKLAND. REPORTED BY A
DERANGED BONANZA PILOT.

OK, I was the deranged Bonanza pilot!
Another read:

BFL UA/ OV EHF/TM 1900/FL100/TP UNKN/SK SKC/IC MOD

Could that be an example of "clear air icing"? No. A strong weather system was forecast to
move rapidly into central California. As it happened, the system stalled off the coast, with
weather advisories for mountain obscuration and icing continuing in effect. When it
became apparent that the system had stalled, I called the Aviation Weather Center and
asked the forecaster to amend the advisory for mountain obscuration. I didn't mention
icing; I assumed it would be amended as well. Silly me! Sure enough, the forecaster
amended the mountain obscuration but left the icing advisory in effect. I stopped issuing
the icing advisory, but one of my coworkers vented some frustration in the form of the
PIREP above.

Many years ago I provided a PIREP to the Los Angeles Flight Watch. Afterward, the
flight watch specialist commented to another controller, "That pilot really knows how to
give a PIREP!" The other controller responded, "He should; he works here."

PIREPs from air carriers, the military, and corporate aircraft tend to be more accurate because of the pilot's training and experience. Few student pilots fly MD11s or F18s. New and low-time pilots (inexperienced) tend to overestimate intensities of turbulence and icing, then they may think they've experienced severe conditions and might not heed reports or forecasts. This is not to say PIREPs from pilots of Cessna 150s or Piper Tomahawks are never accurate or that they should be ignored.

Pilots must evaluate PIREPs within the context of surface reports, forecasts, and other PIREPs (part of the complete picture). A single report of severe turbulence from a Beech Sundowner under clear skies and light winds should be viewed with skepticism. On the other hand, a report of severe turbulence from a Cessna 172 with conditions favorable for a mountain wave and advisories in effect should be taken seriously. PIREPs that are not objective are worse than useless. Not only do they give a false impression to other pilots, but forecasters must take them as fact and issue advisories, which undermine forecast credibility.

PIREPs are an essential part of the observational program. This can be especially true at ASOS stand-alone D sites. Consider, for example, the following METAR and PIREP for Gunnison, Colorado:

METAR KGUC 201436Z AUTO 00000KT 10SM CLR M23/M28 A3018 RMK A01

GUC UA /OV GUC/TM1520/FLUNKN/TP C310/RM PILOT REPORTS AWOS WRONG-THERE IS A LOW CIG-NEED IFR TO DEPART

The fog layer drifted over the laser CHI the next hour, and a broken layer was reported at 500 ft. (Don't think this can't happen at a manual or augmented site; it does, especially at night.)

Every time we fly, we become observers, but our reports must be timely. Some pilots have a tendency to wait until the latter portion or end of a flight to provide a report. A pilot on a flight from Seattle to Los Angeles contacted Oakland Flight Watch and reported conditions departing Seattle two and a half hours earlier. A somewhat overzealous briefer instructed the student to make a pilot report at the conclusion of the flight. The student calling the FSS the following day meekly apologized for failing to provide the report, then proceeded to recount in detail the conditions encountered. I'll bet this pilot doesn't forget on the next flight. Get into the habit of routinely providing timely reports. Keep in mind that reports confirming the forecast are as important as those for unforecast weather. To be of most value, reports of turbulence and icing must accurately contain location, time, altitude, type aircraft, sky cover, and temperature, as well as turbulence and icing. Here again, negative reports are as valuable as those reporting severe conditions.

At present most domestic PIREPs are manually entered by FSS controllers, Center Weather Service Unit (CWSU) meteorologists, Air Route Traffic Control Center (ARTCC) weather coordinators, and military base operations personnel. To clarify any miscoded reports, pilots have only one option: Contact the local FSS. But if the local FSS did not submit the report, there may be no way to verify the information. In that case the only option is to ignore the report.

The next time someone complains about the lack of weather information or forecast accuracy, ask if he or she routinely provides objective pilot reports. If you fly and just don't get around to making a report, consider this:

BFL UA/OV BFL/TM 1450/FLUNK/TP ALL/RM WISH I HAD A TOP REPORT FROM BFL TO ONT.

CHAPTER 10
SATELLITE IMAGES

Weather observations consist of surface observations, observations from pilots—either on the ground or during flight (PIREPs)—radar reports, upper-air soundings, and satellite images. Our three-dimensional observational system begins with surface observations—the lower layer; next comes PIREPs, radar, and upper-air observation—the middle layer; finally satellite images provide a look from the top down.

Since April 1960 weather satellites have been orbiting over the Earth. There are two main types of meteorological satellites: *polar orbiters* and *equator orbiters*. Pilots usually have access to the equator orbiters, which operate as *geostationary operational environmental satellites* (GOESs). Our discussion will be limited to GOESs. GOESs orbit directly over the equator at approximately 19,000 nautical miles (nm). They circle the Earth once every 24 hours. From the satellites' view, the Earth appears to remain stationary—hence the name "geostationary." Satellite images are available on many Internet sites. Most of the images used in this book appear through the courtesy of the National Oceanic and Atmospheric Administration (NOAA) at www.goes.noaa.gov.

Satellite interpretation is a complex science in itself. Therefore, we will limit our discussion to two basic images: visible and infrared (IR). A *visible image,* as the name implies, is a "snapshot" of conditions on the Earth. Resolution of visible images range between 0.5 and 2 nm. An *infrared* (IR) *image* is a temperature picture. That is, the satellite senses the temperature of an area with a resolution of approximately 5 nm. (Most of the pictures used in this chapter and those available from the NOAA Web site have a resolution of approximately 5 nm.) Resolution deteriorates with distance from the satellite, both north and south of the equator and west or east of the satellite's position. This error is known as *parallax.* As a result of the limitations of resolution and parallax, some clouds may be displaced several miles from their actual location, and certain objects will not show accurate brightness values.

GOES east is located at approximately 75° W longitude, and GOES west, at approximately 135° W longitude. GOES east covers the eastern two-thirds of the United States, the Caribbean, and Atlantic. GOES west covers the western third of the United States, Hawaii and the eastern Pacific, and Alaska.

When interpreting visible imagery, note that various types of clouds and terrain reflect different amounts of sunlight. The best reflectors are large cumulonimbus clouds. Thin clouds or areas of very small clouds appear darker because they reflect less sunlight. Below is a list of various surfaces arranged in order of their relative reflectivity, beginning with the brightest:

- Large thunderstorms
- Fresh new snow
- Thick cirrostratus
- Thick stratocumulus

- Snow 3 to 7 days old
- Thin stratus
- Thin cirrostratus
- Sand
- Sand and brushwood
- Forest
- Water surfaces

IR images begin by portraying different temperatures as black, shades of gray, and white; black is the warmest, white the coldest. Typically, black represents a temperature of about 33°C and white, −65°C, with the gray shades representing decreasing temperature toward the white end. In fact, there are 256 distinct shades from black to white. But basically warm temperatures are dark, cool temperatures gray, and cold temperatures are light.

Computer technology enables the enhancement of IR images. This technique allows the operator to highlight areas of interest. Colors can be assigned to specific temperature values within the 256 shades. This results in the variety of color satellite images seen on television and available on various Internet sites. However, without knowing the exact enhancement curve, specific interpretation is difficult. *Enhancement curves* allow for greater detail of certain phenomena, such as snow and ice, fog and thunderstorms, and haze, dust, and volcanic ash. These typically are not available for pilot use in an operational environment. Therefore, our discussion will be limited to those satellite products and phenomena to which pilots have operational access.

As in chart interpretation, best results are obtained by comparing the visible image with the same-time-frame IR image. Often what may be misinterpreted or ambiguous on one image can be resolved by comparing it with the other image. Unfortunately, this doesn't work at night, when only the IR image is available. (I had a pilot ask me for the visible satellite image one morning at about 6 a.m. In a feeble attempt at humor, I replied, "It's not available. The flash bulb has burned out." The pilot replied, "Yes, you've been having a lot of trouble with that satellite lately." Oh well.) Optimum interpretation results from a comparison of chart and satellite information—part of the complete picture.

If we have the capability to view several images in succession—a satellite loop—we can often get a sense of weather movement, development, and dissipation. Viewing a loop will also help us distinguish terrain and surface features from cloud cover.

The next two sections will discuss the identification of geographical and weather features. The last section of this chapter will discuss interpretation and application. This section will apply the principles of geographical and weather features identification, and it will include practical examples of application. Various satellite terms will be defined. Pilots can expect to see this terminology on various weather products, such as Center Weather Advisories, Alert Weather Watches, and the Convective Outlook.

GEOGRAPHICAL FEATURES

Typically a distinct boundary occurs between land masses and oceans on both visual and IR satellite imagery. On visible images, water is dark, land is gray; thick clouds are white, thin clouds are gray. Water surfaces are the least reflective of the various surfaces, and there is often a distinguishable temperature difference, especially during the time of maximum daytime heating. Large lakes, bays, and rivers can usually be identified. Major mountain ranges and valleys can also be identified. Here again, these features usually have different reflectivities and temperature ranges. Deserts are also distinguishable because of the low reflectivity of sand compared to adjacent wooded areas and mountains, and temperature contrasts, especially during the time of maximum daytime heating.

Most satellite images contain a grid showing cultural boundaries, such as state and international borders; and often the grids depict large lakes for reference. A knowledge of terrain features within these boundaries is very helpful in determining the location of specific terrain features. Appendix B, Graphs and Charts, along with aeronautical charts, provides the location of various geographical features.

Snow cover is often difficult to identify. Both low clouds and snow reflect about the same amount of sunlight. This is especially true over relatively flat terrain. Differentiating between low clouds and snow can often be done by recognizing terrain features, such as unfrozen rivers and large lakes. Clouds normally obscure terrain features, but snow cover does not. Snow in mountainous country is usually easier to identify because it often forms a dendritic pattern. Mountain ridges above the tree line are essentially barren and snow is visible. In the tree-filled valleys, most of the snow is hidden beneath the trees. This branchy, sawtooth, dendritic pattern identifies areas of snow cover.

Figure 10-1, a visible image, shows a dendritic pattern over the southern Sierra Nevada mountains of California and the central mountains of Idaho. Surface observations and PIREPs are helpful, especially when they confirm clear skies. Also, a comparison of earlier photos of a loop may be used. Cloud patterns typically change over time; snow cover normally changes very little from day to day. Snow cover is normally not identifiable on IR images because there is not enough temperature contrast between snow cover and adjacent surfaces.

FIGURE 10-1 Branchy, sawtooth, dendritic patterns identify areas of snow cover over mountainous regions.

WHY DOES LAKE ERIE FREEZE?   Lake Erie often freezes in winter while the other Great Lakes remain ice free. In the autumn, surface lake temperatures decrease as air temperatures lower. Just as in the atmosphere, cooler, denser surface water sinks as warmer water from the depths rises. This mixing continues through midwinter when water throughout the lake reaches about 4°C. With further cooling, surface water becomes less dense, which sets the stage for freezing. The other Great Lakes are much deeper and almost never reach 4°C throughout their depths. Upwelling of warmer water prevents their freezing.

WEATHER FEATURES

In visible satellite imagery a thick cloud deck appears white, with large thunderstorms the brightest. Thin or small clouds, in themselves, do not present an aviation hazard. In an area of small or thin clouds, part of the reflected sunlight sensed by the satellite is from the tops of the clouds and part from the surface below. The resultant image is the average of the two reflectivities. The image is darker than a thick cloud and lighter than the normal surface shade. On IR imagery thin or small clouds also create errors in shading. The imager averages surface temperature and cloud top temperature. Therefore, satellite-depicted cloud height is lower than actual cloud tops. This accounts for some apparent errors between surface observation and satellite imagery.

Texture is one means of identifying cloud types. Texture appears lumpy, caused by shadows. Stratiform clouds appear flat and sheetlike because they are formed in stable air. Stratiform clouds normally show no texture. Cumuliform clouds appear rounded, billowy, and puffy on visible imagery because they are formed in unstable air. Cumuliform clouds have a lumpy texture. Cirroform clouds often have a fibrous texture. Cloud appearance is an excellent indicator of atmospheric stability.

Low cloud tops, stratus and fog, are characterized by a flat, smooth, white appearance and a lack of an organized pattern or texture. Boundaries are often sharp and defined by topography and may exhibit a dendritic pattern. It may be difficult to distinguish fog from low stratus and snow. On visible imagery they appear bright when thick; on IR images they appear dark to medium gray and may be difficult to distinguish from the surface since there is little temperature difference between the surface and cloud tops. Most types of fog show up well on visible satellite imagery. Often, the satellite is better at determining the extent of fog and stratus than surface observations.

Radiation fog is usually clearly discernible on visible imagery and at times on IR imagery, depending upon the contact between surface and cloud top temperatures. Figure 10-1 shows an extensive area of radiation fog in California's central valley. Note the bright, flat appearance, with sharp edges. The edges follow the contour of the valley. Although barely discernible in the picture, there is what appears to be a small hole in the fog at its northern end. This is, in fact, the Sutter Butte, elevation 2140 ft. This formation often pokes through the fog. Advection, upslope, and steam fog are usually well defined on visible imagery except when there is a high cloud cover. In Fig. 10-3, an afternoon visible image, advection fog has cleared most of the land areas, but it remains along central California's coastline. Since rain-inducted fog is caused by precipitation from a higher layer, it does not show up on satellite imagery. Fog dissipates from the edges inward.

As we shall see, weather advisory phenomena usually lie well within their designated boundaries. Satellite imagery can often further delineate these areas. Additionally, imagery may expose the fact—sorry I couldn't help it—of weather phenomena are developing or dissipating as forecast. From Fig. 10-1, this would seem to be a good day for VFR flight along the coast and in the mountains. In the central valley even IFR flights may have difficulty with extensive areas of near zero-zero conditions.

Stratocumulus clouds often appear in sheets or lines of clouds. Sometimes individual cloud elements are seen. A *cloud element* is the smallest cloud that can be seen on a satellite image. These clouds sometimes form in narrow bands in which individual cloud cells are connected, and they are known as *cloud lines,* or they are not connected and are known as *cloud streets.* Stratocumulus sometimes exhibit a cellular cloud pattern—that is, a more-or-less pattern of cloud cells. They may form a closed-cell or open-cell pattern. A *closed-cell pattern* refers to cloud cover that is solid, with individual convective elements rising through the layer. An *open-cell pattern* indicates clear air surrounding each individual convective cell. On visible imagery, cumuliform clouds typically appear rounded.

Middle clouds may appear in cellular (altocumulus) or sheet (altostratus) patterns. Like stratocumulus, altocumulus may exhibit stationary lines (altocumulus standing lenticular) indicating mountain wave activity. In IR imagery these clouds appear lighter than low-level clouds because of their colder tops. Figure 10-2 illustrates a mountain wave as seen on visible imagery. Wave activity has developed over the northern Sierra Nevada mountains and the mountains of northeast California and southern Oregon. The wave extends hundreds of miles downstream, as shown in Fig. 10-2, with wave clouds throughout northern Nevada and into northern Utah. Could wave activity be occurring south of Lake Tahoe, east of the southern Sierra Nevada mountains? Probably. Without sufficient moisture, clouds would not develop.

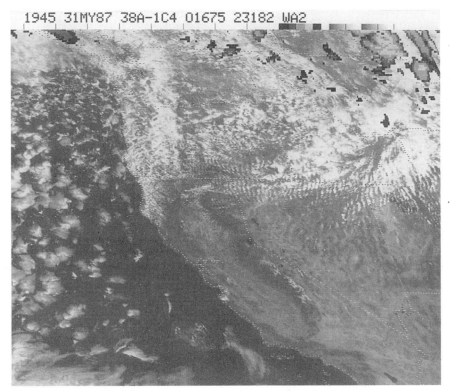

FIGURE 10-2   A satellite image, in itself, cannot be used as a weather briefing; it is merely one piece of the complete picture.

Figure 10-2 is a higher resolution than available from the NOAA Web site. In this mid-day photo, terrain features are clearly visible. In fact, all of the terrain features in App. B, Charts and Graphs, can be identified. Also, note that many smaller lakes are visible. Dendritic snow patterns of the southern Sierra Nevada mountains are easily identifiable. It would seen to be a good flying day, except for the turbulence. All too often this is the case—clear skies and unlimited visibilities mean strong winds and lots of turbulence. Mother Nature never gives you anything for nothing. This also illustrates that the satellite image, in itself, cannot be used as a weather briefing. It is merely one piece of the complete picture.

High clouds (cirroform) form where temperatures are very cold. Thin cirrostratus appear on visible imagery as a medium gray; on IR imagery as light or very light gray. Recall the discussion of thin clouds. Thick cirrostratus appear on visible and IR imagery as almost white. Thick cirrostratus are among the highest in reflectivity, with very cold tops. As we shall see, one of the best techniques to identify cirrus is to compare both visible and IR images of the same time frame.

Clouds with vertical development (cumulus and cumulonimbus) show up well on visible imagery. They tend to be thick, with large thunderstorms the highest on the list of reflectivity. IR imagery is a good indicator of vertical development. The closer to white, the colder, and therefore the higher the tops. Overshooting cirroform tops are also an indicator of development and severe weather potential. High clouds and those with vertical development also tend to cast shadows, especially during morning and afternoon on visible imagery.

Figure 10-3 is a late-summer, afternoon visible image. As is typical this time of year, air mass thunderstorms have developed over the Great Basin and the mountains of central and southern California. There are no overshooting tops, indicating storms are not severe. Typically winds at mid and upper levels are light. This is shown by no distinct cirrus blowoff, indicating little upper wind flow and little, if any, thunderstorm movement. Usually these storms are circumnavigable for both VFR and IFR operations. Notice that cirrus tops cover the storms along the southern Sierra Nevada and central California coastal mountains. Without access to real-time weather radar (storm detection equipment) or visual contact with the cells—the ability to maintain visual separation—no pilot should attempt to penetrate these areas either VFR or IFR. (*Storm detection equipment* refers to airborne access to real-time weather radar or lightning detection equipment. More about this subject in Chap. 11.) These rules apply to any area of air mass thunderstorms.

Coastal advection fog has dissipated land areas, but remains along central California coastal sections. As previously mentioned, the visible satellite image is usually an excellent indicator of the extent and location of advection fog.

Air masses tend to show up well on satellite imagery. Large high-pressure areas are often cloud-free zones. Low-pressure areas, with moisture present, contain large, organized cloud patterns, with the ultimate example a hurricane. Boundaries between air masses—fronts, again with moisture present, show up well on satellite imagery. Large weather systems often appear in the shape of a "comma." A *comma cloud* indicates an area of low pressure with occluded, warm, and cold fronts, and the jet stream. Maximum vorticity—upward vertical motion—occurs in the center. The surface location of fronts, especially warm and occluded, may be masked by higher clouds in the form a *cloud shield*—a broad cloud pattern—or overrunning cirrus. Fronts may be indicated by cloud bands, a nearly continuous cloud formation. However, under extremely dry conditions, frontal boundaries may be cloud free.

The location of the jet stream is frequently identifiable on satellite imagery. The jet stream usually crosses an occluded system just to the north of the point of occlusion. Cirrus is often associated with the jet stream and high-altitude turbulence. Cirrus that form as transverse lines or cloud trails perpendicular to the jet stream indicate moderate or greater turbulence. These clouds might be reported as cirrocumulus. *Cirrus streaks,* parallel to the jet, are long narrow streaks of cirrus frequently seen with jet streams. Typically, the sharp

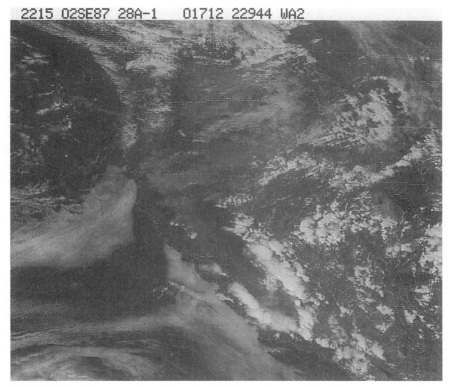

2215 02SE87 28A-1   01712 22944 WA2

FIGURE 10-3 Without access to real-time storm avoidance equipment or the ability to maintain visual separation, no pilot should attempt to penetrate an area of air mass thunderstorms either VFR or IFR.

northern edge of a cirrus cloud shield indicates the location of the subtropical jet stream, and the clouds are called *jet stream cirrus.*

Convective activity appears very bright on both visible and IR imagery. Often individual cells can be identified with air mass thunderstorms, previously discussed. Organized lines, produced by fronts and squalls, can also be seen. Once the cells or lines have developed, the exact location of the thunderstorms may be masked by overrunning cirrus tops. Overshooting tops are most evident on visible imagery with a low sun angle. As air flows rapidly out of the storm at the surface, a line of clouds—called an *arc cloud*—forms along the leading edge. *Arc lines* (clouds) are a good indicator of severe or extreme turbulence. Outflow boundaries are sometime depicted on the surface analysis chart. Often, new thunderstorms develop along the outflow boundary. At a point where two or more arc lines converge, convection is strongest, producing severe weather.

Another term used with satellite imagery to identify severe convective activity is *enhanced V.* On enhanced IR imagery, a severe thunderstorm often exhibits a V-shaped notch in the top of the cloud. The narrow end of the V points upwind. This signature frequency, but not always, is associated with tornadoes, hail, strong winds, and extreme turbulence.

Land and sea breezes and lake effects can be detected on satellite imagery. Like mountain waves, these phenomena appear on IR, but they are typically easier to see on visible imagery. Along a sea breeze front there often exist cumulus clouds, while on the seaward side of the sea breeze clear skies are typically present. Especially in the southern and eastern United States, where abundant moisture is available along with unstable air, thunderstorms can develop. These conditions also occur with large lakes. Refer to Fig. 10-4. The top image was made at 1515Z, or 8:15 a.m. Eastern Daylight Time (EDT). Both land and sea areas are relatively cloud free. As the land heats during the day, convection begins. Cool air over the water rushes in to replace the rising air. The middle image of Fig. 10-4 occurred at 1815Z, or 2:15 p.m. EDT. Note that cumulus clouds have developed over most of Florida, with cumulonimbus occurring along the coast, especially east of Lake Okeechobee. The bottom image in Fig. 10-4 occurred at 2015Z, or 4:15 p.m. EDT. Thunderstorms have developed over the southern three-quarters of the Florida land mass. This is an excellent example of weather development as seen on satellite imagery.

Widespread areas of haze, smoke, dust, sand, and volcanic ash can be seen on visible imagery, and at times on IR imagery. Haze on visible imagery is most easily seen over dark ocean surfaces and rarely shows up well on IR imagery. During the summer, haze boundaries may indicate frontal or air mass boundaries. Haze and smoke are most easily seen on visible imagery during the early morning or late afternoon. Smoke appears as a light gray area. In Fig. 10-3 the light gray areas of the Sacramento Valley, northern California, and southern Oregon are the result of smoke caused by the devastating fires of September 1987. (Recall in Chap. 8 the example of smoke producing an indefinite ceiling over Redding, California. This event was produced by these fires.) Dust often shows up better on IR than it does on visible imagery; and like haze and smoke, dust is best seen during early morning or late afternoon. How can we determine if a particular image is showing haze, smoke, or dust? By comparing a satellite image with a current METAR, along with reading the surface analysis and weather depiction charts, the exact phenomenon can be identified.

The NWS uses various enhancements to highlight areas of haze, dust, sand, and volcanic ash for weather advisory and forecast purposes. These enhancements are normally not available to pilots in an operational environment.

IMAGERY INTERPRETATION AND APPLICATION

When using any satellite product, a pilot's first task is to identify the nature of the image and the date and time it was made. The image can be visible, IR, moisture, or enhanced IR. Refer to Fig. 10-5. From the bottom margin, this is a visible image with a resolution of 8 km—about 5 nm. The top margin reveals the date and time: "3 10 1999 1815Z,"—March 10, 1999, at 1815 UTC. This information is very important when comparing visible and IR images or using several images to determine weather movement, development, or dissipation. Especially for visible imagery, the time of day is important. In Fig. 10-4, 1815Z is early afternoon on the east coast and midmorning over the Rocky Mountains. We won't expect to see as much shadowing as early-morning or late-afternoon images or smoke, haze, or dust.

The grid outlines state and international boundaries and the larger lakes. Even without the grid, the ocean and large lakes appear in distinct contrast to land areas. (Before GOES 8 became operational in 1995, the grid was often hundreds of miles off—at least for FSS users of Kourvoras products.) Even at this relatively poor resolution, mountains and valleys can be seen in the southwestern states.

The upper midwest, especially Wisconsin, Illinois, and Indiana, are snow covered. Note the area south of Green Bay and Lake Winnebago in Wisconsin and the Illinois River. Even at this resolution these land features, along with the Great Lakes, help confirm that this is a region of snow, rather than clouds. Additionally, the surface analysis and weather depic-

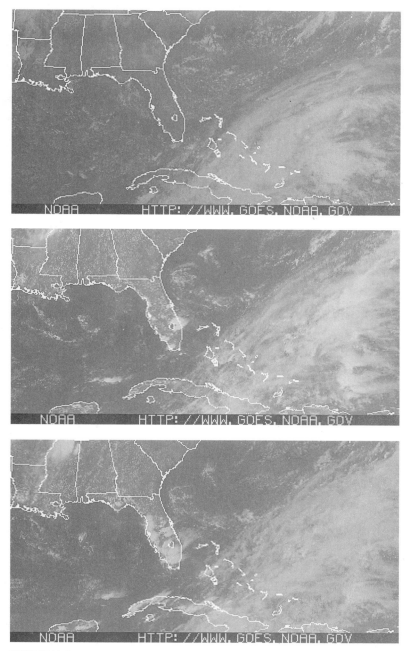

FIGURE 10-4 A sea breeze is often depicted on visible imagery as scattered cumulus over land areas, with relatively clear skies over the water.

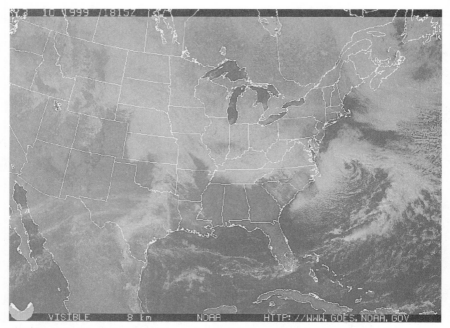

FIGURE 10-5   When using any satellite product, a pilot's first task is to identify the type of image it is and the date and time it was taken.

tion charts can be used to confirm that this is a cloud-free area and that the image, in fact, depicts snow.

Figure 10-5 shows relatively bright clouds throughout the western portion of the Mississippi Valley, central Texas, and the Ohio River Valley. What are the cloud bases? Cloud bases for this system cannot be determined from the satellite image alone. For that information we would have to consult the METARs and the surface analysis and weather depiction charts. A disorganized storm system is over the Atlantic. It still retains its more-or-less comma cloud appearance, with some texturing. To determine precipitation and precipitation intensity, we would have to consult radar observations or the radar chart. Over northern Mexico, western Texas, and the Gulf of Mexico are darker cloud formations. From this information alone, the only deduction that can be made is that the bright clouds are relatively thick and the darker clouds, relatively thin.

To continue the analysis, refer to Fig. 10-6, and note that this is an IR image of the same time frame as the visible image in Fig. 10-5. In the vicinity of the Great Lakes, the surface and water temperatures are the same, as indicated by the same shade of gray. From the IR image alone, it's difficult to distinguish the snow cover of the upper midwest from low clouds. However, by using the two images together, we can conclude it is indeed snow cover. In Florida and Mexico, land temperatures are much warmer than the oceans, as revealed by much darker—warmer—land masses.

The relatively bright area over the Dakotas indicates high, cold tops. For Nebraska, Kansas, and the Ohio River Valley, darker gray represents lower, warmer tops. Notice the high, cold clouds over Mexico and Texas. These clouds have an anticyclonic curvature. Since they appear dark on the visible image, we can conclude they are high, thin, cirrus clouds and are associated with the jet stream. The jet stream runs just north of the band from

FIGURE 10-6   Recall that on IR images, high clouds are cold and bright, low clouds, warm and gray.

central Baja California through southern New Mexico and the Texas panhandle. The clouds over the Gulf of Mexico are dark, indicating they have low, warm tops. This, along with their cellular appearance in the visible image, indicates stratocumulus.

Refer to the storm system off the Atlantic. The IR image shows high, cold tops along the cold front boundary, in the regions of texturing on the visible image. This indicates considerable vertical development, possibly thunderstorms. High, cold, thick clouds also exist in the area of low pressure at the comma's head. Low, thick, warm clouds are present to the southwest of the low center. From the visible image they appear to be closed-cell stratocumulus, becoming open celled to the southeast. Between the stratocumulus and the cold front is an area known as the dry slot. A *dry slot* is an area of sinking air beneath the jet stream caused by the intrusion of dry, relatively cloud free air.

From these satellite images we can conclude there is an upper ridge of high pressure over the southwest United States, with an upper trough of low pressure along the southeast Atlantic coast. Conditions are relatively clear in the southern tier of the states in the area of ridge-to-trough flow. However, the area of trough-to-ridge flow supports the weather system off the Atlantic coast. The jet stream runs from the Texas panhandle, then weakens over an area of relatively high pressure at the surface in the southeast United States, then curves northeast through the dry slot in the comma cloud over the Atlantic, and exits northeast over the Atlantic Ocean. (We hope the previous discussions on surface and upper-level weather and the jet stream are beginning to come together.)

Figures 10-7 and 10-8 are visible and IR images from GOES west. Major mountain ranges and valleys are clearly distinguishable, even at this resolution. Note California's central valley, the deserts of southern California, and the dendritic patterns over the Sierra Nevada mountains. Take a look at the white spot in southern New Mexico, just north of El Paso. You guessed it! It's White Sands.

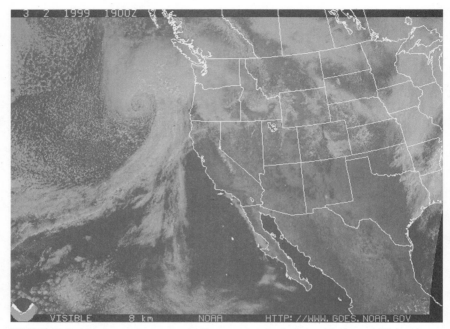

FIGURE 10-7   Major mountain ranges and valleys are often clearly distinguishable on visible imagery.

The main feature on these images is the well-developed comma cloud off the Pacific Northwest. The low-pressure area at the center of the spiral is clearly visible. From the visible image, thick clouds are associated with the comma cloud. Distinct texturing is occurring in the region of the occluded and warm fronts. The exact location of these fronts is covered by the cirrus shield. However, the cold front position is distinctly visible, along with its surface frontal boundary. Ahead of the front are typical cloud formations associated with this phenomenon. Behind the front is an area of open-cell stratocumulus. How do we know these are stratocumulus? The IR image shows dark, warm, low tops. There are considerable, relatively thick clouds along the front in the Pacific Ocean. However, south of the latitude of San Francisco the front is weak, as revealed by the low tops in this region.

Note the clouds along the southern California and northern Baja coasts. By comparing the visible and IR images, we can determine they are coastal stratus. From the visible image, the clouds are relatively thick; however, the IR image reveals their tops have temperatures the same as sea surface temperatures. Therefore, their tops are very low.

Figure 10-9 contains midday satellite images, with the visible image on the left and a same-time-frame, IR image on the right. On this day lake effect snow was falling over the Great Lakes, considerable cloud cover existed over the northeast, and a line of thunderstorms had developed over the Atlantic, east of the Florida, Georgia, and South Carolina coasts.

The visible image shows thick cloud cover over the Great Lakes. Caused by a northeasterly wind, there is a clear area along the northern coast of Lake Superior, and to a lesser degree the west coast of Lake Michigan. The IR image reveals a distinct temperature—gray shade—difference between the surface water temperature of Lake Superior and the cloud deck. This indicates some depth to the cloud layer. We would expect the greatest snowfall to the southeast of the lakes. This could be confirmed by surface observations and the radar chart. In fact, this is a weak event as shown by the radar summary chart in Fig. 10-10. The

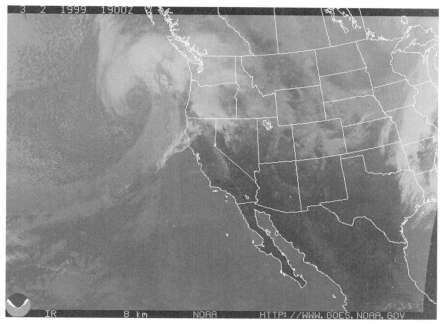

FIGURE 10-8   Cloud tops often reveal the strength of weather, which can be determined from IR imagery.

radar chart shows an area of snow falling along the southern shore of Lake Michigan.

What is occurring over Illinois, Indiana, and central portions of Kentucky and Tennessee? The visual image indicates relatively bright reflectivity. The IR image reveals temperatures colder than surrounding clear areas to the west. The radar chart is not depicting any precipitation echoes in these areas. From this information alone, it appears this is an area of snow cover.

Cloud cover begins over Ohio, and eastern Kentucky and Tennessee. This is revealed by the distinctly colder temperature on the IR image. The radar chart also supports this conclusion, showing light to moderate rain in these areas. The activity is relatively benign, except in the eastern third of North Carolina and along the central Atlantic coast where *thunderstorm activity* (TRW) is reported.

Refer to the activity along the southeast Atlantic coast. There is a narrow, organized line with textured tops indicating thunderstorms with overshooting tops, probably into the lower stratosphere. The cirrus anvils are clearly visible on the visible image, indicating upper-level winds are from the southwest. The IR image shows the highest, coldest tops along the line depicted in the visible image. The exact location, intensity, and movement of this activity can be determined from the radar chart. Refer to Fig. 10-10, the radar chart. This is in fact a *solid* (SLD) *line* of thunderstorms, with precipitation tops to 41,000 ft. The overshooting tops could be several thousand feet above this level. No aircraft, regardless of equipment, should attempt to penetrate this line. Even south of the solid line, penetration will be difficult, if possible at all. The radar chart confirms the thunderstorms are moving from the southeast at between 28 and 37 kn.

We have touched on the concept of the "complete picture." Notice throughout this discussion we continually refer to other aviation weather products for additional information and verification. Satellite imagery must be used in conjunction with the surface observa-

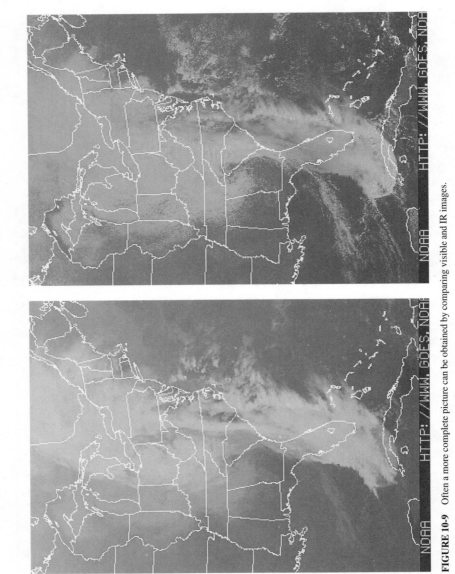

FIGURE 10-9 Often a more complete picture can be obtained by comparing visible and IR images.

10-14

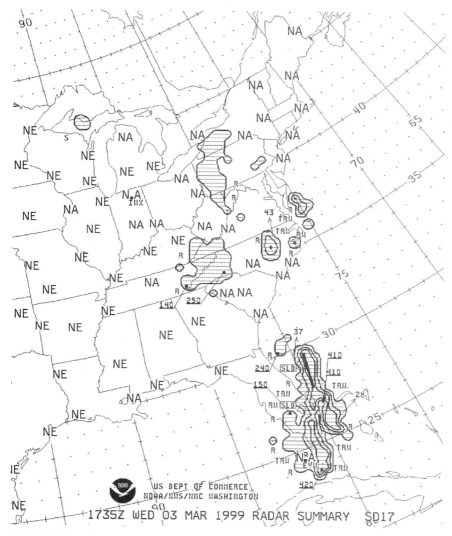

FIGURE 10-10 The radar chart can usually confirm the intensity and movement of weather depicted on satellite imagery.

tions, and the surface analysis, weather depiction, and radar summary charts. Satellite imagery can often reveal the extent of weather phenomena, such as stratus and fog layers, location and intensity of some types of fronts and convective activity, and regions of snow or clear skies. The combined application of satellite and observational products can confirm or refute both manual and automated weather observations.

CHAPTER 11

RADAR AND CONVECTIVE ANALYSIS CHARTS

Radar development was one of many projects initiated in the early 1950s to gain understanding of thunderstorm behavior. Early airborne radars had many technical problems, but by the beginning of the 1960s most airliners were equipped with airborne weather radar. Ground-based radar was also developed.

Access to radar information has increased over the years. Until the mid-1990s, NWS radars in the east and FAA radars west of the Rockies, supplemented by a few NWS sites, were used to compile a national radar summary chart. An NWS network in the west was originally thought unjustified because severe weather is relatively rare in that area. *Radar weather reports* (RAREPs) and convective analysis charts are routinely transmitted on NWS and FAA circuits, and they are available also through many private services. Many twin-, and, increasingly, single-engine aircraft are being equipped with airborne weather radar or lightning detection systems, and there have been several attempts to place ground-based radar displays in aircraft.

The last of the old WSR-57 weather radars have been decommissioned and have been replaced by the *next-generation* (NEXRAD) *weather radar* (WSR-88D). NEXRAD is a Doppler radar system that, with a few minor exceptions, provides coverage from coast to coast in the contiguous United States, Hawaii, Alaska, and the Caribbean.

Like weather observation and forecasts, each system, product, or service has its own particular application and limitations that must be thoroughly understood for safe and efficient flight.

RADAR

Radar is an electronic device that is used to detect and locate moving or fixed objects. Radar works by emitting radio waves and then recording data from those radio waves that are reflected back by solid objects. Radar displays an image created by the reflected energy, or *backscatter*. The intensity of the image depends on several factors; among them are particle (or droplet) size, shape, composition, and quantity. NEXRAD radars are capable of displaying both precipitation as well as cloud-size particles. However, *radar weather reports* (SDs) and radar summary charts display precipitation-size particles only. Therefore, a precipitation-free area on these products does not translate into a cloud-free sky.

Pilots have access to NWS radar and airborne weather radar, and to a limited extent, ATC radar. The FAA is currently providing air traffic controllers with either separate or overlay NEXRAD products. Each system has a specific purpose and its own applications and limitations.

ATC radar is specifically designed to detect aircraft; a narrow fan-shaped beam reaches from near the Earth's surface to high altitudes. ATC radars have a wavelength of 23 cm, and thus they are ideal for detecting aircraft, but they are not as proficient at seeing weather. For example, they reduce the intensity of detected precipitation.

To efficiently detect aircraft and eliminate distracting targets, ATC radars use *circular polarization* (CP), *moving-target indicators* (MTIs), and *sensitivity time control* (STC).

CP results in a low sensitivity to light and moderate precipitation. MTIs displays only moving targets; thus unless droplets have a rapid horizontal movement, they remain undetected, and even rapidly moving precipitation will not be detected when advancing perpendicular (tangentially) to the radar beam. STC further eliminates light and decreases the intensities of displayed precipitation. Naturally, controllers, especially at approach facilities, engage these features during poor weather to accomplish their primary task—aircraft separation.

NWS radars, on the other hand, with a narrow linearly polarized beam, are ideal for detecting precipitation-size particles. Sensitivity time control on NWS radars compensates for *range attenuation,* which is the loss of power density due to the distance between the radar and the echoes. STC-displayed intensity remains independent of range; therefore, targets with the same intensity but at different ranges appear the same to the radar specialist. NWS radars can detect targets up to 250 nautical miles (nm) away; however, due to *range and beam resolution*—which is the ability of the radar to distinguish individual targets at different ranges and azimuth—an effective range of 125 nm is used.

NEXRAD has been a quantum leap in providing early warning of severe weather. NEXRAD will be the standard for the next 20 to 25 years with a wavelength of 10 cm. The WSR-88D network will consist of up to 195 units, 113 NWS sights in the contiguous United States, with additional sites in Alaska, Hawaii, the Caribbean, and western Europe, and 22 Department of Defense (DOD) units. The NEXRAD network fills most of the radar gaps in the western United States.

Since NEXRAD is a Doppler radar, it detects the relative velocity of precipitation within a storm. It has increased the accuracy of severe thunderstorm and tornado warnings, and it has the capability of detecting wind shear. Figure 11-1 shows a NEXRAD weather radar transmitter site. The radar specialist using a WSR-88D will have *radar data acquisition, radar product generation,* and *display units,* as shown in Fig. 11-2. NEXRAD provides hazardous and routine weather radar data typically above 6000 ft east of the Rockies and above 10,000 ft in the western United States. It can also be made available in the aircraft through mode S transponders with automatic data link capability.

Airborne weather radars are low power, generally with a wavelength of 3 cm. Precipitation attenuation, which is directly related to wavelength and power, can be a significant factor. Precipitation attenuation results from radar energy being absorbed and scattered by close targets, which renders the display unreliable in close proximity to heavy rain or hail. Intensity might be greater than displayed, with distant targets obscured. An accumulation of ice on the aircraft's radome causes additional distortion.

Figure 11-3 illustrates the effects of precipitation attenuation, showing how a heavy precipitation pattern with a very strong gradient might appear on an NWS 10-cm radar, compared with the same weather system as seen on 3-cm aircraft units. A pilot seeing the pattern on a 3-cm set might elect to penetrate the weather at what appears to be the weakest point only to find the most severe part of the storm, or find additional severe weather where the radar indicated clear.

CASE STUDIES   According to the National Transportation Safety Board (NTSB), precipitation attenuation was a contributing factor in the crashes of a Southern Airways DC-9 in 1977 and an Air Wisconsin Metroliner in 1980. Precipitation attenuation is not significant with NWS 10-cm high-power units; however, it can be a serious problem with units of 5 cm or less, especially in heavy rain. The NTSB recommends: "In the terminal area, comparison of ground returns to weather echoes is a useful technique to identify when attenuation is occurring. Tilt

FIGURE 11-1   National Weather Service and military NEXRAD radars cover most of the continental United States.

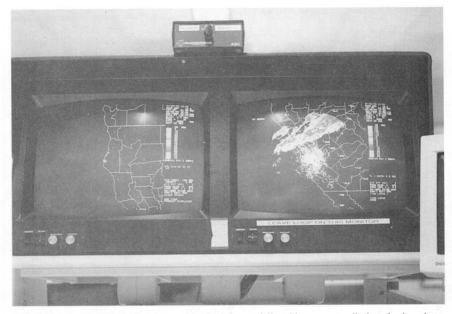

FIGURE 11-2   The NEXRAD radar provides the radar specialist with a computer display of radar echoes, including the relative motion of a storm system.

the antenna down and observe ground returns around the radar echo. With very heavy intervening rain, ground returns behind the echo will not be present. This area lacking ground returns is referred to as a shadow and may indicate a larger area of precipitation than is shown on the indicator. Areas of shadowing should be avoided."

In August 1985 a Delta Air Lines L-1011 crashed at the Dallas/Fort Worth Airport. The NTSB was unable to determine if the crew had been using airborne weather radar at the time of the crash. The NTSB report did state, however, the following: "The evidence concerning the use of the airborne weather radar at close range was contradictory. Testimony was offered that the airborne weather radar was not useful at low altitudes and in close proximity to a weather cell…," although "at least three airplanes scanned the storm at very close range near the time of the accident." The accident was probably caused by a microburst from a single severe storm cell, which illustrates how weather can develop rapidly, often without any severe weather warning.

When using an airborne weather radar, it is imperative to understand the particular unit, its operational characteristics, and limitations. "Just reading through the brochure that comes with the equipment is certainly not enough to prepare a pilot to translate the complex symbology presented on the [airborne] scope into reliable data. A training course with appropriate instructors and simulators is strongly recommended," according to the March–April 1987 *FAA Aviation News.*

Flight service stations (FSSs) and center weather service units (CWSUs) have access to real-time weather radar. Controllers and specialists are specially trained to interpret information for preflight briefings and inflight weather updates. FSS and Flight Watch specialists translate radar echo coverage into the following categories:

Widely scattered   Less than 1/10
Scattered   1/10 to 5/10

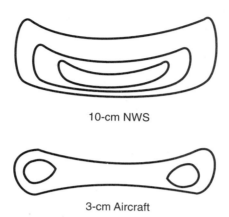

10-cm NWS

3-cm Aircraft

FIGURE 11-3 Precipitation attenuation, caused by close targets absorbing and scattering the radar's energy, can be a serious problem with low-power, short-wavelength sets.

| Broken | 6/10 to 9/10 |
| Solid | More than 9/10 |

Limitations exist. West of the Rockies, sites suffer from extreme *ground clutter*—interference of the radar beam due to objects on the ground. However, the implementation of the NEXRAD network has helped to eliminate many limitations.

Private vendors also have access to radar data. Information from radars is coded and transmitted. When using radar information, it's important to know how individual vendor units display information, such as intensity, and if it is indeed a real-time observation, or a freeze or memory display. The display may be in the precipitation or clear air mode. The *clear air mode* is used for light snow or clear weather events. In the clear air mode such targets as clouds, dust, bugs, birds, or other small objects may be displayed. Therefore, the radar display will show large aerial coverage that will not be precipitation.

The violent nature of thunderstorms causes gust fronts, strong updrafts and downdrafts, and wind shear in clear air adjacent to the storm out to 20 mi with severe storms and squall lines. Precipitation, which is detected by radar, generally occurs in the downdrafts, while updrafts remain relatively precipitation free. Clear air or lack of radar echoes does not guarantee a smooth flight in the vicinity of thunderstorms.

Lightning Detection Equipment

Lightning detection equipment, trade named Stormscope, was invented in the mid-1970s by Paul A. Ryan as a low-cost alternative to radar. Stormscope and similar lightning detectors sense and display electrical discharges in approximate range and azimuth to the aircraft.

TABLE 11-1   Radar Precipitation Intensity Levels

Precipitation intensity	Symbol	Rainfall rate (stratiform)	Rainfall rate (cumuliform)	Associated weather
1. Light	–	<0.1	<0.2	LGT-MOD TURBC PSBL LTNG
2. Moderate	No symbol	0.1–0.5	0.2–1.1	LGT-MOD TURBC PSBL LTNG
3. Heavy	+	0.5–1.0	1.1–2.2	SEV TURBC & LTNG
4. Very heavy	++	1.0–2.0	2.2–4.5	SEV TURBC & LTNG
5. Intense	×	2.0–5.0	4.5–7.1	SEV TURBC LTNG WIND GUST HAIL
6. Extreme	××	>5.0	>7.1	SEV TURBC LTNG EXTENSIVE WIND GUST HAIL

Stormscope also has limitations. One misconception proclaims that in the absence of dots or lighted bands there are no thunderstorms. However, NASA's tests of the Stormscope differed. Precipitation intensity levels of 3 and occasionally 4 (heavy to very heavy; see Table 11-1) would be indicated on radar without activating the lightning detection system. A clear display indicates only the absence of electrical discharges. This does not necessarily mean convective activity and associated thunderstorm hazards are not present. Even tornadic storms have been found that produced very little lightning. The lack of electrical activity, as with the absence of a precipitation display on radar, does not necessarily translate into a smooth ride.

Many authorities agree that a combination of radar and Stormscope is the best thunderstorm detection system. It cannot be overemphasized that these are avoidance, not penetration, devices. Thunderstorms imply severe or greater turbulence, and neither radar nor Stormscope, at the present, directly detect turbulence.

Automated Radar Weather Reports

The National Weather Service routinely takes radar observations at 35 minutes past the hour. These observations are coded and transmitted over the FAA's weather distribution system and available on DUAT and the Internet from the AWC's Web site. Locations are contained in App. B, Graphs and Charts. Aerial coverage is graphically depicted in the *Aeronautical Information Manual* (AIM) and AC 00-45, *Aviation Weather Services*. The radar report, or SD *storm detection* (SD) *report* is now automatically generated by NEXRAD radars, which explains some apparent inconsistencies. For example, in the west ground clutter is sometimes reported as weather echoes. SDs contains the following information:

1. The location of the radar
2. The time of observation
3. The configuration of echoes
4. The coverage of echoes
5. The type of precipitation
6. The intensity of precipitation
7. The location of echoes
8. The movement of echoes
9. Height of echoes

The following illustrates a coded SD report. All reports begin with the radar site. This is a report for Davenport, Iowa (DVN). Note that location identifiers do not necessarily correspond to LOCIDs used with airports and navigational aids (NAVAIDs). This is the 1535Z observation:

    DVN 1535 LN 9TRWXX 95/72 119/119 10W C2237
    AREA 1TRW++4R-288/127 91/114 155W C2239 MT 450 AT 108/99
    AUTO
    ^IJ10112 JI32222222 KI322213332 LI324212344 MH12220001456
    NH4302 NQ266 OH45 OR14 PI1 RN1

Echo Configuration and Coverage.   Echo configuration falls into three categories: cell, area, and line. A single isolated area of precipitation, clearly distinguishable from surrounding echoes, constitutes a *cell.* The following illustrates how cells are indicated on a RAREP:

    FWS 1035 CELL TRWXX 323/95 D30 C2730 MT 550 HOOK 321/91...

This Fort Worth, Texas (FWS), 1035Z special observation reports a cell with a 30-nm diameter (D30) exhibiting a hook echo. *Cell diameter* refers to precipitation, not necessarily the diameter of the cloud, which could be considerably larger. A *hook echo* is the signature of a mesolow, often associated with severe thunderstorms that produce strong gusts, hail, and tornadoes. An *area* consists of a group of echoes of similar type that appear to be associated. A *line* (LN) defines an area of precipitation more or less in a line—straight, curved, or irregular—at least 30 mi long, five times as long as it is wide, with at least 30 percent coverage. Echo coverage is reported in tenths. In the DVN example, the line has 9/10 (LN "9"TRWXX) coverage, and the area 1/10 (AREA "1"TRW++4R−) coverage of thunderstorms with very heavy rain and 4/10 (AREA 1TRW++"4"R−) light rain.

Precipitation Type and Intensity.   SDs, at least at this writing, use the old, pre-1996, aviation weather symbols. These are contained in Table 11-2. In the DVN example the line and the area contain thunderstorms and rain showers (TRW). Following precipitation type, one of the six standard levels describe intensity. Intensity level definitions and descriptions are contained in Table 11-1. The line is extreme (TRW"XX"), and the area very heavy (TRW"++"), with light rain (R"−").

NOTE   Prior to the commissioning of the WSR-88D, radar intensity levels were commonly referred to as "VIP levels." This was because precipitation intensity, and therefore the digits, were derived using a video integrator processor. The WSR-88D does not use a video integrator processor to determine precipitation intensity. Therefore, "VIP level" is no longer a valid term when describing precipitation intensity. Although pilots may still here the term, like "heavy icing," it is a relic of the past. Intensity levels 1 through 6 used on the SD correspond to the precipitation intensity levels contained in Table 11-1.

TABLE 11.2   RAREP/Radar Summary Chart Plotted Data

Symbol	Meaning	Symbol	Meaning
PPINE	No echoes	PPINA	Not available
NE	No echoes	NA	Observation unavailable
OM	Out for maintenance	PPIOM	Out for maintenance
LM	Little movement	RW	Rain showers
T	Thunderstorm	S	Snow
R	Rain	SW	Snow showers

Echo Location and Movement.   The RAREP defines the location of precipitation by points, azimuth (true), and distance (nm) from the reporting station. The line in the DVN report extends from a point 95° at 72 nm (95/72) to a point 119° at 119 nm (119/119), 10 nm wide (10W). The points 288/127 91/114 155W encompass the area. Cell movement indicates short-term motion of cells within the line or area. Cell movement within the line is from 220° at 37 kn (C2237) and area is 220° at 39 kn (C2239).

Echo Height.   Maximum heights are reported in relation to azimuth and distance from the reporting station, with approximate elevation in thousands of feet MSL (MT 450 AT 108/99). Tops within a stable air mass are usually uniform, indicated by the letter *U* (MT U120, uniform tops to 12,000 ft MSL). It's important to remember these are precipitation tops, not cloud tops. Precipitation tops will be close to cloud tops within building thunderstorms. However, precipitation in dissipating cells will normally be several thousand feet below cloud tops. The abbreviation MTS will be used to indicate that satellite data as well as radar information were used to measure precipitation tops.

RAREP digital data appear at the bottom of the report. A grid centers on the reporting station. Each block, 22 nm on a side, is assigned the maximum intensity level observed. When 20 percent of a block contains light intensity, that level is assigned. Therefore, from digital data alone, all that can be concluded from intensity level 1 is that at least 20 percent of that grid contains light precipitation.

Letters represent coordinates; numbers indicate the maximum intensity level for that and succeeding coordinates to the right. These numbers are a prime ingredient in convective SIGMETs. From the Davenport example, we see that strong activity is relatively isolated. It appears that most significant cells can be circumnavigated with storm detection equipment, with only isolated thunderstorm activity within the AREA. We should be able to confirm this with actual real-time radar imagery and the radar summary chart.

Below is another example of a RAREP. This observation came from the Sacramento, California (DAX), NWS site. Sacramento radar was not showing any significant weather prior to the following observation:

DAX 0135 CELL TRW++ 349/80 D9 C2920 MT 250
AREA 1R-10/115 46/30 80W C2920 MT 130 AT 33/79
IN41 KM101 =

Look what popped up, an intensity level 4 (very heavy) cell. Thunderstorms were not forecast. This single cell with a diameter of 9 mi and approximate precipitation tops to 25,000 ft "didn't read the forecast." The cell should have been easily circumnavigable, presenting a hazard only if it had occurred in the vicinity of the departure or destination airport. A pilot should not fly into, close to, or under this cell.

DAX 0235 CELL RW 346/81 D6 C2820 MT 170
AREA 1R-6/95/87/15 80W C2820 MT 120 AT 22/89
IN21 LO1 =

One hour later the cell has deteriorated to moderate rain showers, with maximum precipitation tops down to 17,000 ft MSL. Convective activity can develop and dissipate, often unforecast, at an alarming rate.

Radar Summary Charts

The radar summary chart graphically displays a computer-generated summation of RAREP digital data. The date and time of the observation—time is important because the transmission system might make the report several hours old—appear on the chart. Figure 11-4 illustrates a May 17, 1999, summary, based on 1435Z data. [Note that this is the same day and same time frame as the Davenport, Iowa (DVN), RAREP in the previous section.] Similar to the RAREP, the chart contains information on precipitation type, intensity, configuration, coverage, tops and bases, and movement.

Echo movement and tops are depicted using symbology similar to the RAREP. An arrow with the speed printed at the arrowhead represents echo or cell movement. Echoes within the line in Illinois (Fig. 11-4) are moving from the southwest at 27 kn at the northern edge and from the southwest at 48 kn central and southern portions. Maximum tops are to 46,000 ft. When bases can be determined, the height MSL will appear below the line.

Echo configuration is graphically depicted. Echoes reported as a line are drawn and labeled solid (SLD) when at least 8/10 coverage exists, such as the line in Illinois. From the RAREP data we can determine that this line has 9/10 coverage. The computer plots lines of equal value to indicate echo coverage and intensity. However, unlike the RAREP, the chart displays only intensity 3 levels. The first contour includes intensity levels 1 and 2 (echoes in central Nebraska and eastern Washington state), the second contour depicts levels 3 and 4 (echoes in western Washington state) and the third contour levels 5 and 6 (echoes in Iowa, northeast Missouri, and around the line in Illinois).

AC 00-45 *Aviation Weather Services,* states: "When determining intensity levels from the radar summary chart, it is recommended that the maximum possible intensity be used." Precipitation type uses the same symbology as the RAREP.

When a *tornado* (WT) or *severe thunderstorm* (WS) *watch* is in effect, it is listed on the chart to the right of the legend. On this chart under WEATHER WATCH AREAS, NONE is indicated; there are no watches in effect at the time of observation. An active watch will also be depicted as a dashed-line box on the chart.

Using NEXRAD, RAREPs, and Radar Summary Charts

Pilots can expect to find holes in what the RAREP or radar summary chart portrays as an area of solid echoes. This apparent inconsistency is due to several factors. Targets farther from the antenna might be smaller than depicted due to range and beam resolution. NWS weather radars, at a range of 200 mi, cannot distinguish between individual echoes less than 7 mi apart. A safe flight between severe thunderstorms requires 40 mi, so this provides adequate resolution to detect a safe corridor. Recall that as little as 20 percent coverage of intensity 1 requires the entire grid to be encoded, so holes also occur with isolated and scattered precipitation. On the radar summary chart large areas might be enclosed by relatively isolated echoes.

For example, in Washington state west of the Cascades the chart depicts a large area of light to moderate rain and rain showers, precipitation tops to 19,000 ft, moving to the east northeast at 23 kn. We would suspect this area to only contain scattered precipitation, except along the west slopes of the Cascades where orographic effect has resulted in heavy to very heavy precipitation; however, the chances of very heavy precipitation are low. How far south does this activity extend? Be careful. The chart shows that the Portland, Oregon, radar is *not available* (NA).

Assumptions can never be made with RAREPs or the radar summary chart. Can a pilot fly from Oklahoma City and Kansas City avoiding severe weather by at least 20 mi? Not without weather avoidance equipment, contact with a facility with real-time weather radar, or visual contact with the convective activity. Why? RAREPs and the radar summary chart are observations, not forecasts. They report what occurred in the past, and convective activity, as we've seen, can develop and move at an astonishing rate. Time of observation is an important consideration. RAREPs might be as much as 2 h old, and the radar summary chart from 2 to 4 h old. On the other hand, reports from air traffic controllers with access to NEXRAD observations are usually real time.

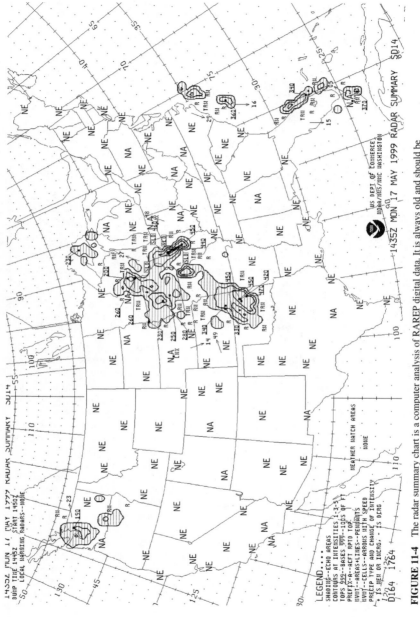

FIGURE 11-4 The radar summary chart is a computer analysis of RAREP digital data. It is always old and should be used for preplanning purposes only. The radar summary should always be updated with current observations.

11-10

Even though the NEXRAD radar network is complete, there are still a few gaps in the west, coverage is only designed for precipitation above 10,000 ft, and ground clutter is still a problem. Although coverage should be adequate for most severe weather, there is no guarantee of complete coverage. Additionally, pilots must be careful to observe the notation NA, which means radar data are not available. This is especially true when using products from commercial vendors. What appears to be a hole in convective activity might be only missing data. When using such services, be sure you know how the vendor displays missing data.

A pilot can never assume there are holes in the solid line in Illinois. The chart can be interpreted only as 8/10 or more coverage within the line. In fact, as we've seen, the Davenport RAREP for about the same time frame showed 9/10 coverage! By reviewing RAREP data, a pilot can quickly assess echo intensity and coverage. For example, review the digital data for the Davenport RAREP. Most of the activity is intensity level 2 or lower. But, again without storm detection equipment, access to real-time radar information, or visual contact with convective activity, penetrating the area should not even be considered.

The radar summary chart provides general areas and movement of precipitation for planning purposes only, and it must be updated by hourly RAREPs or real-time weather radar. Chart notations, such as *out for maintenance* (OM) or NA must be considered. The chart must always be used in conjunction with other charts, reports, and forecasts—the complete picture. Once airborne, inflight observations—visual or electronic—and real-time weather radar information from air traffic controllers must be used. As stated in AC 00-45E, *Aviation Weather Services:* "Once airborne...pilots must depend on contact with Flight Watch, which has the capability to display current radar images, airborne radar, or visual sighting to evade individual storms."

Precipitation should not of itself be cause to cancel a flight. Consider, for example, the returns in Washington and Oregon in Fig. 11-4. Mostly light to moderate activity. Chances are the Cascades and high mountains will be obscured. Again, without further information a flight decision cannot be made with the information available. A flight decision must be made with the following considerations. What is the coverage and intensity of precipitation? What is the weather expected to do (improve or deteriorate)? What is the pilot's experience level, the capability of the aircraft, and the time of day? It can be difficult to see clouds at night; although lightning flashes are visible, distances can be difficult to judge. How familiar are we with the terrain and weather patterns over the intended route? Is an alternate available if the planned flight cannot be completed? Are we mentally prepared to divert, should it become necessary? What about the our physical condition? Tired and anxious to get home is a potentially fatal combination.

CONVECTIVE ANALYSIS CHARTS

Convective Outlooks and Convective Outlook Charts

The *convective outlook* (AC) is prepared by the Storm Prediction Center (SPC). It consists of two forecasts: day 1, the first 24 hours, and day 2, the next 24 hours. Times of issuance for day 1 are 0600Z, 1300Z, 1630Z, 2000Z, and 0100Z. The initial day 2 issuance is at 0830Z, which is later updated at 1730Z. The AC describes the potential for thunderstorm activity and describes areas where thunderstorms might approach severe limits. (A *severe thunderstorm* is defined as wind of 50 kn or hail three-quarters inch or greater at the surface.) Risk categories are defined: slight risk, moderate risk, or high risk for severe thunderstorms. The AC also forecasts areas of general thunderstorms—nonsevere.

Slight (SLGT) *risk* implies well-organized severe thunderstorms are expected, but few in number with limited (circumnavigable) aerial coverage. *Moderate* (MDT) *risk* implies a greater concentration of severe thunderstorms, with larger aerial coverage; thunderstorms may not be circumnavigable. *High* (HIGH) *risk* usually means a major severe weather outbreak, with greater aerial coverage. The notation SEE TEXT is used for those situations where a slight risk was considered, but at the time of the forecast, it was not warranted. The text of the AC will provide information about the potential for a threat to develop if some particular conditions do develop:

```
MKC AC 222001
CONVECTIVE OUTLOOK...REF AFOS NMCGPH94O.
VALID 222000Z - 231200Z
THERE IS A SLGT RISK OF SVR TSTMS TO THE RIGHT OF A LINE FROM 40 E IND 30
NE OWB PAH 35 ESE FYV MKO 45 WSW TUL 25 SE PNC2030 ENE EMP 30 NE MKC BRL
25 NE MMO 15 WNW SBN 30 W FWA 40 E IND. GEN TSTMS ARE FCST TO THE RIGHT
OF A LINE FROM 20 ESE NEL 25 WSW ABE PIT BWG JBR 10 SSE FSM 25 NNW MLC 30
NE OKC PNC 10 NNW EMP 20 W 3OI 10 NNW DBQ 30 SE OSH 20 ENE OSC 30 SSW ART
15 SSE PVD.
GEN TSTMS ARE FCST TO THE RIGHT OF A LINE FROM 30 S MTJ GUP PRC 30 NE DAG
40 SSE BIH 50 ESE NFL 55 S EKO 20 SSE VEL 30 S MTJ.
...SEVERE THUNDERSTORM FORECAST DISCUSSION...

SURFACE LOW CENTER REMAINS OVER SERN KS AT THIS TIME...WITH STATION-
ARY FRONT EXTENDING NEWD INTO NERN MO AND NRN IL AND TRAILING COLD
FRONT MOVING ACROSS WRN OK/NWRN TX. EXTENSIVE SURFACE HEATING HAS
OCCURRED    INTO    NERN    OK/FAR    SERN    KS/SWRN    MO    EARLY    THIS
AFTERNOON...THOUGH AMOUNT OF LOW LEVEL WAVE CLOUDS ON VISIBLE
IMAGERY SUGGESTS STRONG LID/CAP REMAINS OVER THIS REGION. HOWEVER
LARGE SCALE LIFTING ALONG NOSE OF MID/UPPER LEVEL SPEED MAX AND ASSO-
CIATED WITH SHORTWAVE TROUGH LIFTING ACROSS ERN KS...SHOULD ALLOW
STORMS TO INCREASE ALONG FRONT INTO MO OVER THE NEXT FEW HOURS. AXIS
OF MODERATE INSTABILITY WITH SURFACE-BASED CAPES TO1500 J/KG SHOULD
ALSO CONTINUE AHEAD OF FRONT AND...COMBINED WITH INCREASING DEEP
LAYER SHEAR...WILL SUPPORT A THREAT OF ISOLATED SEVERE THUNDER-
STORMS. FORECAST SOUNDINGS INDICATE WIND FIELDS WILL REMAIN MORE
THAN SUFFICIENT FOR SUPERCELLS...HOWEVER STRONG CAP MAY LIMIT ACTIV-
ITY TO A LINE OF STORMS FORCED ALONG COLD FRONT WITH MAIN THREATS OF
ISOLATED LARGE HAIL/DAMAGING WINDS. OTHER STORMS MAY INCREASE
INVOF STATIONARY FRONT OVER NRN MO INTO NRN IL DURING THE EARLY
EVENING...THOUGH 18Z SOUNDING FROM ILX SUGGESTS BEST SEVERE POTEN-
TIAL INTO THIS REGION MAY WAIT UNTIL LATER THIS EVENING WHEN STRONGER
MID/UPPER LEVEL SPEED MAX LIFTS NEWD AND SURFACE LOW MOVES ACROSS.
ACTIVITY SHOULD INCREASE IN COVERAGE INTO THE EVENING ALONG LOW
LEVEL JET AXIS INTO THE MID MS RIVER VALLEY...THOUGH SEVERE POTENTIAL
WILL LIKELY DIMINISH LATER TONIGHT AS INSTABILITY AND LAPSE RATES
WEAKEN.
..EVANS.. 04/22/99
```

The preceding AC was issued on the 22d day of the month and was valid from 222000Z until 231200Z. The time frame for this AC is the same as the example WST and WW discussed in Chap. 14, Weather Advisories. There is a slight risk of severe thunderstorms to the right of a line described by the location identifiers. General thunderstorms are forecast to the right of a line specified by the location identifiers. General thunderstorms are those not expected to reach severe limits. This is followed by a severe thunderstorm forecast discussion. This narrative falls into the same category as the outlook portion of convective SIGMETs, providing synoptic details in meteorological terms directed toward forecasters more than pilots. However, the discussion often provides insight into the overall weather picture.

A surface low center remains over southeastern Kansas at this time, with a stationary front extending northeastward into northeastern Missouri and northern Illinois, with a trailing cold front moving across western Oklahoma and northwestern Texas. Extensive surface heating has occurred into northeastern Oklahoma, far southeastern Kansas, and southwestern Missouri early this afternoon, although the amount of low-level wave clouds on visible imagery suggests a strong lid or cap remains over this region. However, there is large-scale lifting along the nose of a mid- to upper-level speed max, and associated with short wave trough lifting across eastern Kansas should allow storms to increase along the front into Missouri over the next few hours. An axis of moderate instability with surface-based caps to 1500 J/kg should also continue ahead of the front and combined with increasing deep layer shear, will support a threat of isolated severe thunderstorms. Forecast soundings indicate wind fields will remain more than sufficient for supercells. However, strong cap may limit activity to a line of storms forced along the cold front with the main threats of isolated large hale and damaging winds. Other storms may increase in the vicinity of the stationary front over northern Missouri into northern Illinois during the early evening, although the 18Z sounding from ILX suggests best severe potential into this region may wait until later this evening when stronger mid- to upper-level speed max lifts northeastward and the surface low moves across. Activity should increase in coverage into the evening along low-level jet axis into the mid-Mississippi River Valley, although severe potential will likely diminish later tonight as instability and lapse rates weaken.

The convective outlook chart provides a preliminary 2-day thunderstorm probability potential. Issuance times are similar to the textual AC. The left panel gives the day 1 outlook for general and severe thunderstorms. Figure 11-5 is valid for the same time frame as the AC in the previous discussion. From Fig. 11-5, there is a slight risk of severe thunderstorms in the central Mississippi and Ohio River valleys. The right panel, provides an outlook for day 2. It shows the severe activity moving southeastward.

This chart indicates areas where conditions are right for the development of convective activity sometime during the period. The manually prepared chart is basically a pictorial display of the AC; however, updates are limited.

A potential for thunderstorms exists within depicted areas. This does not necessarily mean thunderstorms will develop. The chart, like the AC, is strictly for advanced planning to alert forecasters, pilots, briefers, and the public to the possibility of future storm development. Appropriate FAs, TAFs, WSTs, and severe weather watches must be consulted just prior to and during flight for details on convective activity.

Moisture/Stability Charts

Moisture/stability charts consists of the lifted index analysis, precipitable water, freezing level, and average relative humidity charts. Available twice daily, the charts are computer generated from radiosonde data. Notice in Figs. 11-6 and 11-7 that the analysis is based on the 0000Z observation of Thursday, April 22, 1999. This is the same general time frame as the AC previously discussed. Due to computation and transmission times this chart is about four-and-a-half hours old by the time it becomes available.

The *lifted index analysis* provides an indication of atmospheric moisture and stability—thunderstorm potential at the time of observation. The lifted index compares the temperature that a parcel of air near the surface would be if it were lifted to the 500-mb level and cooled adiabatically, with the observed temperature at 500 mb. The index indicates stability at the 500-mb level. The index can range from $+20$ to -20, but generally it remains between $+10$ and -10. (The figures are strictly an index, not a representation of temperature.) A positive index indicates a stable condition, and high positive values, very stable air. A zero index indicates neutral stability. Values from 0 to -4 indicate areas of potential con-

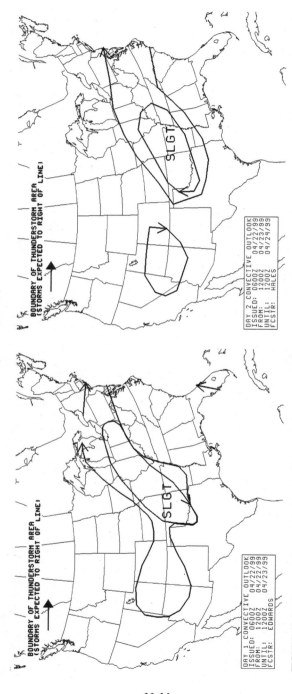

FIGURE 11-5 The convective outlook chart provides a preliminary look at general and severe thunderstorm potential for 48 hours.

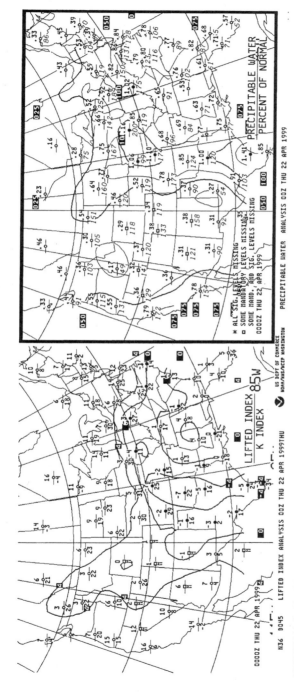

FIGURE 11-6 The lifted index analysis provides an indication of atmospheric moisture and stability and stability—thunderstorm potential. The percepible water chart analyzes water vapor content between the surface and the 500-mb level.

11-15

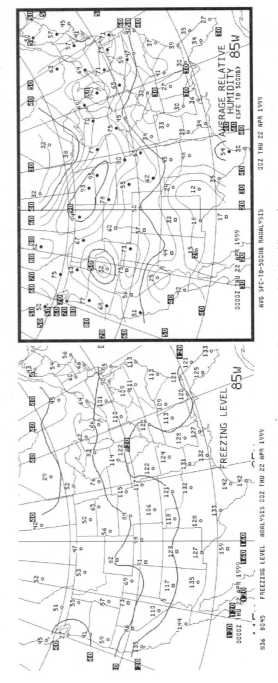

FIGURE 11-7 The freezing level panel plots the lowest observed freezing level. The average relative humidity panel indicates large-scale moisture content in the lower part of the troposphere.

vection. Values of $+2$ to -2 indicate the potential for general thunderstorms that will likely not be severe. Large negative values from -5 to -8 represent very unstable air, which could result in severe thunderstorms—should convection develop.

The *K index* evaluates moisture and temperature. The higher the K index, the greater potential for an unstable lapse rate and the availability of moisture. Values range from -20 to $+40$, with some higher values associated with precipitation. The K index must be used with caution; it is not a true stability index. Large K indices indicate favorable conditions for air mass thunderstorms, during the thunderstorm season. K values can change significantly over short periods due to temperature and moisture advection.

NOTE Thunderstorm prediction involves more than just the lifted index and K index. Consideration must include available moisture, winds aloft, potential lift, and deep convection. These indices indicate that there is a potential for thunderstorms; they are not an absolute prediction that they will occur.

Refer to the lifted index panel, left panel, Fig. 11-6. The lifted index appears above the K index in the plotted data. *Isopleths*—lines equal in number or quantity—of stability are plotted beginning at zero then for every four units (plus and minus). Negative lifted indices and large K values exist over the southern Mississippi Valley and into the Ohio River Valley. The chart would, therefore, indicate a potential for thunderstorms in those areas—no surprise. Conversely, positive lifted indices and small K values were observed over the southwest, indicating little moisture and a stable lapse rate.

The *precipitable water panel,* right panel of Fig. 11-6, analyzes water vapor content from the surface to 500 mb. Darkened station circles indicate large amounts of available water. Isopleths of precipitable water are drawn at one-quarter intervals. The panel is more useful for meteorologists concerned with flash floods. The chart can be used to determine if the air is drying out or increasing in moisture with time by looking at the wind field upstream from a station. Therefore, a pilot can get an excellent indication of changes in moisture content. For example, considerable moisture exists over the southern Mississippi Valley and into the Ohio River Valley. Elsewhere in the United States the air is relatively dry this day.

By applying wind flow from observed and forecast charts, we can obtain a general sense of movement. For example, from the thunderstorm discussion in the AC, we would expect the moisture and instability to be advected to the southeast—which, in fact, the convective outlook chart indicates.

The *freezing level chart,* left panel of Fig. 11-7, plots the lowest observed freezing level. These values are plotted in hundreds of feet mean sea level above the station circle. Lines connecting equal values of freezing level are drawn at 4000-ft intervals. Below-freezing temperatures at the surface are indicated by a dashed line labeled 32F. For this observation, surface temperatures across the United States are above freezing. (Below-freezing surface temperatures are indicated by the abbreviation BF.) Freezing levels range from around 4000 ft in the north to well above 12,000 ft in the south. Multiple entries indicate inversions, and above-freezing temperatures, aloft (none exist on this particular observation). Keep in mind this is observed data and must be used with forecast information contained in the AIRMET Bulletin, SIGMETs, and other forecasts for flight planning.

Two crossings of the freezing level can occur with surface temperature below freezing. For example, should a station report 68 above 39, it would mean that temperatures are below freezing from the surface to 3900 ft MSL, with a layer of above-freezing air from 3900 to 6800 ft.

Figure 11-8 pictorially illustrates three crossings of the freezing level. Note first surface temperature is above freezing. The first crossing occurs at 2200 ft MSL (22 below the station circle), the second crossing occurs at 3200 ft MSL (32), and the third at 7600 ft MSL

MULTIPLE FREEZING LEVELS

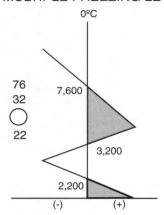

FIGURE 11-8   Plotting RADAT, or freezing
level chart data, shows areas of observed above-
freezing temperatures aloft.

(76). In this example, air temperature that is above freezing occurs from the surface to 1900 ft, and again between 3200 and 7600 ft—the gray shaded area to the right of the 0° isotherm. This could be significant, should an altitude be required to shed ice. *A word of caution.* Don't bet your life that this condition will exist. Hoping an inversion aloft will be available should never be your only escape plan.

Freezing-level data (RADAT) in the past appeared in remarks on 00Z and 12Z surface observations associated with radiosonde (upper-air) observation sites. The RADAT has been removed from these observations in the contiguous United States. At present, RADAT information is available only from an FSS. In the future the DUAT or the Internet may make this data available. Radiosonde release sites can be found in the back of the appropriate *Airport/Facility Directory.*

The data consist of the freezing level in hundreds of feet MSL, and relative humidity at that level. For example,.../RADAT 32078, the first two digits represent relative humidity (32 percent) at the freezing level of 7800 MSL. With multiple freezing levels the highest relative humidity is indicated by the letter *L* lowest, *M* middle, or *H* highest crossing of the zero degree temperature. For example, the RADAT for the station in Fig. 11-8 might appear: .../RADAT98M022032076. In this example the highest relative humidity is 98 percent, which occurred at the middle (M) 3200-ft crossing. High relative humidity indicates moisture and the possibility of icing at and above the freezing level. This is similar to the information available on the freezing level chart, except that relative humidity is not plotted on the chart.

Other RADAT and chart symbols indicate that the sounding was missed (M), due to equipment trouble or other factors, or delayed (NIL). If the entire sounding is below 0°C, RADAR ZERO or BF will be reported.

The average relative humidity panel, right panel of Fig. 11-7, analyzes the average relative humidity from the surface to 500 mb. Darkened station plots indicate high humidity—50 percent or greater. *Isohumes*—lines of equal relative humidity—are drawn at 10 percent intervals. The chart indicates large-scale moisture content in the lower part of the troposphere. Frequently clouds and precipitation are indicated by relative humidity of 50 percent or greater. Clouds and precipitation are indicated in the northern Mississippi Valley and Great Lakes regions with a lifting mechanism available and 70 to 90 percent relative humidity. Few, if any, clouds or little precipitation would be expected in the desert south-west with humidity of 40 percent or less.

Figure 11-7 indicates several likely areas for significant convective activity. The central Mississippi Valley and Ohio River Valley are prime candidates, due to abundant moisture and instability.

These charts are useful in determining the characteristics of a particular weather system in terms of stability, moisture, and possible aviation weather hazards. Even though these charts are hours old by the time they become available, the weather systems will tend to move these characteristics with them. However, caution should be exercised as characteristics may be modified through development, dissipation, or the movement of weather systems.

The lifted index analysis is only one element used to develop the AC, convective out-look charts, and other forecasts. Instability below 500 mb might not be indicated. The chart is several hours old when received, and the lifted index does not consider a lifting mecha-nism, nor can it consider modifications by the development, dissipation, and movement of systems. AC 00-45, *Aviation Weather Services* states: "It is essential to note that an unsta-ble index does not automatically mean thunderstorms."

Certain pilots have attached great significance to the AC and lifted index analysis. They infer some additional insight into thunderstorm activity and severity from these products. The lifted index is an observation and only one element used to prepare the AC and other products. The convective outlook and convective outlook charts are just that, outlooks. They merely provide a statement of potential and must never be used in place of weather advisories, area forecasts, or terminal aerodrome forecasts.

CHAPTER 12
AIR ANALYSIS CHARTS

Weather occurs at all altitudes within the troposphere. The surface analysis chart often cannot solely explain the weather, even weather occurring at or near the surface.

Surface and upper-air analysis charts graphically display a three-dimensional view of the atmosphere based on selected locations at the time of observation. These charts provide a primary source for locating areas of moisture and vertical motion. The following products are normally available to aviation:

- Surface analysis
- Weather depiction
- 850-mb constant pressure chart
- 700-mb constant pressure chart
- 500-mb constant pressure chart
- 300-mb constant pressure chart
- 200-mb constant pressure chart

Each provides details on phenomena occurring at that level. The most complete description of the atmosphere can be obtained only from a combined analysis, which should include the radar summary chart.

SURFACE ANALYSIS CHARTS

The surface analysis chart provides a first look at weather systems. Sea level pressure is the key element. Observed station pressure, converted to sea level, allows analysis from a common reference. The sea level conversion introduces errors, especially in mountainous areas. The data are computer analyzed for lines of equal sea level pressure. The chart also contains wind flow, temperature, and moisture patterns, providing a primary source for the synopsis.

The surface analysis chart has come a long way since its inception. These maps were first drawn by hand by individual weather stations; then they were transmitted by facsimile. Now charts are computer generated and distributed by the Hydrometeorological Prediction Center (HPC), near Washington, D.C.

Often the exact location, and sometimes even the presence, of fronts is a matter of judgment (a front can be a zone several hundred miles across). Additionally, fronts do not necessarily reach the surface; they might be found within layers aloft. This is especially true in the western United States and the Appalachians where mountain ranges break up fronts. Therefore, there might be differences between the charted position of fronts and their location as described in the *area forecast* (FA) or TWEB route forecast synopsis. In such cases it would be advisable to compare chart, FA, and TWEB analysis, and the time of each product.

The surface analysis chart is prepared and transmitted every 3 h and is available at flight service stations and through most commercial vendors with graphics capability. The amount of detail available depends on the vendor's software. Observed data must be plotted and analyzed, so the chart is always old, sometimes several hours, by the time it becomes available. The chart should always be updated with current reports.

An overall perspective of the history of system movements can be obtained by reviewing previous charts. However, care must be used with the apparent movement of low-pressure centers, fronts, and troughs, especially across the western United States. Their movement is subjectively analyzed by the meteorologist.

Surface Analysis Station Models

Most pilots have forgotten how to read station models. But, with commercial weather vendors, pilots might need to brush up on this skill. FSS Flight Watch specialists, trained in their interpretation, use surface analysis chart station models when METARs are not available. Pilots don't need to decode the entire model, just the details significant to aviation.

Figure 12-1 provides an abridged explanation of the station model. Information on the right of the model is primarily for the meteorologists. Above and below the model are cloud types. On the left is information most significant to aviation. This includes temperature and dew point (degrees Fahrenheit—we just can't let go of the past!), and present weather. (Note that the number of precipitation symbols represents intensity—one symbol light, two moderate, etc.) Unknown light precipitation from automated sites is indicated by "?-." Total sky cover and wind direction and speed are contained in the center of the model. The symbols in Fig. 12-1 are fairly standard; they are used on most National Weather Service's charts.

With the introduction of automated observations, the amount of data displayed has significantly decreased. Stations without augmentation or backup are indicated by a square sky cover symbol or with a bracket to the right of the station model. Total sky cover is the summation of all cloud layers.

Notice that in Fig. 12-1 Cu and Cb are listed with low clouds, associated with their bases. The following discussion, based on the cloud types in Fig. 12-1, sometimes refers to a code number (1 through 9) to the left of the abbreviation and symbol.

The code starts with low clouds with vertical development. Cumulus (1) describes fair weather Cu, seemingly flattened, with little vertical development. Cumulus (2) contains considerable vertical development, generally towering. Cumulonimbus (3) exhibit great vertical development with tops composed, at least in part, of ice crystals. Tops no longer contain the well-defined cauliflower shape. Cumulonimbus (9) have clearly fibrous (cirroform) tops, often anvil shaped. (You may have heard the saying "I'm on cloud nine." This was its origin.)

Stratocumulus (4) and (5) and cumulus and stratocumulus (8) represent a moist layer with some convection. Stratocumulus (4) forms from the spreading of Cu. Stratus indicates low-level stability. Fractostratus and fractocumulus (scud) are normally associated with bad weather.

Altostratus (1) is thin, semitransparent, while altostratus (2) is thick enough to hide the sun or moon. When the cloud layer thickens and lowers, it becomes nimbostratus (Ns).

Altocumulus (3) is thin, mostly semitransparent. Altocumulus numbers (5), (6), and (7) describe cloud thickness, development, and altocumulus associated with other cloud forms.

Cirrus (1) consists of filaments, commonly known as *mares' tails*. Cirrus (2) and (3) are often associated with cumulonimbus clouds. Cirrus (4) usually indicates a thickening layer.

STATION MODEL

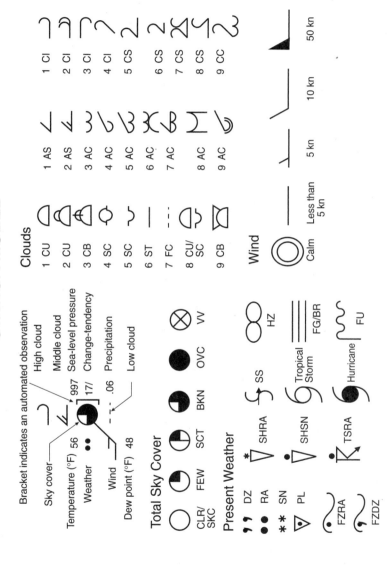

FIGURE 12-1 Surface analysis chart station models contain much valuable information significant to aviation.

Cirrostratus (5 through 8) describe sheets or layers of cirrus. Sun or moon halos often appear in these layers.

Temperature and dew point, in degrees Fahrenheit, appear to the left of the station model. Wind is indicated as the direction (true) from which the wind is blowing, with barbs, half barbs, and flags representing speed in knots.

Information to the right of the station model is less significant to the pilot. Sea-level pressure, in millibars, is decoded the same as for METARs. Figure 12-1 shows 997; therefore, the sea-level pressure is 999.7 mb. The pressure change, during the past 3 h, in tenths of millibars, and its tendency—increasing, decreasing, steady—appear below sea-level pressure. Any precipitation during the past 6 h, to the nearest hundredth of an inch, appears in the lower right.

Let's decode, translate, and interpret the station model in Fig. 12-1. Total sky cover is broken. Cloud bases are not provided, but they can be inferred from cloud type. The low clouds consist of fractostratus, or scud. Thick altostratus and cirrus are also being reported, so the scud can't be too extensive. The higher cloud types might indicate an approaching front. Wind is out of the southwest at 15 kn. Visibility, which does not directly appear, is probably good in spite of a relatively close temperature/dew point spread (8°F). Moisture is being added by rain, however. If the wind remains constant, additional moisture could increase the amount of fractostratus. But if the wind subsides, fog and reduced visibility might result. Except for low-level mechanical turbulence due to wind, the atmosphere appears stable from the cloud types reported. Therefore, mostly smooth flying conditions and light icing in clouds and precipitation above the freezing level are indicated.

Surface Chart Analysis

Figure 12-2 contains front symbols, provides a description of the front, and defines suggested hues used for color displays. Semicircles and triangles are normally omitted with a color display. Open semicircles and triangles indicate the location of a front aloft. *Frontogenesis,* the initial formation of a front or frontal zone, is depicted by a broken line with the appropriate front symbol. *Frontolysis,* the dissipation of a front or frontal zone, is represented by a broken line with the appropriate front symbol on every other line. As previously mentioned, frontal zones can occur in layers aloft. Hollow semicircles and triangles indicate the location of a frontal boundary aloft.

Refer to Fig. 12-3, which is the 1500Z, February 24, 1998, surface analysis chart.

Pressure patterns indicate areas at the surface that are under the influence of high or low pressure, troughs, or ridges. The terms *high* and *low* are relative. A *high* is defined as an area completely surrounded by lower pressure. Conversely, a *low* is an area surrounded by higher pressure. Lines known as *isobars* connect areas of equal sea level pressure. Beginning with 1000 mb, lines are drawn at 4-mb intervals (00, 04, 08, etc.). A weak pressure gradient may be drawn at 2-mb intervals using dashed isobars. Pressures at each pressure center are indicated by a three- or four-digit number. For example, the high center over Iowa has a central pressure of 1019 mb, the low over southern Nevada, 997 mb.

A *trough* consists of an elongated area of low pressure. Figure 12-3 shows a trough extending southwest from the low in Nevada through southern California. A *ridge,* the opposite of a trough, is an elongated area of high pressure, almost always associated with anticyclonic (clockwise in the northern hemisphere) wind flow. Figure 12-3 shows that a ridge extends from the high center over Iowa into south, central Canada.

Wind blows across isobars due to friction between the wind and surface causing convergence and divergence; *convergence*—upward vertical motion—destabilizes the atmosphere, which increases relative humidity, clouds, and precipitation. *Divergence*—downward vertical motion—stabilizes the atmosphere, which decreases relative humidity and clouds.

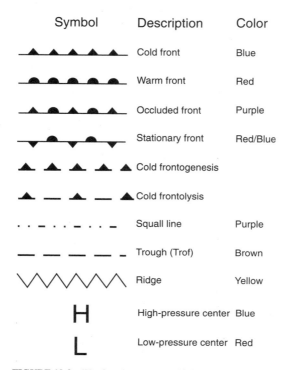

Symbol	Description	Color
	Cold front	Blue
	Warm front	Red
	Occluded front	Purple
	Stationary front	Red/Blue
	Cold frontogenesis	
	Cold frontolysis	
	Squall line	Purple
	Trough (Trof)	Brown
	Ridge	Yellow
H	High-pressure center	Blue
L	Low-pressure center	Red

FIGURE 12-2 Weather charts use standard symbols or colors to describe front type and other weather features.

Convergence in itself does not necessarily produce poor weather, nor does divergence produce good weather. Other factors, such as moisture, must be considered. Surface convergence and divergence affect only the atmosphere below about 10,000 ft, but they are a major factor in the "weather machine"—vertical motion.

Convergence occurs along curved isobars surrounding a low or trough. Maximum convergence takes place at the low center or along the trough line. Figure 12-3 shows surface winds blowing into the low in northern Utah.

Troughs are not fronts, although fronts normally lie in troughs. A front is the boundary between air masses of different temperatures, whereas a trough is simply a line of low pressure. Both phenomena produce upward vertical motion.

Note the anticyclonic, outward flow from both high centers in Fig. 12-3. Maximum divergence takes place at the high center. The symbol for a ridge is a continuous zigzag line, normally not depicted on the surface analysis. A high or ridge implies surface divergence.

There is a misconception that high pressure always means good flying weather. Although good weather often occurs, there are exceptions. Strong pressure gradients at the edge of high cells can cause vigorous winds and severe turbulence. Near the center of a high, or with weak gradients, moisture and pollutants can be trapped at lower levels,

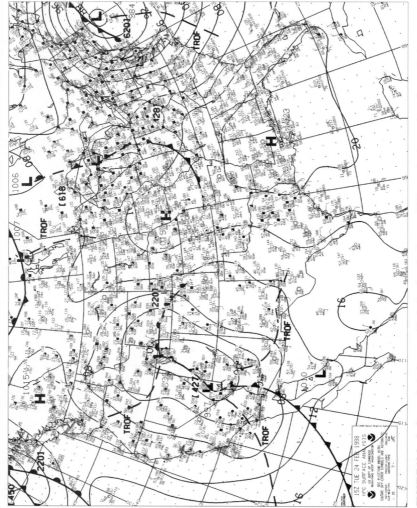

FIGURE 12-3 Surface analysis charts provide the location of moisture and lifting mechanisms at the surface and in the lower atmosphere.

12-6

TABLE 12-1 Front Type, Intensity, and Character

Code	Type (first digit)	Intensity (second digit)	Character (third digit)
0	Quasi-stationary surface	No specification	No specification
1	Quasi-stationary aloft	Weak, decreasing	Frontal area activity, decreasing
2	Warm front surface	Weak, little or no change	Frontal area activity, little change
3	Warm front aloft	Weak, increasing	Frontal area activity, increasing
4	Cold front surface	Moderate, decreasing	Intertropical
5	Cold front aloft	Moderate, little or no change	Forming or existence expected
6	Occlusion	Moderate, increasing	Quasi-stationary
7	Instability line	Strong, decreasing	With waves
8	Intertropical front	Strong, little or no change	Diffuse
9	Convergence line	Strong, increasing	Position doubtful

causing reduced visibilities and even producing zero-zero conditions in fog for days or even weeks.

A three-digit code entered along the front specifies its type, intensity, and character (Table 12-1). For example, the type, intensity, and character of the front over Lake Superior is 618. The type is 6 (occlusion), intensity 1 (no specification), and character 8 (diffuse). Two short lines crossing a front indicate a change in classification.

Frontal intensity is based on frontal speed. That is the temperature gradient in the cold sector, the region of colder air at a frontal zone. A front with waves indicates weak low-pressure centers or portions of the front moving at different speeds. A front with waves needs to be watched. The weak low-pressure areas can intensify and cause significant weather.

The position of the isobars represents pressure patterns or gradients that determine wind flow. Surface wind blows at an angle to the isobars from high to low pressure. Station models show this flow for moderate or strong gradients. However, in mountainous areas, and with weak gradients, the pattern might be confused by terrain or local surface temperature differences.

The isobar pattern represents the relative strength of the wind. Closer spacing of the isobars means stronger pressure gradient forces and, therefore stronger wind. Wider spacing of the isobars means weaker wind. Figure 12-3 shows a weak gradient in New Mexico accompanied by light winds and a stronger gradient over New England with stronger winds.

The surface analysis chart can be used to determine vertical motion at and near the surface. Convergence and divergence have been discussed. Fronts also produce vertical motion. Figure 12-2 and Table 12-1 describe how these phenomena are depicted on the surface analysis chart.

The surface analysis chart may depict a dry line or an outflow boundary. An outflow boundary is a surface boundary left by the horizontal spreading of thunderstorm-cooled air. The boundary is often the lifting mechanism needed to generate new thunderstorms. An outflow boundary will be labeled OUTBNDY; a DRYLINE may be indicated as a line with open semicircles—color brown.

Upslope and downslope flow causes vertical motion. This can be determined from the chart when the interpreter is familiar with the terrain. Upslope produces the same characteristics as convergence and downslope divergence.

Forecast synopses often include "upslope" and might contain "Chinook" or "Santa Ana" when these conditions occur. In Fig. 12-3 upslope is occurring over the eastern Rockies of Colorado and Wyoming, associated with circulation around a low, and eastern Texas, associated with circulation around a high. Note that around the low, upslope with convergence is producing precipitation; around the high with divergence aloft, fog has developed. Conversely, over Alabama and Georgia there is a downslope, offshore flow, with associated clear skies.

Onshore and offshore flows of moderate or greater intensity can be determined from the chart. An onshore flow can translate into advection fog, upslope, or convection with the development of thunderstorms depending on conditions, and an offshore flow clear skies. An onshore flow can be seen in Fig. 12-3 along the Texas Gulf Coast.

Temperature and moisture patterns are determined by analyzing station model temperatures and dew points. Considerable moisture at the surface exists along the Texas Gulf Coast with an onshore flow that has caused fog to develop.

WEATHER DEPICTION CHARTS

Figure 12-4 shows the 1600Z February 14, 1998, weather depiction chart. The weather depiction chart is computer generated, analyzed, and transmitted every 3 h, and it provides a record of observed surface data. Frontal positions are manually analyzed. The information is hours old by the time the chart becomes available; data should always be updated with current reports.

Refer to Table 12-2. The chart is analyzed into three categories—IFR (instrument flight rules), MVFR (marginal visual flight rules), and VFR (visual flight rules). These are not the definitions contained in 14 CFR Part 91. However, these are the same categories used in the area forecast outlook and the *low-level significant weather prognosis* (prog) *chart.*

The chart is computer analyzed, so it cannot consider terrain; nor can it represent conditions between reporting locations. Gross errors between depicted categories and actual weather can occur. It may be helpful to compare the weather depiction chart with a current satellite image. The rather significant differences between the coverages are due to the limitations of the weather depiction chart. Conditions could improve or deteriorate. The weather depiction chart is not a substitute for current observations.

Weather Depiction Station Models

Station models on the weather depiction chart plot cloud height in hundreds of feet AGL, beneath the station circle. When total sky cover has few or scattered clouds, the base of the lowest layer appears. Visibilities of 6 mi or less and present weather are entered to the left of the station. Sky cover and present weather symbols are the same as used on the surface analysis.

Refer to the inset in Fig. 12-4. The station model in south central Texas shows the sky obscured (X, in the station circle), cloud base at 100 ft AGL (1, underneath the station circle). Visibility is reduced in fog to one-quarter mile (1/4, left of the fog symbol). The bracket to the left of the station model indicates this is an automated observation. Because the number of stations analyzed exceeds the number plotted, contoured areas might appear without station models—for example, the depicted area of MVFR in southern California.

Weather Depiction Chart Analysis

The weather depiction chart provides a big, simplified, picture of surface conditions. It alerts pilots and briefers to areas of potentially hazardous low ceilings and visibilities. The chart is a good place to begin looking for an IFR alternate.

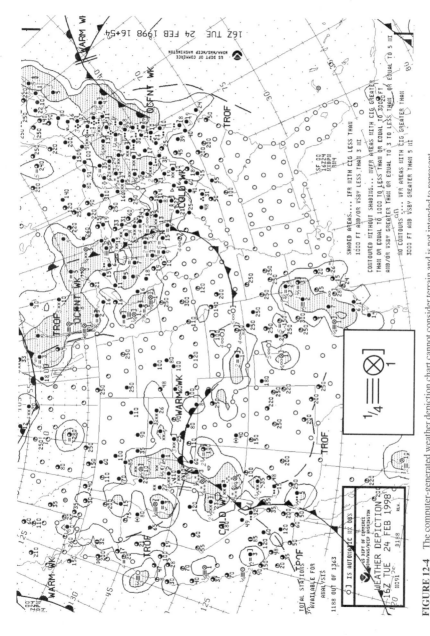

FIGURE 12-4 The computer-generated weather depiction chart cannot consider terrain and is not intended to represent conditions between reporting locations.

TABLE 12-2   Weather Categories

Category	Ceiling, ft		Visibility, mi
IFR	Less than 1000	And/or	Less than 3
MVFR	1000–3000	And/or	3–5
VFR	More than 3000	And	More than 5

Figure 12-4 was generated 1 h after the surface analysis chart in Fig. 12-3. Widespread IFR conditions exist over the Great Lakes and northeast. The chart could be used to determine likely locations for a suitable alternate for an IFR flight into the New York area. With an occluded front moving into New England, an alternate to the south, such as Virginia, would be indicated.

Caution is indicated along the east slopes of the Rockies in Colorado and Wyoming. Even though the weather depiction indicates only isolated areas of IFR, with upslope occurring, conditions could deteriorate rapidly. Appropriate TAFs must be consulted. To find a suitable IFR alternate, a pilot may have to go all the way to central Nebraska, an area not under the influence of upslope conditions.

The weather depiction chart confirms our analysis of the surface chart. IFR and MVFR due to upslope fog is depicted in eastern Texas; clear skies and unrestricted visibilities are reported over the southeast United States.

UPPER-AIR ANALYSIS CHARTS

Weather exists in the two lower layers of the atmosphere—the troposphere and the stratosphere—and the boundary between them, the tropopause.

Pilots fly and weather occurs in three dimensions, so a need exists to describe the atmosphere within this environment. The National Weather Service prepares several constant pressure charts. These computer-prepared charts are transmitted twice daily based on 0000Z and 1200Z upper-air observations. Each level has a particular significance. Table 12-3 describes the general features of each level.

To decode station height, prefix 850 mb with a 1, prefix a 2 or 3 to the 700-mb height, whichever brings it closer to 3000 m, add a 0 to the 500-mb and 300-mb heights, and, for 200-mb, prefix with a 1 and add a 0. For the 300- and 200-mb charts, the temperature/dew point spread is omitted when the air is too cold to measure dew point (less than $-41°C$).

Each chart represents a constant pressure level, so it is analyzed for altitude or height in meters above sea level. Lines, known as *contours,* connect areas of equal pressure surface height. Contours are analyzed in the same way as isobars; the closer the spacing of the contours, the stronger the wind. However, wind blows parallel to the contours, due to the lack of friction; only pressure gradient and Coriolis forces are present.

Constant Pressure Analysis Station Models

Constant pressure analysis charts depict upper-air data. Figure 12-5 illustrates standard station plots. Wind direction and speed use standard symbology, except for the abbreviation LV, which indicates light and variable. A square station symbol indicates the data were obtained from aircraft or satellite data.

TABLE 12-3 Constant Pressure Chart Analysis

Pressure level, mb	Pressure altitude, ft	Temp/dew point spread	Isotachs	Contour interval, m	Height plotted/decode, m	Primary uses
850	5,000	Yes	No	30	585/1,585	Synopsis; advection; convergence; divergence
700	10,000	Yes	No	30	928/2,928	Synopsis; advection
500	18,000	Yes	No	.60	572/5,720	Synopsis; advection; troughs/ridges
300	30,000	Yes	Yes	120	911/9,110	Synopsis; jet stream
200	39,000	Yes	Yes	120	192/11,920	Synopsis; jet stream

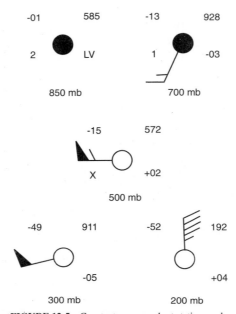

FIGURE 12-5 Constant pressure chart station models depict radiosonde data of height, wind, temperature, and moisture for various pressure levels.

Temperature is plotted in degrees Celsius. However, unlike surface station plots, the temperature/dew point spread, or dew point depression, appears instead of the dew point temperature. For example, in Fig. 12-5, the 850-mb station model has a temperature of -1, and a dew point depression is 2; therefore, the dew point temperature is -3. Darkened station circles are plotted when the temperature/dew point spread is 5° or less, indicating a moist atmosphere at that level. This alerts pilots and forecasters to potential clouds, precipitation, and icing depending on temperature. An X indicates a temperature/dew point spread greater than 29°—for example, in Fig. 12-5, the 500-mb plot. With temperatures less than -41, the air is too dry to measure dew point, and dew point depression is omitted on the 300- and 200-mb plots in Fig. 12-5.

The upper right corner of the plot contains the height of the constant pressure level. Table 12-3 decodes these values. The number in the lower right corner represents height change during the past 12 h, in 10s of meters. For example, in Fig. 12-5, the height of the 700-mb surface has lowered 30 m (-03). In general, lowering heights indicate deteriorating weather and rising heights indicate improving weather. The greater the fall or rise, the more rapid the change. Weather systems tend to move in the direction of greatest height change.

Note that the constant pressure charts in the following sections correspond to the same time frame and the surface analysis and weather depiction charts previously discussed.

850- and 700-mb Constant Pressure Charts

Table 12-3 shows that the 850- and 700-mb charts represent the lower portion of the troposphere, approximately 5000 and 10,000 ft, providing a synopsis for these levels. For surface conditions west of the Rockies, the 850-mb chart might be more representative than the surface analysis. In the west, areas of frictional convergence/divergence can be located. For example, from the 850-mb chart in Fig. 12-6, a downslope condition exists over South Carolina, Georgia, and Alabama; where the air is dry as indicated by the open station models. However, in the extreme southwest and New England, the air at this level is moist, as indicated by the solid station models. A weak upslope can be seen along the eastern slopes of the Rockies.

These charts are particularly useful in monitoring cold and warm air advection. When *isotherms*—dashed lines of equal temperature—cross contours at right angles, the temperature properties of the air mass are advected (moved) in the direction of the winds. At the 850- and 700-mb levels, warm air advection produces upward motion, and cold air advection downward motion. Therefore, warm air advection destabilizes conditions, whereas cold air advection tends to stabilize the weather. Figure 12-6 shows warm air advection on the east side of the low in Nevada. Conversely, cold air advances toward Florida. Note that these patterns fit the analysis of the surface and weather depiction charts. Warm air advection might be all that's needed to trigger thunderstorms.

Areas of moisture can be determined by examining station models for temperature/dew point spread. Icing is implied in areas of visible moisture with temperatures between 0°C and −10°C. The 700-mb chart, Fig. 12-7, shows temperatures within this range, at this level, over the most of the United States. The air is moist in the western states and New England, implying a potential for icing; however, over the southeast there is little moisture present, indicating little icing potential.

These charts can be used to determine the potential for turbulence and mountain wave activity. The 700-mb chart is usually the reference level for mountain waves. Winds in excess of 40 kn imply moderate or greater mechanical turbulence. When winds of these speeds blow perpendicular to a mountain range, accompanied by cold air advection (a stabilizing condition), a strong potential for mountain waves and associated turbulence exists.

Air mass thunderstorms tend to move with the 700-mb winds. In midlatitudes, a 700-mb temperature of 14°C or greater tends to inhibit convection at this level. If convection occurs below, clouds tend to stop rising and spread out at about the 700-mb height.

500-mb Constant Pressure Charts

Probably the most important and useful chart—maybe even more important to meteorologists than the surface analysis—the 500-mb chart describes the atmosphere in the middle troposphere, which is an altitude of approximately 18,000 ft. This chart provides important pressure, wind flow, temperature, and moisture patterns, and it can be used to determine areas of vertical motion at this level.

Troughs and ridges are easily seen in Fig. 12-8. Unlike the surface analysis chart where upward vertical motion takes place along a trough line, upward vertical motion at the 500-mb level takes place between the trough and ridge line. In Fig. 12-8 upward vertical motion is occurring from the trough over California to the ridge over the midwest. Troughs transport cold air down from the north and warm air up from the south. Warm air rides northward on the east side of the trough—trough-to-ridge flow—so the air is lifted as it moves northward, producing upward vertical motion. When moisture is present, such as the area

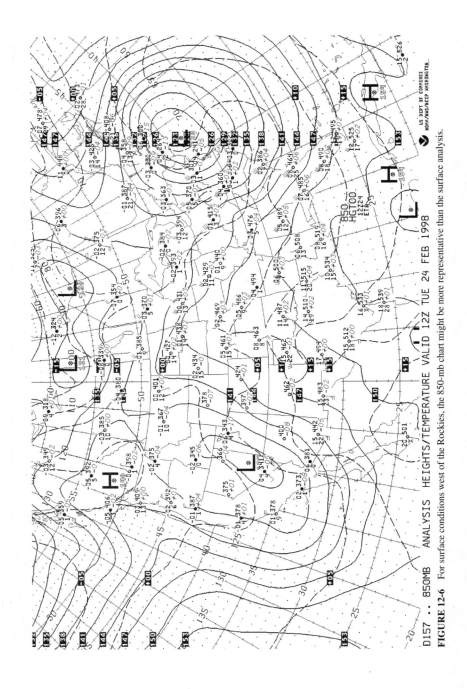

D157 .. 850MB   ANALYSIS   HEIGHTS/TEMPERATURE VALID 12Z TUE 24 FEB 1998

FIGURE 12-6 For surface conditions west of the Rockies, the 850-mb chart might be more representative than the surface analysis.

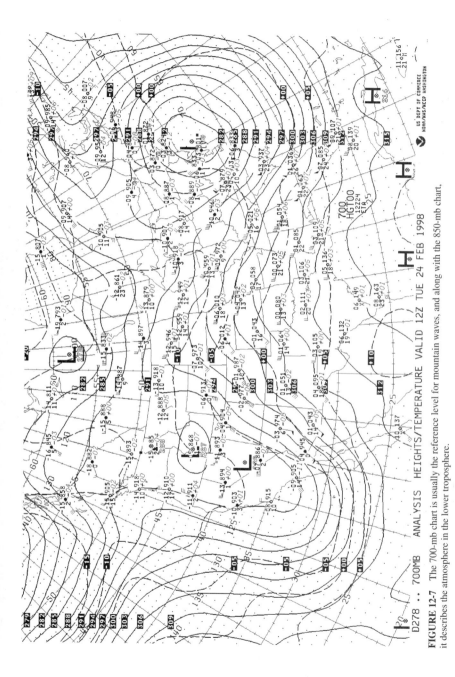

D278 .. 700MB   ANALYSIS   HEIGHTS/TEMPERATURE VALID 12Z TUE 24 FEB 1998

FIGURE 12-7 The 700-mb chart is usually the reference level for mountain waves, and along with the 850-mb chart, it describes the atmosphere in the lower troposphere.

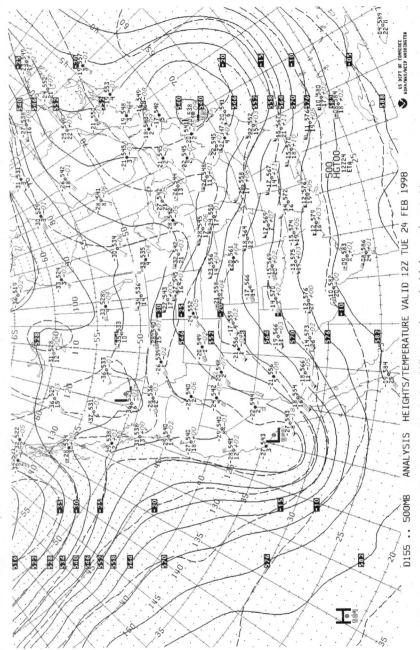

FIGURE 12-8 The 500-mb chart describes the atmosphere in the middle troposphere and might be more important to meteorologists than the surface analysis.

over the Great Basin and northern Rockies, clouds and precipitation in the midtroposphere develop. Conversely, in the ridge-to-trough flow from the midwest to the east coast, cold air sinks southward, producing downward vertical motion with clear, dry conditions in the midtroposphere. Clouds and precipitation frequently accompany upper-level lows and troughs, even without surface frontal or storm systems.

Figure 12-8 illustrates a short wave trough from the low over Idaho to the Texas panhandle. This feature enhances the vertical motion at this level. Although usually best seen on the 700-mb chart, significant waves usually extend to the 500-mb level. Upper troughs are a key to the evolution of weather systems.

When a short wave trough moves through a long wave trough, upward vertical motion is amplified. Short wave troughs can even create upward vertical motion as they move through a ridge. Short waves can be strong vertical motion producers. Information on short waves often appears in FA and TWEB synopses.

Cold air advection destabilizes and warm air advection stabilizes the atmosphere at the 500-mb level, opposite to the effects of cold and warm air advection near the surface. Warm air advection at this level strengthens high-pressure ridges and diminishes low-pressure troughs. In Fig. 12-8, cold air advection is occurring in the trough-to-ridge flow from the west coast to the upper midwest.

Moisture at the 500-mb level can be determined by darkened station models. In Fig. 12-8, darkened station models occur over the Great Basin and northern Rockies and over New England. These indicate areas where the temperature/dew point spread is 5° or less. This chart is a good indicator of high-level icing in summer months and with storms that are either well developed or contain tropical moisture.

Surface weather systems tend to follow the 500-mb flow, and organized thunderstorms tend to move in the direction of the 500-mb winds. The 12-h height change provides a general trend for system movement. Rising heights indicate a building ridge or weakening trough, and lowering heights indicate a deepening trough or weakening ridge.

At the 500-mb level, the shape of the contours rather than wind speed determines the potential for turbulence. At this level, wind shear turbulence occurs as an airplane flies through an area of changing wind direction or speed. Therefore, the greater the curvature, or direction change, the greater the potential for, and intensity of, turbulence. The horizontal distance where this change occurs is critical. The greater the curvature of contours, the greater the probability of turbulence. Therefore, more turbulence potential exists in the trough over the west coast than over the Plains states (Fig. 12-8). The probability of turbulence also exists in the sharp ridge over western Canada. Developing low-pressure troughs moving from the northwest are particularly dangerous.

Areas of potential turbulence occur in merging flows or the neck of a cutoff low. In Fig. 12-8, a merging flow exists over south central Canada. Turbulence could also be expected in the neck of the cutoff lows over Nevada and New England.

FSS controllers have been criticized for not appreciating the importance of, or providing information from, the 500-mb chart and for even discarding the product. Every FSS has access to the 500-mb chart. However, like pilots, some controllers are better schooled than others. Translation and interpretation of the chart over the phone is difficult. The best answer is direct access, which is available through DUAT and other commercial vendors' software with graphics capability.

300- and 200-mb Constant Pressure Charts

The 300- and 200-mb charts provide details of pressure, wind flow, and temperature patterns at the top of the troposphere and occasionally into the lower stratosphere. (A 250-mb chart is also available, representing the atmosphere at approximately 34,000 ft. It is analyzed in the same manner as the 300- and 200-mb charts.) The charts indicate the strength of features in the lower atmosphere. Strong storm systems on the surface are reflected in the 300- and 200-mb patterns, whereas weaker systems lose their identity at these levels. The 500-mb low over Idaho in Fig. 12-8 has lost its identity at the 300- and 200-mb level. (Rather than a closed low—surrounded by a closed contour—it has weakened into a trough.) But the low over New England retains its identity through the 300-mb level (Figs. 12-9 and 12-10). This indicates that the relative strength of the low over New England is greater than the low over Idaho.

At midlatitudes, such as the United States, the jet stream can usually be found on the 300- and 200-mb charts. Wind speeds and curvature of contours provide a clue to *clear air turbulence* (CAT). Because wind speed and direction is of primary importance, areas with observed wind speeds of 70 to 110 kn are indicated by hatching. A clear area within the hatching identifies speeds of 110 to 150 kn. If speeds exceed 150 kn, a second hatched area appears. Areas of potential turbulence occur in the following:

- Sharp troughs (Fig. 12-9, along the west coast and over New England)
- In the neck of cutoff lows (Fig. 12-9, over New York State)
- In a divergent flow (Fig. 12-9, over the northern Rockies)

Turbulence in these areas can exist despite relatively low wind speeds.

Remember that these are observed data charts, not forecasts. The forecast locations of upper-air phenomena, such as turbulence, thunderstorms, and the location and strength of the jet stream must come from the FA, high-altitude significant weather prog, and the tropopause wind and wind shear prog discussed in Chap. 15.

A major factor associated with the jet is wind shear turbulence (CAT). With an average depth of 3000 to 7000 ft, a change in altitude of a few thousand feet will often take the aircraft out of the worst turbulence and strongest winds. Maximum jet stream turbulence tends to occur above the jet core and just below the core on the north side. Additional areas of probable turbulence occur where the polar and subtropical jets merge or diverge, such as off the southeast coast of the United States as illustrated in Figs. 12-9 and 12-10. In this area the polar jet, as seen in Fig. 12-9, merges with the subtropical jet depicted in Fig. 12-10.

CASE STUDIES   In December 1998 a passenger aboard a United Air Lines Boeing 747 died as the result of head injuries suffered when the aircraft flew into an area of severe CAT over the Pacific Ocean. The passenger was not wearing a seat belt. Other less severe injuries have resulted from CAT, again mostly due to the failure to wear seat belts.

In November 1997 I was riding "jump seat" on a Delta Air Lines Boeing 737 from Oakland to Salt Lake City. At flight level (FL) 370, the ride was smooth to occasional light turbulence. Over the radio we could hear a Delta Boeing 757 constantly complaining to ATC about the moderate turbulence at FL390. As chance would have it, I later flew on that airplane, with that crew, from Salt Lake City to Washington's Dulles Airport. It turned out that they had flown their first leg from San Jose to Salt Lake City, almost the same route I had flown at FL370 in 73'. This illustrates how the intensity of jet stream CAT can change significantly over a relatively short distance.

OBSERVED WINDS ALOFT CHARTS

The observed winds aloft chart, which is transmitted twice daily, plots upper-air data. This four-panel chart provides observed winds at four levels:

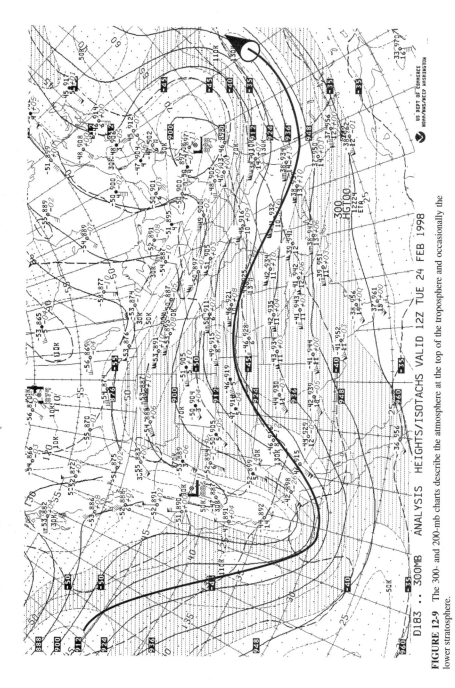

FIGURE 12-9 The 300- and 200-mb charts describe the atmosphere at the top of the troposphere and occasionally the lower stratosphere.

D183 .. 300MB   ANALYSIS   HEIGHTS/ISOTACHS VALID 12Z TUE 24 FEB 1998

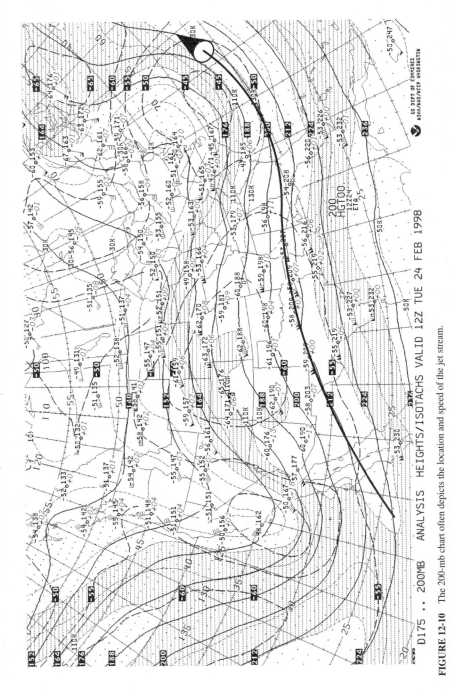

FIGURE 12-10 The 200-mb chart often depicts the location and speed of the jet stream.

D175 . . 200MB   ANALYSIS   HEIGHTS/ISOTACHS VALID 12Z TUE 24 FEB 1998

- Second standard level
- 14,000 ft (600 mb)
- 24,000 ft (400 mb)
- 34,000 ft (250 mb)

The second standard level (lower left panel, Fig. 12-11, pp. 12-22 and 12-23) occurs between 1000 and 2000 ft AGL. The chart provides observed winds above the surface but within the frictional layer. This chart supplements the constant pressure charts by providing observed wind and temperatures between constant pressure levels. Observed winds are, therefore, available for the following heights:

- Second standard level
- 5000 ft*
- 10,000 ft*
- 14,000 ft
- 18,000 ft*
- 24,000 ft
- 30,000 ft*
- 34,000 ft
- 39,000 ft*

(*Obtained from constant-pressure charts.)

Station models are similar to those on the winds and temperatures aloft forecast chart, Chap. 18. Refer to the San Diego plot at the 34,000-ft level in Fig. 12-11. The small number 6 adjacent to the wind pennant indicates the second digit of the wind direction (260°); wind barbs indicate the speed is 45 kn; the temperature is −49°C. Note the station circles for Las Vegas and Salt Lake City; they indicate the wind speed is less than 5 kn or light and variable.

The chart portrays the state of the atmosphere in the past; like constant pressure charts, the charts are about two-and-a-half hours old by the time they becomes available. Observed winds should not be substituted for winds aloft forecasts. However, should gross differences occur between observed and FD forecasts, a pilot may wish to consult an FSS or Flight Watch; both have direct access to NWS forecasters.

UPPER-LEVEL WEATHER SYSTEMS

In Chap 7 and the introduction to this chapter, the effects of upper-level weather systems have been discussed. Figure 12-12 on pp. 12-24 and 12-25 illustrates the effects of an upper-level weather system. The 1300Z weather depiction chart shows no major weather systems, but almost one-quarter of the country is covered by snow producing IFR and MVFR weather. The 1200Z 500-mb chart reveals the culprit: a strong upper-level low over the Ohio Valley. The FA synopsis read:

DEEP UPR LOW OVER SRN IL AT 10Z MOVG SEWD. LARGE TROF AND
CYCLONIC FLOW FROM THE IL LOW SWD THRU GLFMEX....

This weather event closed airports for days, and it was blamed for the deaths of dozens of people.

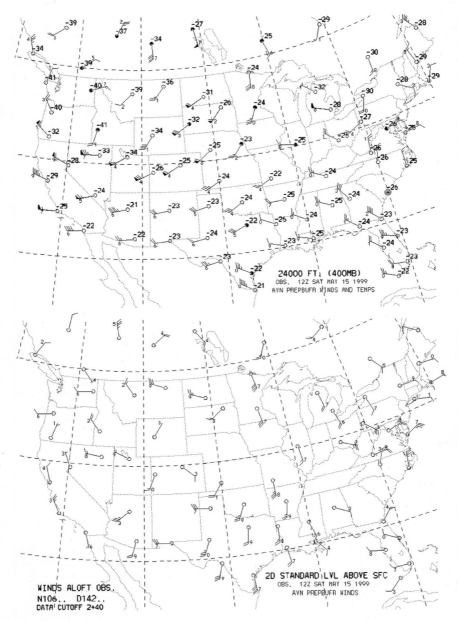

FIGURE 12-11 The observed winds aloft chart describes conditions that existed in the past; it should not be substituted for winds aloft forecasts.

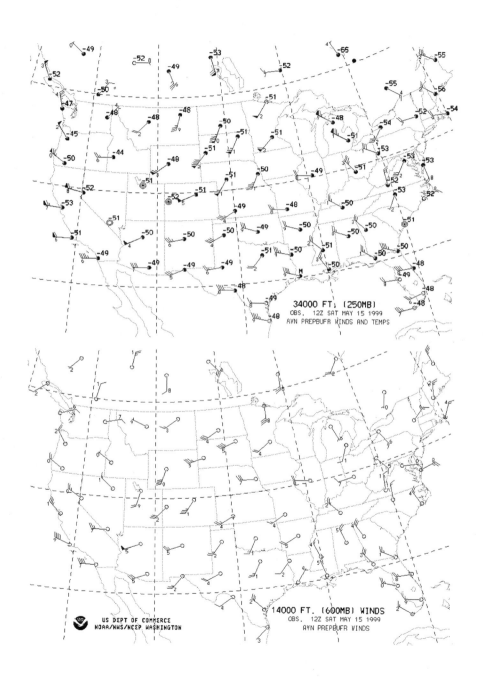

34000 FT. (250MB)
OBS. 12Z SAT MAY 15 1999
AVN PREPBUFR WINDS AND TEMPS

14000 FT. (600MB) WINDS
OBS. 12Z SAT MAY 15 1999
AVN PREPBUFR WINDS

US DEPT OF COMMERCE
NOAA/NWS/NCEP WASHINGTON

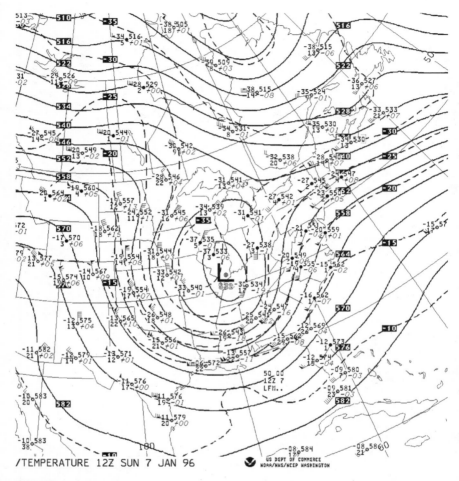

FIGURE 12-12 Upper-level weather systems can cause devastating surface weather, often without a clue about the cause on the surface analysis chart.

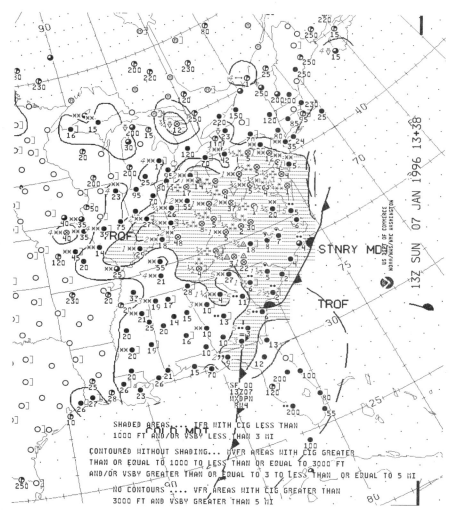

FIGURE 12-11 The observed winds aloft chart describes conditions that existed in the past; it should not be substituted for winds aloft forecasts.

CHAPTER 13
INTRODUCTION TO
FORECASTS

This chapter applies to all forecast products and includes a discussion of accuracy, specificity, and expectations. When it comes to weather forecasts, Sir William Napier Shaw's *Manual of Meteorology* published in the late 1920s nicely sums it up: "Every theory of the course of events in nature is necessarily based on some process of simplification of the phenomena and is to some extent therefore a fairy tale."

Aviation weather forecasts began with the Wright brothers' request for surface winds at Kitty Hawk, North Carolina, in 1903. In the early 1920s aviation forecast centers were established at Washington, D.C., Chicago, and San Francisco. Establishment of lighted airways in the mid-1920s encouraged the Weather Bureau to begin night forecasting. Several airlines had established their own forecasting systems by the mid-1930s. Because meteorologists could be, and sometimes were, held responsible for weather-related accidents, forecasts tended to be pessimistic.

Forecasts were limited by lack of observational data and the complexity of the atmosphere. Meteorologists were plagued with unexpected thunderstorms, the transient nature of icing and turbulence, and the unanticipated development of fog.

Advances were made in the late 1930s and 1940s with upper-air observations and facsimile transmission of weather charts. By the late 1950s weather radar was added to the observational arsenal. Today, satellites and computers help produce forecasts. However, due to the lack of observational data and the complexity of the atmosphere, computer programs can generate only approximations.

Refer to Fig. 13-1. Observational data come from surface, upper-air, and radar reports, satellite imagery, and PIREPs. Data processing—analyzing and forecasting—is accomplished at the National Centers for Environmental Prediction (NCEP) in Washington, D.C., the Storm Prediction Center (SPC) in Norman, Oklahoma, the Aviation Weather Center in Kansas City, Missouri, center weather service units collocated at Air Route Traffic Control Centers (ARTCCs), and local weather forecast offices (WFOs). Forecasts are disseminated through flight service stations, direct user access terminals, and commercial vendors. Figure 13-1, the National Aviation Weather System chart, illustrates the importance of PIREPs and how they fit into the overall picture.

Meteorologists develop a forecast based on the equation:

Existing weather + weather trend = expected weather

A *trend* is the rate of change in the weather. *Advection*—the movement of an atmospheric property from one location to another—and *development* create the weather trend. Fronts and upslope fog are examples of an atmospheric property moving from one location to another, producing weather. Development is the growth of air mass thunderstorms, increase in afternoon thermal turbulence, or the dissipation of radiation fog.

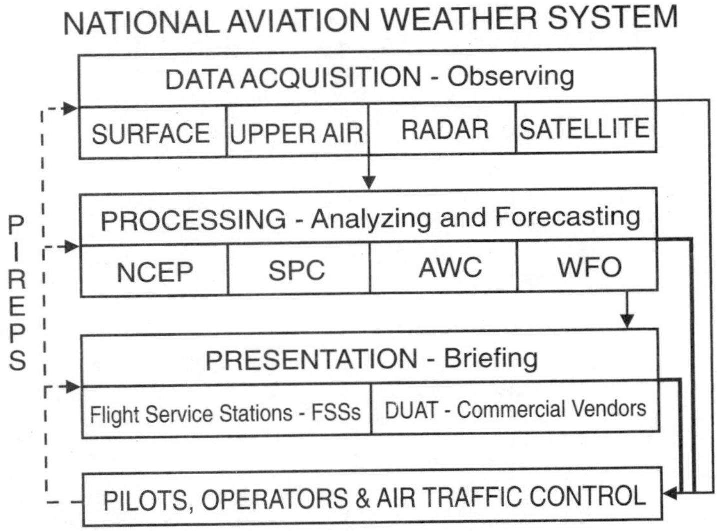

NATIONAL AVIATION WEATHER SYSTEM

```
┌──────────────────────────────────────────────┐
│        DATA ACQUISITION - Observing            │
│  ┌─────────┬──────────┬────────┬───────────┐  │
│  │ SURFACE │ UPPER AIR│ RADAR  │ SATELLITE │  │
│  └─────────┴──────────┴────────┴───────────┘  │
└──────────────────────────────────────────────┘

┌──────────────────────────────────────────────┐
│   PROCESSING - Analyzing and Forecasting       │
│  ┌─────────┬──────────┬────────┬───────────┐  │
│  │  NCEP   │   SPC    │  AWC   │    WFO    │  │
│  └─────────┴──────────┴────────┴───────────┘  │
└──────────────────────────────────────────────┘

┌──────────────────────────────────────────────┐
│        PRESENTATION - Briefing                 │
│  ┌──────────────────────────┬──────────────┐  │
│  │ Flight Service Stations - │ DUAT -       │  │
│  │ FSSs                      │ Commercial   │  │
│  │                           │ Vendors      │  │
│  └──────────────────────────┴──────────────┘  │
└──────────────────────────────────────────────┘

┌──────────────────────────────────────────────┐
│   PILOTS, OPERATORS & AIR TRAFFIC CONTROL      │
└──────────────────────────────────────────────┘
```

P
I
R
E
P
S

FIGURE 13-1   The National Aviation Weather System chart illustrates the importance of PIREPs and how they fit into the overall picture.

ACCURACY

Are aviation forecasts accurate? The FAA admits that needs "cannot be met by an immediate application of existing technology. The need for accurate short-term forecasts exists in every phase of flight operations and is critical to an efficient, smoothly operating air traffic control system." To this end the FAA and National Weather Service established the Enroute Flight Advisory Service, including the implementation of the high-altitude Flight Watch and center weather service units.

Forecast accuracy begins—or maybe begins to deteriorate—with observational data. The limited number of observations hinders forecast accuracy. The observational network dwindled until the introduction of automated observations. However, automated observations have their own limitations, and certain sensors (freezing rain, lightning) have yet to be fully implemented. Upper-air observations, radar, and satellites help, but extensive areas remain outside the observational network. Fully half of the forecast is based on existing weather.

Available data are computer processed and analyzed. Computer equations represent the atmosphere at points approximately 125 mi apart, depending on the computer model, and at various heights. Large-scale (synoptic) weather systems are detected, but smaller-scale (mesoscale) weather systems, such as individual thunderstorms, might not be detected. Additionally, because of computer model limitations, factors such as the interaction of water, ice, and local terrain cannot be adequately taken into account. These are major limitations to the second factor in the weather equation, weather trend.

The National Weather Service monitors forecasts, but other than amendment criteria, no specific agency exists to determine forecast accuracy. Who defines accuracy? C. Donald Ahrens wrote, in *Meteorology Today* (St. Paul, Minn.: West Publishing Co., 1985), "At present, there is no clear-cut answer to the question of determining forecast accuracy." This subject was recently addressed by John E. Jones, Jr., Deputy Director, National Weather Service, in the *AWCommunications* newsletter, Fall 1999. He states that the NWS is "developing an improved national verification program for the TAF and other aviation products." Few pilots or FSS controllers are aware of forecast limitations or amendment criteria. Many have developed their own perceptions that are erroneous, more often than not, due to misconceptions and misunderstandings.

SPECIFICITY

Each forecast is written for a specific purpose in accordance with specific criteria. *Area forecasts* cover entire states, *TWEB route forecasts* cover routes 50 mi wide, and *terminal aerodrome forecasts* (TAFs) relate conditions basically within 5 mi of an airport. Differences are to be expected due to scale, interpretation of the weather situation, issuance times, and starting conditions. For example, localized areas of fog predicted in TWEB route forecasts or TAFs might not appear in the area forecasts. Or, the area forecast might contain a prediction for thunderstorms that might not appear in individual TAFs when the forecaster does not expect the phenomena to occur at that airport. Forecasters might legitimately differ in their interpretations of data. The area forecast might predict frontal passage at one time and the TAF at another time. Forecasts are issued at different times. Therefore, information available to the forecaster on which he or she bases the forecast will differ.

A thorough understanding of format, limitations, and amendment criteria are required to adequately apply a forecast, especially when using a self-briefing media. The FAA and NWS have said: "There probably is no better investment in personal safety, for the pilot as well as the safety of others, than the effort spent to increase knowledge of basic weather principles and to learn to interpret and use the products of the weather service." Then there's the legal requirement. Each pilot in command is required by regulations to become familiar with all available information concerning a flight. This includes: "For a flight under IFR or a flight not in the vicinity of an airport, weather reports and forecasts...." Even student pilots must receive instruction in the "use of aeronautical weather reports and forecasts..." before venturing solo cross-country.

From Table 13-1 the following conclusions are apparent. Forecasts for good weather are more likely to be correct than forecasts for poor weather; this should be no surprise—good weather occurs more often than poor weather. Forecasts are most accurate during the first few hours of the period. Accuracy deteriorates below 80 percent beyond 4 h when less than VFR conditions are forecast. Accurate forecasts of specific values beyond 3 h is not yet possible. Forecast issuance and valid times and amendment criteria are often based on the these limitations. Forecasts are most reliable for distinct weather systems (fast-moving cold fronts, squall lines, or strong high-pressure areas). Synoptic-scale systems are detected within the forecast models.

Phenomena such as the time freezing rain will begin, severe or extreme turbulence, severe icing, the movement of tornadoes, ceilings of 100 ft or zero before they exist, the onset of thunderstorms that have not yet formed, and low-level wind shear are difficult to predict with accuracy. These phenomena are often caused by mesoscale systems, or they are transitory and remain undetected within the normal observational system. Computer models are of limited use. The most hazardous weather is the most difficult to forecast.

Outlook forecasts for good weather are more likely to be correct than forecasts for poor weather. Errors in timing are more prevalent than errors of occurrence. One forecaster put it this way: "We're never wrong, our timing's just off sometimes." Or forecasts are 100 percent correct—90 percent in the summer and 10 percent in the winter!

Unwarranted pessimism is a major forecast complaint. In a 1981 National Aeronautics and Space Administration (NASA) study, pilots complained of canceling flights based on forecasts when the weather turned out to be VFR. One pilot took off in spite of the forecast and completed the flight. The report did say, however, "in this latter case...the forecast was substantially correct and the pilot was fortunate enough to

TABLE 13-1   Limitations on Aviation Weather Forecasts

---

1. Forecasts 12 hours and beyond for good weather (ceiling 3000 ft or more, visibility 3 mi or greater) are more likely to be correct than forecasts for poor weather (ceiling below 1000 ft, visibility below 1 mi).
2. Poor weather forecast to occur within 3 to 4 hours has a better than 80% probability of occurrence.
3. Forecasts for poor weather within the first few hours of the period are most reliable with distinct weather systems.
4. Errors occur with attempts to forecast a specific time poor weather will occur. Errors are less frequent forecasting poor weather within a time frame.
5. Surface visibility is more difficult to forecast than ceiling.

Forecasters Can Predict the Following with 75% Accuracy

---

1. Within 2 hours, the passage of fast-moving cold fronts or squall lines up to 10 hours in advance
2. Within 5 hours, the passage of warm fronts or slow-moving cold fronts up to 12 hours in advance
3. Within 1 to 2 hours, the onset of thunderstorms, with radar available
4. Within 5 hours, the time rain or snow will begin.

Forecasters Cannot Predict the Following with an Accuracy That Satisfies Operational Requirements

---

1. The time freezing rain will begin
2. The location and occurrence of severe or extreme turbulence, or severe icing
3. The location of the initial occurrence of a tornado or low-level wind shear
4. Ceilings of 100 ft or zero before they exist
5. The onset of a thunderstorm that has not yet formed

---

find breaks in the overcast...." As we will see in subsequent chapters, forecasters are making positive efforts to improve their products. Both the NWS and FAA are working on improving forecast accuracy.

EXPECTATIONS

Pilots complain equally about pessimistic forecasts and unforecast weather. Pilots, often, have an overexpectation of forecasts. Each situation is different, with many variables and local factors. The limitations in Table 13-1 remain, and forecasts are going to be missed. Errors fall into two categories: timing and the "Big Bust." The 1965 edition of *Aviation Weather* said it best: "The weather-wise pilot looks upon a forecast as professional advice rather than as the absolute truth."

> **PERSONAL OBSERVATION**   Periodically in aviation periodicals, articles are written bashing the FAA and NWS forecasters and products. I have, along with others, attempted to get details of such occurrences in order to analyze the event, only to be met with silence. It's easy to criticize, especially when there is no responsibility attached. If pilots wish, they can bring specific events to the attention of the National Weather Association's Aviation Meteorology Committee through www.nwas.org or the Aviation Weather Center at www.awc-kc.noaa.gov.

It's essential to remember these concepts during the discussion of forecasts. Forecast issuance times, purpose, conditional terms, amendment criteria, and regulations reflect the limitations on aviation forecasts. Meteorologists sometimes produce a *strategic forecast*—that is, a forecast with values and conditional terms that will allow conditions to go from zero-zero to ceiling and visibility unlimited without a requirement to amend. Well, that's a bit of an exaggeration, but it has occurred.

The National Weather Service's modernization and restructuring is well underway. Approximately 250 weather service forecast offices have been reduced to about 115 weather forecast offices. A new computer system known as the Advanced Weather Interactive Processing System (AWIPS) is coming online. Automated weather observing systems (AWOS/ASOS) and next-generation weather radar (NEXRAD) observations are expanding the observational network. Terminal Doppler Weather Radar (TDWR) will detect microbursts and alert pilots to low-level wind shear.

The next-generation weather satellites have been launched and are operation. The wind profiler, a Doppler radar system, will replace radiosonde balloon observations early in the new century for upper-air wind data. These technological advances promise to improve the accuracy and reliability of weather forecasts. But, for now, pilots will have to deal with today's system and its limitations.

There is no question that weather forecasts have improved dramatically since Sir William's observation some 80 years ago. But the science of meteorology is still not exact. Several years ago during a frustrating period of unsettled weather, a NWS forecaster wrote: "Finally figured out the difference between a ridge and trough in California this month. A trough gives us cold rain; a ridge give us warm rain."

FACT February 5 is National Weather Forecaster's Day. Have you hugged your meteorologist today?

From the FAA publication *The Weather Decision:* "At the weather briefing keep in mind that: Meteorologists tend to be optimists." Say again, please....

Those who struggle with forecasting the chaos of the weather can take solace from Galileo: "I can foretell the way of celestial bodies, but can know nothing about the movement of a small drop of water."

CHAPTER 14
WEATHER ADVISORIES

During the latter part of the 1950s, the U.S. Weather Bureau issued warnings of potentially hazardous or severe aviation weather in the form of Flash Advisories. These were subsequently divided into AIRMETs (WA) and SIGMETs (WS). AIRMETs and SIGMETs alerted pilots that significant, previously unforecast, weather had developed. AIRMETs advertise conditions less severe than SIGMETs (less than VFR conditions and phenomena of moderate intensity). SIGMETs warn of severe conditions that apply to all operations. Twenty-one Weather Bureau offices routinely issued advisories. By 1970 the number of Flight Advisory Weather Service offices was reduced to nine. According to the Weather Service Operations Manual at the time, "the Inflight Weather Advisory program is intended to provide advance notice of potentially hazardous weather developments to enroute aircraft.... " However, advisories were issued even when conditions were accurately portrayed in the *area forecast* (FA). To paraphrase, during this period: There were AIRMETs by the number, SIGMETs by the score. It was not uncommon to have four or more AIRMETs continuously in effect, and to this end the Continuous AIRMET (WAC) was developed. Weather advisories issued by adjacent offices were not always consistent and often overlapped. This led to confusion for pilots and briefers alike.

The area forecast changed in 1978 to include a HAZARDs or flight precautions section, ostensibly to reduce the number of advisories. The FA was still issued by local NWS offices.

FACT When the San Francisco Weather Service Forecast Office (WSFO) issued the FA, there was invariably a flight precaution for occasional moderate turbulence within 5000 ft of rough terrain. Pilots and briefers became equally disgusted with this generalization.

Responsibility for issuing AIRMETs and SIGMETs for the 48 contiguous states was centralized in 1982 in Kansas City at the National Aviation Weather Advisory Unit (NAWAU).

In 1991 AIRMET criteria phenomena were removed from the area forecast and issued separately as the AIRMET Bulletin. The NWS Aviation Services Branch working with FAA Headquarters, industry, user groups, and to a somewhat limited extent, input form NWS and FAA field offices initiated the change. The change was initiated to reduce the redundancy of AIRMET/Flight Precautions and correct the disappearance of AIRMETs at the next FA issuance.

SIGMETs were issued for convective activity until a DC-9 crashed in a severe thunderstorm near Atlanta in 1977. From this accident the convective SIGMET (WST) evolved. The National Aviation Weather Advisory Unit in Kansas City had WST responsibility. Specifically assigned meteorologists issue these advisories.

In October 1995, NAWAU became the Aviation Weather Center (AWC). On Tuesday, March 23, 1999, after 110 years in downtown Kansas City, AWC moved to its new quarters near the Kansas City International Airport.

Center weather service units (CWSUs) were established at Air Route Air Traffic Control Centers (ARTCC) in 1980. The purpose of the CWSU is to assist controllers and flow control personnel and alert pilots of hazardous weather through a *center weather advisory* (CWA). As is often the case with government bureaucracy, the cart came before the horse. There were no instructions for ATC personnel when CWAs first appeared. Distribution went all the way from immediate broadcast to the trash can. The FAA took months to decide that the CWA had the weight of a SIGMET and apply identical distribution and broadcast procedures.

Alert weather watches (AWWs) and *severe weather watch bulletins* (WWs), public forecasts, are also produced. The Severe Local Storms office (SELS) in Kansas City issued AWWs and WWs for severe thunderstorms and tornadoes until 1997. At that time SELS was relocated to the Storm Prediction Center (SPC) in Norman, Oklahoma.

The AIRMET Bulletin, SIGMETs, convective SIGMETs, center weather advisories, and alert weather watches and bulletins are issued when phenomena reach specified criteria, and, like urgent PIREPs (UUA), receive priority handling and distribution.

With the number of advisories, it would seem impossible to fly into an area of hazardous weather without warning. But this is not necessarily the case. An advisory cannot be issued for each thunderstorm, instance of turbulence, strong surface winds, icing, mountain obscuration, or IFR condition. Severe weather can develop before an advisory is written and distributed. The absence of an advisory is no guarantee that hazardous weather does not exist or will not develop.

AIRMET BULLETINS

The AIRMET Bulletin (WA) is issued on a regular basis four times a day (0300Z, 0900Z, 1500Z, 2100Z) for each of the six contiguous area forecast (FA) areas (MIA—Miami, BOS—Boston, OAK—San Francisco, etc.) by the AWC. Issuance times change 1 h during Daylight Savings Time (0200Z, 0800Z, 1400Z, 2000Z). The Honolulu WFO issues the AIRMET Bulletin for Hawaii, with the first issuance at 0400Z. In Alaska the Anchorage, Fairbanks, and Juneau WFOs issue the AIRMET Bulletin for their respective FA areas; the product is issued with, and times coincides with FA issuance. FA areas are depicted in App. B, Graphs and Charts. The forecast is valid for 6 h, with an additional 6-h outlook. Each subsequent issuance is indicated by an update (UPDT) number, beginning with the second of the Zulu date "UPDT 1." Should an amendment be issued between normal update times, it will be reflected in the header (SFOS WA 251730 AMD) and given the next update number.

The bulletin is divided into three sections designated AIRMET SIERRA, AIRMET TANGO, and AIRMET ZULU. AIRMET SIERRA describes areas of IFR conditions and mountain obscurement. AIRMET TANGO forecasts turbulence, nonconvective (thunderstorm) low-level wind shear (LLWS), and strong surface winds. AIRMET ZULU shows the location and intensity of icing and includes the freezing level.

Conditions must be widespread for inclusion as an AIRMET in the AIRMET Bulletin. That is, occurring or forecast over an area of at least 3000 mi^2, approximately three times the size of Rhode Island. Localized occurrences do not warrant the issuance of an AIRMET. Failure to consider this fact has led many a pilot and FSS briefer to unwarranted criticism of this product.

WHAT'S LOCAL? According to Dick Williams, NAWAU forecaster, AIRMET Bulletins "written on the scale of whole states do not endeavor to describe every single occurrence of IFR, icing or turbulence. The forecaster, wishing to indicate that there may be isolated observations, pilots reports, or hazardous weather, may use the term 'local.' No hard and fast rule exists for determining when 'local' becomes widespread. The forecaster relies on observations, pilot reports, satellite imagery, and his or her own judgment in determining the extent of weather features." Federal aviation regulations recognize the fact that every occurrence of adverse weather cannot be forecast. Regulation 14 CFR 61.93 requires that even student pilots receive instruction in "the recognition of critical weather situations" and "estimating visibility while in flight...." In other words, the excuse, "They didn't tell me," is just that, an excuse, not a reason.

AIRMETs are issued when the following phenomena occur or are expected to develop:

- Moderate icing
- Moderate turbulence
- Sustained wind of 30 kn or more at the surface
- Ceilings less than 1000 ft and/or visibility less than 3 mi affecting over 50 percent of the area at any time
- Extensive mountain obscurement

Because hazards often affect more than one FA area and to provide an overview, the AIRMET Bulletin is issued using the common set of VORs depicted on the Weather Advisory Plotting Chart found in App. B. This appendix also contains a list of location identifiers used on the chart. Affected areas start with the most northern location and continue clockwise.

From a forecast point of view, phenomena usually lie well within the delineated area. Since the AIRMET Bulletin covers a period of 6 h, the phenomena may move through the area, develop, or dissipate during the forecast period. Therefore, the conditions advertised may affect only a portion of the area at any particular time. Details appear in the text of the bulletin. For example, an icing advisory may cross several FA boundaries. Due to differing temperatures, humidity, and frontal locations, specific altitudes may vary. When a phenomenon is peculiar to a specific mountain range, coastal area, river basin, or valley, the geographical area may be included (SNAKE RIVER VALLEY, TEXAS WEST OF THE PACOS, etc.). Appendix B contains forecast designators.

Conditional terms describe widely varying conditions over large areas. Definitions are contained in Table 14-1 on weather advisory, area forecast conditional terms. They're self-explanatory, with the exception of *occasional* (OCNL). *Occasional* describes a better than 50/50 chance of occurrence during less than half of the forecast period. OCNL often describes turbulence, icing, mountain obscurement, and IFR conditions, and it reflects the transitory nature of these phenomena. A pilot may or may not encounter the condition flying through a forecast area of OCNL, but the pilot has been warned. Look at it this way: If you run into it, they're right, if you don't, they're still right! We'll discuss how to interpret and apply conditional terms throughout this chapter and Chap. 15, Area Forecasts.

TABLE 14-1 Weather Advisory, Area Forecast Conditional Terms

Term	Abbreviation	Definition
Occasional/occasionally	OCNL/OCNLY	Greater than a 50% probability of occurrence/occurring for less than 1/2 of the forecast period
Isolated	ISOLD	Affecting less than 3000 mi^2/events widely separated in time
Widely scattered	WDLY SCT	Less than 25% of area affected
Scattered/areas	SCT	25–50% of area affected
Numerous/widespread	NMRS	More than 50% of area affected

To further define the transitory nature of phenomena, the forecaster will use remarks concerning the development and dissipation of conditions. For example:

- CONDS BGNG BY...
- CONDS CONTG...
- CONDS IPVG BY...
- CONDS ENDG AFT...
- CONDS SPRDG OVR...
- CONDS MOVG...

Some of these remarks may be combined to better describe expected conditions. For example, "CONDS BGNG BY 1200Z AND CONTG BYD 1400Z."

To alert all users that a SIGMET is in effect, the AIRMET Bulletin will reference the SIGMET in the appropriate section. For example:

...SEE SIGMET XRAY SERIES FOR SEV TURBC...

Details—location, severity, and altitudes—on the phenomena would be obtained from the SIGMET text.

IFR and Mountain Obscuration

AIRMET SIERRA describes the location of ceilings and visibilities below 1000 ft and 3 mi and areas where the mountains are obscured by clouds or precipitation. The AIRMET SIERRA portion of the AIRMET Bulletin looks like this:

SFOS WA 261345
AIRMET SIERRA UPDT 2 FOR IFR AND MTN OBSCN VALID UNTIL 262000

AIRMET IFR...CA
FROM 70 WSW OED TO ENI TO SNS TO 20W SNS TO 30W ENI TO FOT TO 70WSW OED
OCNL CIG BLO 10 VSBY BLO 3F. CONDS ENDG 17Z-19Z.

AIRMET MTN OBSCN...CA
FROM 40 W SAC TO 20 NNW RZS TO 50WNW TRM TO 40E MZB TO 10S MZB TO LAX
TO 40W RZS TO 40NW OAK TO 40W SAC
CSTL MTNS OBSCD CLDS/FOG. CONDS ENDG CNTRL CA PTN AREA 19Z-20Z..CONTG
SRN CA BYD 20Z THRU 02Z.

This San Francisco AIRMET SIERRA Bulletin (SFOS WA) was issued on the 26th day of the month at 1345Z. This is the third issuance for this Zulu day (UPDT 2). It is valid until the 26th at 2000Z.

IFR conditions are forecast for all or portions of California (AIRMET IFR...CA). Specifically, IFR is expected within the area from 70 nautical miles (nm) west-southwest of Medford (OED), to Ukiah (ENI), to Salinas (SNS), to 20 west of Salinas, to 30 west of Ukiah, to Fortuna (FOT), and back to 70 west-southwest of Medford.

Within the delineated area, occasional ceilings below 1000 ft AGL and visibilities below 3 mi in fog are expected. These conditions should end between 17Z and 19Z. Even if some localized areas of IFR remain beyond 19Z, the forecast is correct. To indicate that other areas within the FA coverage are not expected to contain widespread areas of IFR, the following statement may appear: RMNDR...NO WDSPRD IFR EXP.

The mountain obscuration is expected to end in the central California portion between 19Z and 20Z but continue in southern California beyond 20Z through 02Z—the outlook period.

Turbulence, Winds, and Low-Level Wind Shear

AIRMET TANGO describes the location, intensity, and height of nonconvective-related turbulence. Localized or isolated moderate turbulence, not requiring the issuance of an AIRMET, also appears in this section (LGT ISOLD MOD). When turbulence is forecast that does not meet AIRMET criteria, only the general geographical area is indicated (WRN WA AND OR, etc.) Light or no turbulence is indicated by the statement: NO SGFNT TURBC EXPCTD or NO SGFNT TURBC OUTSIDE CNVTV ACTVTY.

Turbulence forecasts are based on wind flow, winds aloft, evaluation of terrain, and PIREPs. Generally, moderate intensity is forecast when the winds reach 25 to 30 kn, and severe when winds exceed 40 kn. High-level turbulence is difficult to forecast.

Refer to the Chicago Area Forecast Synopsis and AIRMET TANGO portion of the AIRMET Bulletin below for the following discussion:

```
CHIS FA 201040
SYNOPSIS VALID UNTIL 210500
COLD FNT ALG LN MOT-LBF-ABQ AT 11Z WL MOVE TO A DLH-DSM-ALI LN BY 23Z
AND DISIPT BY 210500Z.


CHIT WA 200745
AIRMET TANGO UPDT 1 FOR TURBC VALID UNTIL 201400
FROM ISN TO MQT TO COU TO ACT TO SJN TO FMN TO DEN TO ISN
OCNL MOD TURBC BLO 150 WI 100 ML OF COLD FNT. CONDS CONTG BYD 14Z THRU
20Z.
```

The turbulence forecast must cover the 6-h forecast period—08Z through 14Z—and the outlook period—14Z through 02Z. A cold front is expected to produce occasional moderate turbulence. The CHI FA synopsis and AIRMET TANGO illustrate how a hazard can move through the delineated area, only affecting portions at any one time during the forecast period. Unfortunately, some pilots and FSS briefers fail to understand and consider this, and they accept or issue the advisory whether or not it applies, undermining the credibility of both forecast and briefing.

The probability of moderate turbulence is better than 50 percent. However, it is only expected to occur for less than half of the forecast period. A pilot could expect a greater probability in the vicinity of the frontal zone, with its approach, and shortly after passage. This is not inconsistent but rather, reflects the dynamic character of weather.

Strong surface winds, for the purpose of AIRMET TANGO, are defined as sustained winds 30 kn or greater. For example:

```
AIRMET STG SFC WIND...MT...UPDT
FROM YXH TO BIL TO DLN TO 50N FCA TO YXH
SUSTAINED SFC WINDS GTR THAN 30 KT EPCD. CONDS CONTG BYD
21Z...ENDG BY 00Z.
```

Nonconvective low-level wind shear (LLWS) will be included in a separate paragraph. The paragraph will state LLWS potential, location, and cause—for example, LLWS POTENTIAL OVER MOST OF NEW ENGLAND AFTER 03Z DUE TO STG NWLY FLOW BHND CSTL LOW PRES SYS. There is a potential for LLWS over most of New England after 0300Z due to a strong northwesterly flow behind the low-pressure system over the coast. Nonconvective LLWS can be caused by:

- Fronts
- Low-level jet streams
- Terrain
- Valley effects
- Sea breezes
- Lee-side effects
- Inversions
- Santa Ana or similar foehn-like winds

The occurrence, exact location, and intensity of turbulence and LLWS is difficult to predict. The forecaster must consider the often widespread and transitory nature of turbulence. Here again, PIREPs are the only means of validating the forecast.

> **JOHN 3:8** "The wind bloweth where it listeth, and thou hearest the sound thereof, but canst not tell whence it cometh, and whither it goeth:…."

Icing and Freezing Level

AIRMET ZULU describes the location, intensity, and type (rime, clear, or mixed) of non-convective icing. Layers, where significant icing can be expected, are expressed as specific values or ranges with bases and tops. This section includes forecasts for light or local moderate icing. Like AIRMET TANGO, light or local moderate icing will be described using geographical areas (NE AZ AND NW NM, etc.). Trace or no icing is indicated by the statement NO SGFNT ICING EXPCD or NO SGFNT ICING EXPCD OUTSIDE CNVTV ACTVTY. A separate paragraph contains forecast freezing levels. Terms such as *sloping* or *lowering* describe varying levels.

The AIRMET ZULU portion of the AIRMET Bulletin looks like this:

```
SFOZ WA 141345
AIRMET ZULU UPDT 2 FOR ICG AND FRZLVL VALID UNTIL 142000
AIRMET ICG…WA OR CA
FROM YQL TO GGW TO BFF TO ALS TO 120W OAK TO 120W FOT TO
120W TOU TO YQL
OCNL MOD RIME/MXD ICGICIP BTWN 040 TO 140 WA BTWN AND 060 TO 160 OR/CA.
CONDS CONTG BYD 20Z THRU 02Z.

FRZLVL..WA W OF CASCDS..045 LWRG BY 20Z TO 035.
    WA CASCDS EWD..AT/NEAR SFC WITH MULT FRZLVLS 30-35.
    OR W OF CASCDS..55 NORTH TO 65 SOUTH. AFT 18Z 50
        NORTH TO 75 SOUTH. OR CASCDS
        EWD..AT/NEAR SFC WITH MULT FRZLVLS TO 40-50
    CA..AT/NEAR SFC SIERRAS AND NE PTN TIL 18Z. ELSW
        NEAR 70 NORTH SLPG TO 90 SOUTH. AFT 18Z 70
        NORTH SLPG TO 90 CNTRL AND 100 SOUTH.
```

The specific area extends from Lethbridge, Alberta (YQL), to Glasgow (GGW) to Alamosa (ALS) to 120 nm west of Oakland (OAK) to 120 nm west of Fortuna (FOT) to 120 nm west of Tatoosh (TOU) and back to YQL. This area includes the coastal waters.

Intensities, type, and altitudes may differ significantly over the areas affected. Occasional moderate rime or mixed icing in clouds and precipitation is forecast from 4000 to 14,000 ft MSL over Washington, and from 6000 to 16,000 ft MSL over Oregon and

California. Conditions are expected to continue beyond the end of the forecast period 20Z, through the outlook period of 0200Z. Pilots planning flights beyond the forecast period of 20Z can expect this AIRMET to be in effect at least through 02Z.

It seems a bit redundant to include the base of the icing in the icing paragraph and freezing level in the subsequent paragraph. This resulted from a pilot obtaining a briefing with icing from the FRZLVL. The base of the freezing level was not specified. Now the forecaster must enter a specific altitude in the icing paragraph. This may be in the form of BTWN as in the previous example or FRZLVL-150. Pilots can expect the most significant icing within the layer specified in the icing paragraph. Trace or light icing information can be expected between the freezing level and the altitude data in the icing paragraph.

Next appears the freezing level paragraph (FRZLVL). Areas are defined using area forecast designators provided in App. B. In Washington, west of the Cascades the freezing level is expected to lower from 4500 to 3500 ft MSL by 2000Z. East of the Cascades the freezing level is at or near the surface with multiple freezing levels between the surface and 3000 to 3500 ft MSL. Multiple freezing levels are caused by overrunning warm air, such as occurs with a warm front. Freezing rain occurs in these areas. Similar conditions are expected in Oregon, but at different levels and times.

In California the freezing level is forecast at or near the surface in the Sierra Nevada mountains and northeast portion until 1800Z. Elsewhere the freezing level is expected near 7000 ft MSL in the north sloping to 9000 ft MSL in the south. After 1800Z the freezing level is forecast to remain around 7000 in the north and to rise in the central and southern portions to 9000 to 10,000 ft.

In this section the forecaster has described an icing layer 6000 to 10,000 ft deep, that slopes upward about 2000 ft from north to south. The type of ice forecast, mixed, and depth of the anticipated icing layer indicate an unstable air mass. IFR pilots with aircraft certified for flight in icing conditions should have little trouble in these areas, assuming performance will allow them to climb out of the icing layer. However, they must be prepared to contend with freezing rain east of the Cascades and icing to the surface in parts of California. For IFR pilots of aircraft without ice protection equipment, this forecast would be a very strong no-go indicator, especially east of the Cascades and the mountains of California. The VFR pilot flying in Washington or Oregon east of the Cascades will be just as susceptible to icing as an IFR pilot because of multiple freezing levels and possible freezing rain. But, as we've seen, any flight decision cannot be based on one forecast, especially taken out of context, as in this case.

Pilots planning flights below the freezing level can normally not expect to receive this advisory during an FSS preflight briefing since icing will not affect their proposed flight. Some briefers fail to understand and consider this, and they issue the advisory even though it is not a factor. This practice undermines the credibility of both the forecast and the briefing. A pilot planning a flight and briefed for low altitudes should keep this point in mind, should he or she elect, or be instructed by ATC, to climb to a higher altitude. Pilots might well consider the advisability of accepting the clearance without additional information on icing and freezing level. This point also applies to rerouting. Should the pilot or ATC reroute the aircraft, advisories that were not pertinent during the briefing may then apply. In Chap. 19, we'll discuss various means to obtain updated weather status information.

When AIRMET phenomena are expected to develop beyond the valid time of the AIRMET Bulletin—first 6 h, but within 12 h of issuance—the details appear in the outlook section. For example:

```
CHIS WA 021945
AIRMET SIERRA FOR IFR VALID UNTIL 030200

NONE XPCD.

OTLK VALID 0200-0800Z...IFR NE KS IA MO
```

AFT 05Z OCNL CIGS BLO 10 VSBYS 3-5 IN PCPN/BR DVLPG NE NRN
KS NRN MO SW IA. CONDS CONTG BYD 08Z.

SIGMETs

SIGMETs, like the AIRMET Bulletin, often cover large areas due to the scattered and tran-sitory nature of the phenomena they report. Therefore, the term *occasional* (OCNL) fre-quently appears. Additionally, phenomena may move through or only affect certain geographical features within the advisory area. In the example, strong updrafts and down-drafts are expected only in the vicinity of the mountains. Failure to thoroughly read and understand an advisory has led many pilots and briefers to unjustly criticize this product.

A SIGMET will be issued only when the phenomenon is widespread. Local occurrences of severe turbulence and icing will appear in AIRMET Bulletins. (*Note:* At this writing, AWC has an agreement with certain users not to include locally severe conditions in the AIRMET Bulletin. The reason for that policy is the fact that some users are prohibited from flying in areas of advertised severe conditions.) SIGMETs are issued when the following phenomena occur or are expected to develop:

- Severe icing
- Severe or greater nonconvective turbulence
- Moderate or greater clear air turbulence (CAT)
- Widespread duststorms, sandstorms, or volcanic ash reducing visibility below 3 mi over an area at least 3000 mi^2

For the purposes of a SIGMET, CAT is defined as nonconvective turbulence occurring at or above 15,000 ft, although it usually refers to turbulence above 25,000 ft. Because it is difficult to forecast these phenomena, the term *moderate or greater* (MOGR) indicates a threat of CAT. SIGMETs contain the qualifiers *severe* or *extreme* only with actual reports.

In Alaska and Hawaii, WFOs responsible for FAs issue SIGMETs for these conditions. These WFOs also issue SIGMETs for tornadoes, hail greater than 3/4 in in diameter, and embedded or lines of thunderstorms.

SIGMETs are identified by forecast area, and alphabetic and product designators. The forecast area specifies within which FA the advisory applies. Next appears the alphabetic designator for the phenomenon being described. (To avoid confusion with international SIGMETs, domestic SIGMET names now run NOVEMBER through YANKEE—excluding SIERRA, TANGO, and ZULU, which are reserved for AIRMETs.) The product designator (1, 2, 3, etc.) indicates the number of successive times the advisory has been issued. For example, a cold front causing severe turbulence may begin as San Francisco SFO OSCAR 1, as the front moves into the Rocky Mountains, become Salt Lake City SLC OSCAR 2, and into the Plains, Chicago CHI OSCAR 3. To assure continuity and alert pilots, briefers, and controllers that OSCAR 3 is the first CHI issuance, a referencing remark may be appended to the message (FOR PREVIOUS ISSUANCE SEE SLC OSCAR 2). Updates often contain changes; they must be reviewed for affected areas, altitudes, and times. It's important to note both phe-nomenon and product designators.

SFO N WS 152130
SIGMET NOVEMBER 3 VALID UNTIL 160130
CA
FROM FOT TO 50NW FMG TO 50NE EHF TO RZS TO 40W RZS TO 30W OAK TO FOT
OCNL SEV TURBC BLO 100 XCP BLO 150 VCNTY SIERRAS. STG UDDFS VCNTY MTNS
AND LLWS POTENTIAL BLO 20 AGL. CONDS CONTG BYD 0130Z.

In the above example, the FA designator is San Francisco (SFO N). The issuance date and time group follows the product designator (WS). This SIGMET was issued on the 15th day of the month at 2130Z (152130). The alphabetic phenomenon designator is N (SFO N). SIGMET NOVEMBER is spelled out on the following line along with the product designator (3). WSs are valid for 4 h as indicated by the VALID UNTIL time, the 16th day of the month at 0130Z (160130). San Francisco SIGMET NOVEMBER 3 affects all or part of California (CA). Like the AIRMET Bulletin, specific geographical areas are described using VORs on the Weather Advisory Plotting Chart, App. B, starting with the most northern and continuing clockwise. The phenomenon will usually lie well within the delineated area.

Figure 14-1 shows SIGMET NOVEMBER 3 laid out—the gray area—on a Weather Advisory Plotting Chart. Figure 14-1 also displays a portion of the Geographical Area Designators map. The gray area again shows the area covered by the SIGMET. The advisory affects a portion of northern California and most of central California. Plotting may be required to determine the extent of an advisory, and it is extremely helpful visualizing affected areas. Occasional severe turbulence is expected below 10,000 ft MSL, except below 15,000 ft MSL in the vicinity of the Sierra Nevada mountains. Strong updrafts and downdrafts in the vicinity of the mountains and low-level wind shear are anticipated. Conditions are expected to continue beyond the end of the advisory period (0130Z). This means SIGMET NOVEMBER 4 should be issued prior to 0130Z. However, should factors, such as transmission trouble, delay the updated advisory, pilots should consider the advisory still in effect, unless a cancellation message is received.

Because of the hazards associated with widespread duststorms, sandstorms, and volcanic ash, a SIGMET is issued for these phenomena. To meet SIGMET criteria, duststorms and sandstorms must cause surface or inflight visibilities to be below 3 mi over an area of at least 3000 mi^2. In the case of volcanic ash, a SIGMET is issued regardless of the area affected as soon as possible after receipt of notification of an eruption. The following is an example:

```
ANCA WS 231604
PAZA SIGMET ALFA 39 VALID 231600/230200 PANC-
POSSIBLE VOLCANIC ASH BLW FL300 PAVLOP VOLCANO (55.4N/161.9W) WITHIN AN
AREA FM 57.0N/166.0W TO 55.6N/161.7W TO 55.0N/162.0W TO 55.0N/167.0W TO
57.0N/166.0W MOVING WNW.

SATELLITE INDICATES A NEW ERUPTION OF PAVLOP VOLCANO AT ABOUT 1330
UTC ON 23 NOV 96. PLUM INITIALLY MOVING WNW.
```

International SIGMETs are issued for oceanic areas adjacent to the United States by a *meteorological watch office* (MWO). The National Weather Service has MWOs at Anchorage, Alaska, Guam Island, and Honolulu in the Pacific, Kansas City, Missouri, and the Tropical Prediction Center in Miami, Florida. Criteria for domestic and international SIGMETs are similar; however, the format, abbreviations, and wording used are different. (Abbreviations are contained in App. A.)

For inclusion in a SIGMET, phenomena must be widespread—affecting an area of at least 3000 mi^2. International SIGMETs are issued for the following phenomena:

- Thunderstorms

- Tornadoes

- Lines of thunderstorms

- Embedded thunderstorms

- Large areas of thunderstorms

- Large hail

- Tropical cyclone

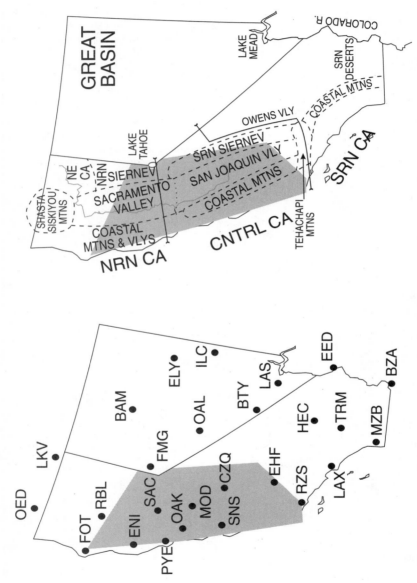

FIGURE 14-1 Plotting may be required to determine the extent of advisories, and it is extremely helpful in visualizing affected areas.

- Severe icing
- Severe or extreme turbulence
- Duststorms and sandstorms lowering visibilities to less that 3 mi
- Volcanic ash

International SIGMETs are issued for 12 h for volcanic ash events, 6 h for hurricanes and tropical storms, and 4 h for all other phenomena. If conditions persist beyond the forecast period, the SIGMET is updated and reissued. Below are two examples of international SIGMETs:

```
KNHC 021349 CCB KZNY KZMA
SIGMET ALFA 1 VALID 021335/021735 UTC KNHC-
EMBD TS OBS BY SATELLITE AND LIGHTNING DATA WI AREA BOUNDED
BY 30N77W 30N75W 25.5N75.5W 25.5N77W 30N77W. CB TOPS TO
FL410. MOV NE 30 KT. NC.
DGS
```

```
KNHC 021356 TJZS
SIGMET BRAVO 1 VALID 021355/021955 UTC KNHC-
VA CLD OBS FM SOUFRIERE HILLS VOLCANO MONTSERRAT 16.7N62.2W.
THIN VA CLD OBS BY SATELLITE BLW 050 10 NM EITHER SIDE OF
LINE 17.1N63W 17.2N63.7W. MOV NW 10-15 KT. NC.
DGS
```

CONVECTIVE SIGMETs

Convective SIGMETs (WST) provide detailed, specific forecasts for thunderstorm-related phenomena. Since thunderstorms are accompanied by severe or greater turbulence, severe icing, and low-level wind shear (LLWS), these conditions are not specifically addressed in the advisory or in a weather briefing. The Aviation Weather Center's WST unit makes extensive use of radar data to analyze thunderstorm systems. Meteorologists compare radar data with satellite imagery, lightning information, and other conventional sources to determine the need for a SIGMET. WSTs are issued when the following phenomena occur or are expected to develop and continue for more than 30 min within the valid period:

- Severe thunderstorms
- Embedded thunderstorms
- A line of thunderstorms
- An area of active thunderstorms affecting at least 3000 mi^2

WSTs for severe thunderstorms may include specific information on tornadoes, large hail, and wind gusts of 50 kn or greater. An embedded thunderstorm occurs within an obscuration, such as haze, stratiform clouds, or precipitation from stratiform clouds. Embedded thunderstorms alert pilots that avoidance by visual or radar detection could be difficult or impossible. A line of thunderstorms must be at least 60 mi long, with thunderstorms affecting at least 40 percent of its length. Active thunderstorms must have an intensity level of 4 or greater, or they must affect at least 40 percent of an area. (Table 11-1 contains a description of radar precipitation intensity levels.)

The 48 contiguous states are divided into three areas for convective SIGMET issuance: west (MKCW WST, west of 107° W longitude), central (MKCC WST, between 107° W and 87° W longitude), and east (MKCE WST, east of 87° W longitude). These areas are shown on the Weather Advisory Plotting Chart, App. B. Issued on an unscheduled basis as needed, beginning with number 1 at 0000Z, WSTs contain a forecast for up to 2 h and an outlook from 2 to 6 h:

```
MKCC WST 222055
CONVECTIVE SIGMET 63C
VALID UNTIL 2255Z


MO KS OK
FROM 40WNW IRK-20WNW BUM-30N TUL
LINE TS 20 NM WIDE MOV FROM 23025KT. TOPS TO FL450.
HAIL TO 1 IN...WIND GUSTS TO 50 KT POSS.


OUTLOOK VALID 222255-230255
FROM TVC-ASP-FWA-SJT-CDS-60WSW PWE-TVC
TS CONTG INVOF CDFNT THAT EXTDS FM CNTRL GRTLKS SWWD TO DPNG SFC
LOW OVER N CNTRL OK. STGST STMS RMN OVER KS/MO IN AREA OF STGST LOW
LVL WARM ADVCTN AND MOST FVRBL MSTR/INSTBY. STMS EXPD TO INCR IN
COVERAGE AND INTSTY NEXT SVRL HRS. ADDNL STG/PSBLY SEV TS DVLPMT
LKLY INVOF DRYLN THAT EXTDS SWD FM OK LOW.
PDS
```

This central (MKC C) WST was issued on the 22d day of the month at 2055Z (222055). It is the 63d central issuance for this ZULU day (CONVECTIVE SIGMET 20 C). It affects portions of Missouri, Kansas, and Oklahoma, and it is valid for 2 h (VALID UNTIL 2255Z).

Specific areas are described using the VORs on the Weather Advisory Plotting Chart. In this case from 40 mi west-northwest of Kirksville, Missouri, to 20 mi west-northwest of Butler, Missouri, to 30 mi north of Tulsa, Oklahoma. The advisory warns of a line of thunderstorms 20 nm wide moving from 230° at 25 kn, tops to 45,000 ft, hail to 1 in, and wind gusts to 50 kn possible. The black line in Fig. 14-2 encloses this area; it's nice to have the chart to visualize locations.

The WST outlook was designed primarily for preflight planning and aircraft dispatch. It normally includes a meteorological discussion of factors considered by the forecaster. It is supplemental information not required for weather avoidance, but it is useful to CWSU and FSS specialists for analysis and background information. Normally the outlook will not be included in broadcasts nor provided during a briefing.

The example outlook is valid for an additional 4 h, covers an area from Traverse City, Michigan, to Oscoda, Michigan, to Ft. Wayne, Indiana, to San Angleo, Texas, to Childress, Texas, to 60 mi west-southwest of Pawnee City, Nebraska, back to Traverse City. The broken line in Fig. 14-1 encloses this area.

The outlook translates: Thunderstorms continuing in vicinity of cold front that extends from central Great Lakes southwestward to deepening surface low over north central Oklahoma. Strongest storms remain over Kansas/Missouri in area of strongest low-level warm advection and most favorable moisture/instability. Storms expected to increase in coverage and intensity next several hours. Additionally, strong/possibly severe thunderstorm development likely in vicinity of dry line that extends southward from Oklahoma low. Standard abbreviations are contained in App. A.

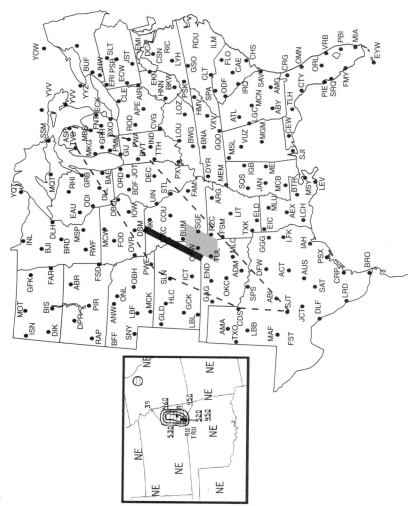

FIGURE 14-2 Both the convective SIGMET and alert weather watch are perfectly consistent within the scope and purpose of each product.

The outlook interpretation is that thunderstorms will continue due to the cold front. The strongest storms will be produced in the area of strongest low-level warm air advection (upward-moving air—a lifting mechanism), and most favorable area of moisture and instability—the three elements required for thunderstorms. Strong, possibly severe thunderstorms are likely to develop along a dry line (a lifting mechanism) that extends from the Oklahoma low-pressure area. Most meteorological terms used in the outlook discussion are contained in the Glossary.

SEVERE WEATHER BULLETINS

The National Weather Service produces three severe weather bulletins that apply directly or indirectly to aviation operations. The *alert weather watch* (AWW) warns of severe convective weather. The *severe weather watch bulletin* (WW) provides details on the AWW. The *hurricane bulletin* (WH) advertises hazards associated with this phenomenon. Both the AWW and WH are written in plain language and are primarily public forecasts.

Alert weather watches alert forecasters, briefers, pilots, and the public to the potential for severe thunderstorms or tornadoes. Subsequent to the AWW, a severe weather watch bulletin is issued. The WW contains details on the phenomenon described in the AWW. These unscheduled bulletins are primarily a public forecast, whereas the WST is a combination observation and aviation forecast.

Although Storm Prediction Center and Aviation Weather Center meteorologists coordinate their products, criteria and time frames differ. Therefore, aerial coverage might not coincide. The issuance of an AWW might precede or coincide with a WST. AWWs are numbered sequentially beginning each January 1st. The following AWW was issued just after the WST in the previous example (refer to the shaded area in Fig. 14-2):

```
MKC AWW 222036
WW 166 TORNADO OK KS MO AR 222100Z - 230200Z
AXIS..70 STATUTE MILES EAST AND WEST OF LINE..
10S MKO/MUSKOGEE OK/ - 65NNE JLN/JOPLIN MO/
..AVIATION COORDS.. 60NM E/W /45SSE TUL - 25SE BUM/
HAIL SURFACE AND ALOFT..2 INCHES. WIND GUSTS..70 KNOTS. MAX TOPS TO 500.
MEAN STORM MOTION VECTOR 25025.
```

When the area is described with locations not on the Weather Advisory Plotting Chart, a separate line titled ..AVIATION COORDS.. will be added. The mean wind vector is the direction and magnitude of the mean winds from 5000 ft AGL to the tropopause. It can be used to estimate cell movement—in the example, 250° at 25 kn.

The WST describes a developing area of interest to aviation. Inset in Fig. 14-1 is a portion of the radar summary chart observed at 2035Z. Thunderstorms have already developed along the southern portion of the line described in the WST. The area is moving toward the northeast at almost 40 kn. The WST forecaster expects the line to develop toward the northeast. Additionally, the AWW forecasts expects tornadic activity to develop along the southern portion of the line and move eastward. Expect later WSTs to cover the area toward the northeast and southwest, into the AWW, and along the area described in the WST outlook. All of this is perfectly consistent within the scope and purpose of these products.

Below is the severe weather watch bulletin corresponding to the alert weather watch in the previous example:

MKC WW 222036
URGENT — IMMEDIATE BROADCAST REQUESTED
TORNADO WATCH NUMBER 166
STORM PREDICTION CENTER NORMAN OK
336 PM CDT THU APR 22 1999
THE STORM PREDICTION CENTER HAS ISSUED A
TORNADO WATCH FOR PORTIONS OF
    NORTHEAST OKLAHOMA
    EXTREME SOUTHEAST KANSAS
    SOUTHWEST MISSOURI
    NORTHWEST ARKANSAS
EFFECTIVE THIS THURSDAY AFTERNOON AND EVENING FROM 400 PM UNTIL
900PM CDT.
TORNADOES...HAIL TO 2 INCHES IN DIAMETER...THUNDERSTORM WIND GUSTS
TO 80 MPH...AND DANGEROUS LIGHTNING ARE POSSIBLE IN THESE AREAS. THE
TORNADO WATCH AREA IS ALONG AND 70 STATUTE MILES EAST AND WEST OF A
LINE FROM 10 MILES SOUTH OF MUSKOGEE OKLAHOMA TO 65 MILES NORTH
NORTHEAST OF JOPLIN MISSOURI.
REMEMBER...A TORNADO WATCH MEANS CONDITIONS ARE FAVORABLE FOR
TORNADOES AND SEVERE THUNDERSTORMS IN AND CLOSE TO THE WATCH AREA.
PERSONS IN THESE AREAS SHOULD BE ON THE LOOKOUT FOR THREATENING
WEATHER CONDITIONS AND LISTEN FOR LATER STATEMENTS AND POSSIBLE
WARNINGS.
DISCUSSION...THREAT FOR SEVERE THUNDERSTORMS AND POSSIBLY SUPER-
CELLS SHOULD INCREASE ALONG COLD FRONT THIS AFTERNOON AS AIR MASS
CONTINUES TO DESTABILIZE AND CAP WEAKENS. TORNADO THREAT APPEARS
GREATEST IN OK IF ISOLATED CELLS CAN DEVELOP. AVIATION...TORNADOES
AND A FEW SEVERE THUNDERSTORMS WITH HAIL SURFACE AND ALOFT TO 2
INCHES. EXTREME TURBULENCE AND SURFACE WIND GUSTS TO 70 KNOTS. A FEW
CUMULONIMBI WITH MAXIMUM TOPS TO 500. MEAN STORM MOTION VECTOR
25025.
...VESCIO

Like the severe weather watch bulletin, the hurricane bulletin is essentially a public forecast product issued by the National Hurricane Center and written in plain language. When hurricanes are the cause for the issuance of aviation weather advisories, the hurricane will be referenced in the advisory. Below is an example of a hurricane bulletin:

HURRICANE GEORGES FORECAST/ADVISORY NUMBER 46
NATIONAL WEATHER SERVICE MIAMI FL AL0798
2100Z SAT SEP 26 1998

A HURRICANE WARNING IS IN EFFECT FROM MORGAN CITY LOUISIANA TO
PANAMA CITY FLORIDA. A HURRICANE WARNING MEANS THAT HURRICANE CON-
DITIONS ARE EXPECTED IN THE WARNED AREA WITHIN 24 HOURS. PREPARATIONS
TO PROTECT LIFE AND PROPERTY SHOULD BE RUSHED TO COMPLETION.

A TROPICAL STORM WARNING AND A HURRICANE WATCH ARE IN EFFECT FROM
EAST OF PANAMA CITY FLORIDA TO ST. MARKS FLORIDA. A HURRICANE WATCH
IS IN EFFECT FROM WEST OF MORGAN CITY TO INTRACOASTAL CITY LOUISIANA.

HURRICANE CENTER LOCATED NEAR 26.6N 86.2W AT 26/2100Z POSITION ACCU-
RATE WITHIN 30 NM

PRESENT MOVEMENT TOWARD THE NORTHWEST OR 305 DEGREES AT 9 KT

ESTIMATED MINIMUM CENTRAL PRESSURE 968 MB
MAX SUSTAINED WINDS 95 KT WITH GUSTS TO 115 KT
64 KT.......100NE 30SE 0SW 25NW
50 KT.......150NE 75SE 30SW 50NW
34 KT.......175NE 150SE 100SW 80NW
12 FT SEAS..300NE 150SE 150SW 275NW
ALL QUADRANT RADII IN NAUTICAL MILES

REPEAT...CENTER LOCATED NEAR 26.6N 86.2W AT 26/2100Z AT 26/1800Z CENTER
WAS LOCATED NEAR 26.3N 85.8W

FORECAST VALID 27/0600Z 27.1N 87.2W

MAX WIND 95 KT...GUSTS 115 KT
64 KT...100NE 45SE 30SW 30NW
50 KT...150NE 75SE 45SW 60NW
34 KT...175NE 150SE 100SW 90NW

FORECAST VALID 27/1800Z 28.1N 88.5W
MAX WIND 100 KT...GUSTS 120 KT
64 KT...100NE 45SE 30SW 45NW
50 KT...150NE 75SE 45SW 60NW
34 KT...175NE 150SE 100SW 100NW

FORECAST VALID 28/0600Z 28.9N 89.3W
MAX WIND 100 KT...GUSTS 120 KT
64 KT... 90NE 45SE 30SW 45NW
50 KT...120NE 75SE 45SW 60NW
34 KT...150NE 150SE 100SW 90NW

STORM SURGE FLOODING OF 10 TO 15 FT...LOCALLY HIGHER...ABOVE NORMAL
TIDE LEVELS IS POSSIBLE IN THE WARNED AREA AND WILL BE ACCOMPANIED BY
LARGE AND DANGEROUS BATTERING WAVES.

SMALL CRAFT FROM INTRACOASTAL CITY TO HIGH ISLAND TEXAS SHOULD
REMAIN IN PORT. SMALL CRAFT ALONG THE WEST COAST OF THE FLORIDA
PENINSULA SHOULD REMAIN IN PORT UNTIL WINDS AND SEAS SUBSIDE.

REQUEST FOR 3 HOURLY SHIP REPORTS WITHIN 300 MILES OF 26.6N 86.2W

EXTENDED OUTLOOK...USE FOR GUIDANCE ONLY...ERRORS MAY
BE LARGE

OUTLOOK VALID 28/1800Z 29.7N 89.7W...INLAND
MAX WIND 100 KT...GUSTS 120 KT
50 KT... 50NE 50SE 50SW 50NW

OUTLOOK VALID 29/1800Z 30.5N 90.0W...INLAND
MAX WIND 65 KT...GUSTS 80 KT
50 KT... 30NE 50SE 45SW 30NW

NEXT ADVISORY AT 27/0300Z
RAPPAPORT

CENTER WEATHER ADVISORIES

Center weather advisories (CWAs), unscheduled inflight advisories, are issued when conditions are expected to significantly affect IFR operations and help pilots avoid hazardous weather. The advisories update or expand the AIRMET Bulletin, SIGMETs, or the area forecast, and they might be issued when conditions meet advisory criteria. In such cases, the center weather service unit will coordinate with aviation weather center forecasters for the issuance of the appropriate advisory. CWAs are also issued when local hazardous conditions develop that do not warrant other advisories. Because they often report localized phenomena, the area might be described using locations other than those on the Area Designators map, or VORs on the Weather Advisory Plotting Chart.

The CWA numbering system was somewhat complex but has since been simplified. CWAs have a three-digit number. The first digit is a phenomenon number—that is, a specific weather event that required the issuance of the CWA. A separate phenomenon number will be assigned each distinct condition (turbulence, icing, thunderstorms, etc.). For example, 101 may forecast a turbulence event; 201 might be issued for icing. The second and third digits indicate the number of times a specific phenomenon event has been updated. For example, 101 (first issuance), 102 (second issuance), and so on.

Below is an example of a Denver Center CWA:

ZDV CWA 101 132050-132250
CAUTION FOR MOD-SEV TURBC/MTN WAVE ACTVTY PSBL ALL FLGT LVLS OVER
AND NEAR MTNS IN WY AND CO. STRONG CLD FNT MOVG THRU AREA THIS AFTN
AND EVE WITH PSBL SFC WND GUSTS TO 60KT. HIGH WND WARNINGS ARE IN
EFFECT FOR FNT RANGE AREA BEGINNING 14/0000Z. THIS ADVRY SUPPLEMENTS
SIGMET NOVEMBER 2.

This Denver Center CWA describes phenomenon number 1. It is the first issuance for this phenomenon (ZDV CWA 101), valid from the 13th day of the month at 2050Z until 2250Z. If the advisory requires updating at 2250Z, it will become ZDV CWA 102. The CWA advises caution for moderate to severe turbulence associated with mountain wave activity for all flight levels over and near the mountains of Wyoming and Colorado. Additionally, a strong cold front moving through the area during the afternoon and evening might bring surface winds with gusts to 60 kn. High-wind warnings are in effect for the Front Range area beginning on the 14th at 0000Z. This advisory supplements SIGMET NOVEMBER 2.

This example illustrates many uses of the CWA. The advisory expands on SIGMET NOVEMBER 2 by indicating the phenomenon is due to mountain wave activity and mechanical turbulence. The CWA also mentions high-wind warnings for the Front Range. The Front Range, normally a local geographical reference, refers to the mountains just west of Denver, from Fort Collins to about 40 mi southwest of Denver.

ZHU CWA 101 061355-061455
FM A BPT to 40SE LFT LN..S 150 MI INTO GULF...AREA SCT INTST 3-5 TSTMS MOVG
N 15 KTS. NMRS TOPS ABV 450.

The Houston Center CWSU has issued this advisory for an area of scattered thunderstorms, intensity level 3 to 5, moving north at 15 kn, with numerous tops to above 45,000 ft MSL. The area extends along a line from Beaumont, Texas, to 40 mi southeast of Lafayette, Louisiana, south 150 mi into the Gulf of Mexico. These locations are not on the Weather Advisory Plotting Chart. The condition has not yet met the criteria for a WST.

CWSUs also issue *meteorological impact statements* (MISs). Strictly an in-house product, the MIS alerts controllers of weather that might affect the flow of IFR traffic. The MIS describes conditions already contained in other advisories and forecasts. From time to time overzealous FSS briefers might refer to an MIS or tower controllers might record it on the *automatic terminal information service* (ATIS).

DISSEMINATION

Advisories are routinely provided during FSS standard, and they are offered during abbreviated briefings. (FSS weather briefings will be discussed in detail in Chap. 19.)

During routine FSS radio contacts, advisories within 150 mi will be offered when they affect the pilot's route. It's important to note SIGMET series and number to ensure receipt of the latest information.

In the contiguous United States, Hazardous Inflight Weather Advisory Service (HIWAS) has been commissioned. Advisories and urgent PIREPs are broadcast continuously over selected VORs. The availability of HIWAS can be determined from aeronautical charts and the *Airport/Facility Directory*.

When a WA, WS, WST, AWW, or CWA affects an area within 150 mi of a HIWAS outlet or an ARTCC sector's jurisdiction, an alert is broadcast once on all frequencies—except flight watch and emergency. Approach controls and towers also broadcast an alert, but it may be limited to phenomena within 50 mi of the terminal. When the advisory affects operations within the terminal area, an alert message will be placed on the ATIS. Here again, overzealous controllers have been known to place SIGMET alerts for conditions hundreds of miles away on the ATIS.

In spite of criticism that advisories cover too much area, their issuance has become more conservative. Ironically, some pilots and briefers now criticize the forecast for not containing enough precautions. Virtually all criticism, however, is due to misconceptions and misunderstanding about the product.

The existence of an advisory, or lack thereof, does not relieve the pilot from using good judgment and applying personal limitations. Like all pilots, I have had on occasion to park my turbo Cessna 150 and take one of American's Boeing 757s. These instances lend credence to the aviation axiom: "When you have time to spare, go by air, more time yet, take a jet." When you don't have the equipment or qualifications to handle the weather, don't go! This doesn't mean every time we hear an advisory we cancel; but we do take a close look at all available information—to develop the complete picture.

CASE STUDY   For a flight from Las Vegas, Nevada, to Van Nuys, California, I was told by the briefer: "Well, you aren't going today!" My jaws locked up, and I replied: "Oh yes, I am!" I hadn't looked at the weather yet; my statement was a gut reaction to this individual's horrible technique. Advisories for turbulence, mountain obscurement, and rain showers were in effect. As is often the case in this part of the country, a direct flight was out. But, by choosing a course over lower terrain, VFR is frequently possible. My decision was based on my experience, knowledge of the terrain, a thorough review of all available weather reports and forecasts, and always having an out should the weather ahead become impassable. When developing the complete picture, a knowledge of the terrain is just as important as the weather.

The lack of an advisory does not guarantee the absence of hazardous weather. An unfortunate pilot learned this lesson the hard way.

CASE STUDY The synopsis described a moist unstable air mass. Thunderstorms were not forecast for the time of flight but were expected to develop; thunderstorms, however, were already being reported along the route. The pilot, without storm detection equipment, encountered extreme turbulence inadvertently entering a cell. The pilot, with three passengers, filed an IFR flight plan based on the fact that there were no advisories. About a half hour into the flight, according to the pilot's statements to the FAA and National Transportation Safety Broad (NTSB) from the NTSB report, "We noticed a heavy layer of clouds at and below our altitude and some 20 mi ahead....The layer in front of us seemed to be light cumulus with a heavier layer behind it (not ominous looking)." After the encounter, the pilot could not understand why he was "never given a precaution or advisory regarding that system!" He went on to say that the accident "would not have happened if [the] pilot had been aware of weather conditions...." There were no advisories in effect because, at the time of the briefing, none were warranted. The pilot had the clues—moist unstable air and thunderstorms already reported—but he put complete trust in a forecast that included no flight precautions or advisories.

The preceding examples illustrate two go decisions. One resulted in a routine flight, the other in an almost fatal accident. My intent is not to brag about my skills or criticize another individual. Instead, I hope to emphasize the decision-making process as based on available information, a knowledge of weather products, and limitations.

CHAPTER 15
AREA FORECASTS

In this chapter we will discuss various written and graphic *area forecast* (FA) products designed primarily for the enroute phase of flight. Written products include domestic FAs, along with those for Alaska and Hawaii, and forecasts available for the Gulf of Mexico and Caribbean. These products are available from FAA flight service stations and DUATs and via the Internet. One Internet site is the Aviation Weather Center (AWC) at www.awc-kc.noaa.gov. Graphic products include significant weather prognostic charts and the tropopause wind and wind shear and volcanic ash forecast transport and dispersion chart. These products are available on the Internet at http://weather.noaa.gov/fax/nwsfax.shtml.

Pilots, briefers, and meteorologists share misconceptions and misinterpret the purpose and scope of the area forecast.

> **CASE STUDY**   A forecast for the desert portion of a fuel proficiency air race predicted scattered thunderstorms and rain showers, with wind gusts to 35 kn (SCT TSRA G35KT). Pilots criticized the forecast, complaining they didn't encounter any gusts; by avoiding the thunderstorms they remained clear of the winds. The thunderstorms were there, and you can bet gusty winds could be found in the vicinity of the cells. The forecast was perfectly correct.

Some people will criticize an FA for being too lengthy, then ironically, in the next breath, for not containing enough detail. A west coast FSS manager was quoted (April 1986, *Pacific Flyer,* Lance Stalker) as saying, "Before, you had people that were familiar with the local conditions and put that into their forecasting." That's still true. Local NWS offices issue TWEB route and terminal aerodrome forecasts (TAFs). The area forecast does not now, nor has it ever been intended to, cover every single condition.

The area forecast predicts conditions over an area the size of several states. Due to limitations on size, computer storage, and communications equipment, the forecast cannot be divided into smaller segments, nor can it provide the detail available in other forecast products. Widely varying conditions over relatively large areas must be included; therefore small-scale events are often described using conditional terms, such as occasional, isolated, and widely scattered (Table 14-1). The FA provides a forecast for the enroute portion of a flight and destination weather for locations without TAFs. This contradicts a widely held notion that without a TAF there is no destination forecast. Conditions are forecast from the surface to 70 mb (approximately 63,000 ft).

Area forecasts in the 1960s were issued every 6 h, and they remained valid for 12 h, with a 12-h outlook. This was time-consuming for the forecaster and, therefore, expensive. Area forecasts in the 1970s were issued twice a day, and they were valid for 18 h with an additional 12-h categorical outlook (IFR, MVFR, or VFR). Today the FA is issued three times a day, and it is valid for 12 h, with a 6-h outlook. The increased number of issuances and reduced valid times directly reflect forecast limitations.

TABLE 15-1   Area Forecast Issuance Times

SFO and SLC	CHI and DFW	BOS and MIA
1045/1145Z	0945/1045Z	0845/0945Z
1945/2045Z	1845/1945Z	1745/1845Z
0245/0345Z	0145/0245Z	0045/0145Z
HNL	JNU	ANC and FAI
0345Z	0645Z	0645Z
0945Z	1345Z	1445Z
1545Z	2245Z	2245Z
2145Z		

Area forecasts for the 48 contiguous states were reduced to six in 1982. Rather than being issued by local offices, responsibility was transferred to the National Aviation Weather Advisory Unit (NAWAU) in Kansas City. During this period the area forecast was divided into five sections: HAZARDS, SYNOPSIS, ICING AND FREEZING LEVEL, TURBULENCE AND LOW-LEVEL WIND SHEAR, and SIGNIFICANT CLOUDS AND WEATHER.

In 1991 with the introduction of the AIRMET Bulletin, the FA was reduced to a SYNOPSIS and VFR CLDS/WX (VFR clouds and weather). Alaska WFO offices in Anchorage (ANC), Fairbanks (FAI), and Juneau (JNU) issue FAs three times a day; the Honolulu (HNL) WFO issues FAs four times a day. Alaskan and Hawaiian FAs are depicted in App. B and use a similar format as those in the contiguous states. The Miami WFO issues a Gulf of Mexico FA for the area west of 85° west longitude and north of 27° north latitude, which includes the coastal plains and waters from Apalachicola, Florida, to Brownsville, Texas.

Appendix B, Graphs and Charts, depicts FA coverage areas. Table 15-1 contains FA issuance times. Note that UTC or ZULU issuance times within the 48 contiguous states change twice a year with daylight savings.

The area forecast begins with a heading describing the coverage of the product and valid times:

DFWC FA 210945
SYNOPSIS AND VFR CLDS/WX
SYNOPSIS VALID UNTIL 220400
CLDS/WX VALID UNTIL 212200...OTLK VALID 212200-220400
OK TX AR TN LA MS AL AND CSTL WTRS

This is the Dallas Ft. Worth FA issued on the 21st day of the month at 0945Z (DFWC FA 210945). (The C in DFW C is left over from the FAs prior to 1991. In those days DFWC was the SIGNIFICANT CLOUDS AND WEATHER section of the FA.) The next line identifies this as the synopsis and VFR clouds and weather. The synopsis is valid until the 22d at 0400Z (18 h). The clouds and weather section is valid until the 21st at 2200Z (12 h), with an outlook from the 21st at 2200Z until the 22d at 0400Z (6 h). This FA covers the states of Oklahoma, Texas, Arkansas, Louisiana, Mississippi, and Alabama, and it includes the adjacent coastal waters (OK TX AR TN LA MS AL AND CSTL WTRS). *Coastal waters* are defined as the area from the coastline to the *domestic flight information region* (FIR) boundary—typically a distance of 100 nautical miles.

Following the heading is the disclaimer paragraph:

SEE AIRMET SIERRA FOR IFR CONDS AND NTM OBSCN.
TS IMPLY SEV OR GTR TURBC SEV ICG LLWS AND IFR CONDS.
NON MSL HGTS DENOTED BY AGL OR CIG.

The first sentence refers the user to AIRMET SIERRA for IFR conditions and areas of mountain obscuration. Often when IFR is forecast, the body of the FA will contain only a tops forecast. For example, AIRMET SIERRA may forecast ceilings and visibilities below 1000 ft and 3 mi. The FA may contain only: ST TOPS 030 (stratus tops 3000 ft). The bases of the clouds and visibilities are contained in the AIRMET Bulletin.

> **FACT** Since 1991 the area forecast has not been a standalone product. For ceilings and visibilities, AIRMET SIERRA must be consulted. This has led to confusion for pilots and briefers alike.

From AC 00-6, *Aviation Weather*: "A thunderstorm packs just about every weather hazard known to aviation into one vicious bundle." To eliminate redundancy and serve as a "we told you so," the following statement appears on every FA: TS IMPLY SEV OR GTR TURBC SEV ICG LLWS AND IFR CONDS (Thunderstorms imply possibly severe or greater turbulence, severe icing, low-level wind shear, and IFR conditions). A report or forecast of thunderstorms implies these and other hazards associated with thunderstorms (hail, lightning, gusty winds, and altimeter errors). The body of the FA will not specifically address these hazards, nor will briefers normally include this statement. The fact that thunderstorms are reported or forecast infers all associated thunderstorm hazards!

NON MSL HGTS DENOTED BY AGL OR CIG: This statement simply means all heights are above *mean sea level* (MSL), unless noted as *above ground level* (AGL) or *ceiling* (CIG). This distinction can be significant, especially in mountainous areas. Forecasts for mountain states (the west and Appalachian states) will normally reference cloud bases to MSL, while forecasts for flat terrain (midwest and east coast) will normally reference bases to AGL or CIG. When comparing the FA with METARs, this differentiation must be considered.

> **CASE STUDY** One evening several FSS controllers brought the east of the Cascades portion of the SFO FA to my attention. They contended the observations and the forecast had nothing in common. However, after converting the METARs to MSL heights, the observations and forecast were perfectly consistent. This example emphasizes the point that to apply a forecast, a pilot or briefer must have a thorough knowledge of terrain.

SYNOPSES

The synopsis describes the location and movement of pressure systems and fronts and weather patterns, usually as a brief, generalized statement. The following example is quite detailed:

COLD UPR SYS OFF WA CST AT 19Z INVOF 48N 127W MOVG EWD AT ABT 10-15 KTS. PVA INDUCED/ENHANCED CNVTV CLDS WERE ROTG ONSHR FROM NRN CA THRU SWRN WA. AMS THIS RGN APPRS QUITE MOIST AND UNSTBL AND WL SPRD ACRS NRN HLF OF FCST AREA THRU THE PD.

A cold upper-level low-pressure system is off the Washington coast at 11 a.m. PST about 100 mi west of Seattle moving east at about 10 to 15 kn. *Positive vorticity advection* (PVA) is inducing and enhancing convective clouds that were rotating onshore from northern California through southwestern Washington. The air mass in this region appears quite moist and unstable and will spread across the northern half of the forecast area through the period.

I prefer a detailed synopsis because my training and experience allow extra insight regarding the weather situation. This is especially important if the weather improves or deteriorates more rapidly than forecast. This synopsis would normally be summarized during

FSS briefings and on broadcasts: "An upper-level low off the Pacific Northwest is bringing moist unstable air over Washington, Oregon, and Northern California." Translating and summarizing in this manner is a prime function of FSS weather briefers. Pilots using DUAT will have to decode, translate, and interpret the synopsis on their own.

Based on the preceding synopsis, a pilot might conclude that on a flight from northern California to Washington, the best weather lies to the east, ahead of the system. The example also contradicts a widely held misconception that fronts are the only weather producing systems.

A synopsis describes the cause of the weather; therefore, language and detail will depend on the situation. The synopsis will vary from the lengthy detail in the previous example to HI PRES OVR THE ERN GLFALSK WL MOV OVR THE PNHDL AND WKN (high pressure over the eastern Gulf of Alaska will move over the panhandle and weaken).

The importance of the synopses cannot be overemphasized. For example, the significance of a forecast for IFR conditions will depend on whether the IFR situation is due to a coastal marine layer, upslope fog covering several states, a frontal system, or a tropical storm. Here's another example:

```
CHIS FA 300940
SYNOPSIS VALID UNTIL 310400
AT 10Z CDFNT FROM LS SWWD THRU NWRN IA INTO NWRN KS THEN WWD THRU
CO. HI PRES OVR OH VLY AND MT. THE CDFNT WL CONT EWD AND BY 00Z WL
EXTEND FROM LWR MI SWWD INTO SRN KS AS HI PRES BLDS OVR DKTS. MRNG
FOG/ST OVR ERN GRTLKS WL IPV BY 16Z. AFTN/EVE TSTMS MOST ACTV ALG FNT
FROM IL NEWD THRU MI...WILLIAMS...
```

This Chicago FA SYNOPSIS (CHIS FA) was issued on the 30th day of the month at 0940Z (300945), and it is valid until the 31st at 0400Z. The synopsis covers the entire 18-h forecast period. At 1000Z a cold front extends from Lake Superior southwestward through northwestern Iowa into northwestern Kansas, then westward through Colorado. High pressure dominates the Ohio Valley and Montana. The cold front will continue eastward and by 0000Z will extend from lower Michigan southwestward into southern Kansas as high pressure builds over the Dakotas. The morning fog and stratus over the eastern Great Lakes will improve by 1600Z. Afternoon and evening thunderstorms will be most active along the front from Illinois northeastward through Michigan. This forecast was prepared by Williams. The CHIS synopsis has been plotted in Fig. 15-1. Notice the ease of visualizing conditions.

VFR CLOUDS AND WEATHER

The VFR clouds and weather section includes the following:

- Sky conditions
- Non-IFR cloud heights
- Visibility
- Weather and obstructions to visibility
- Surface winds
- Outlook

Sky conditions contain cloud height, amount, and tops. Heights are normally MSL, with AGL and CIG generally limited to layers within 4000 ft of the surface. (Recall from

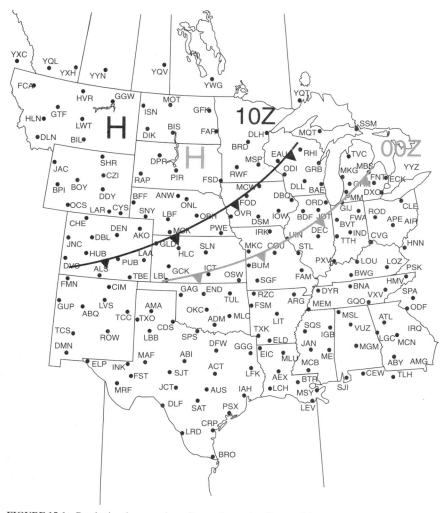

FIGURE 15-1 By plotting the synopsis, a pilot can better visualize conditions and the expected movement of weather systems.

Chap. 9, Pilot Weather Reports, how it took several years to standardize the format for cloud bases and tops. Now all aviation forecast products report cloud bases and tops using the same format.) Cloud tops are always referenced to MSL. Since tops of building cumulus, towering cumulus, and cumulonimbus are quite variable, only upper limits appear (CB TOPS FL300). Note that tops at or above 18,000 ft are referenced to pressure altitude or *flight level* (FL). When multiple or merging layers are forecast that would not permit VFR flight between layers, only the top of the highest layer appears (BKN-OVC080-100 LYRD TOPS TO FL200, MEGG/NMRS LYRS TOPS TO FL180). Because of its scope, FA tops cannot be more precise. TWEB route forecasts may contain more detail.

Surface visibility appears in the forecast when expected to be 6 mi or less. To be included, the area affected must be at least 3000 mi^2. For example, 3-5SM -RA BR (visibility 3 to 5 mi in light rain and mist), or VIS 3-5SM -SHRA AND WDLY SCT - TSRA (visibility 3 to 5 mi in light rain showers and widely scattered light rain showers and thunderstorms). Table 14-1, Weather Advisory, Area Forecast Conditional Terms, defined WDLY SCT as less than 25 percent of the area. Because of the scope of this product, the forecast cannot be more precise. A pilot can interpret this forecast to mean the thunderstorms should be circumnavigable.

The absence of a visibility forecast only implies general visibilities greater than 6 mi. Since they are not within the scope of this product, widespread visibilities of greater than 6 mi, or local conditions less than 6 mi, may exist and not be included in the FA. TWEB route and terminal aerodrome forecasts may contain greater detail, especially on local conditions.

Weather and obstructions to visibility use standard ICAO weather abbreviations. As in the preceding examples, weather and obstructions may be included with visibility.

Widespread areas of surface winds that are expected to be operationally significant appear in the forecast (20G30KT). Often associated with convective activity, TSRA G40KT translates to wind gusts of 40 kn expected to accompany thunderstorms. Direction is true, referenced to the eight points of the compass (N, NE, E, etc.). The lack of a wind forecast only implies widespread sustained speeds less than 20 kn. TWEB route and terminal aerodrome forecasts can often be used to determine winds of lesser speeds and local conditions. Below is an example of how surface winds appear in the FA:

```
NRN CA—STS-SAC-TVL LN NWD
NERN CA/NRN SIERNEV...20Z BKN100. WNDS NWLY G20KT.
OTLK...VFR.

CNTRL CA
MTNS...15Z SCT150. WND NLY G25KT. OTLK...VFR.
```

Note that these winds apply only to northeastern California and the northern Sierra Nevada mountains and the mountains of central California. These winds are not expected in the central valley or coastal sections.

The forecaster divides the forecast area using standard geographical designators as found in App. B. The extent and detail will depend on the weather situation. The example below illustrates a standard division of the Dallas Ft. Worth FA:

```
SWRN TX
WEST OF PECOS RVR...SKC OR SCT CI.
EAST OF PECOS RVR...AGL SCT030 SCT100. ISOLD -RA/-TSRA.
```

This portion of the DFW FA covers southwestern Texas (SWRN TX). The forecaster has further divided the area into west and east of the Pecos River. This is a common feature in this FA. (If the Pecos River ever dries up, I don't think they'll be able to write a DFW FA.)

Below is another example of a VFR CLDS/WX section. The synopsis indicates midlevel moisture, with a stable air mass in the valleys. The following is a typical situation during the winter months for northern and central California:

```
NRN/CNTRL CA
CNTRL VLYS AND CSTL VLYS SFO NWD..ST TOPS 020-025 CNTRL VLYS AND 010-020
CSTL VLYS. CONDS IPVG CSTL PTN 18-21Z.
ELSW.. SCT150 CI ABV. OCNL BKN100 TOPS 150 NRN CSTL WTRS AND CSTLN WITH
SCT -RA.
```

In this forecast the meteorologist has divided California, specifying this portion for northern and central sections. The forecaster has further divided the area within the text of the FA. Figure 15-2 graphically displays the forecast. Below each excerpt from the area designators map in Fig. 15-2 is a portion of the written forecast. The gray shaded area represents the affected area. The forecast for the central valleys (Sacramento and San Joaquin) and coastal valleys San Francisco northward—ceilings below 1000 ft, visibilities below 3 mi in fog—are contained in AIRMET SIERRA. Therefore, it is omitted from the FA. Stratus tops are expected between 2000 and 2500 ft in the central valleys and 1000 and 2000 ft in coastal valleys; coastal valleys are forecast to improve between 18Z and 21Z. Notice how the forecaster specifies a period (18Z–21Z) rather than an exact time—another reflection on the limitations of forecasts.

Radiation fog has formed from the moisture of previous storms trapped in valleys and cooled at night under stable air. The relatively shallow layer in the coastal valleys is expected to improve by midday. Conditions in the central valleys will continue. VFR flights will be delayed until afternoon in the coastal valleys, and they most probably will not be possible at any time in the central valleys. IFR operations will be possible assuming the pilot has takeoff, landing, and alternate minimums. Since ceilings of less than 1000 ft and visibilities of less than 3 mi are the lowest values normally found in the FA, TWEB route and terminal aerodrome forecasts should be consulted. It is not within the scope of this product to provide more detail. A pilot planning an IFR flight into an airport without a TAF must specify an alternate. Additionally, airports within this area, without a TAF, do not satisfy alternate requirements. Pilots must specify an airport with a TAF forecasting alternate minimums, or an airport out of the affected area. VFR operations above the valley fog will not be restricted. However, both VFR and IFR pilots may be flying above extensive areas of zero-zero conditions. This will pose a risk to single-engine operations in case of engine failure. Pilots should weigh this danger carefully.

Elsewhere, the forecast for the coastal mountains (including northern coastal mountains above the stratus), northern mountains, Sierra Nevada mountains, and the coast south of San Francisco is 15,000 MSL scattered with cirrus above. There will be no restrictions to either VFR or IFR operations in these areas. Landing fields in these areas would be suitable IFR alternates for valley airports. Additionally, over the northern coastal waters and coastline, due to sufficient moisture at midlevels, occasional 10,000 MSL broken tops 15,000 MSL with a chance of light rain is expected.

Note that the area forecast may use the conditional term TEMPO, the international code for occasional; forecasters may also use BECMG to indicate a change period (BECMG 1720, becoming between 17Z and 20Z).

OUTLOOKS

A 6-h categorical outlook (OTLK) appears at the end of each 12-hour VFR CLDS/WX statement—18 hours for Alaskan FAs. These are the same categories described in Table 12-2, Weather Categories. The OTLK consists of IFR (instrument flight rules), MVFR (marginal visual flight rules), and VFR (visual flight rules). (Note that these categories do not necessarily correspond to 14 CFR Part 91 definitions.) Additionally, when sustained surface wind or gusts are expected to be 20 kn or more, the contraction WND appears.

References to IFR or MVFR explain the phenomenon causing the condition. For example:

- *IFR CIG.*   Ceiling less than 1000 ft.
- *IFR CIG BR.*   Ceiling less than 1000 ft and visibility less than 3 mi.

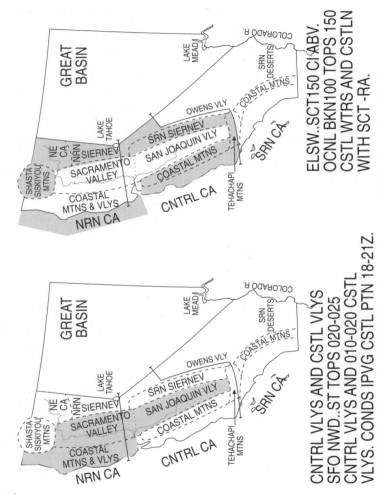

CNTRL VLYS AND CSTL VLYS
SFO NWD..ST TOPS 020-025
CNTRL VLYS AND 010-020 CSTL
VLYS. CONDS IPVG CSTL PTN 18-21Z.

ELSW..SCT150 CI ABV.
OCNL BKN100 TOPS 150
CSTL WTRS AND CSTLN
WITH SCT -RA.

FIGURE 15-2 Without a knowledge of geographical area designators, it is difficult to apply a written forecast.

15-8

- *MVFR HZ FU.* Visibilities between 3 and 5 mi in haze and smoke.
- *VFR WND.* Ceiling greater than 3000 ft and visibility greater than 5 mi; sustained surface wind or gusts 20 kn or greater for 50 percent or more of the outlook period.

A *categorical outlook* is another direct reflection on the limitations of aviation forecasts. The outlook is based on synoptic-scale events and might not contain local conditions. Pilots must carefully consider an outlook. VFR cannot be interpreted as clear, although conditions might actually be clear. VFR translates as a ceiling greater than 3000 ft AGL and visibility greater than 5 mi; how much "greater" is not specified. Often, mountain obscurement is not considered. VFR might be expected in valleys, while VFR flight through mountainous areas might not be possible. This category is an indicator that airports within the area will not require an IFR alternate.

MVFR cannot necessarily be interpreted as allowing VFR operations. If conditions are at the lower limit of the category, IFR, most often, would be required. This category is an indicator that airports will be above instrument minimums but will require an alternate. IFR indicates that VFR flight is out. The IFR pilot, however, cannot interpret this category as indicating an airport will be above instrument minimums. It indicates an IFR alternate will be required.

A word of caution to the IFR pilot: Outlooks are merely indicators. Categorical outlooks can never be used to determine IFR alternate requirements or suitable IFR alternate airports. Regulatory requirements must be met based on the latest forecasts prior to departure.

AREA FORECAST UPDATE CRITERIA

Weather advisories (AIRMETs, SIGMETs, CWAs, etc.) automatically amend the area forecast. When updated, the contraction AMD (amendment) appears in the forecast header (SLCC FA 141220 AMD), and the amended section of the forecast is indicated by the contraction UPDT (update). The SYNOPSIS will normally be updated with a significant change in the synoptic pattern. The CLDS/WX section is updated whenever the weather improves or deteriorates—that is, when current or expected weather changes significantly from what was forecast, making the FA, in the judgment of the forecaster, unrepresentative. Updated paragraphs are indicated by the contraction UPDT in the body of the forecast (ID MT...UPDT). The OTLK will be amended when a change from one or more of the categories is expected (VFR to MVFR, IFR to MVFR, etc.). A change in the wind outlook alone, however, is not sufficient for an amendment.

These are broad statements, and pilots must remember the FA describes conditions over large areas. We should not expect amendments for local or localized changes, which might be reflected in TWEB routes or TAFs, or their updates. When wide-scale changes do occur, the FA will be amended.

The following appeared in the FAA's June 1988 *Air Traffic Bulletin* written by forecaster Paul Smith:

A DAY ON THE FA DESK "A typical day shift begins at 6:45 a.m. with a briefing from the midnight forecaster. We keep this as brief as possible, explaining current conditions and any expected trouble spots for the upcoming day.

"Three forecasters start the shift together, one each for the East, Central, and West. (Each forecaster is responsible for two FAs: East-BOS and MIA, Central-CHI and DFW, and West-SLC and SFO.) The three forecasters' work areas are adjacent to one another to allow easy coordination.

"Several things are routine on every shift. For instance, surface maps are analyzed every two hours to keep up with current weather conditions over the area. PIREP collectives alarm at our consoles twice an hour and are displayed both in text form and graphically on a computer-generated map. [This emphasizes the importance of PIREPs.] Surface and upper-level guidance material is received from NCEP in Washington, DC.

"Many other things are not routine and occur as weather conditions warrant. When SIGMETs are in effect, or are being considered, coordination with the CWSUs occurs regularly. AIRMETs and amendments must be issued if forecast conditions go sour. When things of this type occur on a shift, the forecaster may become rushed to meet product deadlines.

"Product composition is accomplished on computer terminals. Once composed and checked, our products are transmitted to WMSC [FAA's Weather Switching Center in Kansas City, MO] for nationwide dissemination. While we do not have backup procedures in case of computer failure, delayed FAs can and do occur due to computer or communications failure.

"Each forecaster has a different routine in preparing a forecast. Usually, the meteorologist spends the first hour or so on shift analyzing the current weather situation and reviewing computer guidance concerning the evolution of weather systems over the following six to 24 hours. SIGMETs, AIRMETs, and forecast amendments have high priority and are issued as needed to keep briefers up-to-date.

"Approximately three hours prior to FA issuance, the forecasters begin work in earnest on the two FAs being written. Most forecasters draw up tentative outlines for their flight precaution areas, then consult with the person writing the adjacent forecast to develop a common set of VOR points to describe the entire area. Transmission times are staggered; the BOS and MIA FAs are transmitted first, followed an hour later by CHI/DFW and finally SLC/SFO. The issue time differences make early coordination a necessity. Thus, for flight precaution areas extending from the Rockies to the East Coast, the west forecaster must make decisions very early on the aerial outline of expected weather conditions. IFR areas, because of their changeable nature, are often the last VOR outline to be "nailed down." Negotiation and point changes are made right up to our transmission deadline.

"The SGFNT CLDS AND WX [now VFR CLDS AND WX] portion of the FA is usually the last section to be composed. The forecaster makes use of many separate sources of information to develop the FA including METARs, PIREPs, TAFs, TWEBs, satellite imagery, radar, radiosonde data, prognostic charts, standard level charts, and CSIS [the NWS] interactive computer system. During periods of extensive adverse weather conditions with multiple flight precautions and frequent conversations with CWSU meteorologists, the FA forecaster is particularly busy as transmission deadline approaches."

ALASKAN, HAWAIIAN, AND GULF OF MEXICO/CARIBBEAN AREA FORECASTS

In addition to the differences already mentioned between the Alaskan and contiguous states' FAs, the Alaskan FAs contain brief forecasts for designated major mountain passes. This statement will reference any appropriate AIRMET Bulletin or SIGMET and include a remark whether the passes are open (VFR), marginal (MVFR), or closed (IFR).

Below is an example of an Alaskan FA. Area designators can be found in App. B:

ANCH FA 071745 AAB
AK SRN HLF EXCP SE AK...

AIRMETS VALID UNTIL 072100
TS IMPLY POSSIBLE SEV OR GREATER TURB SEV ICE LLWS AND IFR CONDS.
NON MSL HEIGHTS NOTED BY AGL OR CIG.

SYNOPSIS VALID UNTIL 080900
LOW PRES OVER THE NRN GLFALSK WKNS SLOWLY THRU THE PERIOD. A 1002 MBLOW 120 NW OF PAMY MOVES OVER NRN BRISTOL BAY BY 09Z. THE HIGH PRES-RDG OVER THE CNTRL ALUTNS MOVES E TO NR PADU BY 09Z AND STARTS TOWKN. A NEW OCFNT MOVES OVER THE WRN ALUTNS BY 06Z.

COOK INLET AND SUSITNA VLY AB...VALID UNTIL 080300
...CLOUDS/WX...
FEW050 FEW100. SFC WNDS N G25 KT. VCY MT PASSES..HIER G40KT.
OTLK VALID 080300-082100...VFR WND.
PASSES...ALL PASSES TURBT.
LK CLARK..MERRILL..WINDY..VFR.
RAINY..VFR. WRN APCH ISOL MVFR CIG SHSN.
PORTAGE...MVFR CIG SHSN OCNL IFR CIG SHSN BLSN.
...TURB...
AIRMET TURBOCNL MOD TURB BLW 080. VCY CHANNELED TRRN ISOL SEV
TURB WI 020 AGL. NC...
...ICE AND FZLVL...
NIL SIG. FZLVL SFC.

COPPER RIVER BASIN AC...VALID UNTIL 080300
...CLOUDS/WX...
FEW045 SCT-BKN100 TOPS 120. ISOL -SHSN. ISOL CIGS BLW 010 VIS BLW 3SM
BR/FZFG. OTLK VALID 080300-082100...VFR.
PASS...TAHNETA...VFR TURBT.
...TURB...
NIL SIG.
...ICE AND FZLVL...
NIL SIG. FZLVL SFC.

CNTRL GLF CST AD...VALID UNTIL 080300
...CLOUDS/WX...
AIRMET MT OBSCMTS OCNL OBSC IN CLDS/PCPN. NC...
SCT015 BKN060 TOPS 090. OCNL BKN012 OVC040 TOPS 160 3SM -SHSN. ISOL CIGS
BLW 010 VIS BLW 3SM SN BLSN. SFC WNDS CHANNELED TRRN N 25G45KT. OTLK
VALID 080300-082100...MVFR CIG SHSN WND.
...TURB...
AIRMET TURBOCNL MOD TURB BLW 080. ISOL SEV TURB VCY CHNLD TRRN.
NC...
...ICE AND FZLVL...
AIRMET ICEOCNL MOD RIME ICEIC 0415-120. FZLVL SFC....

KODIAK IS AE...VALID UNTIL 080300 AAA
...CLOUDS/WX...
AIRMET MT OBSCMTS OCNL OBSC IN CLDS/PCPN. NC...
SCT015 SCT060 5SM BLSN. ISOL BKN-OVC015 TOPS 120 VIS BLW 3SM -SHSN BLSN.
SFC WNDS NW 25G40 KT. HIER G65 KT OFSHR.
OTLK VALID 080300-082100...MVFR CIG SHSN BLSN WND.
...TURB...AAA
AIRMET TURBOCNL MOD TURB BLW 080 AND FL300-360.
ISOL SEV TURB WI 020 AGL. NC...
...ICE AND FZLVL...
NIL SIG. FZLVL SFC.

KUSKOKWIM VLY AF...VALID UNTIL 080300
...CLOUDS/WX...
GENLY SKC.
ISOL CIGS BLW 010/VIS BLW 3SM IC BR/FG TOPS 015 VCY SETTLEMENTS.
OTLK VALID 080300-082100...VFR.
...TURB...NIL SIG.
...ICE AND FZLVL...NIL SIG. FZLVL SFC.

YKN-KUSKOKWIM DELTA AG...VALID UNTIL 080300 AAA
...CLOUDS/WX...
***AIRMET MT OBSC ***SLOLY DVLPG FM W..

MTS OCNL OBSC IN CLOUDS/PCPN ABV 015. NC...
ALL SXNS..FEW015 SCT030 BKN060 TOPS 100 FEW LYRS ABV TOPS FL250. W PAEM-
PAEH LN AND SPRDG SLOLY E ACRS AREA BY 03Z.. OCNL BKN015 OVC030 SCT 3SM
-SHSN. OFSHR AND ALG CST ALL PDS..SFC WND SW-NW 20G35 KTS.
OTLK VALID 080300-082100...MVFR CIG SHSN.
...TURB...AAA
AIRMET TURBPABE S..OCNL MOD TURB FL300-360. WKN...
ISOL MOD TURB SFC-050 VCY RUF TRRN.
...ICE AND FZLVL...
NIL SIG. FZLVL SFC.

.

BRISTOL BAY AH...VALID UNTIL 080300 AAB
...CLOUDS/WX...AAB
AIRMET MT OBSCPADL W SPRDG E THRU ALL AREAS BY 21Z..MT OBSC IN
CLDS AND SN. NC...
OTRW..SCT020 SCT-BKN070 TOPS 100. SPRDG FM W THRU ALL AREAS BY 21Z..
BKN015 OVC050 TOPS 130 3SM -SN. OCNL VIS BLW 3SM
-SN ISOL CIG BLW 010.

.

OTLK VALID 080300-082100...MVFR CIG SN. FM 15Z VFR.
...TURB...AAA
AIRMET TURBOCNL MOD TURB FL300-FL360. WKN...
...ICE AND FZLVL...AAB
SPRDG FM W THRU ALL AREAS BY 21Z..OCNL LGT RIME ICEIC 0115-130. OTRW..NIL
SIG. FZLVL SFC.

.

AK PEN AI...VALID UNTIL 080300 AAA
...CLOUDS/WX...
AIRMET IFR/MT OBSCBERING SIDE..OCNL CIGS BLW 010 VIS BLW 3SM -SN
BLSN. BOTH SIDES..OCNL MT OBSC IN CLDS/PCPN. NC... OTRW..SCT005 SCT025
BKN-OVC040 TOPS 090. OCNL SCT005 BKN-OVC015 3SM -SN BLSN. SFC WNDS NW
G215-30 KT.
OTLK VALID 080300-082100...MVFR CIG SHSN WND.
...TURB...AAA
AIRMET TURBOCNL MOD TURB BLW 060. NC...
...ICE AND FZLVL...
OCNL LGT RIME ICEIC 0115-090. FZLVL SFC.

.

UNIMAK PASS TO ADAK AJ...VALID UNTIL 080300
...CLOUDS/WX...
SCT025 SCT-BKN050 TOPS 070. ISOL SCT005 BKN-OVC020 3SM -SN BECMG OCNL AFT
21Z. SFC WNDS W G25 KT.
OTLK VALID 080300-082100...MVFR CIG SHSN.
...TURB...
VCY RUF TRRN..ISOL MOD TURB BLW 060.
...ICE AND FZLVL...
NIL SIG. FZLVL SFC.

The following is an example of a Hawaiian FA. Area designators can be found in App. B:

HNLC FA 290940
SYNOPSIS AND VFR CLDS/WX
SYNOPSIS VALID UNTIL 300400
CLDS/WX VALID UNTIL 292200...OTLK VALID 292200-300400

.

SEE AIRMET SIERRA FOR IFR CLDS AND MT OBSCN.
TS IMPLY SEV OR GTR TURB SEV ICING LLWS AND IFR CONDS.
NON MSL HGTS DENOTED BY AGL OR CIG.

SYNOPSIS...SFC RDG 400 NM N PHNL MOVG SOUTH SLOWLY.

ENTIRE AREA.
BKN-SCT250. LWR CLDS AND WX FLW.

WNDWD CSTL/MTN SXNS AND ADJ WNDWD CSTL WTRS OF THE BIG ISLAND.
SCT020 BKN-OVC040 TOPS 080 TEMPO BKN020 VIS 5SM -SHRA.
21Z SCT020 BKN-SCT040 TOPS 070 TEMPO BKN020 -SHRA. OTLK...VFR.

WNDWD CSTL/MTN SXNS AND ADJ WNDWD CSTL WTRS OF THE RMNG ISLANDS.
SCT025 SCT-BKN045 TOPS 070 ISOLD BKN025 -SHRA. OTLK...VFR.

KONA...KAU AND LEEWARD KOHALA SXNS.
SCT-BKN040 TOPS 080. 21Z FEW030 BKN050 TOPS 080. OTLK...VFR.

REST OF AREA.
SCT025 SCT045 ISOLD BKN040 TOPS 070 -SHRA. OTLK...VFR.

Two forecasts have been developed to support aviation operations in Gulf of Mexico and Caribbean. These are the Gulf of Mexico FA and the Atlantic, Caribbean, and Gulf of Mexico FA.

The Gulf of Mexico FA is issued twice a day at 1040/1140Z and 1740/1840Z depending on whether it's standard or daylight savings time. Aerial coverage is depicted in App. B. The forecast has been specifically developed to support helicopter operations. It is unique in that it is a single forecast combining the FA, weather advisories, and marine precautions. Each section describes the phenomenon impacting the respective area.

The following is an example of the Gulf of Mexico FA:

TTAA00 KNHC 071830
FAGX01 KNHC 071830
FCST...071900Z-080700Z
OTLK...080700Z-081900Z
AMD NOT AVBL 0200Z-1100Z

TROPICAL ANALYSIS AND FORECAST BRANCH
TROPICAL PREDICTION CENTER MIAMI FL

GLFMEX N OF 27N W OF 85W...CSTL PLAINS AND WTRS AQQ-BRO...HGTS MSL UNLESS NOTED.

TSRA IMPLY POSS SEV OR GTR TURB...SEV ICE...LOW LVL WS AND STG SFC WND...HIGH WAVES...CIG BLW 010...AND VIS BLW 3SM.

01 SYNS...
HI PRES OVR GLFMEX THRU FCST AND OTLK PD.

02 FLT PRCTNS...
NONE.

03 MARINE PRCTNS...
NONE.

04 SGFNT CLD/WX...
CSTL PLAINS CSTL WTRS BRO-LCH...
BKN/OVC0115-025 TOPS 040-060. 21Z BKN/SCT020-030. AFT 03Z AREAS VIS 3-4SM

BR...AFT 06Z BKN/OVC010 WDSPRD VIS 3-4SM BR...LOC VIS BLW 3SM BR/FG.
OTLK...LIFR CIG FG. 16Z MVFR CIG BR. 18Z VFR.

CSTL PLAINS CSTL WTRS LCH-AQQ...
SCT/BKN020-030 BKN050-060...ISOL -SHRA. AFT 04Z SCT/BKN010-015 AREAS VIS 3-
4SM BR...AFT 06Z BKN/OVC010 WDSPRD VIS 3SM BR...LOC VIS BLW 3SM BR/FG.
OTLK...IFR CIG BR. PNS-AQQ AFT 15Z VFR. LCH-PNS AFT 16Z MVFR CIG BR.

OFSHR WTRS W OF 90W...
FEW/SCT020-030. OTLK...VFR.

OFSHR WTRS E OF 90W...
SCT/BKN020-030...ISOL -SHRA. OTLK...VFR.

05 ICE AND FZ LVL BLW 120...
NONE. FZ LVL ABV 120.

06 TURB BLW 120...
NONE.

07 WND BLW 120...
CSTL PLAINS CSTL WTRS BRO-CRP...
SFC-010 S 10-15 KT. 010-080 SW 115-20 KT. 080-120 W 15 KT. OTLK...SFC-080 SLGT
INCR. 080-120 NOSIG.

CSTL PLAINS CSTL WTRS CRP-LCH AND OFSHR WTRS W OF 93W...
SFC-010 SW 10-15 KT. 010-080 SW 20-30 KT...AFT 00Z DCR SW 115-20KT. 080-120 W-NW
30-40 KT...AFT 00Z DCR NW 115-20 KT.
OTLK...NOSIG.

CSTL PLAINS CSTL WTRS LCH-AQQ AND OFSHR WTRS E OF 93W...
SFC-010 SW 10-15 KT. 010-060 SW 20-30 KT...OFSHR WTRS AFT 00Z DCR SW-W 115-20
KT. 060-120 SW-W 215-35 KT...AFT 00Z DCR W 115-25KT. OTLK...NOSIG.

08 WAVES...
CSTL WTRS BRO-PSX.....2-4 FT. OTLK...NOSIG.
CSTL WTRS PSX-BVE.....2-3 FT. OTLK...NOSIG.
CSTL WTRS BVE-AQQ.....3-4 FT. OTLK...SLGT DCR.
OFSHR WTRS............3-4 FT. OTLK...NOSIG.

The following is an example of the Atlantic, Caribbean, and Gulf of Mexico area
forecast. The forecast covers the western Atlantic, Caribbean, and Gulf of Mexico and
adjacent coast. Produced by the NWS's Tropical Prediction Center (TPC), this forecast
covers conditions from the surface to 400 mb (24,000 ft). Refer to App. B for a graphical depiction of coverage:

TTAA00 KNHC 071505
TROPICAL ANALYSIS AND FORECAST BRANCH
TROPICAL PREDICTION CENTER MIAMI FLORIDA
071800-080600

ATLANTIC S OF 32N W OF 57W...CARIBBEAN...GULF OF MEXICO AND ADJCOAST N
OF 23N...AND FLORIDA SFC TO 400 MB.

SYNOPSIS...

ATLANTIC DISSIPATING FRONT 21N58W 19N70W THROUGH 06Z. WEAK COLD FRONT WILL MOVE RAPIDLY ACROSS AREA N OF 30N E OF 70W THROUGH06Z. HIGH PRESSURE OVER ATLANTIC/GULF OF MEXICO NEAR 25N.

SIGNIFICANT CLD/WX...
ATLANTIC WI 30NM OF DISSIPATING FRONT
BKN/SCT0215-030 BKN060-080. WIDELY SCT SHRA.

ATLANTIC N OF 30N E OF 78W
SCT/BKN030-035 BKN/OVC050-070. ISOL -SHRA.

ATLANTIC ELSEWHERE N OF 25N BTN 68W-76W
SCT/LOC BKN030 BKN/OVC060. 00Z SCT/LOC BKN0315-045.

REMAINDER ATLANTIC
SCT020-030 SCT/BKN050-070.

CARIBBEAN S OF 15N E OF 75W
SCT/LOC BKN020-030 BKN/SCT050-070. ISOL -SHRA.

REMAINDER CARIBBEAN
SCT/LOC BKN030-040. ISOL -SHRA.

GULF OF MEXICO ADJ COAST W OF 93W
SCT/LOC BKN020-030. AFT 03Z AREAS BKN/OVC010-015 VIS 3SM BR.
GULF OF MEXICO ADJ COAST E OF 93W/GULF OF MEXICO N OF
28N/FLORIDA W OF 85W BKN/SCT0215-030 BKN/LOC OVC070-090. ISOL -SHRA.

REMAINDER GULF OF MEXICO
SCT/LOC BKN020-030. ISOL -SHRA.

FLORIDA N OF 28N E OF 85W
FEW/SCT0315-040. 00Z FEW030 SCT/BKN080-100.

REMAINDER FLORIDA
FEW/SCT0315-050...EXC OVER FLORIDA KEYS SCT/LOC BKN030 WITH ISOL -SHRA.

ICE AND FZ LVL...
MOD OR GREATER IN TCU/CB TOPS ABV FZ LVL.
NONE. NE OF 32N69W 26N61W LINE FZ LVL 100-120 SLOPING TO 1215-145 ELSE-
WHERE N OF 20N AND 150-160 S OF 20N.

TURB...
MOD OR GREATER NEAR TSRA OR +TSRA. NONE.

OUTLOOK 080600-081800...
NEXT ATLANTIC COLD FRONT WILL APPROACH FORECAST AREA W OF 75W BY18Z.
HIGH PRESSURE WILL REMAIN NEAR 24N/25N. OTHERWISE LITTLE CHANGE.

USING THE AREA FORECAST

Below is an example of a Boston AIRMET Bulletin and Boston area forecast used for the following discussion. It may be helpful to refer to Fig. 15-3 to better visualize conditions:

FIGURE 15-3 Forecasts for distinct and fast-moving weather systems are typically more accurate than those for diffuse and slow-moving systems.

BOSZ WA 301345
AIRMET ZULU UPDT 3 FOR ICG AND FRZLVL VALID UNTIL 302000
.
ME CSTL WTRS
LGT TO LCL MOD RIME ICGIC ABV FRZLVL BTW 080 AND 150 OVER ME AND NEW
ENG CSTL WTRS.
OTRW...NO SGFNT ICING EXPCD.

FRZLVL...50-80 ME. 80-100 NH VT MA RI CT NY LO LE OH. 100-110 NJ PA WV MD DC
DE VA.
.
BOST WA 301345
AIRMET TANGO UPDT 3 FOR TURBC VALID UNTIL 022000
.
AIRMET TURBC...NY PA OH LE LO WV MD
FROM MQT TO YOW TO HNK TO EKN TO LOZ TO ARG TO MCW TO MQT
MOD TO LCL SEV TURBC 200-400. CONDS DMSHG BY 20Z.

BOSS WA 301345
AIRMET SIERRA UPDT 3 FOR IFR AND MTN OBSCN VALID UNTIL 302000

AIRMET IFR...ME NH MA RI AND CSTL WTRS
FROM MLT TO YSJ TO 150SE ACK TO ORF TO PVD TO CON TO MLT
OCNL CIG BLO 10 VIS BLO 3F. CONDS CONTG NEW ENG AND CSTL WTRS BYD 20Z
THRU 02Z.

AIRMET MTN OBSCN...ME NH VT MA NY
FROM PQI TO CON TO 40SE ALB TO HNK TO YOW TO PQI
MTNS OSBCD IN CLDS. CONDS CONTG BYD 20Z IPVG BY 01Z.

OTLK VALID 2000-0200Z...MTN OBSCN WV VA
AFT 00Z MTNS OBSCD IN CLDS AND PCPN. CONTG THRU 02Z.

BOSC FA 301745
SYNOPSIS AND VFR CLDS WX
SYNOPSIS VALID UNTIL 011200
CLDS/WX VALID UNTIL 010600...OTLK VALID 010600-011200
ME NH VT MA RI CT NY LO NJ PA OH LE WV MD DC DE VA AND CSTL WTRS

SEE AIRMET SIERRA FOR IFR CONDS AND MTN OBSCN.
TSTMS IMPLY SEV OR GTR TURBC SEV ICG LLWS AND IFR CONDS.
NON MSL HGTS DENOTED BY AGL OR CIG.

SYNOPSIS...AT 18Z LOW WAS OVER LH WITH CDFNT CURVG THRU LO ERN PA
CNTRL NC NRN AL BCMG STNRY TO SRN TX. BY 12Z CDFNT WILL MOVE OFF NRN
AND MID ATLC CST CURVG ACRS SC NRN GA BCMG STNRY TO SRN TX.

ME NH
W OF MLT-CON...BKN030 BKN100 TOPS 150. AFT 01Z SCT-BKN050. OTLK...VFR.
E OF MLT-CON...OVC010-020 TOPS 150. VIS 3-5SM BR. SCT -RA. AFT 00Z SCT-BKN020
TOPS 040. VIS 3-5SM BR. OTLK...IFR CIG BR.

MA RI CT
CT WRN MA...BKN020-030 TOPS 060. AFT 00-03Z SKC. OCNL VIS 3-5SM BR.
OTLK...IFR BR.
RI ERN MA...OVC020 TOPS 050. VIS 3-5SM BR. AFT 21-00Z SCT-BKN020. VIS 3-5SM BR.
AFT 00-03Z SCT020 TOPS 040. VIS 3-5SM BR. OTLK...IFR CIG BR.

VT
BKN030 TOPS 080. AFT 01Z SCT040. OCNL VIS 3-5SM BR AFT 03Z. OTLK...MVFR BR.

NY LO
LO WRN AND NERN NY...BKN030 TOPS 060. AFT 01Z SCT040. OCNL VIS 3-5SM BR AFT
02Z. OTLK...MVFR BR.
SERN NY...SCT-BKN030 TOPS 060. AFT 22Z SKC. AFT 02Z SKC. VIS 3-5SM BR.
OTLK...IFR CIG BR.

NJ PA
SCT-BKN040 TOPS 060. ISOLD TSRA SRN NJ SERN PA TIL 00Z. CB TOPS FL350. VIS 3-
5SM BR. AFT 00Z SKC. VIS 3-5SM BR. OTLK...MVFR BR.

OH LE
SCT040. AFT 00Z SKC. OCNL SCT-BKN100 WRN AND SRN PTNS OH. TOPS 150.
OTLK...VFR.

WV
NRN 2/3...SCT050. AFT 00Z SKC. AFT 03Z SCT-BKN100 TOPS 150. VIS 3-5SM BR.
OTLK...VFR BCMG MVFR CIG BR BY 10Z.
SRN 1/3...SCT-BKN050. AFT 00-03Z BKN030-040 BKN100 TOPS 180. VIS 3-5SM BR.
WDLY SCT -TSRA. CB TOPS FL350. OTLK...IFR CIG TSRA BR.

MD DC DE VA
MD DC DE NRN VA...SCT-BKN040-050. WDLY SCT -TSRA. OCNL BKN-OVC020 CSTL
SXNS. CB TOPS FL350. OTLK...MVFR CIG BR.
SRN VA...BKN040-050. WDLY SCT TSRA. OCNL OVC020 WDLY SCT TSRA ERN PTN
WITH SCT SEV TSTMS. AFT 00-03Z BKN020-030. WDLY SCT -TSRA. CB TOPS FL450.
OTLK...IFR CIG TSRA BR.

CSTL WTRS
N OF ACK...OVC010-020 TOPS 150. VIS 3-5SM BR. SCT -RA. AFT 00-03Z OVC010-020
TOPS 040. VIS 3-5SM BR. OTLK...IFR CIG BR.
S OF ACK...OVC020. VIS 3-5SM BR. WDLY SCT -TSRA. AFT 00-03Z BKN020-030 TOPS
060. OTLK...MVFR CIG.

A pilot's first task is to determine the valid time of the forecast. This Boston FA was
issued on the 30th day of the month at 1745Z, and it becomes effective at 1800Z and is valid
until the 1st at 0600Z. Old forecasts, due to late revisions or computer trouble, occasionally
remain in the system. Next, does the product cover the proposed flight? Notice the Boston
FA includes Lake Erie (LE), Lake Ontario (LO), and the District of Columbia (DC). The
synopsis paragraph follows the standard reference to AIRMET SIERRA, thunderstorm,
and height statements. The main weather feature is a cold front that has been forecast to
move through the FA area. Notice the 18-h valid time, through the outlook period.

Refer to AIRMET ZULU. Light to locally moderate rime icing in clouds from 8000 to
15,000 ft MSL over Maine and New England coastal waters is forecast. The forecaster does
not expect extensive moderate icing. This is an example of icing that does not meet
"AIRMET criteria." Otherwise, no significant icing—trace or no icing—is expected. From
the freezing level paragraph, we see the freezing level is expected to slope from around
5000 in northern Maine to 8000 ft in the south. This condition would be a definite consideration for aircraft without ice protection equipment.

Next refer to AIRMET TANGO. Moderate to locally severe turbulence is forecast
between 20,000 and 40,000 ft within the delineated area (see Fig. 15-3). Notice that only
portions of New York and Pennsylvania are affected, with turbulence expected to diminish
by 2000Z. This is an example of a condition dissipating during the forecast period. Since
severe turbulence will be localized, a SIGMET has not been issued. Should a SIGMET be
in effect, a statement following the paragraph will refer to the advisory (...SEE SIGMET
OSCAR 1 FOR SEV TURBC...). Is this the only area where significant turbulence can be
expected? No, remember thunderstorms imply moderate or greater turbulence and low-
level wind shear, which are not addressed separately.

Now refer to AIRMET SIERRA. Occasional ceiling below 1000 ft AGL and visibilities below 3 mi in fog are forecast. The outlined section in Fig. 15-3 over New England
and the coastal waters represents this area. Conditions will continue beyond 2000Z
through 0200Z. The gray shaded areas of Fig. 15-3 depict mountain obscurement. The
mountains of New York and New England are expected to be obscured in clouds. Here,
conditions are expected to improve by 0100Z. A second mountain obscuration paragraph
provides details on conditions in West Virginia and Virginia. Mountains are expected to
become obscured in clouds and precipitation after 0000Z, which will continue through
0200Z—the end of the outlook period. This is an example of a condition developing during the period.

Generally, advisories for IFR and mountain obscurement are synonymous. They represent low ceilings and visibilities that will preclude VFR flight within all or part of the affected area. Pilots, especially those used to flying over flat terrain, must use caution with mountain obscurement forecast. Weather reports from valley stations may indicate VFR flying conditions, but VFR flight through passes and over the mountains may be impossible. Again, it's imperative for the pilot to have a sound knowledge of the terrain. When the forecast indicates the phenomenon will be occasional—for example, the IFR paragraph in the BOS AIRMET Bulletin—the probability of occurrence is greater than 50 percent, but it is expected to occur for less than half the forecast period. A forecast for occasional should not in itself warrant the cancellation of VFR flight. All available information, however, must be considered and caution exercised, with suitable alternates available—the complete picture. The absence of a conditional term indicates the phenomenon will be widespread—for example, the BOS AIRMET SIERRA mountain obscurement paragraphs. This means a greater than 50 percent probability occurring for more than half the forecast period. VFR flight probably will not be possible, but there could be areas where the phenomenon does not exist. This is not an error but a limitation of the forecast product.

VFR CLDS-WX paragraphs describe expected weather for the forecast area. Areas are usually described using area designators, as contained in App. B. In Maine and New Hampshire west of a MLT CON line—the mountains—BKN030 BKN100 TOPS 150 (these heights are all MSL) are expected from 1800Z until after 0100Z. With mountain peaks in the 3500- to 5000-ft range, they will be obscured as advertised in AIRMET SIERRA; where *minimum enroute altitudes* (MEAs) are above 5000, icing in clouds can be expected. After 0100Z conditions are forecast to improve to SCT-BKN050, consistent with AIRMET SIERRA. General visibilities through the period will be 6 mi or greater, with no significant precipitation. The outlook is VFR, but remember its definition.

East of MLT-CON line—the coastal plain—OVC010-020 TOPS 150, VIS 3-5SM BR, SCT -RA (these are again MSL), with occasional IFR conditions (AIRMET SIERRA). The forecaster expects conditions to begin to improve, but not until after 0000Z. VFR will be iffy. MEAs are lower and, with flights below 5000, ice should not be a factor. After 0000Z conditions will improve somewhat, but they will remain marginal to IFR. Tops will lower significantly in the coastal areas. The outlook indicates the ceilings below 1000, and visibilities below 3 mi in mist will persist.

Moving to the New Jersey and Pennsylvania paragraph (NJ PA), notice that isolated thunderstorms and rain showers are forecast for southern New Jersey and southeastern Pennsylvania. Since they're expected to be isolated, circumnavigation should be possible. However, we wouldn't want to be poking around in clouds without storm detection equipment. Under the circumstances, climbing above the lower layer where visual separation from convective activity can be maintained would be a prudent procedure. What about turbulence and icing? They're implied with any forecast of thunderstorms. The outlook, MVFR BR, indicates visibilities will be between 3 and 5 mi in mist. The ceiling? Well, greater than 3000 ft.

Refer to the Maryland, District of Columbia, Delaware, Virginia (MD DC DE VA) paragraph. Notice for southern Virginia the forecasts indicate OCNL OVC020 WDLY SCT TSRA ERN PTN WITH SCT SEV TSTMS. Over the eastern portion of southern Virginia, there is a chance of severe thunderstorms. The outlook, IFR CIG TSRA BR, indicates ceiling less than 1000 ft, visibilities less than 3 mi in mist, rain showers, and thunderstorms.

The final paragraph forecasts conditions for the CSTL WTRS. The area is divided into north and south of Nantucket, Massachusetts (ACK).

This FA describes a distinct weather system. Notice how the AIRMET Bulletin, synopsis, and VFR clouds and weather are all tied together. FAs will not always be this detailed or consistent, but the example has provided most variables used in this product.

Both the area forecast and AIRMET Bulletin must be carefully reviewed to determine conditions enroute.

The following case study is based on data from the 1145Z FA and the 1445Z AIRMET Bulletin. The following discussion reflects the limitations and scope of these products, which must be understood for correct interpretation and applications.

CASE STUDY
FA
SAN JOQUIN VLY...BKN010-020 SCT-BKN120. VIS 3-5SM BR. 17-
20Z BKN025-035 BKN100. TOPS FL180....
WA:
AIRMET IFR...CA
FROM 30NE CZQ TO 40SE EHF TO 40SW EHF TO 40W CZQ TO 30NE CZQ
OCNL CIGS BLW 010 VIS BLW 3SM BR. CONDS ENDG 18-20Z....

At first glance it may seem as if these forecasts are contradictory—it did to the FSS controllers that brought it to my attention. The FA appears to predict VFR, the AIRMET IFR. Let's analyze these products. The FA forecasts conditions for the San Joaquin Valley, which consists of California's central valley Stockton southward. The WA forecasts conditions for the San Joaquin Valley Fresno/Madera southward. This distinction will be important.

For the San Joaquin Valley the FA predicts a greater than 50 percent probability for greater than 1/2 of the forecast period (12Z to between 17-20Z) for broken clouds based between 1000 and 2000 ft MSL. This translates to between 500 to 1000 and 1500 to 2000 ft AGL; because of terrain elevations the lower conditions will be in the southern portion of the valley. A higher scattered to broken layer based at 12,000 ft MSL, visibilities 3 to 5 statute miles (SM) in mist is also forecast. Between 17 and 20Z conditions are forecast to become: 2500 to 3500 ft broken, 10,000 ft broken.

For the southern half of the San Joaquin Valley, the WA predicts a greater than 50 percent probability for less than 1/2 of the forecast period (15Z to between 18 and 20Z) ceilings below 1000 ft, visibilities below 3 SM in mist.

These forecasts are perfectly consistent within the scope and purpose of each product.

SIGNIFICANT WEATHER PROGNOSTIC CHARTS

On October 27, 1997, the Significant Weather (SIGWX) Graphics Unit of the Aviation Weather Center took over the production of the 12–24-h low-level significant weather prognoses chart and all high-level charts for the Atlantic and Pacific oceans. The low-altitude charts provide outlook guidance for flights below 24,000 ft over the contiguous United States. They are not a substitute for a standard preflight briefing. High-altitude charts are designed to provide the enroute portion of international civil aeronautical operations above 24,000 ft; these charts must be used along with any valid domestic or international weather advisories.

The 12–24-h low-level significant weather prognostic (prog) chart is issued four times a day, valid at 0000Z, 0600Z, 1200Z, and 1800Z, depending on issuance time. Because it takes approximately 6 h to prepare and distribute the chart, by the time it becomes available, valid times are only up to 18 h. If a pilot calls just prior to the next issuance, only a 12-h forecast would be available. The 12–24-h prog consists of four panels. The two upper panels forecast significant weather from the surface to 400 mb (24,000 ft). These panels are produced by AWC forecasters responsible for the domestic FAs. The lower panels depict the location and coverage of precipitation, and the location of surface features—fronts and pressure systems. The lower panels are produced by the Hydrometeorological Prediction Center (HPC), in Camp Springs, Maryland.

Forecast weather categories (VFR, MVFR, and IFR) have the same definitions and limitations as the weather depiction chart and area forecast outlook categories. These are defined in Table 12-2, Weather Categories.

Turbulence is depicted within dashed lines, with turbulence symbols indicating intensity. In Fig. 15-4 at 0000Z, over the southern Rockies, moderate turbulence is forecast below 24,000 ft MSL. Moderate to severe turbulence below 14,000 ft is expected over portions of California, Arizona, Utah, Colorado, and New Mexico. (Note the turbulence symbols representing moderate and severe intensities in the inset of Fig. 15-4.)

Although icing is not directly forecast, it's implied in clouds and precipitation above the freezing level. Supercooled large drops are inferred in areas of forecast freezing precipitation. Freezing level is represented by short, dashed lines, and the freezing level at the surface is denoted by a zigzag line and the contraction SFC. During the forecast period in Fig. 15-4, freezing levels range from around 4000 ft in the north to 12,000 ft in the south. In the early morning hours of April 23, freezing temperatures at the surface are expected over the northern Rockies and Great Basin.

Refer to the lower panels in Fig. 15-4. Fronts and pressure systems use standard symbology contained in Chap. 12. Central pressure is also noted. For example, in the 12Z panel the central pressure of the high over southern Canada is 1037 mb, and the low over Texas is 1003 mb. Front type, intensity, and character are also depicted (Table 12-1), for example, the frontal system in southern California [425. This decodes as a cold front at the surface, weak, little or no change, forming or existence expected. Note the symbol in Texas used to depict a dry line, or temperature/dew point front.

Lines on the lower panels enclose areas of precipitation. Hatching used with stable precipitation (rain and snow) indicates continuous precipitation. Nonhatched areas represent intermittent precipitation. Hatching used with unstable precipitation (showers and thunderstorms) indicates the precipitation is expected to cover half or more of the area.

In Fig. 15-4 there are large enclosed areas, with an area of hatching within the overall area. In the outer—nonhatched—area mostly intermittent precipitation is forecast. The exception is in the vicinity of the low in the midwest where intermittent rain, with scattered rain showers and thunderstorms, is forecast. However, from an operation point of view, we would consider all nonhatched areas as forecast for scattered precipitation. The hatched area, again from an operation point of view, forecasts precipitation covering more than half the area, meaning that it will be widespread.

Precipitation type is depicted using standard symbols. In general, one precipitation symbol indicates light intensity, two symbols light to moderate, and three symbols moderate. Refer to Fig. 15-4, 00Z panel. The widespread area of continuous snow over the central Rockies is forecast to be light to moderate, to moderate (symbols in Arizona). The widespread precipitation in extreme eastern Colorado is expected to be a combination of light to moderate rain and snow (symbols in New Mexico). Note the dashed line that separates the area of snow, from the area of rain and snow. The scattered precipitation in Kansas consists of light rain showers and thunderstorms, with light rain showers.

In preparing this chart, forecasters cannot consider mesoscale features; the chart is a synoptic depiction. Local conditions might not be accurately portrayed. These progs tend to underestimate convective activity in the west, along with the intensity of Pacific storms, and the forecaster tends to smooth over local variations due to terrain. Therefore, these charts are most useful in the midwest and east.

The chart in Fig. 15-4 is valid at 0000Z Friday, April 23, and 1200Z Friday, April 23. Be careful converting to local time—for example, 0000Z Friday translates to 5 p.m. Thursday, Pacific Daylight Time.

Let's say we're planning a flight from Kansas City to St. Louis. Refer to the Thursday evening (00Z FRI APR 23 1999) panel of Fig 15-4. A surface low and associated stationary front are forecast over Missouri. MVFR ceilings and visibilities in widespread light to

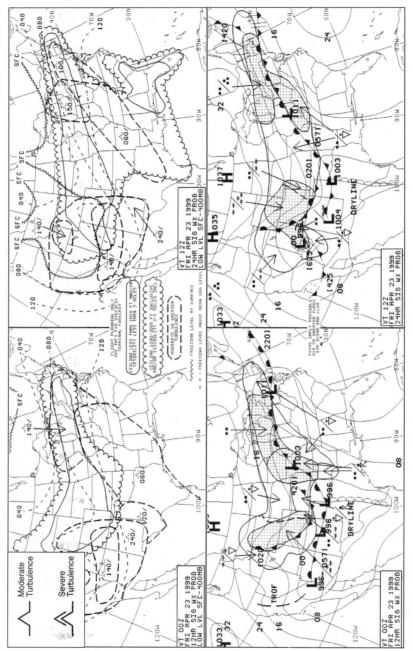

FIGURE 15-4 The 12- to 24-h significant weather prog describes the location and movement of synoptic-scale weather systems. Small-scale and local events cannot be depicted.

moderate rain and thunderstorms are expected. Moderate turbulence is forecast below 14,000 ft, with the freezing level high at 12,000 ft.

VFR flight may be possible, and an IFR flight below 12,000 ft should not have to contend with icing. However, both VFR and IFR flights will have to deal with thunderstorms. With an organized weather system and thunderstorms, it does not appear that flight above the clouds for light aircraft will be possible. Without storm detection equipment, an IFR flight does not appear practical. Even though VFR ceilings and visibilities are forecast, avoiding thunderstorms will be chancy.

Refer to the Friday morning (12Z FRI APR 23 1999) panel of Fig 15-4. The low-pressure system and associated front are forecast to move eastward. MVFR is still forecast, with scattered rain or rain showers. No significant turbulence is expected, the freezing level remains around 12,000 ft. Both VFR and IFR operations appear practical under these conditions. However, a flight decision must also be based on our personal minimums. More about personal minimums in Chap. 27, Risk Assessment and Management.

Strictly a surface prog, the 36–48-h significant weather prog is issued twice daily, valid at 0000Z and 1200Z. By the time the chart becomes available, valid times only provide a 30- and 42-h forecast. Should a pilot call a flight service station or download the product just prior to the next issuance, only a 30-h forecast would be available.

Refer to Fig. 15-5, the 0000Z and 1200Z Saturday forecast. The front is forecast to continue to move slowly southward as high pressure builds into the upper midwest. Scattered rain and rain showers are expected to persist through the forecast period.

What can be concluded about clouds, visibilities, turbulence, and icing? Drawing on our knowledge, we can make the following deductions. High pressure generally means fair weather. Weak pressure gradients and the absence of convective activity indicate no significant turbulence. These conditions are not conducive to icing. However, conditions could be right for extensive areas of radiation fog causing IFR weather.

Using progs, remember the limitations on aviation forecasts, especially timing. Then, when requesting outlooks, review issuance and valid times to obtain the latest forecasts. Keep in mind that these forecast progs are not available beyond about 42 h.

The High-Altitude Significant Weather Prog

The high-altitude significant weather prog provides a graphic view of forecast weather from 400 mb to about 70 mb, approximately 24,000 through 60,000 ft. All heights are pressure altitudes (flight levels). These charts depict the following:

- Thunderstorms and cumulonimbus clouds
- Tropical cyclones
- Severe squall lines
- Moderate or severe turbulence
- Moderate or severe icing
- Widespread sandstorms or duststorms
- Surface fronts
- Tropopause height
- Jet streams
- Volcanic eruptions

Charts are valid at 0000Z, 0600Z, 1200Z, and 1800Z.

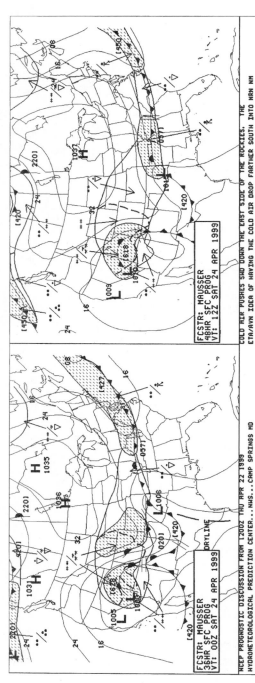

FIGURE 15-5 The 36- to 48-h significant weather prog provides only the location and movement of weather system, and areas of precipitation.

Scalloped lines enclose areas of forecast cumulonimbus development, divided into three categories: *isolated* (ISOL), or less than one-eighth coverage; *occasional* (OCNL), or one-eighth to four-eighths coverage; and *frequent* (FRQ), five-eighths to eight-eighths coverage. Cumulonimbus clouds imply hail and moderate or greater turbulence and icing. Bases and tops are shown by numerical figures below and above a short horizontal line. Bases below 24,000 ft are indicated by XXX below the line. For example, in Fig. 15-6, through northeast Mexico, Texas, and Louisiana, the forecast indicates an area of isolated (less than one-eighth coverage), embedded cumulonimbus with tops to 42,000 ft, bases below 24,000 ft (XXX). Clear air turbulence, not associated with cumulonimbus clouds, is indicated by a dashed line containing a turbulence intensity symbol and forecast height. In Fig. 15-6, off of Chile, South America, moderate turbulence is expected between 28,000 and 40,000 ft.

The expected location of jet streams is indicated by long black lines. Figure 15-6 implies the turbulence off Chile is due to a jet stream at 40,000 ft (FL400), with a core speed of 90 kn. The double-hatched line to the right or left of the maximum speed indicates a change in speed of 20 kn. For example, the jet stream over the North Atlantic has a core speed of 90 kn, then increases to 120 kn east of Newfoundland.

The chart also contains the expected height of the tropopause—the boundary between the troposphere and stratosphere. The height of the tropopause over the west coast is indicated by numerals enclosed in a box and forecast around 50,000 ft off northern California. A five-sided polygon indicates areas of high or low tropopause heights. For example, in Fig. 15-6, off the southern California coast is a high point in the tropopause, forecast to be at 54,000 ft.

Significant frontal boundaries are depicted using standard symbols, with system movement indicated by an arrow and speed printed by the arrowhead. For example, the cold front in the central Mississippi Valley depicted in Fig. 15-6, is expected to move to the south at 10 kn.

The names of tropical cyclones are entered next to the symbol. In Fig. 15-6 hurricane Danielle is located off the Florida coast, tropical storm Bonnie off New Brunswick, and tropical depression Howard in the central Pacific.

Severe squall lines are depicted within areas of cumulonimbus clouds by the symbol - V - V - V. Widespread sandstorms or duststorms are enclosed by scalloped lines, labeled with the appropriate symbol.

Volcanic activity is indicated by a trapezoidal figure depicting an eruption (see inset Fig. 15-6). The symbol and any known information concerning the name of the volcano, its latitude and longitude, the date and time of the first eruption, and a reminder to check SIGMETs will be included in the legend of the chart. Figure 15-6 shows the Popocatepetl eruption in central Mexico.

The chart is often an excellent place to begin planning a high-altitude flight (above FL 240). For example, a pilot might plan to follow the jet stream to take advantage of tail winds. If turbulence—for example, a lifeguard flight—were a significant factor, the pilot might plan a route to avoid wind shear turbulence. A pilot planning a flight from east to west might also select a route to avoid maximum jet stream head winds. The chart can also be used to determine areas of cumulonimbus, cloud tops, and tropopause heights. If aircraft equipment (storm detection equipment) or performance does not allow the pilot to climb above thunderstorms, the pilot may wish to avoid areas where coverage is expected to be occasional or frequent. The pilot should note the expected height of the tropopause. If thunderstorms develop, the isothermal layer of the tropopause caps all except severe activity.

Pilots, like FSS briefers, can use the chart for the enroute portion of a high-altitude weather briefing, but pilots, as well as briefers, must not substitute the chart for the area forecast and weather advisories. The chart is not amended, and severe weather might develop that is not depicted. The area forecast is also needed to provide weather details for climbout and descent.

Additional data related to the jet stream may be obtained from the domestic tropopause wind and wind shear prog. This two-panel chart depicts the forecast winds at the tropopause, and tropopause height, and vertical wind shear. An example is contained in Fig. 15-7.

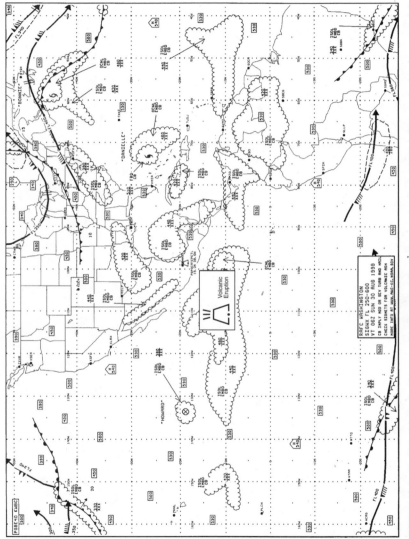

FIGURE 15-6 The high-altitude significant weather prog pictorially displays weather above 24,000 ft. Because the chart is not amended, the area forecast and weather advisories must be consulted to complete the forecast weather picture.

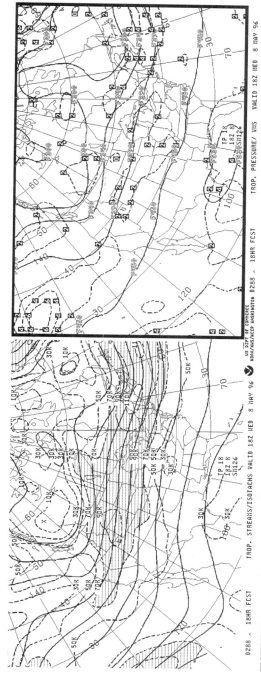

FIGURE 15-7 Data related to the jet stream may be obtained from the domestic tropopause wind and wind shear prog.

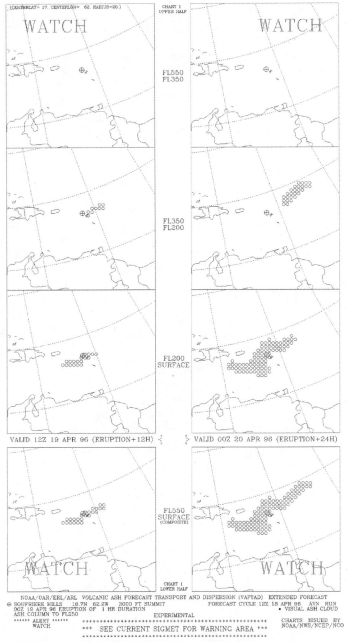

FIGURE 15-8 The National Weather Service has developed the experimental *Volcanic Ash Forecast Transport and Dispersion chart* (VAFTED) to assist pilots in avoiding ash hazards.

The left panel in Fig. 15-7 forecasts winds at the tropopause. Wind speed is shown by isotachs at 20-kn intervals. Areas of wind speeds between 70 and 110 kn, and between 150 and 190 kn are shaded to help locate the jet core. Horizontal wind shear can be determined from the spacing of the isotachs. Expect moderate or greater turbulence when winds exceed 18 kn per 150 mi. In Fig. 15-7 significant horizontal wind shear turbulence is indicated north of the jet core over New England and southeastern Canada.

The right panel in Fig. 15-7 depicts tropopause height and vertical wind shear. Heights are pressure altitudes or flight levels (F450, 45,000 ft). Vertical wind shear is in knots per thousand feet and depicted by dashed lines at 2-kn intervals. Wind shear is averaged through a layer from about 8000 ft below to 4000 ft above the tropopause. Vertical wind shear can be determined directly from the dashed lines in Fig. 15-7. The vertical wind shear critical for probable turbulence is from 4 to 6 kn/1000 ft. This value is shown from eastern Pacific, north of 30° latitude; with another area over Nova Scotia, Canada.

The National Weather Service has produced the experimental *Volcanic Ash Forecast Transport and Dispersion chart* (VAFTED). Figure 15-8 contains an example.

Each complete chart consists of eight panels. The four panels in any column are for a single valid time after eruption. In the example, valid times are 12Z 19 APR 96 (eruption+12H) and 00Z 20 APR 96 (eruption+24H). Individual panels are for layers applicable to aviation operations and are identified at the side of a panel with upper and lower flight levels (FL) in hundreds of feet—in the example, surface to FL200, FL200 to FL350, and FL350 to FL550. The bottom panel is a composite from the surface to FL550. For each column, the forecast valid time separates the upper three panels from the composite panel. Volcano eruption information is at the lower left. A description of the input meteorology is at the lower right with a message to SEE CURRENT SIGMET FOR WARNING AREA. The visual ash cloud symbol and run description are at the lower center. Upper half and lower half panels are included for viewing detail. Volcanic ash cloud charts for hazards alert volcanoes are always listed first. Charts for volcanoes of interest (TEST-volcanoes) will then be listed. Only upper half and lower half panels are included for TEST-volcanoes. TEST-volcanoes are usually updated on a daily basis. (If the same volcano is listed first as a hazards alert volcano and then below as a TEST-volcano, the hazards alert volcano listed first takes precedence.)

There's no question that the FA has improved. The NWS's decision to consolidate in Kansas City and the commitment of Aviation Weather Center forecasters has resulted in standardization and consistency. The FA is a valuable and useful tool as long as pilots, briefers, and dispatchers understand its purpose, format, conditional terms, and limitations.

The forecasters at AWC request positive user feedback—that is, constructive criticism of their products. But they need to know the date and time of the occurrence and specific user comments. These can be sent to AWC via its Internet site or by writing to them at the following address: Aviation Weather Center, 7220 NW 101st Terr., Rm. 105, Kansas City, MO 64153-2371. Consider, for example, the following case study:

CASE STUDY An AIRMET Bulletin forecast IFR conditions in stratus and fog for the San Joaquin Valley. Rather than forecasting stratus tops, which were only about 2000 ft, and a high layer based about 12,000 ft in the area forecast, it simply read ST BR TOPS FL200. This would indicate that VFR flight was not possible below 20,000 ft! The problem? There was about 2 mi of clear air between the tops of the stratus and the next layer. We had a chat about this, and it hasn't occurred since.

Never rely on a single piece of data. The FA is only one product, one piece of the complete picture. Localized conditions may not be included. Flight decisions should never be based solely on this or any other product. The FA must be used along with all available information. Forecast limitations cannot be over emphasized along with using all available sources.

"Whatever may be the progress of the sciences, never will observers who are trustworthy and careful of their reputations venture to forecast the state of the weather." [Dominique Argo, French astronomer-physicist (1786–1853)].

CHAPTER 16
TWEB ROUTE FORECASTS

TWEB route forecasts were developed as scripts for the FAA's Transcribed Weather Broadcast (TWEB) and Pilots Automatic Telephone Weather Answering Service (PAT-WAS). With few exceptions, TWEB broadcasts over low-frequency radio beacons, and *very high frequency omnidirectional radio ranges* (VORs) have been decommissioned. The PATWAS has also been decommissioned. TWEBs may be used for the enroute forecast portion of the Telephone Information Briefing Service (TIBS). These forecasts can also be used during flight service station weather briefings. Telephone numbers for TIBS can be found in the telephone numbers section of the *Airport/Facility Directory*. As of this writing TWEBs are not available through DUATs or other commercial weather providers. TWEBs are, however, available from the Aviation Weather Center's Internet page (www.awc-kc.noaa.gov) and many local weather forecast office sites. TWEB route numbers and locations are contained in App. B, Graphs and Charts.

Like most forecast products, TWEBs have been revised over the years, and issuance and valid times have been standardized. The most recent revisions occurred in 1997 when the FAA and NWS eliminated many routes in the southeastern United States. TWEBs are issued four times a day and are valid for 12 hours, as shown in Table 16-1.

TWEB forecasts are divided into route forecasts and local vicinity forecasts. TWEB route forecasts cover a corridor 25 nm either side of the route centerline, and a 25-nm-radius semicircle around the end points. Local vicinity forecasts are, typically, valid for a 50-nm radius, except for certain irregularly shaped areas. Additionally, designated weather forecast offices write TWEB synopses.

Refer to the 420 TWEB route forecast heading below for the following discussion:

420 TWEB 061402 KOAK-KACV. ALL HGTS MSL XCP CIGS.

TWEB route forecasts are numbered as shown in App. B. Following the route numbers are the valid period and the route end anchor points. In the example this is the 420 TWEB route, valid on the sixth day of the month from 14Z until 02Z, and the route end anchor points are Oakland and Arcata in California (420 TWEB 061402 KOAK-KACV). This is followed by a height statement similar to that used on the area forecast: All heights are relative to mean sea level except ceilings.

The loss of reporting stations has caused revisions to the TWEB network, resulting in the issuance of certain routes on a part-time basis. In such cases the notation NIL TWEB is used:

426 TWEB 250520 TSP MTNS SOLEDAD-CAJON-BNG PASSES AND ADJ
MTNS. NIL TWEB.

This evening issuance of the 426 route, for the Tehachapi Mountains, Soledad, Cajon, Banning Passes and adjacent mountains, will be delayed due to lack of observations.

TABLE 16-1   TWEB Issuance Times and Valid Periods

Issuance time (Z)	Valid period (Z)
0200	0200–1400
0800	0800–2000
1400	1400–0200
2000	2000–0800

Additionally, amendments might not be available for selected routes that will contain the remark NO AMDTS AFT 04Z. Significant changes can occur without an amendment, so this serves the same purpose as NOSPECI on METAR. It's a warning that caution must be exercised, especially during marginal and deteriorating conditions.

Pilots, FSS controllers, and forecasters have been instrumental in revising TWEB routes. For example, the 420 route at one time forecast conditions from Concord to Arcata to Crescent City, California (KCCR-KACV-KCEC). However, with the loss of 24-h observations at Crescent City and only part-time observations at Concord, the forecast was suspended during the night. But by changing the anchors—end points on the route—to Oakland and Arcata (KOAK-KACV), with 24-h observations, the route once more became available full-time. Other routes in the area have also been revised as a direct result of pilot input. This is an excellent example of how pilots, FSS controllers, and forecasters, working together, can improve the system. In recent years automated weather observations have almost eliminated this type of problem.

SYNOPSES

The synopsis contains a brief description of fronts, pressure systems, and local climate or terrain factors affecting the routes, valid for the same period as the route forecasts. The TWEB synopsis often contains more detail than is possible in the area forecast. Therefore, the TWEB synopsis might be of more value describing local conditions. More detail is possible because the TWEB synopsis covers only about one-fifth the area of the FA. Refer to App. B for TWEB synopses locations.

LAX SYNS 191402 UPR RDG ALG W CST WITH NLY FLO ALF. SFC HI
CONTS TO BLD OVER GT BASIN FOR CONTG SANTA ANA CONDS.

This Los Angeles (LAX) synopsis covers the routes issued by the Los Angeles WFO. An upper-level ridge is along the west coast with a northerly flow aloft. Surface high pressure continues to build over the Great Basin—the high plateau generally consisting of Nevada, Utah, and southern Idaho—for continuing Santa Ana conditions—strong, sometimes warm, northeasterly winds over southern California.

The detail of a TWEB synopsis depends on the weather pattern:

ATL SYNS 302008 STNRY FNT NR MYR-ABY-PAM. HI PRES CNTRD OH DRFTG SEWD.

This Atlanta (ATL) synopsis describes a stationary front from Myrtle Beach, South Carolina (MYR), to Albany, Georgia (ABY), to Panama City, Florida (PAM). High pressure centered over Ohio is drifting southeastward.

A more complex weather system is reflected in a longer synopsis:

IND SYNS 042008 CD FNT XTNDS FRM NRN LK ERIE THRU CNTRL OH INTO NRN AL. WK TROF EXTNDS FRM SFC LO SW LWR MI THRU W CNTRL IN. UPPR LVL LOW OVER NRN IL WL BRNG SCT TRW MAINLY OVR WRN AND NRN RTES THIS AFTN AND EVE. PRZYBYLINSKI

In this Indianapolis synopsis, a cold front extends from northern Lake Erie through central Ohio into northern Alabama. A weak trough extends from a surface low in southwest lower Michigan through west central Indiana. The upper-level low over northern Illinois will bring scattered rain showers and thunderstorms mainly over the western and northern routes this afternoon and evening. The synopsis was prepared by Przybylinski.

CASE STUDY   Pilots using the Internet will have to contact an FSS for clarification of contractions or phrases they do not understand or are not decoded through the system's decode function. One evening I had a pilot call the FSS with such a question. He didn't understand the last word of the synopsis. I explained it was the name of the meteorologist who wrote the forecast.

Earlier it was mentioned how personal comments sometimes get into weather products:

SFO SYNS 091402 SFC HI PRES OFSHR WITH LWR PRES OVER INTR CA. THIS IS MY LAST SET OF TWEBS.....EVER. SHORTY THOMAS.

On one synopsis the forecaster signed his name SENYOR CHEWBACCA. Then, every once in a while something gets transmitted that shouldn't have:

SFO SYNS 032121 THE QUICK BROWN FOX JUMPED OVER THE LAZY DOG BECAUSE THERE WAS NO WEATHER OF ANY MERIT TO TALK ABOUT.

I'll leave the names off this one.

SIGNIFICANT CLOUDS AND WEATHER

TWEBs use the same terminology and abbreviations as the area forecast. Forecasts contain *significant* clouds and weather, with significant defined as phenomena affecting at least 10 percent of a route, or in the forecaster's judgment important to flight planning.

In the body of the forecast all time references are UTC, abbreviated Z. The forecast is in the following order: locations, time, conditions. Conditions may be modified using the conditional terms given in Table 16-2.

TWEB forecasts consist of the following:

- Surface winds
- Surface visibility
- Weather and obstructions to visibility
- Sky conditions
- Mountain obscurement
- Nonthunderstorm low-level wind shear

Sustained surface winds are normally forecast when expected to be 25 kn or greater. Format is the same as used on the METAR and TAF. For example, SFC WND 34025G35KT

TABLE 16-2 TWEB Route Forecast Conditional Terms

Term	Abbreviation	Definition
Isolated	ISOLD	Single cells or localized conditions (no percentage); implies circumnavigable
Widely scattered or local	WDLY SCT/LCL	Less than 25% of area or route affected
Scattered or large areas	SCT/AREA	25–54% of area or route affected
Numerous or widespread	NMRS/WDSPRD	More than 54% of area or route affected

(surface winds 340° at 25 kn, gusting to 35 kn). Wind gusts associated with thunderstorms may also be included (SFC WND G40KT). The lack of a wind forecast implies only sustained speeds less than 25 kn over 90 percent of the forecast area.

Surface visibilities are always included. To indicate unrestricted visibility, P6SM is used (greater than 6 statute miles). Care is taken to avoid crossing categories in the forecast—IFR-MVFR, or IFR-VFR, and so on.

Significant weather is included using standard weather abbreviations described in Chap. 8. Obstructions to visibility are included when visibility is expected to be less than 6 mi.

Cloud heights are based on a standard reference: ALL HGTS MSL XCP CIGS (all heights are relative to mean sea level except ceilings) or ALL HGTS AGL XCP TOPS (all heights are above ground level except tops). The use of AGL and CIGS is normally limited to layers within 4000 ft of the surface. Originally, TWEBs were basically low-altitude forecasts—below 15,000 to 18,000 ft. That no longer necessarily applies, but forecasters emphasize cloud conditions below 12,000 ft. With no clouds, or scattered clouds above approximately 12,000 ft, the forecaster might use the phrase NO SGFNT CLDS/WX (no significant clouds or weather); this means that any clouds present should be easily circumnavigable. Like other reports and forecasts, sky cover uses standard abbreviations (SKC, FEW, SCT, BKN, OVC). Again, the forecaster will attempt to avoid crossing categorical boundaries. CLR BLO 120 or AOA BKN120 may be used based on automated observations when the forecaster considers them representative for the route.

When forecast, tops will normally only be included for layers with tops below 15,000 ft. Like the FA, for multiple or merging layers, which would not permit VFR flight between layers, only the top of the highest layer appears. However, because of the local nature of the product, tops are often more specific and, therefore, useful. Cloud tops may also be indirectly specified. For example, CIGS BKN020 OBSCG TRRN BLO 030.... Cloud tops are expected to be 3000 ft.

Mountain obscurement has been added (MTNS OBSCD ABV..., MTN RDGS OBSCD, or ALL PASSES OBSCD), and geographical features might be specified (TSP—Tehachapi—MTNS, CSTL MTNS). Because of the local nature of the forecast, not all locations appear on the area designators maps in App. B. Pilots unfamiliar with locations on TWEB forecasts will have to consult an FSS.

Nonconvective low-level wind shear is included when expected. The extent of coverage will be included.

Icing and turbulence information are not included in TWEB forecasts. This information is available from appropriate weather advisories.

The following is the complete 420 TWEB previously introduced:

420 TWEB 061402 KOAK-KACV. ALL HGTS MSL XCP CIGS. KOAK-KUKI P6SM SCT015

AREAS CIGS BKN015...AFT 20Z P6SM SKC-FEW250. KUKI-KACV P6SM SCT020 BKN040 LCL 5SM -SHRA BR CIGS BKN020 MTNS OBSCD TOPS BLO 080...AFT 20Z P6SM SCT-BKN040 BKN150

The route forecast begins with the segment from Oakland to Ukiah: Visibility unrestricted; 1500 ft MSL scattered, areas of ceilings 1500 ft broken; after 20Z conditions improving to clear to a few clouds at 25,000 ft. The forecast for the segment from Ukiah to Arcata: Visibility unrestricted; 2000 ft MSL scattered, 4000 ft MSL broken; local areas of visibility 5 mi in light rain showers and mist, with ceilings 2000 ft broken, mountains obscured, tops below 8000 ft MSL; after 20Z 4000 ft MSL scattered to broken, 15,000 ft broken.

CASE STUDY As an FSS supervisor, I counsel specialists to keep briefings clear and concise. One area of excess verbiage is high thin cirroform clouds; all cirroform clouds are high! But, leave it to the National Weather Service: Sure enough, one day I found 426 TWEB...HI THIN CIFM CLDS. What can you do?

The length and detail of TWEB forecasts vary widely. Some are one-liners, others take up a half of a page. Just try getting that and a half-dozen more like it on a 3-min tape for the broadcast and you get some idea why FSS specialists speak so fast.

LOCAL VICINITY FORECASTS

Local vicinity forecasts have been developed to cover metropolitan areas; locations are contained in App. B. They normally cover a radius of 50 nm:

358 TWEB 061402 PGTSND WITHIN 25NM RDS KSEA. ALL HGTS MSL XCP CIGS. P6SM SCT050 BKN100 OVC200...17Z P6SM -RA SCT025 OVC040...19Z P6SM -RA SCT-BKN015 BKN025 OVC040...LCL VIS 3-5SM -RA...21Z P6SM -SHRA SCT-BKN030 BKN050. VC KPAE AFT 21Z AREAS VIS 3-5SM -RA CIGS BKN-OVC009.

This is an example of a Seattle local vicinity forecast. It covers the Puget Sound area within a 25-nm radius of the Seattle-Tacoma (KSEA) Airport.

Exceptions to the 50-mi radius occur in southern California and Arizona where large, irregular, homogenous geographical areas exist. These routes are depicted in Fig. 16-1. The 431 TWEB covers the Los Angeles Basin, from the San Fernando Valley along the San Gabriel and San Bernardino Mountains to Hemet, and back to Santa Ana. The 432 route from Santa Ana to San Diego also predicts conditions for the adjacent mountains—shown in gray in Fig. 16-1. The 426 route forecasts conditions over the Tehachapi Mountains, and mountains and passes north and east of Los Angeles—shown in gray in Fig. 16-1. This includes a specific forecast for the SOLEDAD, CAJON, and BNG (Banning) PASSES. The 429 route includes the southern California deserts, south of a Palm Springs–Needles line, and the 425 route the southwestern Arizona deserts south of a Phoenix–Needles line—shown in gray in Fig. 16-1.

AMENDMENT CRITERIA

Table 16-3 contains TWEB amendment criteria. When a phenomenon occurs, is expected to develop, or is no longer anticipated, an amendment will be issued. Amendments are also required for thunderstorms and low-level wind shear. Routes will also normally be

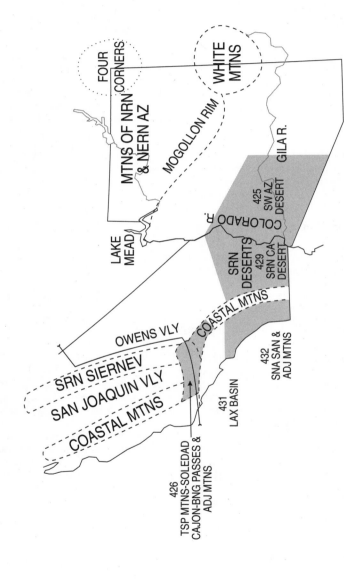

FIGURE 16-1 TWEB local vicinity forecasts normally cover a 50-mi radius of large metropolitan areas, except in southern California and southwest Arizona where large, irregular homogeneous geographical areas exist.

TABLE 16-3   TWEB Forecast Amendment Criteria

Ceiling, ft		Visibility, mi	
With a forecast of:	Amend if:	With a forecast of:	Amend if:
>3000	<3000	<5	>5
≤3000	≥3000	≥5	<5
3000–1000	<1000	3–5	<3
<1000	≥1000	<3	≥3

amended when trends indicate, in the forecaster's judgment, that the forecast will be substantially in error or unrepresentative. FSS controllers make inquiries when discrepancies develop or significant differences occur between FAs and TWEBs. Amendments can originate in this manner. Pilots have no direct way of resolving FA and TWEB discrepancies. Their only option would be to consult with the flight service station.

Forecasters consider issuance times in their decision to amend. Because it takes time to write and distribute an amendment, if the next issuance is in less than 2 h, the forecaster might elect to delay and issue the new forecast rather than amend. As can be seen from Table 16-3, amendments are directed at low-altitude changes. If these criteria are not met, the forecast is considered accurate and will not be amended.

USING THE TWEB ROUTE FORECAST

Flight service station controllers in the western third of the United States use TWEB forecasts more often than controllers in other parts of the country. This is due to the effect of terrain on weather. Often the TWEB is more representative because it covers a smaller area in more detail than is possible with the area forecast. With the Internet these TWEB forecasts are becoming more available to pilots. Pilots should not overlook this valuable, but underused product.

CASE STUDY   A student at our local FBO wanted to fly from Livermore to Salinas. The area forecast predicted conditions below student pilot minimums. (More about personal minimum in Chap. 27.) Since the Oakland, California, FSS uses TWEB forecasts on their TIBS, we were able to get a "second opinion." The TWEB forecast better conditions, which were more consistent with reported weather and TAFs. Our student was able to make the flight based on the TWEB forecasts, along with other reports and TAFs along the route.

Below are examples of the San Francisco area forecast synopsis and Seattle TWEB synopsis for the same period:

SYNOPSIS AND VFR CLDS/WX
SYNOPSIS VALID UNTIL 070500 CLDS/WX VALID UNTIL 062300...
OTLK VALID 062300-070500 WA OR CA AND CSTL WTRS.

SEE AIRMET SIERRA FOR IFR CONDS AND MTN OBSCN.
TS IMPLY SEV OR GTR TURB SEV ICE LLWS AND IFR CONDS.
NON MSL HGTS DENOTED BY AGL OR CIG..

SYNOPSIS...AT 10Z CDFNT WAS OFF WA OR CST. BY 05Z CDFNT WILL LIE FROM
XTRM NRN ID TO NWRN NV. ASSOCD UPR TROF OFFSHR WILL MOV TO WA OR..

SEA SYNS 061402 CDFNT OFSHR RCHG WA CSTLN ARND 17Z...KBLI-KSEA-KPDX LN
19Z-21Z...IDAHO BDR NR 02Z. SWLY FLOW ALF. AMS MOIST STBL AHD OF FNT WRN
WA...STBL WITH INCRG MID/HI LVL MSTR ERN WA AFT 18Z. VERY COLD/UNSTBL
AMS MOVG OVR WA BHND FNT WITH PSCZ DVLPG VC KPAE20AFT 20Z. OUTLOOK
02Z-12Z UPR LVL TROF OFSHR RCHG CSTLN ARND 12Z.SWLY FLOW ALF CONTG.
AMS MOIST/UNSTBL. PSCZ DSIPTG 06Z-10Z. DF

The FA reports a cold front, with an associated upper trough, offshore moving through
Washington and Oregon. The TWEB synopsis reports the front reaching the Washington
coastline around 17Z, a Bellingham, Seattle, Portland line between 19Z and 21Z, and the Idaho
border near 02Z. This is certainly more precise and detailed than the FA. There is a southwesterly flow aloft. The air mass is moist and stable ahead of the front in western Washington
and stable with increasing mid- to high-level moisture over eastern Washington after 18Z. A
very cold and unstable airmass is moving over Washington behind the front with the Puget
Sound convergence zone (PSCG) developing in the vicinity of Paine after 20Z.

PUGET SOUND CONVERGENCE ZONE   According to Brad Colman, Seattle-Tacoma
National Weather: "A wind-borne rain band often keeps Seattle wet. When a west wind blows
into Washington state and across Puget Sound, Seattle or one of its suburbs more often than not
has a cloudy and rainy day. But the whole area doesn't get soaked. Typically, a narrow band of
clouds and rain often parks over one place while a neighboring locale remains dry, sometimes
sunny. Only those places that fall victim to the Puget Sound convergence zone stay wet."

Since TWEBs are issued 2 h after terminal aerodrome forecasts, an outlook may be
included to coincide with the end of the TAF valid time. This has been done with the Seattle
TWEB synopsis. The outlook (02Z–12Z) forecasts the upper-level trough offshore reaching the coastline around 12Z, the southwesterly flow aloft continuing, the air mass remaining moist and unstable, and the PSCZ dissipating between 06Z and 10Z. What about the
letters DF? In this case they are the forecaster's initials.

The considerable difference between FA and TWEB synopses will not always be this
extreme. But the example dramatically shows how a local synopsis can paint a much more
detailed, specific picture of the weather situation.

Below is an excerpt from the SFO FA for Washington state. Figure 16-2 graphically
compares and contrasts the areas covered by the FA and the TWEB routes.

SFOC FA 061045

WA CASCDS WWD
CSTL SXNS...SCT050 BKN100. 12-15Z BKN015-030 OVC060. OCNL VIS 3-5SM RA. TOPS
FL200. OTLK...MVFR CIG.
INTR SXNS...SCT-BKN120. 14-18Z BKN020-030 OVC060. OCNL VIS 3-5SM RA. TOPS
FL200. OTLK...MVFR CIG..

WA E OF CASCD
WRN PTN...SCT140. CI ABV. 15Z BKN120. 20Z SCT-BKN060 BKN100. ISOL SHRA. TOPS
FL200. OTLK...VFR.
ERN PTN...SKC OR SCT CI. 15Z SCT140. 18Z SCT080 BKN120. 22Z SCT-BKN060 BKN100.
ISOL SHRA. TOPS FL200. OTLK...VFR SHRA..

The forecaster has divided the area west of the Cascades into coastal sections and interior
sections, east of the Cascades into western and eastern portions. The 359 TWEB, Seattle-
Olympia-Portland route corresponds to the interior sections (INTR SXNS) of the FA. The
353 TWEB, the Spokane-Pendelton route, corresponds to the eastern portion section (ERN
PTN) of the FA.

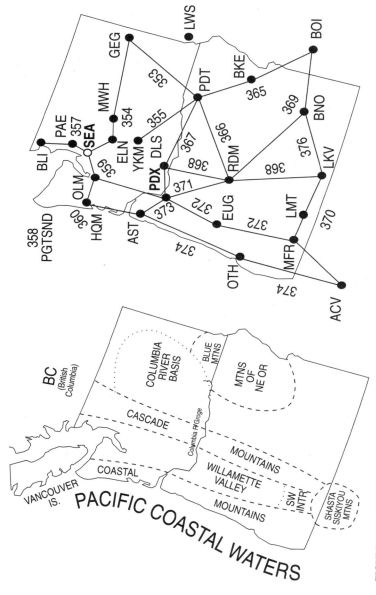

FIGURE 16-2 TWEB forecasts often contain more detail than is possible in an FA, although they are usually perfectly consistent within the scope of each product.

359 TWEB 061402 KSEA-KOLM-KPDX. ALL HGTS MSL XCP CIGS. KSEA-KTDO P6SM SCT050 BKN100 OVC200...17Z P6SM -RA SCT025 OVC040...19Z P6SM -RA SCT-BKN015 BKN025 OVC040...LCL VIS 3-5SM -RA...21Z P6SM -SHRA SCT-BKN030 BKN050. KTDO-KPDX P6SM SCT100 BKN200...16Z P6SM SCT-BKN040 BKN100 OVC200...LCL -RA...20Z P6SM -RA SCT-BKN025 OVC040...00Z P6SM -SHRA SCT025 BKN040.

353 TWEB 061402 KGEG-KPDT. ALL HGTS MSL XCP CIGS. P6SM SCT150 SCT-BKN250...17Z-24Z P6SM SCT080 BKN200...AFT 00Z P6SM SCT-BKN070-080 BKN200 WITH SCT -SHRA.20

The TWEB forecaster has further subdivided the 359 TWEB area into the leg from Seattle to Toledo and Toledo to Portland. To help visualize expected conditions, Table 16-4 breaks down the forecasts into visibility, sky condition, and weather for two specific times, 14Z and 19Z.

CASE STUDY Among the various opportunities that I have had with the FAA, I was once asked to serve as an expert witness. The plaintiff's attorney was quite perplexed that I testified that different forecasts for the same area were essentially the same even though they were not identical. The fact is different forecasts may appear to be substantially different, without any significant operational impact. Weather forecasting is not an exact science.

Back to Table 16-4: At the first time period, 14Z, both visibility and weather are the same for each forecast. Although sky condition appears at first glance to be very different, there is not any significant operational difference. Both forecast midlevel scattered to broken clouds. The SEA-TDO TWEB does predict a local scattered layer at 5000 ft. This should have no significant impact on either VFR or IFR operations.

The second time, 19Z, predicts lowering conditions with both forecasts. The TWEB forecast, because of its scope, can better define the time that conditions will deteriorate. Otherwise, again, forecasts for visibility, sky conditions, and weather are essentially the same.

We come to the same conclusions reviewing the ERN PTN and the FA and the 353 TWEB route forecast. The visibility forecast is the same (P6SM). The FA predicts lower-

TABLE 16-4 Area Forecast Compared to TWEB Forecast

	Area forecast	TWEB SEA-TDO	TWEB TDO-PDX
		14Z	
Visibility	P6SM	P6SM	P6SM
Sky condition	SCT-BKN120 TOPS FL200	SCT050 BKN100 OVC200	SCT100 BKN200
Weather	NIL	NIL	NIL
		19Z	
Visibility	OCNL 3-5SM	LCL 3-5SM	P6SM
Sky condition	BKN020-030 OVC060 TOPS FL200	SCT-BKN015 BKN025 OVC040	SCT-BKN040 BKN100 OVC200
Weather	OCNL RA	-RA	LCL -RA

ing clouds and precipitation earlier than the TWEB. Since the FA must cover a larger area than the TWEB and the TWEB route is in the extreme eastern portion of the FA area, this is perfectly consistent. Both forecasts accurately represent expected weather conditions within the scope and purpose of each product.

TWEBs do not directly forecast turbulence or icing. However, turbulence and strong updrafts and downdrafts are implied by weather associated with these phenomena. For example, strong winds and mountain wave activity indicate turbulence. Icing can be expected above the freezing level where visible moisture exists or in areas of freezing precipitation. Like the FA, thunderstorms imply severe or greater turbulence and icing and low-level wind shear. Specific forecasts for turbulence and icing must be obtained from the FA and weather advisories.

TWEB routes and synopses often provide more precise timing and detail than an FA can provide; TWEBs provide additional specific information on visibility, surface winds, timing, and local conditions. But, like other forecast products, TWEBs cannot cover every instance of hazardous weather. Although the number of TWEB routes is extensive, many areas are not covered. Never extrapolate nor extend the forecast beyond its defined area.

Revised TWEB standards were adopted to impose uniformity on the product. However, as is often the case, interpretation of the new instructions was not consistent. Just after implementation, this message appeared from one weather service office: TWEB ROUTES DELAYED DUE TO RIDICULOUS NEW FORMAT.

CHAPTER 17

TERMINAL AERODROME FORECASTS

Some think the only place a pilot will actually see a *terminal aerodrome forecast* (TAF) is on an FAA exam. This might have been true in the past, but with FSS consolidation, DUAT, and other commercial software, responsibility for decoding, translating, and interpreting this product will rest with the pilot. Incorrect interpretation could lead to anything from an embarrassing chat with a *flight standards district office* (FSDO) inspector to an aircraft accident.

As it did METAR, the United States adopted the international TAF code on July 1, 1996. According to sources within the National Weather Service, "A few of the changes seem to defy logic—in fact, they are illogical." At first glance the format appears confusing, but actually the basic forecast remains the same.

TAFs are issued four times a day (0000Z, 0600Z, 1200Z, 1800Z), and they are valid for 24 h. This has answered the criticism that terminal forecasts be written more often and be valid for a longer period. Forecasts using TAF (World Meteorological Organization—WMO) codes are also issued by local base weather offices for many military locations. Military TAFs are affected by different criteria than those TAFs issued by the NWS; therefore, some differences in format and content will occur.

The meteorologist must, normally, have at least two consecutive METARs before issuing a TAF. These observations must not be less than 30 min or more than 1 h apart. At a minimum, these observations must, usually, contain wind, visibilities and obstructions to visibility, weather, sky conditions, temperatures, dew points, and altimeter settings. However, after analyzing available data, if in the forecaster's judgment, a missing observation or element will have no impact on the quality of the forecast, the TAF may be issued or continued.

Airports with part-time weather observations normally require 3 h after the first observation for the TAF to be written and distributed. The TAF may not be issued if an observation or element is missing. For example, with a missing dew point, pilots could expect to see KXYZ TAF 071919 TAF NIL DUE MSG DEW PT. Likewise when the observation ceases, the forecast will contain the statement AMD NOT SKED AFT 03Z. This alerts the user that amendments are not available after 0300Z.

Automated observations are allowing more TAFs to be written on a 24-h basis. However, should certain sensors be unavailable, a remark will indicate the TAF status. Consider, for example, AMD LTD TO CLD VIS AND WIND (amendments limited to clouds, visibility, and wind). This indicates that weather phenomena such as thunderstorms and freezing precipitation can occur without the issuance of a TAF amendment.

Prepared by local *weather forecast offices* (WFOs), terminal aerodrome forecasts contain specific information for individual airports. TAFs are issued in the following format:

- Type
- Location
- Issuance time
- Valid time
- Forecast

There are two types of TAF issuances: a routine forecast issuance, TAF, and an amended forecast, TAF AMD. TAFs may be *corrected* (COR) or *routinely delayed* (NIL). TAF location is identified by the four-letter ICAO station identifier. Issuance date and time consists of a six-digit group. The first two digits represent the day of the month, and the last four digits, the UTC issuance time. The valid period is a four-digit group, usually 24 h, in UTC. The forecast group is divided into the body and remarks. Remarks amplify or describe conditions that differ from those described in the body of the forecast group. Conditional terms describe variability.

In the United States, TAFs cover a 5-statute-mile (SM) radius from the center of an airport's runway complex. *Vicinity* (VC) is defined as an area from beyond 5 mi to 10 SM from the center of the runway complex.

A word of caution: It is dangerous to extrapolate the TAF beyond 5 mi. Over flat terrain, such as the midwest, chances are that nearby airports might have similar conditions. This does not relieve the pilot, however, of checking appropriate FAs or TWEBs. In hilly or mountainous terrain, this practice can be disastrous. For example, afternoon surface winds during the summer are routinely strong and gusty at the San Francisco International Airport, whereas at other San Francisco Bay area airports, winds remain relatively calm due to wind direction and terrain.

Conversely, an airport in the middle of a valley might have benign surface winds; however, at a nearby airport below a canyon, surface winds can prevent landings completely. This situation is not uncommon in the Los Angeles Basin when Santa Ana winds blow; winds at the Ontario, California, airport might be out of the west and less than 5 kn, but at Rialto, only about 10 mi away, winds might be out of the north gusting to more than 40 kn. Rialto lies just below the Cajon Pass.

Terminal aerodrome forecasts should be written as simply and straightforwardly as possible. To describe significant changes during the forecast period, the forecast is subdivided into one or more smaller time segments. Changes or expected changes in ceiling and visibility that cross a flight category threshold are considered operationally significant. These categories are defined in Table 12-2, Weather Categories, with the addition of low IFR (LIFR, ceiling less than 500 ft and/or visibility less than 1 SM). Additional ceiling and visibility thresholds correspond to IFR alternate requirements and approach and landing minimums. Other elements that are considered operationally significant are:

- Thunderstorms
- Low-level wind shear
- Freezing precipitation
- Ice pellets
- Moderate or greater rain
- Snow expected to accumulate
- Sustained winds greater than 15 kn
- Wind direction changes of 30° or more with speeds of 12 kn or more
- Wind gust value changes of 10 kn or more

The body of the TAF uses the following format:

- Wind
- Visibility
- Weather
- Sky conditions
- Wind shear, when applicable

As with METAR, wind is forecast as a five- or six-digit group when it is considered significant to aviation. The abbreviation KT follows the wind forecast and denotes the units as knots. Gustiness is a forecast for rapid fluctuations of 10 kn or more indicated by the letter *G* following the wind group: 34025G40KT (mean wind speed 25 kn with gusts to 40 kn), or VRBG50 (wind gusts to 50 kn). Unlike the FA or TWEB, the TAF does provide a specific wind forecast.

A popular aviation magazine reported that winds received from a tower were reported in miles per hour and winds on TAFs were reported in knots. Let's settle the matter: Wind direction true or magnetic, blowing from or toward the reported direction, speed knots or miles per hour. All official—for some reason the FAA in Washington hates the word *official*; therefore, I use it as often as possible—aviation wind observations and forecasts are reported in relation to true north, given as the direction from which the wind is blowing, with speed in knots. The only times a pilot will receive winds in relation to magnetic north are in reports from a tower, ATIS, or AWOS broadcast, or as part of a local airport advisory because runway numbers are magnetic.

Prevailing visibility up to and including 6 SM is forecast. Visibility greater than 6 mi is indicated by the letter *P* for plus (P6SM, visibility greater than 6 statute miles). Military and many international TAFs forecast visibility in meters.

Weather and obstructions to vision use the same format and codes as METAR. With no significant weather expected, the weather group is omitted. When significant weather is forecast but expected to change to no significant weather, the abbreviation for *no significant weather*—NSW—appears. NSW will not appear when any of the following phenomena are expected to occur:

- Freezing precipitation
- Moderate or heavy precipitation
- Drifting or blowing dust, sand, or snow
- Duststorms or sandstorms
- Thunderstorms
- Squalls
- Tornadoes
- Phenomena expected to cause a significant change in visibility

Tornadic activity, including tornadoes, waterspouts, and funnel clouds, will normally not appear in terminal forecasts since the probability of occurrence at a specific location is extremely small. On the other hand, volcanic ash will always be forecast when it is expected.

Sky conditions are given in the same format and abbreviations as METAR, and cloud heights are always reported *above ground level* (AGL)—that is, amount, height, cloud type, or vertical visibility. This is important when using a TAF product or comparing it with observations, PIREPs, or other forecasts. The ceiling is always the first broken or overcast layer, or vertical visibility into a complete obscuration. TAFs are specifically intended for arriving and departing aircraft; therefore, cloud layers above 15,000 ft might not be included when a lower ceiling appears. When cumulonimbus clouds are expected, CB is appended to the cloud layer. CB is the only cloud type forecast in TAFs. NWS-prepared terminal forecasts will not include forecasts of partial obscurations.

Low-level wind shear (LLWS) on terminal aerodrome forecasts refers to nonthunderstorm shear, within 2000 ft of the ground. Wind direction, speed, and height are included. LLWS appears when PIREPs report an airspeed gain or loss of 20 kn or more, or vertical shears of 10 kn or more per 100 ft are expected or reported. Low-level wind shear, when forecast, will appear following sky condition on domestic TAFs, in the following format:

...WS015/24035KT...

- *WS.*   Wind shear.
- *015.*   Height in hundreds of feet (above ground level) of the wind shear (1500 ft AGL).
- */24035.*   Wind direction and speed (knots) above the wind shear (240° at 35 kn).

Since turbulence and icing forecasts are contained in the AIRMET Bulletin and SIGMETs, U.S. domestic TAFs will not include these phenomena.

CONDITIONAL TERMS

In the body of the TAF, prevailing conditions have a greater than 50 percent probability of occurrence, are expected to last for more than 1 h during each occurrence, and cover more than half of the forecast period. This explains a major TAF misconception. Conditions may differ in the body of the forecast from actual conditions, but the forecast is considered accurate as long as conditions fall within the prescribed parameters. More about this when we discuss amendment criteria.

TAF conditional terms consists of *temporary* (TEMPO) *conditions* and *probability* (PROB) *forecasts.* TEMPO and PROB are considered remarks to the body of the forecast. TAF terms are defined in Table 17-1. When determining minimums (approach and landing, or alternate), the lowest conditions within the appropriate time period must be considered. Therefore, if conditions are lower in the TEMPO group, those conditions must be applied.

TEMPO indicates that temporary conditions are expected to occur during the forecast period. TEMPO describes any condition with a 50 percent or greater probability of occurrence, expected to last for generally less than an hour at a time, and to cover less than half

TABLE 17-1   Terminal Aerodrome Forecast Conditional Terms

No conditional term	A greater than 50% probability of occurrence, expected to last for more than 1 h during each occurrence and to cover more than half of the forecast period
TEMPO	A 50% or greater probability of occurrence, expected to last for generally less than an hour at a time and to cover less than half of the forecast period
VC (vicinity)	An area from beyond 5–10 SM from the center of the runway complex
Probability of precipitation	
PROB 40	A 40–<50% probability
PROB 30	A 30–<40% probability

of the forecast period. The time during which the condition is expected to occur is indicated with a four-digit group beginning and ending times UTC. For example, SCT030 TEMPO 1923 BKN030; 3000 scattered temporarily between 1900Z and 2300Z ceilings 3000 broken. A ceiling of 3000 broken is expected to exist for periods of less than 1 h during the 1900Z to 2300Z time frame. TEMPO is equivalent to, and may be translated by FSS controllers as, *occasional* (OCNL).

A PROB group indicates the probability of occurrence of thunderstorms or other precipitation events. PROB 40 indicates a 40 to less than 50 percent probability; PROB 30 indicates a 30 to less than 40 percent probability. This is followed by a four-digit time group giving beginning and ending times. PROB 40 is the equivalent of *chance* (CHC), and PROB 30, *slight chance* (SLT CHC).

FORECAST CHANGE GROUPS

Forecast change groups consist of *from* (FM) followed by a *time group* (tttt)—FMtttt—and *becoming* (BECMG) followed by a *time group* (TTtt)—BECMG TTtt. The FMtttt group is used when a rapid change is expected, usually within less than 1 h. For example, BKN020 FM1630 SKC—before 1630Z ceiling 2000 broken, around 1630Z the sky condition will change to clear. The BECMG TTtt group indicates a more gradual change in conditions over a longer period of time. The BECMG group indicates a change in conditions that is expected to occur at either a regular or irregular rate at an unspecified time but within the period.

Figure 17-1 illustrates the operational impact of a BECMG group. If conditions are expected to deteriorate during the BECMG period, those conditions must be considered to exist at the beginning of the period. Conversely, if conditions are expected to improve during the BECMG period, the lower conditions must be considered to exist to the end of the period. The first example in Fig. 17-1 shows the ceiling deteriorating during the BECMG period. From an operational perspective, pilots and dispatchers must consider the lower condition (OVC005) to exist any time after 18Z. The second example in Fig. 17-1 shows the visibility improving during the BECMG period. Again, from an operational perspective the lower condition (1SM BR) must be considered to exist through 22Z.

Although weather typically behaves in this manner, the BECMG group can have significant operational impact. In response to user requirements, the NWS has agreed to use the BECMG group sparingly, and when they do use it, not to exceed 2 h. Additionally, forecasters will avoid using a BECMG group to forecast minimum conditions, especially visibility less than 1/2 SM.

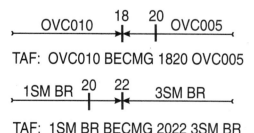

FIGURE 17-1 The BECMG group can have significant operational impact. The NWS has agreed to use the BECMG group sparingly and to not exceed 2 h when they do use it.

Let's decode, translate, and interpret the following TAF for New York's Kennedy International Airport:

```
TAF KJFK 191130Z 191212 08023G33KT 5SM -RA BR OVC007
    WS020/11045KT TEMPO 1212 2SM RA BR OVC005 BECMG 1618
    07028G38KT WS020/11050KT
    BECMG 1921 06023G45KT WS020/10055KT
```

This forecast was issued on the 19th day of the month at 1130Z. It is valid from the 19th at 1200Z until the 20th at 1200Z. Note that the first element contains all the forecast categories. The wind is from 080° at 23 gusting to 33 kn; visibility is 5 statute miles in light rain and mist; ceiling 700 ft overcast; wind shear at 2000 ft, wind at that altitude 110° at 45 kn. (There is a greater than 50 percent probability of occurrence, it is expected to last for more than 1 h during each occurrence, and it is expected to cover more than half of the forecast period.) Occasionally (TEMPO) between 12Z and 12Z—the entire forecast period—visibility 2 statute miles in moderate rain and mist, ceiling 500 ft overcast. Since the wind group is omitted from the TEMPO group, it is expected to remain 08023G33KT. (There is a 50 percent or greater probability of occurrence, and it is expected to last for generally less than an hour at a time and to cover less than half of the forecast period.) Since the remarks in the TEMPO group are lower than the body, for flight planning purposes, they must be applied through the entire period. That is, for determining IFR alternate requirements for the entire period of the forecast (12Z to 12Z), visibility 2SM and ceiling OVC005 must be applied.

Wind and wind shear are expected to become between 16Z and 18Z 070° at 28 gusting to 38 kn. What about the other elements of the forecast? In a BECMG group, only the elements expected to change are included. Therefore, visibility, weather, and sky condition forecast in the body and remarks are expected to continue. This is one difference between a BECMG and an FM group. In an FM group all elements of the forecast, even if they are not expected to change, must be included. Because higher winds are forecast in the second part of the BECMG group, they must be applied beginning at 16Z. For example, if the wind conditions in the BECMG group (07028G38KT) exceed our maximum crosswind component, the forecast would preclude our landing at any time after 16Z. The same interpretation applies to the second BECMG group, where wind and wind shear are expected to increase.

Pilots can expect to see differences in domestic TAF codes and those used on military forecasts. Military forecasts may include temperature, turbulence, and icing, in addition to visibility in meters. Table 17-2 decodes temperature, turbulence, and icing codes. Table 17-3 converts meters to statute miles.

The following is a TAF for Travis Air Force Base, California:

```
KSUU 101111 24010G15KT 9999 BKN070 BKN180 QNH3020INS
    BECMG 1415 22010G15KT 9999 VCRA SCT050 BKN070 OVC180
    620507 QNH3018INS
    BECMG 1718 22010G15KT 9000 -RA BR SCT030 BKN050OV100
    62048 51005 QNH3014INS WND 20012G18KT 21003
    BECMG 0405 20010G15KT 8000 -RA BR BKN030OVC050 62408
    51005 QNH3005INS T11/23Z T08/13Z
```

The forecast was issued on the 10th day of the month and is valid from the 10th at 1100Z through the 11th at 1100Z.

The initial period, 1100Z, forecasts winds 240 at 10 gusting to 15 kn, visibility 9999 m, ceiling 7000 ft broken 18,000 ft broken, and the altimeter setting is 30.20 in Hg. From Table 17-3 we see that 9999 translates to more than 6 statute miles (P6SM). Note that the

TABLE 17-2 Terminal Aerodrome Forecasts

Temperature	TT_1T_1/tt
	T: temperature group
	T_1T_1: temperature Celsius
	tt: time UTC
Turbulence	5ihhhd
	5: turbulence group
	i: turbulence intensity
	hhh: base height, hundreds of feet
	d: thickness, thousands of feet
Turbulence intensity	
0	None
1	Light turbulence
2	Moderate turbulence in clear air, infrequent
3	Moderate turbulence in clear air, frequent
4	Moderate turbulence in cloud, infrequent
5	Moderate turbulence in cloud, frequent
6	Severe turbulence in clear air, infrequent
7	Severe turbulence in clear air, frequent
8	Severe turbulence in cloud, infrequent
9	Severe turbulence in cloud, frequent
Icing	6ihhhd
	6: icing group
	i: icing intensity
	hhh: base height, hundreds of feet
	d: thickness, thousands of feet (0 indicates to top of clouds)
Icing intensity	
0	No icing
1	Light icing
2	Light icing in cloud
3	Light icing in precipitation
4	Moderate icing
5	Moderate icing in cloud
6	Moderate icing in precipitation
7	Severe icing
8	Severe icing in cloud
9	Severe icing in precipitation

altimeter setting (QNH) is also forecast—another difference between domestic and military TAFs.

Conditions are forecast to change between (BECMG) 1400Z and 1500Z to wind 220 at 10 gusting to 15 kn, visibility 9999 m, rain in the vicinity, 5000 ft scattered ceiling 7000 ft broken 18,000 ft overcast, icing 620507, and altimeter setting 30.18 in Hg. Rain in the vicinity translates to rain occurring beyond 5 and up to 10 SM of the airport. Notice this period contains an icing group 620507.

Decode the icing group using Table 17-2. The numeral 6 tells us it is an icing group. The next digit (2) represents icing intensity—light icing in cloud. The next three digits (050) represent the base of the icing layer in hundreds of feet MSL—5000 ft. The last digit (7) indicates the thickness of the icing layer in thousands of feet—7000 ft or tops of the icing layer at 12,000 ft MSL.

TABLE 17-3 Visibility in Meters

Meters	Statute miles	Meters	Statute miles
0000	0	2200	1 3/8
0200	1/16	2400	1 1/2
0300	1/8	2600	1 5/8
0400	3/16	2800	1 3/4
0500	5/16	3000	1 7/8
0600	3/8	3200	2
0800	1/2	3600	2 1/4
1000	5/8	4000	2 1/2
1200	3/4	4800	3
1400	7/8	6000	4
1600	1	8000	5
1800	1 1/8	9000	6
2000	1 1/4	9999	6+

The next change group occurs between 1700Z and 1800Z. The forecast expects wind 220 at 10 gusting to 15 kn, visibility 9000 m in light rain and mist, 3000 ft scattered ceiling 5000 ft broken 10,000 ft overcast, icing 620408, turbulence 51005, altimeter setting 30.14 in Hg, and wind between 21Z and 03Z 200 at 12 gusting to 28 kn. Visibility is expected to decrease to 6 statute miles in light rain and mist. Light icing in clouds is expected between 4000 and 12,000 ft MSL.

Decode the turbulence group using Table 17-2. The numeral 5 tells us it is a turbulence group. The next digit (1) represents turbulence intensity—light turbulence. The next three digits (000) are the base of the turbulence layer in hundreds of feet MSL—0000 ft or the surface. The last digit (5) indicates the thickness of the turbulence layer in thousands of feet—5000 ft or top of the turbulence layer at 5000 ft MSL.

The next change is expected between 0400Z and 0500Z. The forecast expects wind 220 at 10 gusting to 15 kn, visibility 8000 m in light rain and mist, ceiling 3000 ft broken 5000 ft overcast, icing 620408, turbulence 51005, altimeter setting 30.05 in Hg, and temperature T11/23Z T08/13Z. Visibility is expected to decrease to 5 statute miles. Icing and turbulence are expected to continue.

Again, decode the temperature group using Table 17-2. The maximum temperature of 11°C is forecast to occur at 2300Z, minimum temperature of 8°C at 1300Z.

Whether international or military, TAFs suffer from the same limitations as any other forecast. It really can't be said that a domestic TAF or military TAF is necessarily more accurate. However, by comparing a domestic TAF with a nearby military TAF, one could get a "second opinion."

AMENDMENT CRITERIA

The NWS forecast manual states:

"Amendments shall be issued when expected or observed conditions: (1) meet amendment criteria for the specified forecast elements, (2) are expected to persist, and (3) in the forecaster's judgement, there is sufficient, reliable information,...on which to base a forecast."

Amended TAFs contain the contraction AMD. For example:

TAF AMD
PDX 101440Z 101512...

This amended forecast for Portland International Airport was issued on the 10th at 1440Z and is valid from 1500Z on the 10th until 1200Z on the 11th.

Amendment criteria have changed significantly in the 1990s. Even with these changes, reviewing Table 17-4, which gives the terminal aerodrome forecast amendment criteria, we see the reason for some of the criticism of this product. Ceilings above 3000 ft and visibilities of more than 6 mi must decrease below these values before an amendment is required. Amendments are not required, for example, for a forecast ceiling of 10,000 decreasing to 5000 ft. However, the forecaster has the authority to amend the forecast when he or she considers it operationally significant. As conditions lower, becoming more significant, amendment criteria intervals decrease. The forecast is considered accurate as long as the criteria in Table 17-4 are met.

The new amendment criteria reflect operations needs. Note that a decrease or increase from basic VFR (1000 and 3), nonprecision approach minimums (600 and 2), and precision approach minimums (200 and 1/2) require an amendment. As we will see, however, this does not mean that the moment these criteria are met, an amendment will be issued.

TABLE 17-4 Terminal Aerodrome Forecast Amendment Criteria

Forecast	Amend if
Wind, kn	
≥12	Direction differs by 30°; speed differs by 10 kn
Gusts with mean speed ≥12	Gusts 10 kn above forecast
Visibility, SM	
>5	≤5
3–5	>6 or <3
2	≥3 or <2
1–1 1/2	≥2 or <1
1/2–3/4	≥1 or <1/2
<1/2	≥1/2
Weather	
Thunderstorms, freezing precipitation, or ice pellets	Does not occur or no longer expected
No thunderstorms, freezing precipitation, or ice pellets	Occurs or is expected
Ceiling, ft	
>3000	<3000
2000–3000	>3000 or <2000
1000–1900	>1900 or <1000
600–900	>900 or <600
200–500	>500 or <200
>100	≥200
Low-level wind shear	
LLWS	No LLWS expected
No LLWS	LLWS occurs or is expected

As shown in Table 17-4, changes in wind speed of less than 12 kn do not require an amendment. At speeds of 12 kn or greater, the speed must differ by 10 kn or more. Therefore, a forecast of 20 kn would have to decrease to 10 or increase to 30 before an amendment would be required. Basically, TAF wind speeds are considered accurate when they are within 10 kn of forecast. A pilot competent to handle 20 kn must consider, with a forecast of 20 kn, that actual winds could increase to almost 30 kn without a requirement to amend. Within these parameters the forecast is considered correct.

The only weather phenomena requiring amendments are thunderstorms, freezing precipitation, ice pellets, and LLWSs. The unforecast occurrence or ending of rain or snow does not require an amendment. This fact has led both pilots and briefers to erroneously criticize the forecast.

When a phenomenon described in VC moves or is expected to move within 5 mi of the airport, an amendment is required. The phenomenon can appear in the body or remarks of the amendment. But it must be one or more of the phenomena described in the weather section of Table 17-4.

At locations with part-time observations, the remark NIL AMDTS AFT (time)Z will appear. As used in TWEBs, this serves as a warning that significant changes can occur without an amendment. Extra caution must be exercised operating into these airports, especially during marginal or deteriorating conditions.

A forecast of landing minimums, VFR or IFR, is no guarantee those conditions will exist at our estimated time of arrival. An IFR alternate affected by the same weather pattern as the destination may satisfy regulations but leave a pilot on the proverbial limb with a busted forecast. For example, if a pilot's destination was Fresno, in California's central valley, Bakersfield may qualify as a legal alternate. However, with Tule Fog—a local name given to a condition of extensive wintertime fog—in California's central valley, a viable alternate might be along the coast, not affected by this phenomenon. This would also apply to airports affected by upslope fog or frontal systems. If at all possible, a pilot should select an alternate not affected by the weather pattern affecting the destination. At the risk of being redundant, we're back to knowing the synopsis and continually updating weather enroute, always piecing together the complete picture.

USING THE TERMINAL AERODROME FORECAST

Apparent inconsistencies arise because different forecasts serve different purposes. The following example compares FA and TWEB forecasts with TAFs.

Below are excerpts from the Chicago AIRMET Bulletin and FA, the 216 TWEB route forecast from Chicago, Illinois, to Burlington, Iowa, and the Chicago O'Hare International and Burlington Regional TAFs.

CHIS WA 121945
AIRMET SIERRA UPDT 4 FOR IFR VALID UNTIL 130200

AIRMET IFR...MN IA MO WI IL
FROM 100NE GFK TO INL TO BDF TO FAM TO SGF TO 100NE GFK
OCNL CIG BLW 010/VIS BLW 3SM PCPN/BR. LTL MOVMT EXP OVR NRN
SXNS OF THE AREA..SHFTG SLOLY EWD SRN SXNS. CONDS CONTG BYD
02Z THRU 08Z.

CHIC FA 121845
SYNOPSIS AND VFR CLDS/WX
SYNOPSIS VALID UNTIL 131300

CLDS/WX VALID UNTIL 130700...OTLK VALID 130700-131300

IL
NRN...CIG BKN-SCT020-030 BKN080. TOPS 140. OCNL VIS 3-5SM
-SHRA. SCT TSRA. CB TOPS FL400. OTLK...MVFR CIG SHRA BR.

216 TWEB 122008 KCHI-KBRL. ALL HGTS AGL XCP TOPS. KCHI-50SM
SW KCHI P6SM BKN100 WDLY SCT 2SM TSRA OVC040-050...AFT
05Z 4SM -RA BR OVC020-030. 50SM SW KCHI-KBRL 5SM -SHRA
OVC010-020 SCT 2SM TSRA OVC010-020...AFT 06Z 5SM BR
WDLY SCT 2SM TSRA OVC010-020.

TAF
KORD 121730Z 121818 07012KT P6SM BKN100 TEMPO 2024 3SM TSRA
    OVC045CB
FM0000 08015G25KT P6SM OVC040 TEMPO 0004 3SM TSRA
    OVC025CB
FM0400 08012KT 4SM -RA BR OVC022 =

TAF AMD
KBRL 122122Z 122118 32013KT P6SM FEW020 BKN050
FM2130 30012KT 5SM -SHRA OVC035
    TEMPO 2202 3SM TSRA OVC015CB
FM0200 34008KT 5SM BR OVC010 TEMPO 0618 -SHRA
    PROB40 0610 3SM TSRA OVC015CB =

Figure 17-2 depicts these forecasts on excerpts from the weather advisory plotting chart and TWEB route chart. Chicago AIRMET SIERRA has been plotted in gray on Fig. 17-2, occasional ceilings and visibilities below 1000 ft and 3 mi. The AIRMET covers the western half of the 216 TWEB route, including Burlington, Iowa.

The 216 TWEB divides the route into two segments, Chicago to 50 mi southwest of Chicago, and 50 mi southwest of Chicago to Burlington. Note how the TWEB has divided the route into segments and two distinct time periods.

The TAFs are further able to be divided into specific time periods. To help visualize forecast conditions, Table 17-5 provides a comparison of these products. A time frame of 2000Z to 0000Z has been selected with appropriate forecasts covering the 216 TWEB route.

The TAF provides much more detail on surface winds than is available with either the FA or TWEB. Typically the FA does not forecast visibilities or ceilings less than 3 mi or 1000 ft. How much less? We don't know. This makes the FA all but useless for determining IFR approach and landing, or alternate requirements. Therefore, the TWEB is much more helpful in forecasting conditions at airports not served by a TAF but within the area of coverage of the TWEB. This is especially true for the second half of the TWEB route. From Table 17-5 we see that expected weather is consistent within each forecast.

Let's say we're planning an IFR arrival to an airport 5 mi beyond the Burlington airport but within the area served by the TWEB. In accordance with FARs, we must determine if an alternate airport is required, and, if so, select a suitable alternate. Our ETA is 2300Z. From the TWEB we see the forecast for 1 h before until 1 h after our *estimated time of arrival* (ETA) is visibility 2 SM ceiling 1000 ft. Therefore, an alternate is required.

The preceding comparison contradicts a common misconception that in the absence of a TAF there is no forecast on which to base destination, alternate, and fuel requirements. If

216 TWEB 122008 KCHI-KBRL. ALL HGTS AGL XCP TOPS. KCHI-50SM
SW KCHI P6SM BKN100 WDLY SCT 2SM TSRA OVC040-050...
50SM SW KCHI-KBRL 5SM -SHRA OVC010-020 SCT 2SM TSRA
OVC010-020...

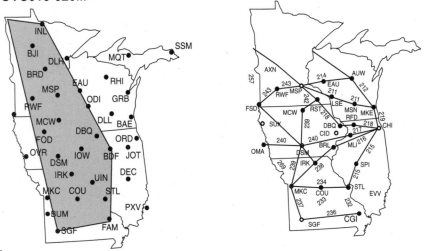

IL
NRN...CIG BKN-SCT020-030 BKN080. TOPS 140. OCNL VIS 3-5SM
-SHRA. SCT TSRA. CB TOPS FL400. OTLK...MVFR CIG SHRA BR.

FIGURE 17-2   Apparent inconsistencies arise because the FAs, TWEBs, and TAFs serve different purposes.

not TWEBs, the FAs are always available; maybe they are not as detailed as we might like, requiring additional alternates and fuel, but nonetheless they are available.

Forecasters consider four elements when writing TAFs:

1. Expected weather

2. Local effects

3. Climatology

4. Amendment criteria

Attention to detail places a premium on the forecaster's time and judgment.

In addition to the new TAF amendment criteria, several other changes have improved the terminal forecast. The NWS recommends that forecasters restrict the use of conditional terms and keep TEMPO restricted to short periods. Forecasters are encouraged to amend when conditions do not materialize, even when amendment criteria have not yet been met.

Due to a number of factors, inconsistencies with reported conditions are to be expected. The forecaster might not believe observations are representative, or the forecaster might expect conditions to change rapidly. Even in the body of the forecast there is only a greater than 50 percent probability of occurrence, during more than one-half the forecast period. Amendment criteria give the forecaster some latitude. The forecaster might be waiting until TAF amendment criteria are reached—reported or expected. The time required to write and distribute an amendment is also a consideration; saturated computer and communications systems can hinder timely updates. If

TABLE 17-5 FA, TWEB, and TAF Forecast Comparison, 2200Z–0000Z

	FA	TWEB CHI-50 SW CHI	ORD TAF
Wind	<20 kn	<25 kn	07012 kn
Visibility	3 SM	2 SM	3 SM
Weather	-SHRA/TSRA	TSRA	TSRA
Ceiling	2000 ft	4000 ft	4500 ft
	FA	TWEB 50 SW CHI-BRL	BRL TAF
Wind	<20 kn	<25 kn	30012 kn
Visibility	<3 SM	2 SM	3 SM
Weather	-SHRA/TSRA	-SHRA/TSRA	-SHRA/TSRA
Ceiling	<1000 ft	1000 ft	3500 ft

there is less than 2 h between the time an amendment is required and a new forecast becomes effective or the next portion of the forecast becomes effective, an amendment might not be issued. Time parameters on domestic TAFs, even with an FM group, usually indicate phenomena will change within 2 h of the specified time. Additionally, forecasters have other duties that might hinder timely amendments.

FAs, TWEBs, and TAFs are usually perfectly consistent given their individual purpose and criteria. Apparent inconsistencies arise because cloud heights in the FA are generally provided relative to MSL, whereas in the TAF cloud heights are always relative to AGL; TAFs can consider local effects not within the scope of other products; or a pilot attempts to extrapolate a TAF beyond 5 mi. When inconsistencies develop, FSS controllers and forecasters work together to resolve differences. Pilots using DUATs will have to consult an FSS for resolution.

Domestic TAFs do not directly forecast turbulence or icing. However, like METAR, these phenomena are implied. Strong surface winds, LLWSs, and thunderstorms indicate turbulence; surface temperatures close to freezing with cloud layers, freezing precipitation, and thunderstorms imply icing.

Pilots continually demand to know exactly when an airport will improve to IFR landing minimums or VFR conditions. This accuracy simply is not yet possible; but both the FAA and NWS are working on these issues. Alternate airport and fuel reserve requirements take these limitations into consideration. However, departing with appropriate alternates and adequate fuel does not relieve the pilot from updating weather information and when necessary, revising the flight plan, enroute.

CHAPTER 18
WINDS AND TEMPERATURES ALOFT FORECASTS

A Navion pilot called flight service with a request for "winds aloft." The specialist asked, "For what altitudes?" The pilot rather indignantly replied, "What ever's best for my direction of flight!" Well, let me tell you, it's 53,000 ft every time.

Winds and temperatures aloft forecasts (FDs) for the contiguous United States, Alaska, and many oceanic areas are computer generated at the National Centers for Environmental Prediction (NCEP) outside Washington, D.C. FDs for the Hawaiian Islands are produced by the Honolulu Weather Forecast Office. FDs are based on the twice-daily *radiosonde balloon* (balloon with a radiosonde attached) *observations*, which are normally released at 1100Z and 2300Z daily. Launch sites can be found in the *Airport/Facility Directory*, on the page opposite the inside back cover. Because these sites and times are published, pilots cannot expect to receive balloon launch warnings in the form of NOTAMs or broadcasts, except for unscheduled releases.

A computer program known as the *Nested Grid Model* (NGM) analyzes data. NGM is one of the latest forecast models, and it has replaced the *Limited Fine Mesh* (LFM) *model* for FD-tabulated forecasts; with a smaller grid, thus better resolution, and terrain taken into account to a greater degree than the LFM, the NGM has improved forecast accuracy. However, the NGM can consider only synoptic (large-scale) weather systems. The computer projects system movements and produces a forecast. Although large-scale terrain, to some extent, is considered, local features are not. Therefore, FDs tend to be less accurate in the western states, especially below 12,000 ft.

Approximately 750 upper-air stations worldwide—about 120 in the United States—take observations. Stations are generally located on land, leaving great expanses of ocean without observations; however, satellite and aircraft reports help fill in the gaps. The computer must interpolate for locations without observations, and sparse observational data hinder the accuracy of the forecast.

Winds and temperatures aloft forecasts can be obtained from a number of sources: flight service stations (FSSs), telephone information briefing services (TIBS), direct user access terminals (DUATs), and commercial vendors.

TABULATED FORECASTS

FDs normally become available after their scheduled transmission times of 0440Z and 1640Z. They consist of three forecast periods: 6, 12, and 24 h. These periods are labeled FD1, FD2, and FD3 for levels through 39,000 ft, and FD8, FD9, and FD10 for 45,000 and 53,000 ft (Table 18-1).

TABLE 18-1 Wind and Temperature Aloft Forecast Schedule

File type	Valid	For use
Forecasts based on 0000Z data, available at 0440Z		
FD1 FD8	0600Z	0500–0900Z
FD2 FD9	1200Z	0900–1800Z
FD3 FD10	0000Z	1800–0500Z
Forecasts based on 1200Z data, available at 1640Z		
FD1 FD8	1800Z	1700–2100Z
FD2 FD9	0000Z	2100–0600Z
FD3 FD10	1200Z	0600–1700Z

CASE STUDY My student and I had planned a cross-country flight using FDs obtained from DUATs. We then called flight service and filed our flight plans. I had my student obtain a standard weather briefing to compare the information from an FSS briefing with that received from DUATs. The FDs provided in the FSS briefing were significantly different from those received from DUATs. Why? Between the time we obtained the DUAT briefing, completed the calculations, and called flight service, new FDs had become available.

FAA automated flight service stations (AFSSs) display levels 3000 through 53,000 ft. However, the 45,000- and 53,000-ft levels are not available for all standard FD locations. Other FSSs (nonautomated Alaskan facilities) usually post levels through 39,000 ft; the two higher levels are available on request. FSSs usually post the 6- and 12-h forecasts (FD1 and FD2), with the 24-h forecasts (FD3) available on request.

FD3s and FD10s are plagued with the same limitations as other forecasts and must be viewed with skepticism and used only for advanced planning, then updated with the latest forecasts prior to departure—recall the previous case study. This requires the pilot to become familiar with the issuance times in Table 18-1. For example, a pilot planning a 1900Z departure might obtain the FD3s, based on 0000Z data, the evening before departure. The FD1s, based on 1200Z data, become available at 1640Z. If the pilot fails to obtain an updated forecast, he or she could be wide open to being cited for a violation in the event of a problem.

Refer to Fig. 18-1. This example contains the FD1 and FD8 forecasts, based on the 30th day of the month 1200Z radiosonde data (DATA BASED ON 301200Z). The DATA BASED ON must always be checked. From time to time old FDs fail to be purged and remain in the system. It's possible to receive data that is 24 h old. The next line states VALID 301800Z FOR USE 1700-2100Z. These FDs are for use between 1700Z and 2100Z. The computer does not forecast an average; the model predicts winds and temperatures for one specific time, in this case 1800Z (VALID 301800Z).

Forecasts based on the expected movement of synoptic systems explains one reason for apparent errors. With rapidly moving or intensifying systems, FDs can change significantly during the FOR USE period. This would be especially true for flights at the beginning or end of the FOR USE period.

Forecast levels are *true altitude*—true height above sea level—through 12,000 ft. From 18,000 through 53,000 ft, levels are *pressure altitude*—height as indicated with an altimeter setting of 29.92 in Hg (1013.2 mb). The example shows FDs for SFO (San Francisco, California) and RNO (Reno, Nevada).

Levels within the area of frictional effect between the wind and the Earth's surface are omitted. Therefore, forecast levels within approximately 1500 ft of the surface and temperatures for the 3000-ft level, or levels within 2500 ft of the surface, do not appear.

DATA BASED ON 301200Z

VALID 301800Z FOR USE 1700-2100Z. TEMPS NEG ABV 24000

FT	3000	6000	9000	12000	18000	24000	30000	34000	39000
SFO	3513	3316+10	3220+06	3224+01	3138-11	3148-24	325539	315846	315954
RNO		0605	3308+02	3217-02	3134-16	3052-28	316542	316848	316853

DATA BASED ON 301200Z

VALID 301800Z FOR USE 1700-2100Z. TEMPS NEG ABV 24000

FT	45000	53000
SFO	314961	304263

FIGURE 18-1 Tabulated winds and temperatures aloft forecasts are available from just above the surface to 53,000 ft, based on the twice-daily radiosonde observations.

Refer to the SFO 12,000-ft winds. The first two digits of a wind group represent true direction, from which the wind is blowing, to the nearest 10° ("32"24+01); the third and fourth digits indicate speed in knots (32"24"+01); and the last two digits are temperature in degrees Celsius (3224"+01"). Temperatures, plus or minus, are indicated through 24,000 ft; all temperatures above 24,000 are below 0°C and the minus sign is omitted. Therefore, 3224+01 is wind blowing from 320° true at 24 kn, temperature +01°C.

Forecast speeds of less than 5 kn are encoded 9900 and translated as "light and variable."

CASE STUDY FSS briefers are periodically asked, "What's the direction and speed of the light and variable winds?" In one extreme case a rather irate pilot demanded to know what was "actually written on the paper." The specialist replied, "niner-niner-zero-zero!"

Pilots must interpolate—compute intermediate values—to determine direction, speed, and temperature between forecast levels and reporting locations. Plan to fly from Oakland to South Lake Tahoe at 13,500 ft; use the SFO and RNO FDs. Average the 12,000-ft and then the 18,000-ft levels. At 12,000 ft, the direction is the same. A difference in speed of 7 kn results in an average of 21 kn (7/2 = 4; 17 + 4 = 21). The average temperature is zero. At 18,000 ft, again the direction is the same. A difference in speed of 4 kn results in an average of 36 kn (4/2 = 2; 34 + 2 = 36). The difference in temperature is 5, resulting in an average of -14 ($-5/2 = -3$; $-11 + (-3) = -14$). Be careful with the algebraic sign. The final result: 12,000 ft, 320° at 21 kn is $-01°C$, and at 18,000 ft, 310° at 36 kn is $-14°C$.

The 13,500 level is one-quarter of the way between 12,000 and 18,000 ft. Therefore, divide the difference between levels by 4 and add the result to the 12,000-ft values. Wind direction is 320°, speed 25 kn (15/4 = 4; 21 + 4 = 25), and temperature $-04°C$ [$-13/4 = -3$; $-1 + (-3) = -4$]. Because direction is to the nearest 10°, speed in whole knots, and temperature in whole degrees Celsius, the result cannot have a value in increments smaller than the original data; values are rounded off. As already noted, be careful of the algebraic sign. FAA exams might require calculating wind direction to the nearest 5°, however, for practical purposes this is not necessary.

Is a forecast of 701548 a misprint, or has it been garbled in transmission? With forecast winds of 100 kn or more, 5 is added to the first digit of the wind direction group. Therefore, to decode, subtract 5 from the first digit of the wind direction, and add 100 to speed. In this example wind direction, speed, and temperature are:

Direction, degrees	Speed, kn	Temperature, °C
70	15	48
-5	$+100$	
200	115	-48

[No mathematical sign (+ or −) was specified, so the temperature must be negative.] Maximum speed for FD tabulated forecasts is 199 kn.

FORECAST CHARTS

Forecast winds and temperatures aloft are also available in graphic form, issued twice daily valid at 1200Z and 0000Z. FD charts are excellent for determining forecast winds for long-distance flights. By visually depicting winds at various levels, favorable routes and altitudes can be determined. The eight panels contain forecast levels from 6000 through 39,000 ft. The two upper panels of the winds aloft chart are illustrated in Fig. 18-2. Because of

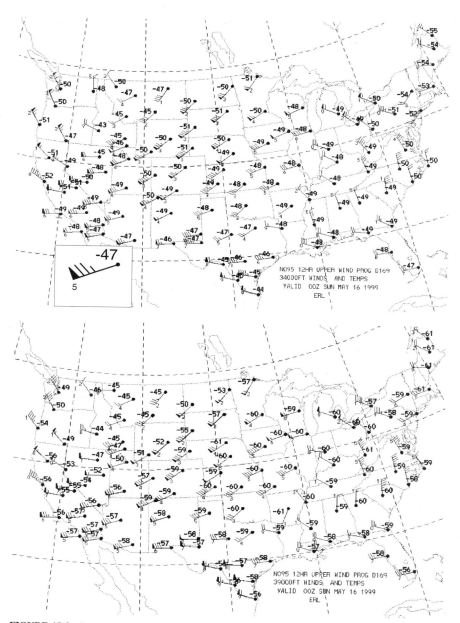

FIGURE 18-2 Because of valid times and computer models, some differences between the winds aloft forecast charts and tabular forecasts are to be expected.

valid times and computer models, some differences between the chart and tabular forecasts are to be expected.

Plotted data are standard. Wind direction is forecast to the nearest 10° and speed, 5 kn. Arrows with pennants and barbs are similar to those on other charts. The first digit of the wind direction is obtained from the general direction of the arrow. The inset in Fig. 18-2 contains the Tucson, Arizona, station model. The wind is from 250° at 75 kn, and the temperature is −47°C.

AMENDMENT CRITERIA

Although FDs are generated in Washington, regional NWS offices are responsible for amendments. FDs are amended when, in the forecaster's judgment, there is a change or an expected change in the wind or temperature that would significantly affect aircraft operations. Amendment procedures are complex. (In some 30 years I can count the number of FD amendments that I have seen on one hand.) Table 18-2 contains amendment criteria.

Reviewing Table 18-2, it can be seen there must be a considerable difference between forecast and actual winds to require an amendment. In general, forecasts within 30° of direction and plus or minus 20 kn are considered accurate. As with turbulence and icing, the only way to verify the forecast is through pilot reports.

USING THE FD FORECAST

FDs provide the pilot with two valuable pieces of information: wind direction and speed, plus temperature. Both significantly affect aircraft operation and performance. Failure to properly consider and apply either can be potentially hazardous.

In spite of its limitations, the FD can never be ignored. Pilots are required by regulations to consider "fuel requirements…;" and are prohibited from beginning a flight either VFR or IFR "unless [considering wind and forecast weather conditions]…" the aircraft will have enough fuel to fly to destination or an alternate if required and still have appropriate fuel reserves. Fuel reserve minimums, which do not necessarily equate to "safe," in no way

TABLE 18-2 Winds and Temperatures Aloft Amendment Criteria

Forecast	Amend if
Wind direction	
Speed 5 kn to <30 kn	Direction change ≥45°
Speed ≥30 kn	Direction change ≥30°
Wind speed	
Speed <70 kn	Speed change ≥20 kn
Speed 70–100 kn	Speed change ≥30 kn
Speed 100–135 kn	Speed change ≥40 kn
Speed >135 kn	Speed change ≥50 kn
Temperature	
Observed or forecast change	≥5°C

relieve the pilot from keeping careful track of ground speed and revising the flight plan accordingly.

When reserves are marginal, good operating practice dictates the careful tracking of position and ground speed. *Marginal* is not necessarily synonymous with *legal*; in sparsely populated areas, a fuel reserve of 30 min with clear weather reported and forecast might be sufficient. But with marginal weather or thunderstorms and the nearest suitable alternate 35 min away, a 30-min reserve doesn't make any sense. Chapter 17 discusses how legal alternates might not be satisfactory with a busted forecast. The same is true for legal fuel reserves.

CASE STUDIES The following situation illustrates how a series of small, at the time seemingly insignificant, factors have the potential to lead to disaster. The flight from Van Nuys, California, to Tonopah, Nevada, was based on 4 h of fuel and a 10-kn head wind, time enroute estimated 3:15. My Cessna 150 was fueled Friday when I arrived at Van Nuys. During the preflight Sunday morning, I noticed the fuel was not at the top of the filler neck—factor 1. This was not unusual because the airplane was parked on a slight incline and some fuel tends to vent overboard. The departure required an IFR climb to on top conditions, which added about 15 min to time enroute—factor 2.

Over Trona, California, about halfway, ground speed checks indicated winds were as forecast. Calculations indicated adequate fuel for Tonopah based on 4 h of fuel and ignoring the extra time required for departure.

The Cessna 150 climbs like a wet mop, so I decided not to land at Trona—factor 3. The fuel gauges were bouncing on zero, and I still had 30 min to destination, and there were no suitable alternates—factor 4. I made a straight-in approach and had everything stowed ready to crash, but I landed safely in spite of some extremely poor planning. By the way, they put 22.6 gal in my 22.5-gal-usable airplane. Never again.

A Grumman Tiger pilot—instructor with student—was not as fortunate. On a flight from Salt Lake City to Tonopah, they crashed short of the airport, out of fuel. The instructor couldn't understand why, after he calculated the airplane had 2:45 fuel, the engine quit after only 2:31. Needless to say, the FAA wanted to have a little chat with this gentleman.

The venturi effect at mountains and mountain passes accelerates winds over ridges and through passes. Stronger-than-forecast winds should be expected in these areas, especially within 5000 ft of terrain.

Low-level jet streams can develop under certain meteorological conditions. A clue to their existence can be obtained by comparing surface winds and winds aloft forecasts. Such a low-level jet occurs during the spring months in California's San Joaquin Valley. On one such occurrence, surface winds were reported as calm, while winds at 3000 ft were 080° at 30 kn! Such reports are strong indicators of moderate or greater nonconvective low-level wind shear.

Conversely, after strong afternoon mixing, surface winds within the mixing layer— a layer several thousand feet deep—can develop to altitudes higher than normal. Under such conditions the lower-level winds aloft forecasts can be erroneous. On low-level flights, especially within about 3000 to 5000 ft of the surface, pilots are usually better off using surface wind observations and forecasts rather than FDs. This would also be a turbulent layer.

Aircraft performance charts are based on the standard atmosphere, or more precisely the *international standard atmosphere* (ISA). Recall that standard atmosphere temperature and pressure at sea level are 15°C and 29.92 in Hg. The standard lapse rate in the troposphere is approximately 2°C per 1000 ft. Temperature decreases to a value of −57°C at the tropopause at approximately 36,000 ft for midlatitudes. An isothermal lapse rate occurs in the stratosphere to about 66,000 ft. This is illustrated in Fig. 1-3, International Standard Atmosphere.

Altitudes on the ISA chart are pressure altitudes. Pilots using the local altimeter setting fly at the indicated altitude for levels through 17,500 ft, and pressure altitude at 18,000 ft and above. Winds aloft forecasts are true altitude through 12,000 ft, so there will be a slight difference between the altitudes in the forecast (true altitude) and those flown by the pilot (indicated altitude). However, unless atmospheric pressure is extremely high or low, the difference is negligible. A variance of 1 in of mercury from standard would result in only 1000-ft altitude difference. Temperature is the biggest factor.

Standard conditions rarely occur in the real world, and performance charts are based on standard conditions; accommodation must be made for a nonstandard environment, usually a temperature correction. Manufacturers sometimes provide an ISA conversion with cruise power setting charts for high, low, and standard temperatures, or they simply note that performance is based on standard conditions. The aircraft doesn't understand any of this and performs based on the environment—pressure altitude and temperature.

Nonstandard conditions affect *true air speed* (TAS) and performance, as well as power settings. (TAS is calibrated, or equivalent, airspeed corrected for air density—pressure altitude and temperature. TAS is used on flight plans and is independent of wind direction and speed.) An aircraft's advertised service ceiling of 13,100 ft is based on standard conditions. Differences are usually not significant unless the pilot is operating at the limit of the aircraft's performance.

Unfortunately, this occurs every year with pilots that attempt to cross the Sierra Nevada or Rocky Mountains in conditions well above standard. Some pilots can't understand why an aircraft with a service ceiling 13,100 ft can't climb above 12,000 ft with an outside air temperature of 0°C. Density altitude, which is pressure altitude corrected for temperature, at 12,000 ft and temperature 0°C is 13,100. And this is an ideal case. A runout engine, poor leaning technique, over gross weight, and the possibility of turbulence and downdrafts would further decrease performance. You can't fool Mother Nature; attempts can be fatal.

CASE STUDIES My 1966 Cessna 150 had a book ceiling of 12,650 ft. We planned to traverse the 9941-ft Tioga Pass in California's Sierra Nevada range. The winds were out of the northeast at only 10 kn, resulting in a slight downdraft from the wind flowing up the east slopes and down the west slopes as we approached the pass. Temperature was slightly above standard, and in combination with the wind, the airplane wouldn't climb out of 9500 ft. We had to proceed north along the west slopes of the mountains to Ebbett's Pass at 8732 ft, where we were able to safely cross the mountains. We needed to gain the required altitude prior to reaching the crest, and we needed sufficient room to make a comfortable 180, if required.

On another occasion we had filed for an IFR flight from Lancaster Fox Field in the Mojave Desert to Ontario, California. I requested 7000 ft and planned to go through the San Fernando Valley because of lower minimum altitudes. The clearance came back, "Cleared via the Cajon two arrival; climb and maintain 11,000." The surface temperature was 30°C, and the Cessna 150 was not going to 11,000 that day! After negotiating with a rather perturbed ground controller, I received my requested routing. Pilots must know their aircraft's performance and not allow ATC, or anyone else for that matter, to push them into an untenable—in this case unobtainable—position.

All aircraft have limitations, including turbojet aircraft that often fly at the edge of their performance envelope where nonstandard conditions critically affect performance. The pilot's task is to determine aircraft performance based on forecast temperatures aloft. This requires the information in Fig. 1-3. From the previous discussion, the average temperature from Oakland to South Lake Tahoe at 13,500 ft was −04°C. Figure 1-3 indicates that the standard temperature for 13,500 ft is −10°C. The forecast temperature is 6° warmer, which is above standard. The air is less dense than standard, so aircraft performance will be less than performance charts advertise for standard conditions. Based on the above conditions, density altitude is about 15,000 ft.

Certain performance charts require the pilot to determine temperature at altitude relative to ISA. For example, the forecast temperature over Reno at flight level 300 (30,000 ft) is −42°C. Figure 1-3 indicates that the standard temperature for that pressure altitude is

$-45°C$. Certain flight computers can be used to determine ISA temperatures; the Jeppesen CR-3 has a true altitude computation window. With the scales aligned (10 on the outer scale with 10 on the inner scale), standard temperature is read under pressure altitude. Under 30,000 ft pressure altitude, -45 appears. Therefore, the forecast temperature is ISA $+3$.

FDs are a source of forecast freezing levels. They should be in general agreement with the FA because they're based on the same data. Differences result from FD freezing levels representing only one particular time, whereas FA forecasts take into account changes during the period. However, if a significant difference occurs, be alert for other possible forecast errors. FSS controllers work with forecasters under such circumstances. Pilots using DUATs would normally have to consult an FSS for resolution. From the example FDs, the expected freezing level at 1800Z over SFO is approximately 12,500 ft, lowering to 11,000 ft in the RNO area.

FDs can be used to determine the approximate height of the tropopause. Winds are strongest just below the tropopause. Checking the example FDs in Fig. 18-1, SFO winds at 39,000 ft are 59 kn, and at 45,000 ft, they have decreased to 49 kn. Also note that speed continues to decrease at 53,000 ft and temperature remains almost constant. Therefore, the tropopause is between the 39,000- and 45,000-ft pressure levels.

FD forecast limitations and amendment criteria must be understood for effective flight planning. Product preparation, plus the advantages—and to some extent, the limitations—of computer models have already been discussed. Surface heating as well as terrain affect winds aloft. Today's technology cannot account for the effects of land/sea and mountain/valley winds, nor frictional effects between wind and the surface. Other forecast problems include the extent and availability of data, and timing.

FDs are based on the expected movement of weather systems, so errors result when weather systems move faster or slower than forecast. This is one reason why flight watch specialists are required to continually solicit reports of winds and temperatures aloft. Specialists are specifically trained to recognize these situations and provide updated information. Careful tracking of position and ground speed will verify the accuracy of the FDs.

Inertial navigation, LORAN, Global Positioning System (GPS), and other computerized navigation systems can provide immediate wind readouts. Anyone remember how to calculate winds with the E6B flight computer? Electronic flight computers make the calculation easy. This is a reminder for pilots to become actively involved in the system with PIREPs. Observed winds aloft, whether they confirm or contradict the forecast, should be routinely passed to flight watch. Only two upper-air observations are made each day, so PIREPs are the only other direct source of observed winds and the only way to verify the forecast.

The situation might change in the future with wind profilers that will automatically and almost continuously provide high-resolution upper-air wind measurements. Naturally, the closer to observation time, the more accurate the forecast. Accuracy normally deteriorates with time—the FD2s and especially the FD3s. Since the evening forecast becomes available after 0440Z, it doesn't make much sense to request winds any earlier for the following day. Flights departing after 1640Z, must consult the new FDs. Weather patterns can change significantly in 12 h.

By regulation, the pilot has no option but to use FDs in flight planning. Local or short flights might mean nothing more than an eyeball interpolation. Exams and flight tests require computer calculations. Flights toward the limit of aircraft range will require a careful interpolation and calculation.

I have on many occasions flown from southern California to destinations in the midwest and east. Most of these flights were completed in Cessna 150s and 172s. The legs through the intermountain region often stretch the range of these airplanes.

CASE STUDIES On a leg from Phoenix to Albuquerque, I selected a point, a little over halfway, to make the decision to divert. On this occasion the promised 10-kn tailwind was as advertised, and the flight was completed as planned.

On another flight from Prescott to Albuquerque, things were just not meant to be. Crossing Winslow, Arizona, the Cessna's ground speed never reached three digits. I changed the flight plan and proceeded to Gallup, New Mexico. Hoping that a stronger-than-forecast headwind will abate is folly.

The theme of the preceding examples is to have a plan, then follow it!

With the general criticism of winds aloft forecasts, it's amazing how many pilots call and must absolutely have winds 2 or even 3 days in the future. Then there's the guy who calls flight watch and can't understand why the winds are 20° and 5 kn off forecast. Oh well, you can't please everyone. By now we should have some insight into FD limitations, the causes of inaccuracies, and perceived errors of this valuable, but often maligned product.

CHAPTER 19
FAA WEATHER BRIEFING SERVICES

In the early days of aviation, because all flights were local, pilots had little need for meteorological information. Nor were the aeronautical *Notices to Airmen* (NOTAMs) necessary because pilots departed and landed at the same field, assuming the engine didn't quit. By the spring of 1918, the U.S. Post Office Department began working on a transcontinental airmail route. A combination rail-air route between New York and Chicago was established by July and a month later extended to San Francisco. Authorization was granted in August 1920 for the establishment of 17 airmail radio stations. Personnel originally were to load and unload mail. However, as traffic increased, the need for weather information became apparent. Airmail radio personnel soon began taking weather observations and developing forecasts. The information was relayed via radio telegraph to adjacent stations. Inflight weather reports were heavily relied upon for the weather briefing.

Postal personnel soon became involved in air traffic as well as postal services, and in July 1927 they were transferred to the Department of Commerce, Bureau of Lighthouses, along with their facilities, now known as *airway radio stations.* The stations were transferred in August 1938 to the Civil Aeronautics Authority, and they became *airway communication stations.* Finally, these facilities became *flight service stations* (FSSs) with the establishment of the Federal Aviation Agency in 1958. The Weather Bureau became increasingly responsible for the collection and distribution of aviation weather and forecasts. Pilots obtained briefings from the Weather Bureau, and they filed flight plans and acquired aeronautical information from the flight service stations.

Due to the increase in air commerce, and other factors, in 1961 the Weather Bureau began the certification of FSS personnel as pilot weather briefers. The Federal Aviation Agency and the Weather Bureau mutually signed a memorandum of agreement in 1965 delegating responsibility for pilot weather briefing to the FAA. FSS briefers had little in the way of guidelines during this period regarding the structure of the briefing; basically, they read weather reports and forecasts verbatim as requested by the pilot.

The Field FSS Pilot Briefing Deficiency Analysis group began a special evaluation of pilot briefing services in 1975. The group cited a deficiency in the use of a standardized briefing format. (The standardized format had been taught at the FAA Academy for some time; however, it had yet to be incorporated in the FSS handbook.) Other areas identified were the reading of weather reports and forecasts verbatim as opposed to interpreting, translating, and summarizing data. A poor level of proficiency in reading, understanding, and employing facsimile charts was noted. Briefers failed to obtain sufficient background information to tailor the briefing to the type of flight planned.

From this study came the agency's emphasis on an extremely rigid briefing format and an ambitious refresher training program. Unfortunately, the agency did little to inform pilots or other offices within the agency of this change in policy. This led to a good deal of friction between briefers and pilots.

Over the years, however, through mostly local efforts, pilots have become acquainted with the standard briefing format. The refresher training was to be a continuing program conducted at least every 5 years. However, with few exceptions, this program has been abandoned presumably due to fiscal constraints.

The National Transportation Safety Board (NTSB) also conducted a special investigation into flight service station weather briefing inadequacies. In 6 of 72 accidents involving fatalities, the safety board determined that pertinent meteorological information was not passed to the pilot during the weather briefing. Basically these deficiencies consisted of failing to pass along important weather advisories and icing forecasts and downplaying forecasts of hazardous weather. The result of the board's determinations has been what many pilots consider overdoing dissemination of these advisories.

A major change occurred in 1983 when the extremely rigid format was relaxed somewhat. Three types of briefings emerged:

- Standard briefings
- Abbreviated briefings
- Outlook briefings

Requirements for inflight briefings were also specified.

Almost from the time the FAA took over pilot briefing responsibility in 1965, the service A (weather) teletype system was obsolete. Since that time, proposal after proposal was made to update weather distribution. Even by the early 1980s, most flight service stations still used the 100-word-per-minute electromechanical teletype equipment. Briefers had to sift through mountains of paper to provide a briefing with weather reports as much as 1 1/2 h old by the time they were relayed and available.

The FAA approved what was termed the *interim service A system* in November 1978 that had been tested at the Chicago FSS. Subsequently the service B teletype system for the transmission of flight plans and other messages was incorporated. Referred to as the *leased A and B system* (LABS), it is still in use at nonautomated FSSs. LABS was designed to update FSS service A until a complete computer system could be developed and installed. LABS eliminates the need for the briefer to sort and post specials, PIREPs, NOTAMs, and most amended forecasts. These housekeeping chores, which took considerable time, have been eliminated. With this system, most weather reports are available within 5 to 15 min of observation. Development of model 1, a so-called completely computerized system, began in 1982 and came online in 1985. *Automated flight service stations* (AFSSs) use this equipment.

According to the FAA, "The primary benefit of the [FSS automation] program is improved productivity through automation of specialist's access to detailed briefing information and flight plan filing. To some extent the improved quality of pilot briefings reduces the need for multiple briefings as in the past." Model 1 is in the same evolutionary category as ARTCC flight data processing was in the early 1970s. It takes care of many of the data processing functions such as flight plan transmission and tracking.

From a weather briefing point of view, however, it presents the same information that was available from teletype and LABS. With model 1, amendments are more timely, but this improvement won't be directly obvious to the pilot. Model 1 does not necessarily improve pilot briefing productivity; in fact, productivity in certain cases is reduced. Model 1 is truly "user hostile" and presents information in much the same way as DUATs and other commercial briefing systems.

In the late 1990s with the loss of the Miami AFSS due to hurricane Andrew and the St. Louis AFSS due to floods, model 1 equipment not only became unsupportable but nonexistent. Model 1's replacement will be the *operational and supportability implementation system* (OASIS). OASIS will provide essentially the same data but now with its own weather graphics system. OASIS, like model 1, will allegedly increase the quality and quantity of FSS briefing services. Although, OASIS does provide some very important ergonomic advances, don't count on improved quality or quantity unless the FAA decides to do some serious FSS controller training. To date the system continues to be plagued with problems, budget cuts, and delays. Project initial installation has now been pushed forward to 2002.

Regulations require each pilot in command, before beginning a flight, to become familiar with all available information concerning that flight. This information must include the following: "For a flight under IFR or a flight not in the vicinity of an airport, weather reports and forecasts...and any known traffic delays of which the pilot has been advised by ATC." Additional regulations specify fuel and alternate airport requirements. The regulations do not, however, require that meteorological and aeronautical information be obtained from the FAA.

STANDARD BRIEFINGS

The standard briefing is designed for a pilot's initial weather rundown prior to departure. Standard briefings are not normally provided when the departure time is beyond 6 h, nor current weather beyond 2 h. It is to the pilot's advantage to obtain a standard briefing, or update the briefing, as close to departure time as possible.

Background Information

1. *The type of flight planned.* Always advise the briefer if the flight can be conducted only VFR or if an IFR flight is planned, or can be conducted IFR. Normally, the briefer will assume a pilot is planning VFR, unless stated otherwise. Student pilots should always state this fact to help the briefer provide a briefing tailored for a student's needs. Also, new or low-time pilots and pilots unfamiliar with the area will receive better service if they advise the briefer of their status. This alerts the briefer to proceed more slowly, and with greater detail.

2. *The aircraft number or pilot's name.* This information is evidence that a briefing was obtained, as well as an indicator of FSS activity. In the absence of an aircraft number, the pilot's name is sufficient. Most briefings are recorded and reviewed in case of incident or accident; it's in the pilot's interest to get "on the record" as having received a briefing.

CASE STUDY At the briefer's request for an aircraft number, the pilot replied, "I don't have one." When asked for a name, the pilot again replied, "I don't have one." One thing for sure, the briefer can't get in trouble on that briefing.

3. *The aircraft type.* Low-, medium-, and high-altitude flights present different briefing problems. This information allows briefers to tailor the briefing to a pilot's specific needs. By knowing the aircraft type, the briefer, many times, can estimate general performance characteristics such as altitude, range, and time enroute.

4. *The departure airport.* Pilots must be specific—they know the airport, but the briefer usually doesn't.

CASE STUDY Some pilots use generalities such as "Los Angeles" when their actual departure airport is Oxnard, more than 50 mi away; and there is always the ever-popular: "Here." The specific departure airport is important with FSS consolidation, 800 phone numbers, and metropolitan areas.

5. *The estimated time of departure.* The estimated time of departure is essential, even if general.

CASE STUDY Briefer: "When are you planning on departing?" Pilot: "Well, that depends on the weather." This response tells the briefer nothing. In such a situation a pilot could respond, "I'd like to go this afternoon, but I can put the flight off until tomorrow."

6. *The proposed altitude or altitude range.* This information is needed to provide winds and temperatures aloft forecasts. If an altitude range is specified—for example, 8000 to 12,000 ft—the briefer can provide the most efficient altitude for direction of flight.

7. *The route of flight.* The briefer will assume a pilot is planning a direct flight unless the pilot states otherwise. If the flight will not be direct, a pilot must provide the exact route or preferred route, and any planned stops. Total time is essential when stops or anything other than a direct flight are planned; for IFR flights, the estimated time of arrival is required to determine alternate requirements. This will assist the briefer in providing weather for the planned route.

8. *The destination airport.* Again, pilots must be specific. If they are not, they might not receive all available weather and NOTAM information.

CASE STUDY A Piper Cub pilot on one occasion requested a briefing to Los Angeles. The briefer asked if his intended destination was Los Angeles International. It was! Another pilot obtained a briefing from Chino, California, to the stated destination of Stockton, California, and was told there were no NOTAMs for the route. At the end of the briefing, the pilot matter-of-factly said the actually destination was going to an airport about 20 mi east of Stockton—Columbia. Now there were a few NOTAMs!—the airport would be closed during certain hours, a temporary tower was in operation, and acrobatic flight and parachute jumping were being conducted.

9. *Estimated time enroute.* Many briefers can estimate time enroute based on aircraft type. This information is needed to provide enroute and destination forecasts. Total time enroute is essential when stops or anything other than a direct flight are planned; for IFR flights, the estimated time of arrival is required to determine alternate requirements.

10. *Alternate airport.* If you already have an alternate in mind, provide it at this time. FSS equipment will automatically display alternate airport current weather, forecast, and NOTAMs to the briefer.

FACT Often the briefer can be of assistance in determining an alternate. But remember it is the pilot's responsibility, not the briefer, to determine whether an alternate is required and if the forecast weather is acceptable for naming it as the alternate.

This might seem like a lot of information, but it really isn't. The briefer must obtain this information before or during the briefing. Providing background information will allow briefers to do a better job of providing pilots with clear, concise, well-organized briefings, tailored to the pilots' specific needs.

Briefing Format

All right, the background information has been provided. What can a pilot expect in return? The briefer is required, using all available weather and aeronautical information, to provide a briefing in the following order. Pilots should be as familiar with this format as the mnemonic CIGAR (controls, instruments, gas, attitude, runup), or the IFR clearance format.

1. *Adverse conditions.* Any information, aeronautical or meteorological, that might influence the pilot to cancel, alter, or postpone the flight will be provided at this time. Items will consist of weather advisories, major NAVAID outages, runway or airport closures, or any other hazardous conditions.

The adverse conditions provided should be only those pertinent to the intended flight. This is one reason why the pilot must provide the briefer with accurate and specific background information. The briefer should then furnish only those conditions that affect the flight. There is, unfortunately, some paranoia among briefers evident in that they provide anything within 200 mi of the flight, whether it's applicable or not.

2. *VFR flight is not recommended (VNR).* Undoubtedly the VNR statement is the most controversial element of the briefing; nevertheless, the FAA requires the briefer to "include this recommendation when VFR flight is proposed and sky conditions or visibilities are present or forecast, surface or aloft, that in [the judgment of the specialist], would make flight under visual flight rules doubtful."

This leaves considerable leeway for the briefer; some use this statement more than others. The inclusion of this statement should not necessarily be interpreted by the pilot as an automatic cancellation, nor its absence as a go-for-it day. Notice that VNR applies to sky condition and visibility only. Such phenomena as turbulence, icing, winds, and thunderstorms, in themselves, do not warrant the issuance of this statement. And it is important to remember that this is a recommendation. Why then such a statement? It's simple. Every year pilots insist on killing themselves and their passengers at an alarming and relatively constant rate by flying into weather that they shouldn't. This statement was instituted in 1974, presumably because the last person a pilot would talk to was usually the briefer.

CASE STUDY A logical, although alarming, result of the VNR statement is the increasing number of pilots who, in the absence of VNR, ask, "Is VFR recommended?" Others criticize a perceived overuse of the statement as reducing its effectiveness. This is certainly true in many cases. But, for now, the bottom line remains: The decision as to whether a flight can be safely conducted rests solely with the pilot.

According to the *Flight Services* handbook, the reason for VNR must be provided. For example, "VFR is not recommended into the Monterey area because of visibilities 1/2 mi and ceilings 200 ft, and conditions are not expected to improve until around noon." Briefers have been known to use some exceedingly poor techniques in this area. One briefer told a pilot, "The San Fernando Valley is still VNR." Oh well.

RULE Some pilots say never say never. Here is the exception that proves the rule. Never let the briefer make the go/no-go decision. The briefer is a resource, and some briefers are better than others. This rule applies equally to optimistic and pessimistic briefers and briefings.

3. *Synopsis.* The synopsis is extracted and summarized from FA and TWEB route synopses, weather advisories, and surface and upper-level weather charts. This element might be combined with adverse conditions and the VNR statement, in any order, when it would help to more clearly describe conditions.

These three elements should provide us with the "big picture," part of the "complete picture." The synopsis should indicate the reason for any adverse conditions, and it should tie in with current and forecast weather. During this portion of the briefing, pay particular attention for clues of turbulence and icing, even if a weather advisory is not in effect. For example, areas of locally moderate turbulence and icing may be overlooked.

4. *Current conditions.* Current weather will be summarized: point of departure, enroute, and destination. Relevant PIREPs and weather radar reports will be included. Weather reports will not normally be read verbatim, and they might be omitted if the proposed departure time is beyond 2 h, unless the report is specifically requested by the pilot. Forecast surface temperatures are not available at this time, but they may be in the future. However, we can extrapolate surface temperatures from current data. Because the METAR database is normally reloaded just after the hour, to obtain the latest reports, avoid, if possible, calling just prior to the hour.

5. *Enroute forecast.* The enroute forecast will be summarized in a logical order (climbout, enroute, and destination) from appropriate forecasts (FAs, TWEBs, weather advisories, and prog charts). The briefer will interpret, translate, and summarize expected conditions along the route.

6. *Destination forecast.* Using the TAF where available, or appropriate portions of the area or TWEB forecasts, the briefer will provide a destination forecast, along with significant changes from 1 h before until 1 h after the ETA.

7. *Winds aloft forecast.* The briefer will summarize forecast winds aloft for the proposed route. Normally, temperatures will be provided only on request. Request temperatures aloft. We want to know if we're going to be below, at, or above the freezing level. Temperature at our flight planned altitude is an indicator of icing severity, as well as aircraft performance.

8. *Notices to Airmen (NOTAMs).* The briefer will review and provide applicable NOTAMs for the proposed flight that are on hand and not already carried in the *Notices to Airmen* publication. This information consists of NAVAID status, airport conditions, temporary flight restrictions, changes to instrument approach procedures, and flow control information. In the briefing, the term *NOTAMs* is all-inclusive.

CASE STUDY I briefed a student one day and, as is my practice, informed this aviator that there were no NOTAMs for the route. There was a pause. I asked if he knew what NOTAMs were; he didn't.

The U.S. NOTAM system (USNS) is computerized and occasionally fails. When this occurs, briefers will include the statement: "Due to temporary NOTAM system outage, enroute and destination NOTAM information may not be current. Pilots should contact FSSs enroute and at destination to ensure current NOTAM information."

9. *Other services and items provided on request.* At this point in the briefing, briefers will normally inform the pilot of the availability of flight plan, traffic advisory, and flight watch services, and he or she will request pilot reports. Upon request, the controller will provide information on *military training route* (MTR) and *military operation area* (MOA) *activity,* review the *Notices to Airmen* publication, check Loran or GPS NOTAMs, and provide other information requested.

It's not necessary to copy all the information provided because much is supplementary and provides a background for other portions of the briefing. Pertinent information should be noted, and it's often advantageous to copy these data. There are many forms available. It's often helpful to have a map containing weather advisory plotting points and jot down significant information, such as the form shown in Fig. 19-1.

ABBREVIATED BRIEFINGS

Briefers provide abbreviated briefings when a pilot requests specific data, information to update a previous briefing, or supplement an FAA mass dissemination system (transcribed

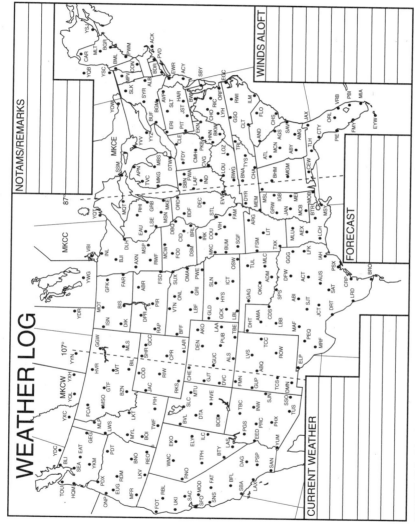

FIGURE 19-1 It's often helpful to note significant items on a form, such as this weather log, during a briefing.

19-7

weather broadcast, telephone information briefing service, or pilot's automatic telephone weather answering service).

Requests for Specific Information

When all that's required is specific information, a pilot should state this fact and request an abbreviated briefing. Because the briefer must normally make a request for each individual item, it's extremely helpful to request all items at the beginning of the briefing, thus reducing delays. The briefer will then provide the information requested. When using this procedure, the responsibility for obtaining all necessary and available information rests with the pilot, not the briefer. Pilots must realize that the briefer is still required to offer adverse conditions. Pilots sometimes become irritated when the briefer mentions weather advisories; the briefer is adhering to a *Flight Services* handbook requirement.

Requests to Update a Previous Briefing

Pilots requesting an update to a previous briefing must provide the time the briefing was received and necessary background information. The briefer will then, to the extent possible, limit the briefing to appreciable changes.

> **FACT**   An alarming number of pilots, when asked the time of their previous briefing, respond, "I got the weather last night." Needless to say; this practice does not comply with regulations. These individuals should be requesting a standard briefing.

Requests for Information to Supplement FAA Mass Dissemination Systems

Again, the briefer must have enough background information and the time the recording was obtained to provide appropriate supplemental data. The extent of the briefing will depend on the type of recording and the time received.

OUTLOOK BRIEFINGS

With a proposed departure time beyond 6 h, an outlook briefing will normally be provided. The briefing will contain available information applicable to the proposed flight. The detail will depend on the proposed time of departure. The further in the future, the less specific. As a minimum, the outlook will consist of a synopsis and route/destination forecast.

Outlooks beyond FA, TWEB, and TAF valid times are available using significant weather prognosis charts. A last word on outlooks: Pilots should not overlook the weather section of the newspaper, the Weather Channel, or local TV programs. Often these sources contain local detail not available in aviation outlook products. Regardless of source, the outlook forecast axiom remains: The weather tomorrow is going to be what the weather is tomorrow, no matter what anybody says.

INFLIGHT BRIEFINGS

Although discouraged, unless unavoidable, briefings once airborne will be conducted as standard, abbreviated, or outlook briefings as requested by the pilot. As with any briefing, sufficient background information must be made available.

CASE STUDY A survey question at the 1997 Oshkosh Fly-in asked: Do you feel that the FSS inflight brief gives you too much, too little, or about the right amount of information during your inflight weather briefing?

Approximately two-thirds of pilots responded that the volume of information was about right from both flight service stations and through flight watch. This is interesting because the FAA requires both FSS and flight watch controllers to "force feed" certain types of meteorological information. In response to a similar question, about the same number indicated that the information received matched what they experienced inflight.

THE FAA PILOT WEATHER BRIEFING SERVICES

Briefings can be obtained in person, over the telephone, or by radio. The preferred methods are to obtain a weather briefing in person or by phone. Initial briefings by radio are discouraged, except when there is no other means. The reasons are simple. The cabin of an aircraft plunging into the wild gray yonder is no place to plan a flight. Attention must be diverted from flying the aircraft to the briefing. Especially with marginal weather, certain pilots have a tendency to push on, regardless of conditions, not to mention the fact that it usually unnecessarily ties up already-congested radio frequencies.

Flight service station controller training begins at the FAA Academy, in Oklahoma City, with the equivalent of a college year in basic meteorology and briefing techniques; the weather portion is taught by NWS meteorologists. At field facilities, "developmentals" receive training in *area knowledge* (local weather, terrain features, weather reporting locations) and must be certified by both the FAA and National Weather Service. Briefers, like pilots, at some point must get hands-on training. From time to time, pilots will encounter this situation; the briefing might not be clear or concise, but pilots should exercise the same patience that briefers use with student pilots.

The briefing is supposed to be presented in a clear, concise manner. Ambiguous terms, such as "looks bad, scuzzy," and even generalizations like "VFR," are to be avoided.

CASE STUDY An FSS friend of mine received a one-liner briefing from Salinas to Sacramento, California. "It's VFR." What does that mean? It could range from clouds almost to the ground and 1 mi visibility, to clear and 100.

The point is made that FSS briefers are not meteorologists—usually by meteorologists. These individuals call attention to the problem that when a forecast goes bad, the quality of the briefing falls apart and the pilot is left on the proverbial limb. This is not often the case. Briefers are trained to recognize *forecast variance,* a difference between the forecast for a given time and existing conditions. A briefer might suggest the pilot wait for a new forecast or coordinate with a forecaster for resolution. In any case, the pilot is made aware of the problem.

Pilots will have less and less access to NWS personnel. The NWS for most practical purposes is out of the weather briefing business. However, where available, telephone numbers are published in the *Airport/Facility Directory* or may be obtained from the local FSS. Pilots will not necessarily talk to a forecaster.

CASE STUDY A rather new flight instructor brought his novice student into a National Weather Service office collocated with an FSS. The instructor explained that if the student wanted a really good briefing, he should always go to the NWS. Unfortunately, for this young instructor, a rather crusty old met tech (meteorological technician) was on duty. The met tech, quite unceremoniously, admonished the instructor, explaining that many NWS specialists are not meteorologists, nor engaged in aviation, and that for aviation, the FSS was the place to go for the weather.

Many FSS specialists are excellent interpreters of the weather, familiar with local weather patterns and terrain, and pass on their knowledge and experience to the pilot.

The biggest complaint about the FAA's pilot briefing service is delays. An Aviation Safety Reporting Service (ASRS) study states: "The inability to reach flight service by telephone was the complaint...." in a number of incidents. "Reporters relate waits as long as 20 to 45 minutes on hold and then being disconnected. Reporters allege that, because of the inability to reach flight service, many pilots in their area depart without preflight weather briefings or take off and contact enroute flight advisory service." Flight watch is not for an initial briefing; pilots who elect to use this procedure must call an FSS on the station's discrete frequency for an initial briefing.

It's not a big mystery why delays are so lengthy. Let's take a large FSS with a flight plan area that serves about 30,000 pilots. At any one time there might be from four to eight briefers during peak periods. Guess what happens when more than eight pilots call? The longest delays occur when weather is marginal; during these periods, the average briefing might take 5 to 8 min, whereas during good weather briefs average only 2 or 3 min. Add to this that after initial checkout, the FAA provides little training to help briefers become more productive.

Pilots complain about the deluge of superfluous information provided by certain briefers and the difficulty of getting information from others. These complaints result from poor briefer training and perceived paranoia about accident investigations.

Pilots are equally guilty of tying up briefing lines. Pilots inhibit the system by not being prepared for the briefing nor prepared to file a flight plan, and some have unrealistic expectations. This begins with the flight instructor who fails to properly prepare a student for the briefing, to pilots—who should know better—who call to file an IFR flight plan but haven't looked at the charts yet. These are usually the people that complain the loudest about delays. I have had, on many occasions, students call for a briefing and to file a flight plan with zero knowledge of how to accomplish either. And instructors still have students call for practice briefings during peak periods. Instructors must take the responsibility to prepare their students.

Pilots need to be specific about the information they require. Ambiguous statements such as "Is it VFR? I'm looking for some soft IFR. Where is it good? Just tell me what I need to know. Is there anything significant?" Such statements have no place in the pilot briefing environment. Does the pilot mean VFR in controlled or uncontrolled airspace? Above or below 10,000 ft? Or special VFR? Try finding soft IFR in the Pilot Controller Glossary. "Good" to one briefer could be "1000-ft ceiling and 3 mi visibility." "Significant" falls into the same category as "good." Some pilots will still simply ask the briefer, "Are there any AIRMETs and SIGMETs for the route?" These individuals might miss significant information.

There are a number of techniques pilots can use when dealing with air traffic and flight service station controllers.

Patience is a virtue. It's certainly no fun holding on the phone for 20 or 30 min. But the briefer may have been at it steadily for up to 6 h. There are very few reasons for departing without a briefing or requesting an initial briefing enroute; there are lots of excuses. A good pilot realize the problems of other pilots, controllers, briefers, and forecasters. They are patient with training controllers and briefers because they remember when they were student pilots. These pilots would no more lose control with a controller or briefer than lose control of the aircraft.

Good pilots seldom blunder into situations and occasionally cancel or discontinue trips because of the weather. Pilots knows the limitations of weather reports and forecasts and plan accordingly. Based on their experience and the capabilities of the aircraft, they know when the weather answer is no go.

Good pilots have their homework done; they have reviewed the routes, and they have completed a partial flight plan. These pilots are ready for the briefing and to file a flight plan; they don't guess at the route trying to file from memory. As required by the practical test standards, these pilots are organized.

Part of the complete picture is having more than one way out. When only one out is left, it's exercised. This might mean canceling a flight, circumnavigating weather, avoiding hazardous terrain, or planning an additional landing enroute. The 180° turn is made before entering clouds. If the situation becomes uncertain, assistance is obtained before an incident becomes an accident.

Good pilots will never be caught on top or run out of fuel. These pilots combine mental attitude and skill to update weather enroute, devise a plan based on this information, and coordinate the action before the situation becomes critical.

Perhaps as important as patience is courtesy. A briefer who has been briefing for 4 to 6 h on a marginal weather day is in no mood for pilot sarcasm. Briefers do not have to put up with obnoxious, rude, or profane pilots; that's the purpose of the telephone release button. Courtesy is a two-way street, however. Pilots don't have to put up with obnoxious or rude briefers. If you don't think you're being treated in a courteous, professional manner, talk to the supervisor or facility manager, or call the FAA's hot line, (800) FAA-SURE.

USING THE FAA WEATHER BRIEFING SERVICES

The weather briefing is a cooperative effort between the pilot and FSS controller. Preliminary planning should be complete, including a general idea of route, terrain, minimum altitudes, and possible alternates. Where available, obtain preliminary weather from one of the recorded services. From the broadcast, determine the type of briefing required—standard, abbreviated, or outlook.

During the briefing, try not to interrupt unless the briefer is going too fast. Often pilots interrupt with a question that was just about to be answered. This can cause the briefer to lose his or her train of thought, resulting in the inadvertent omission of information.

Finally, from the "things that bug briefers the most" category: Some pilots unintentionally engage in a form of Chinese water torture—after every word the briefer says, they interject "ah ha." This is terribly annoying and distracting. Additionally, if at all possible, avoid using a speaker phone. Feedback from these devices is terribly annoying and distracting.

CASE STUDY   I briefed a student one day, and about 15 min later he called back. I recognized the aircraft and said, "Didn't I just brief you?" "Well, I couldn't make heads or tails of my notes. This time I'm recording it." An outstanding idea, especially for those pilots for whom English is a second language. More students and instructors should adopt this practice. This has the added advantage that the pilot can listen without taking his or her attention from the briefing to write or requesting the briefer to repeat information.

Briefers make mistakes, and many are not pilots. At the end of the briefing, don't hesitate to ask for clarification or additional information on any point you do not understand. If conditions are right for turbulence or icing and these phenomena were not mentioned, ask the briefer to verify that there are no weather advisories. Remember that forecasts for light to locally moderate icing do not warrant an advisory, nor does locally severe turbulence warrant a SIGMET. Forecasts for these conditions can be overlooked. On the other hand, don't expect the freezing level on a clear day—gotta watch out for that "clear air icing."

With this as a background and FSS staffing being further reduced, the question becomes, how can a pilot best use the services available?

Become familiar with recorded weather information in your area. These programs have been established to help reduce delays. They provide a general weather picture, with usually enough information to determine if further checking is warranted. If the weather is IFR or beyond a pilot's limits, there's no need to tie up a briefer. Additionally, pilots can determine if, on a particular day, a flight to the coast, the desert, or the mountains would be best. Briefers can be on the line for 10 min or more with pilots looking for a place to fly. This could be eliminated if these individuals would use the recordings. This also applies to student pilots looking for suitable cross-country routes.

Recordings provide much of the information in a standard briefing. Normally these services contain the synopsis, adverse conditions, route, and winds aloft forecasts through 12,000 ft, and selected surface weather reports. Forecasts are normally available 24 h, although, surface weather reports may be suspended between 10 p.m. and 5 a.m. Depending on the broadcast, other information, such as TAFs, NOTAMs, and military training activity, are not available. These broadcasts do not meet regulatory requirements for IFR. However, often the information will be sufficient for a VFR flight. If any clarification or additional information is required, the FSS should be consulted.

There are additional uses of the FAA's recorded weather systems. Knowing when the broadcasts are updated, pilots can obtain an outlook for the following day. Broadcast updates usually coincide with FA and TWEB forecast issuance times. Pilots can check with their FSS for broadcast update times; most FSSs publish these times in *Letters to Airmen* or *Pilot Bulletins*. Student and low-time pilots can use broadcast systems to learn aviation weather terminology. I always recommend this to students. In this way, they can become familiar with the terms and phrases used in weather briefings. Anything they don't understand they can discuss with their instructor.

Pilots have a say in the content of these services. Although the FAA prescribes the general content and format, the exact items of information, such as individual weather reports, are left to the discretion of the facility. Facility managers are supposed to solicit comments from users—pilots—about their content. If pilots want a particular item on the broadcast, they should contact the facility.

Finally, pilots should know where to complain. As far as broadcasts go, this usually means it was unintelligible or read too fast. If there's a problem, contact the supervisor or manager, or use the FAA's hot line, (800) FAA-SURE. If pilots don't make the FAA aware of a problem, whose fault is it if it doesn't get fixed?

UPDATING WEATHER

A pilot's responsibility does not end with an understanding of forecast products and limitations and the means of collecting meteorological and aeronautical information. Due to the dynamic character of the atmosphere, data must be continually updated. Surprisingly, many pilots have not been taught, or learned, the importance of updating weather reports, forecasts, and NOTAMs enroute. Failure to exercise this pilot-in-command prerogative can have disastrous results.

The importance of updating weather and NOTAMs enroute cannot be overemphasized. The primary focal point for these services are the FAA's flight service stations, through FSS communications, broadcasts, and flight watch. Secondary sources of information are *automatic terminal information services* (ATISs) and *automated weather observation* (AWOS/ASOS) *broadcasts,* and *air route traffic control center* (ARTCC) and *terminal* (tower/approach control) *controllers.*

With FSS consolidation, correct, concise, and accurate communications become more important. FSS frequencies are busier than ever, and there are fewer specialists providing communications over larger areas. Correct communications technique, a seemingly simple task, takes on a greater significance. A pilot's first task is to select the appropriate frequency.

A new service called *ground communication outlet* (GCO) has recently been established. This is a remotely controlled, ground-to-ground communications facility. Pilots at uncontrolled airports may contact the FSS via VHF radio to a telephone connection to update a weather briefing prior to takeoff. Pilots use six "key clicks" to contact the FSS. The GCO system is intended to be used only on the ground. Available locations are advertised in the *Airport/Facility Directory.*

The notion of an FSS common frequency 122.2 MHz is a relic of the past. Although this frequency is available at almost all FSSs, it is also used at remote sites. Pilots should use the appropriate frequency published for their vicinity on aeronautical charts.

The *local airport advisory* (LAA) is a terminal service provided by designated facilities located at airports without an operating control tower. However, LAA is another relic of the past. Even where an FSS is located at a nontower or part-time tower airport, few facilities provide this service. Where available, frequency 123.6 MHz (123.62 or 123.65 MHz at some locations) is used. This service provides wind, altimeter setting, favored or designated runway, and known traffic. Local weather conditions can also be included. At airports where part-time towers are collocated with an FSS, the local airport advisory, where available, will be provided on the tower local control frequency when the tower is closed. VFR flights should monitor the frequency when within 10 mi of the airport. IFR flights will be instructed to contact the advisory frequency by the control facility.

The FAA's position remains that automated flight service stations will not provide LAA. This seems to be a waste of a valuable resource. An FAA group has proposed a similar service for AFSSs. An LAA for local or remote airports would be considered on an individual basis. Wind and altimeter would be provided either from direct-reading instruments or the local weather report. If the local weather report is used, the time of observation would be included. The inclusion of time will alert the pilot that wind direction is true, rather than magnetic, as reported from direct-reading instruments.

Routine communications (weather information, flight plan services, position reports, etc.) should be accomplished on the station's discrete frequency. These frequencies are unique to individual facilities and remote locations. Their use will usually avoid frequency congestion with aircraft calling adjacent stations.

FSS frequencies can be found on aeronautical charts as illustrated in Fig. 19-2 and in the *Airport/Facility Directory.* A heavy-line box indicates standard FSS frequencies, 122.2 MHz, and the emergency frequency 121.5 MHz. Other frequencies are printed above the box. If a frequency is followed by the letter *R* (122.15R), the FSS has receiver-capability-only on that frequency. The pilot must receive transmissions from the FSS on another frequency, usually the associated VOR. This duplex communications system requires the pilot to ensure that the volume is turned up on the VOR receiver.

For example, refer to the remote communications outlet/NAVAID portion of Fig. 19-2. Rancho Murietta FSS, noted below the box, has a remote receiver at the VOR site on 122.1 MHz, noted above the box. A pilot wishing to communicate through the VOR would tune the transmitter to 122.1 MHz, and select Friant, 115.6 MHz, on the VOR receiver. The pilot must remember to turn up the volume on the VOR receiver because "Rancho Radio" will transmit on that frequency. A thin-line box indicates a remote communications outlet. The frequency or frequencies available are printed above the box with the name of the controlling FSS below.

After selecting the frequency for the service desired, correct communications technique must be used. By following the procedures below, pilots will realize faster, more efficient service, and the chance for error or delay will decrease.

1. *Monitor the frequency before transmitting.* Monitoring the frequency before transmitting is paramount to effective communications. How many times have we heard someone transmit over someone else? We've all done it—selected the frequency and pressed the

FSS COMMUNICATIONS

HEAVY LINE BOX indicates FSS.
Normally **122.2** and **121.5** are available.

122.35   122.5

┌─────────────────────────────┐
│ **HAWTHORNE HHR**   │
└─────────────────────────────┘

122.35 (Simplex) FSS Primary Discrete
Frequency.
122.5 (Simplex) FSS Secondary
Discrete Frequency.

**REMOTE COMMUNICATIONS
OUTLET/NAVAID**
(Duplex) "FRIANT" is the name of the
RCO and NAVAID. RANCHO
MURIETTA is the controlling FSS.

122.1R

┌─────────────────────────────┐
│ **FRIANT**   │
│ **115.6 Ch 103 FRA**   │
└─────────────────────────────┘
RANCHO

122.1R (Duplex) FSS has receiver
ONLY. *Pilot must transmit on 122.1 and
listen on the VOR frequency 115.6.*

An underlined frequency (i.e. 109.4)
indicates NAVAID only. NO FSS
communications.

HEAVY LINE BOX indicates FSS.
Normally **122.2** and **121.5** are available.
Circle, inside upper right corner,
indicates **HIWAS** available on the VOR
frequency.

122.6
123.65 VOR/DME

┌─────────────────────────────┐
│ **ARCATA**   │
│ **110.2 Ch 39 ACV**   │
└─────────────────────────────┘

122.6 (Simplex) FSS discrete frequency.
123.65 (Simplex) FSS Local Airport
Advisory.

REMOTE COMMUNICATION OUTLET
(Simplex) "BURBANK" is the name of
the RCO with a frequency of **122.35**.

122.35

┌─────────────────────────────┐
│ **BURBANK RCO**   │
└─────────────────────────────┘
HAWTHORNE

HAWTHORNE is the controlling FSS.

FIGURE 19-2 FSSs communication frequencies are depicted on aeronautical charts and can be found in the *Airport/Facility Directory.*

transmit button. All this does is add to the congestion of already crowded frequencies. This basic procedure should be followed when contacting any facility.

2. *Use the complete aircraft identification.* The FSS needs the full aircraft call sign.

3. *Advise the FSS on which frequency you expect a response and your general location.* Most FSSs monitor from between 5 and 10 different frequencies. With FSS consolidation, this practice has become more and more important for efficient communications.

4. *Establish communications before proceeding with your message.* The controller might be busy with other aircraft on other frequencies or other duties. Then listen to what the controller says.

CASE STUDY A classic failure occurs when a pilot calls to file a flight plan. The controller, busy with another aircraft on another frequency, advises the pilot to "stand by." The pilot proceeds with the flight plan. The controller has no option but to mute the receiver and conclude the contact with the other aircraft. The pilot wishing to file is somewhat miffed when the controller advises, "Go ahead with your flight plan."

Enroute Flight Advisory Services (Flight Watch)

The objective and purpose of flight watch is to enhance aviation safety by providing enroute aircraft with timely and meaningful weather advisories. This objective is met by providing complete and accurate information on weather as it exists along a route pertinent to a specific flight, provided in sufficient time to prevent unnecessary changes to a flight plan but when necessary, to permit the pilot to make a decision to terminate the flight or to alter course before he or she encounters adverse conditions.

Flight watch is not intended for flight plan services, position reports, or initial or outlook briefings, nor is it to be used for aeronautical information, such as NOTAMs, center or navigational frequencies, or for single or random weather reports and forecasts. The altimeter setting may be provided only on request. Pilots requesting services not within the scope of flight watch will be advised to contact an FSS.

Using all sources, flight watch provides enroute flight advisories, which include any hazardous weather, presented as a narrative summary of existing flight conditions—real-time weather—along the proposed route of flight, tailored to the type of flight being conducted.

The purpose of flight watch is to provide meteorological information for that phase of flight that begins after climbout and ends with descent to land; therefore, the specialist can concentrate on weather trends, forecast variances, and hazards. Flight watch is specifically intended to update information previously received and to serve as a focal point for system feedback in the form of PIREPs. ARTCC and tower controllers do accept PIREPs, but weather is a secondary duty and, unfortunately, PIREPs aren't always passed along; if at all possible, PIREPs should be reported directly to flight watch. The effectiveness of flight watch is to a large degree dependent on this two-way exchange of information.

CASE STUDY   A Bonanza pilot approached an area of thunderstorms in California's central valley. The pilot received the latest weather radar and satellite information, as well as PIREPs and surface observations from flight watch. The pilot safely traversed the area with minimum diversion or delay.

This is not a very exciting story, but that's the purpose of flight watch, to assist pilots in conducting uneventful flights. *Enroute flight advisory service* (EFAS) has been around for more than 20 years. In spite of this, its function and the best way to use this important service are misunderstood by many pilots.

Enroute flight advisory service, originally "enroute weather advisory service" (radio call "Eee'waas," which no one could pronounce), began on the west coast in 1972 originally as a 24-h service; Flight watch now normally operates from 6 a.m. until 10 p.m. local time. Flight Watch is not available at all altitudes in all areas. The service provides communications generally at and above 5000 ft AGL. However, in areas of low terrain and closer to communication outlets, service will be available at lower altitudes.

The system expanded in 1976, and a network of 44 flight watch control stations became operational in 1979. The common frequency 122.0 MHz immediately became congested, especially from aircraft at high altitudes. To help resolve the problem, a discrete high-altitude frequency was assigned to flight watch stations in the southwest in 1980. With flight service station consolidation, flight watch responsibility has been assigned to the FSSs associated with the air route traffic control centers (Oakland FSS—Oakland Center, Hawthorne FSS—Los Angeles Center, etc.). A discrete high-altitude frequency, for use at and above flight level 180, has been assigned each flight watch control station to cover the associated center's area.

Establishing communications is the first step. Because only one frequency is available for low altitudes, pilots must exercise frequency discipline. In addition to the basic communications technique already discussed, the following procedures should be used when contacting flight watch:

1. When known, use the name of the associate air route traffic control center (ARTCC) followed by "flight watch" (Salt Lake Flight Watch). If not known, simply calling flight watch is sufficient.

2. State the aircraft position in relation to a major topographical feature or navigation aid (in the vicinity of Fresno, over the Clovis VOR, etc.).

Exact positions are not necessary, but the general aircraft location is needed. Flight watch facilities cover the same geographical areas as the ARTCCs. With numerous outlets on a single frequency, the specialist needs to know which transmitter serves the pilot's area. This will eliminate interference with aircraft calling other facilities, garbled communications, and repeated transmissions. Failure to state the aircraft position on initial contact is the single biggest complaint from flight watch specialists.

3. When requesting weather or an enroute flight advisory, provide the controller with cruising altitude, route, destination, and IFR capability, if appropriate. The controller needs sufficient background information to provide the service requested.

Flight watch controllers are required to continually solicit reports of turbulence, icing, temperature, wind shear, and upper winds regardless of weather conditions. This information, along with PIREPs of other phenomena, is immediately relayed to other pilots, briefers, and forecasters. Together with all sources of information, the controller has access to the most complete weather picture possible.

ARTCC controllers are helpful relaying reports of turbulence and icing and providing advice on the location of convective activity, but the information is limited by equipment, and usually to immediate and surrounding sectors. Their primary responsibility is the separation of aircraft. On the other hand, flight watch has only one responsibility, weather. Flight watch—with real-time National Weather Service weather radar displays, satellite pictures, and the latest weather and pilot reports—provides specific real-time conditions, as well as the big picture. Additionally, flight watch controllers have direct communications with center weather service unit personnel and NWS aviation forecasters.

Getting a hold of flight watch is usually a simple matter, even for single-pilot IFR operations. ATC will almost always approve a request to leave the frequency for a few minutes, but don't wait until the last minute. Trying to find an alternate airport in congested approach airspace is no fun for anyone. I routinely use this procedure and have never been denied the request from enroute controllers.

The early days of airline flying were plagued by thunderstorms as well as icing, turbulence, widespread low ceilings and visibilities, and the limited range of the aircraft. Today's jets have virtually overcome these obstacles. More and more pilots of general aviation aircraft, equipped with turbochargers and oxygen or pressurization, are encountering the same problems as yesterday's airline captains. The only difference is a vastly improved air traffic and communication system. Among one of the FAA's best-kept secrets is the implementation of high-altitude flight watch.

Continually updating the weather picture is the key to managing a flight, especially at high altitude in aircraft without ice protection and storm avoidance equipment and with relatively limited range. Winds aloft can be a welcome friend eastbound or a terrible foe westbound. With limited range, even a small change in winds at altitude can have a disastrous result. At the first sign of unexpected winds, flight watch should be consulted if for no other reason than to provide a pilot report. A significant change in wind direction or speed is

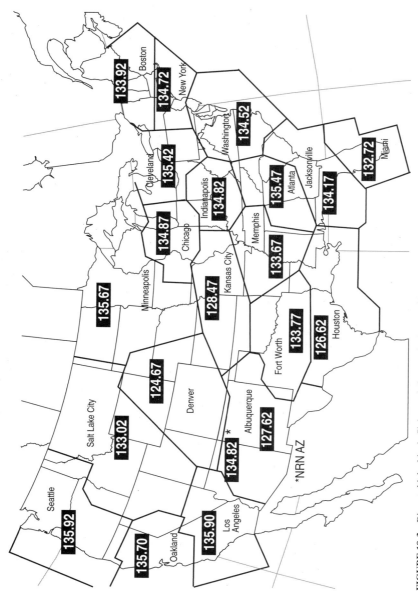

FIGURE 19-3 Discrete high-altitude flight watch frequencies have been commissioned nationwide to eliminate much frequency congestion, especially with aircraft at low altitudes.

Boston 133.92

New York 134.72

Washington 134.52

Cleveland 135.42

Indianapolis 134.82

Atlanta 135.47

Jacksonville 134.17

Miami 132.72

Chicago 134.87

Memphis 133.67

Minneapolis 135.67

Kansas City 128.47

Houston 126.62

Fort Worth 133.77

Denver 124.67

Albuquerque 127.62

*NRN AZ 134.82 *

Salt Lake City 133.02

Seattle 135.92

Oakland 135.70

Los Angeles 135.90

often the first sign of a forecast gone sour. A revised flight plan might be required. Flight watch can provide needed additional information on current weather, PIREPs, and updated forecasts upon which to base an intelligent decision.

A primary reason for high-altitude flying is to avoid mechanical, frontal, and mountain wave turbulence; however, the flight levels have their own problems—wind shear or clear air turbulence. When problems are encountered, flight watch can help find a smooth altitude or alternate route. If the pilot elects to change altitude, an update of actual or forecast winds aloft is often a necessity.

Icing is normally not a significant factor in the flight levels, except around convective activity or in the summer when temperatures can range between $0°$ and $-10°C$. However, icing can be significant during descent, especially when destination temperatures are at or below freezing. Flight watch can provide information on tops, temperatures aloft, reported and forecast icing, and current surface conditions.

Many aircraft are equipped with airborne weather radar and lightning detection equipment. However, these systems are plagued by low power, attenuation, and limited range. A pilot might pick his or her way through a convective area only to find additional activity beyond. Flight watch has the latest NWS weather radar information. Well before engaging any convective activity, a pilot should consult flight watch to determine the extent of the system and its movement, intensity, and intensity trend. Armed with this information, the pilot can determine whether to attempt to penetrate the system or select a suitable alternate. ATC prefers issuing alternate clearances compared to handling emergencies in congested airspace and severe weather.

Finally, there is destination and alternate weather. The preflight briefing provided current and forecast conditions at the time of the briefing. This information should be routinely updated enroute; the airlines do it, often through flight watch. Are updated reports consistent with the forecast? If not, why? Flight watch controllers through their training are in an excellent position to detect forecast variances. Whether the forecast was incorrect or conditions are changing faster or slower than forecast, the pilot needs to know and plan accordingly. A knowledge of forecast issuance times is often helpful. Forecasts might not be amended if the next issuance time is close. Flight watch is in the best position to provide the latest information and suggest possible alternatives.

Updates must be obtained far enough in advance to be acted upon effectively. This must be done before critical weather is encountered or fuel runs low. Hoping a stronger-than-forecast head wind will abate or arriving over a destination that has not improved as forecast is folly. At the first sign of unforecast conditions, flight watch should be consulted, and, if necessary, an alternate plan developed. This might mean an additional routine landing enroute, which is eminently preferable to, at best a terrifying flight or at worst an aircraft accident.

High-altitude flight watch frequencies for individual ARTCCs are provided in Fig. 19-3. Frequencies and outlets can also be found on the inside back cover of the *Airport/Facility Directory*. The standard frequency 122.0 MHz can be used when a pilot is unsure of the discrete frequency.

CHAPTER 20

AUTOMATED WEATHER BRIEFING SERVICES

For our purposes, we will define *automated weather briefing services* as meteorological and aeronautical information that is accessible without human intervention from the provider. Few realize that these efforts, like automated weather observations, began in the late 1950s and early 1960s. Even at that time the FAA and NWS realized that person-to-person weather briefings were time-consuming, expensive, and inefficient. However, the technology of the day, again like automated weather observing systems, did not directly lend itself to providing these kinds of services.

During the decade of the 1960s, the FAA introduced the *transcribed weather broadcast* (TWEB). (Recall that TWEB routes were originally written to support this program.) The service was initially intended to be broadcast over VORs and low- to medium-frequency *nondirection radio beacons* (NDBs). The goal was to replace the *scheduled weather broadcasts* (SWBs), which were manually transmitted on the quarter hour. TWEB not only provided the information of the SWB but also gave additional information on surface weather forecasts and winds aloft, and on a continuous basis. PATWAS also evolved during this period. PATWAS provided telephone access to essentially the same information as TWEB. In many areas pilots eventually gained access to TWEB through dedicated telephone numbers.

During the decades of the 1960s and 1970s, other experiments at self-briefing were conducted. One such effort involved the posting of weather and aeronautical information at a "self-briefing" position in the FSS. Another program used closed-circuit television to display data. Both met with limited success. Most pilots simply bypassed these positions to request briefings directly from FSS controllers.

In the decades of the 1980s and 1990s, the FAA's self-briefing services have evolved into the *hazardous inflight weather advisory service* (HIWAS) and *telephone information briefing service* (TIBS). HIWAS has replaced the SWB and to some extent TWEB. TIBS has replaced TWEB and PATWAS. (Some TWEB locations remain commissioned to support specific user needs.) However, like TWEB and PATWAS, neither HIWAS or TIBS provides all the information contained in a standard weather briefing. Therefore, these services do not meet regulatory requirements for all operations.

During this period various commercial vendors provided weather briefing information and flight plan services to subscribing customers for a fee. With the introduction of computers, subscribers were provided with the best routes and altitudes for specific flights based on weather and winds aloft forecasts. This type of service continues today with customers provided with customs notification, documentation, and aircraft and passenger services. The convenience and extent of these services have changed significantly with the communications revolution.

The FAA's next venture into self-briefing and flight plan services came in February 1990. The agency authorized *data transformation and contel* to provide *direct user access terminal* (DUAT) *service* to pilots within the contiguous United States. This computerized system, available at airports and through personal computers, allows direct access to weather briefing and flight plan services.

Throughout the last decade, various government and commercial organizations have entered the weather provider field through the expanding Internet. Throughout this book we have provided government Internet addresses for various weather products. For those who would like a complete listing of aviation Internet sites, I suggest *300 Best Aviation Web Sites* by John A. Merry (McGraw-Hill, 1999). This book is an essential industry reference tool for management, pilots, aviation-related companies, and enthusiasts.

DUATs, and virtually all other commercial services, as well as government sites, use National Weather Service products. This contradicts the misconception that computer briefings somehow provide a different product than that available through an FSS. Graphic products may differ slightly in appearance. However, again, they are derived from NWS products and are, typically, less detailed.

CASE STUDY   The Monterey WFO evaluation officer was checking the performance of a new FSS controller at the Hawthorne AFSS. On a high-altitude briefing, the evaluator was concerned that the briefer was not presenting the same information as depicted on the evaluator's high-altitude significant weather prog. After further checking, it seems that the AFSS was using a commercial vendor for graphic products rather than NWS charts. And there were some differences between the products.

Even small *fixed-base operators* (FBOs) often have a weather computer available in the flight planning room. Through the Internet, there are literally dozens, if not hundreds, of places a pilot can search for aviation and general weather information. No airplane-related Web site is complete without at least one link to an aviation weather provider. For instance, www.landings.com, which bills itself as "the busiest aviation hub in cyberspace," has over 40 links to weather Web sites, detailing observations and forecasts from government, private industry, university, and commercial sources both in the United States and around the world. The Aircraft Owners and Pilots Association (AOPA) Web site, at www.aopa.org, has an excellent weather flight planning site for members of its organization. Any pilot, from student to ATP, can log into the government-sponsored (and free) direct user access terminal system and get most of the information that's available to flight service station controllers. Many flight planning software packages are bundled to DUATs or similar computer weather servers, automatically downloading winds aloft and other information while compiling a pilot's flight plan; often, the pilot can even file a flight plan directly through the system, without having to call it in manually by telephone.

As important as the weather are *Notices to Airmen* (NOTAMs). Most NOTAMs are provided as part of an FSS standard briefing, with the exceptions noted in Chap. 19. Since NOTAMs are an essential part of the weather briefing, the final section of this chapter contains a discussion of these products.

ADVANTAGES OF AUTOMATED BRIEFING SERVICES

There are a number of advantages to self-briefing. For the weather-wise pilot, experienced in accessing and decoding the hieroglyphics of aviation weather, it's often an easier way to get the complete picture of the weather, as compared to talking with an FSS controller. This is not to criticize; it's just that FSS controllers are trained to distill the weather picture for pilots, and some pilots would rather see the big picture in its entirety for themselves. Computer access to weather information gives pilots all the raw data, often helpful in answering tough go/no-go decisions. Another advantage is that, assuming you can log onto the Internet without being "put on hold," it's often quicker to fire up the computer than it is to reach a live FSS person on the telephone. This is especially true during peak flight planning times and during bad weather. Access to color graphics is another distinct plus with most self-briefing services.

Many pilots like to take a printed hard copy of some of the more critical weather information along in the airplane, for reference inflight. It's nice to be able to point to the expected location of a front, for instance, and compare it to the actual point of encountering frontal weather, to gauge the accuracy of the weather briefing and the severity of predicted conditions. Computers allow you to print out anything you'd like to take with you.

There are a number of locations where experimental prediction charts are posted. For instance, computers let you see the "neural net icing forecast," an excellent, albeit experimental, weather product, that you won't yet hear about from flight service. Through the Internet you have available the very latest technology in predicting aviation weather hazards.

Computerized briefing is a natural for pilots who use flight planning software. As already mentioned, some flight planning programs even use DUAT weather in calculating aircraft performance and filing flight plans. This level of sophistication makes computerized self-briefing and flight planning outstanding for busy executives and hard-pressed charter pilots, who may not have a dispatcher or a lot of time to devote to before-takeoff needs.

With these advantages come the responsibility to decode, translate, interpret, and apply information to a flight. The pilot will have to sift through the mountains of written data, formally reserved for the FSS controller, to determine if a particular flight is feasible under existing and forecast conditions, and aircraft and pilot capability.

DISADVANTAGES OF AUTOMATED BRIEFING SERVICES

The biggest disadvantage of computerized self-briefing is that often pilots don't have enough weather education to make a sound judgment alone. Confounding this is the fact that a typical DUAT printout for even a short cross-country flight will involve the equivalent of a ream of paper, with information valid more than a hundred miles from your intended route. If you key in the wrong response to a question, you might end up with useless-to-you information like temporary flight restrictions over southern Iraq, approach system updates for remote airports, or the inadvisability of flying your Piper Cherokee into Libya. FSS briefers are experts at translating all this information into what they feel you need to make a safe decision; you can always ask for additional information after they've completed their briefing if it doesn't completely answer your questions.

Sometimes it's not convenient or even possible to plug in a computer and access the Internet or DUATs when away on a trip; you may not even have a portable computer despite liking to use DUAT or other Internet weather sources at home.

The sheer amount of information is often overwhelming, especially for long-distance flights. This is especially true with NOTAMs. A pilot might have to study several pages for a single sentence that applies. Flight instructors and pilot examiners might wish to save briefings for training and flight test purposes.

USING AUTOMATED BRIEFING SERVICES

When using these services, it's essential to know what information is available. Pilots using a commercial system must check with the vendor to determine how their system handles aviation products. Certain products, for example, TWEBs, might not be available on some systems, none provide local NOTAMs. Know your service, and check with an FSS for any additional information you require or for clarification of anything you don't understand— remember the disclaimer. Below is one example:

DISCLAIMER "While weatherTAP provides a complete package of aviation weather information, the information available here shall not be used for flight planning or other operational purposes. The weather forecast information provided by this Service should not be relied on in lieu of officially disseminated weather forecasts and warnings."

Let's review a DUAT aviation briefing, for a flight in a Cessna 182 from Livermore, California (LVK), to Salt Lake City, Utah (SLC). The briefing filled nine complete pages— without using the plain-language function—and the weather was good. The briefing is organized as follows:

• The area forecast
• Alert weather watches
• SIGMETs
• Convective SIGMETs
• Center weather advisories
• The AIRMET Bulletin
• METARs
• Pilot reports
• Radar weather reports
• Terminal aerodrome forecasts
• Winds aloft
• Notices to Airmen

Note that the sequence is not the same as used by the FSS controller. The area forecast is first, followed by weather advisories. One advantage of these computer systems is that the winds and temperatures aloft forecasts contain a forecast for the altitude requested.

CASE STUDY I downloaded a DUAT briefing from the San Francisco Bay area to Reno, Nevada. The weather was marginal, and I planned to use the package as a training aid. Even for a such a relatively short distance the printout consumed 17 pages!

The pilot is presented with the same products—except TWEBs, charts, and local NOTAMs—available at the flight service station. The pilot must then decode, translate, and interpret the information to determine which reports and forecasts apply, although a plain-language translation is now available. This briefing took about 15 min to obtain and print, and an additional 10 min to analyze and apply. After a little practice, you should be able to scan the material as it is displayed and only print significant portions for further review.

CASE STUDY Occasionally, incorrect abbreviations are used. One synopsis decoded: "AN UPPER LEVEL LAKE ONTARIO IS OFF NORTHERN BAJA CALIFORNIA." (The forecaster had used the abbreviation "LO," which decodes as "Lake Ontario," to indicate a low-pressure area.)

An FSS briefing would go something like this:

There is an AIRMET for moderate turbulence over California, due to a strong northerly flow aloft over California, with an upper-level trough over the Rockies moving eastward. Livermore's reporting 15,000 broken, visibility two zero, wind calm. Over Stockton at 7500 a Baron reports light turbulence with northerly winds 35 to 40 kn. Enroute, broken to overcast cirroform clouds and unrestricted visibilities becoming, by the Elko, Salt Lake City, portion of the route, scattered clouds based around 6000 to 7000. Salt Lake surface winds zero two zero at one one, with standing lenticular altocumulus southeast through west. A Gulfstream two during climbout of Salt Lake southbound reports smooth, tops of scattered clouds 9000 to 10,000. Conditions forecast to remain the same enroute, with Salt Lake 6000 scattered, a slight chance of ceilings 5500 broken in light rain showers, surface winds three five zero at one five gusts two zero. Winds aloft forecast at 11,500, 330 at two five. There are no NOTAMs for the route.

Throughout the chapters on satellite and graphic weather products, we've stressed the need to understand their purpose, scope, and limitations. This is especially true with DUAT and commercial graphic products.

In our discussion of satellite images, we stressed the need to view visual, as well as IR images for a more complete picture. Before we can effectively use these products, we must know if they are visual or IR images, the time of observation, and, which if any, enhancement curves were employed. A satellite "loop" may be very helpful with a general trend of the movement of weather systems. But, at present, there are no substitutes for written weather reports and forecasts.

Color graphics are "neat" but may be misleading. A generalized weather depiction may indicate VFR conditions along our intended route. However, in reality there may be locations of adverse weather within the general picture, and these charts are always old by the time they become available. Caution must also be exercised with radar depictions. Areas of very light precipitation may be aloft and not reaching the ground (virga), or there may even be cloud particles without precipitation (much like the presentation of intensity levels 1 and 2 on the radar summary chart). However, these systems do accurately portray the location and movement of heavy precipitation.

DTN Kavouras Weather Center has introduced plotted AIRMETs as an added aviation resource. Certainly a graphic depiction of this product makes interpretation much easier. (FSS plots this product.) It is updated hourly. However, a pilot can never be sure of getting the latest information. Weather advisory distribution in an FSS is almost instantaneous. Keep in mind that AIRMETs are only one category of weather advisory.

Undeniably, an advantage of these systems is that the user can "overlay" products—for example, the satellite with the radar summary chart, or the radar chart with a current lightning depiction graphic. We've stressed the need to use all available sources to obtain the complete picture. This is especially true with DUATs and other commercial vendors.

Finally, if you have a technical problem with one of these services, you'll have to contact the vendor. Questions about written reports and forecasts should be easily resolved with the assistance of the FSS because they both use the same products. Graphics are a different matter. At this time neither DUATs nor commercial vendors necessarily use the same graphic products as the FSS. However, by consulting an FSS, you will at very least be able to get a second opinion.

As so elegantly stated by John Hyde, an ex-Army aviator, Kit Fox owner, and retired Oakland FSS controller: "When obtaining a briefing from an FSS or other source, keep in mind that they're in 'sales, not production.'" In other words, don't blame the messenger for the message.

The information you'd like to know may not always be available on a single Web site. For instance, if you'd like to access METARs and TAFs and at the same time know the lifted index for your route of flight, you may have to dial into more than one source. Your flight planning software may work only with one Web weather source, but you like the graphics of another. Or your favorite planning site may be excellent except that it reveals weather only for a narrow corridor along your proposed route and you'd really like to see more for making a "Where will I divert?" or "What weather is blowing in, that may arrive sooner than expected?" sort of evaluation. In these cases, computerized weather briefings may be less convenient than the old-fashioned call on the phone.

Computerized aviation weather is a tremendous tool for making an informed flying decision. It's possible that eventually computerized self-briefing will replace the flight service station network completely. For best results, however, I like to combine all available sources of information, culminating in a call to flight services for the expert analysis of a trained weather specialist. For instance, I planned a flight for yesterday morning by looking at The Weather Channel the night before, then again while I ate breakfast the morning of the trip. That gave me the big picture for the entire country and the region, allowing me to see what could potentially affect my flight. Then I called flight services for a standard weather briefing, asking specific questions (in this case, about the likelihood of clouds and airframe icing along my route) after she completed her presentation. I filed my flight plan by phone, then critically observed the clouds and wind myself while driving to the airport. At the FBO, I used the computer to access some experimental ice prediction sites and for the latest, real-time weather radar and any recent pilot reports for icing, turbulence, and cloud tops along the way. This integrated approach to weather decision making (general information, a specific briefing, personal observation, and a last-minute computer check) combines the best of each source of weather information.

Mode S transponders have the capability of two-way communications, heralding the day when real-time weather information can be displayed directly in the cockpit. A number of pioneering avionics manufacturers are marketing units that combine GPS moving maps, lightning detection, radar, and weather uplinks on a single cockpit cathode-ray tube. I envision a not-too-distant time when you might fly visually with a panel-mounted moving map overlaid with the location of VFR, MVFR, IFR, and LIFR weather, as well as areas and intensities of thunderstorms, turbulence, reduced visibility aloft, and ice. A profile view on the screen might give those systems' locations and altitudes. As "nowcasting" outlets detect hazards over reporting points at specific altitudes, a central processor will broadcast the information to airplanes enroute. A blue-shaded area, for instance, might appear on overhead and profile views to indicate the location of airframe icing; you simply climb, descend, or maneuver your airplane around the display to avoid encountering the hazard. VFR pilots could "see" reduced visibility ahead, perhaps shaded in green, and simply fly around the threat.

Truly usable, real-time aviation weather is in its infancy. Until the time when the types of devices I've described are in widespread use, you'll need to understand what causes aviation weather hazards to form, how they'll move and modify, and what to do if conditions become hazardous. That will enable you to achieve your goals of safety, comfort, convenience, and economy of flight. Weather historically causes 25 percent of all accidents and 40 percent of all fatal accidents in general aviation; knowing how to obtain weather information, including what specialized questions to ask, and comparing that expected weather model to conditions you actually encounter enroute, will allow you to verify or refute the required weather briefing and make better decisions that affect the safe outcome of your trip.

PILOT SURVEYS

This section contains an analysis of two pilot surveys. The first was conducted by an established contractor with experience in this field. The second was more of a "straw poll" conducted by an aviation Web site.

TABLE 20-1   Pilots' Use of Direct User Access Terminals versus Flight Service Stations

Frequency	DUATs, %	FSSs, %	Private or state services, %
Always	12	31	7
Often	24	34	9
Occasionally	15	15	9
Infrequently	14	10	11
Never	20	3	45
No response	15	6	18

The Mitre Corporation (a large government contractor) conducted a survey at the 1997 Oshkosh Fly-in. During the course of the event, they obtained opinions of the National Airspace System from 568 pilots. The survey asked questions concerning many issues including special-use airspace, enroute navigation aids, and inflight and preflight services. The following is an analysis of pilots' responses in the area of preflight meteorological services. Within the group, 70 percent indicated they had access to a computer terminal for preflight planning. Table 20-1 indicates the participants' use of DUATs versus FSSs versus private or state-sponsored services.

If we combine "always" and "often," the following conclusions become apparent. Pilots use the FSS about two-thirds of time, they use DUATs one-third of the time, and they use private or state-sponsored services one-sixth of the time. Many pilots prefer the "human element" when dealing with weather. This is not surprising given the anxiety many pilots feel about weather. Numerous pilots and organizations, including the NWS's Aviation Weather Center, recommend pilots obtain a DUAT or Internet briefing, then contact the FSS, and follow up with an FSS controller's briefing. As is true in the use of TIBS and The Weather Channel, the more familiar a pilot is with the weather situation, the more readily he or she will understand the briefing.

When adverse weather conditions exist, how often during your preflight briefing does DUATs/FSS warn you about the conditions?

Pilots indicated that DUATs provide a warning about one-third of the time, while the FSS issues warnings about three-quarters of the time. One reason for this difference is that during an abbreviated DUATs briefing, pilots are advised that adverse conditions "may" exist; the system then allows the pilot to decline this information. FSS briefings require the briefer to specify adverse conditions (i.e., turbulence, icing, IFR) that exist and then to provide the details on request. Another factor is that the FSS controller is immediately notified of the issuance of adverse weather advisories. This may occur during the briefing. Once the DUAT has downloaded data to the pilot, it cannot be updated.

How often do the conditions forecast in your DUAT/FSS briefing match those that you actually experience in flight?

Pilot response was a little closer than the previous question, with 40 percent for DUATs and 70 percent for the FSS briefings. The FSS briefing was closer to the pilots' actual experiences inflight because the FSS briefer summarizes data, and in doing so, he or she eliminates obviously erroneous or misleading information. The briefer is also able to interpret, translate, and explain the weather, thus, giving pilots a better understanding of the weather picture.

Do you feel that the DUAT/FSS briefing gives you too much, too little, or about the right amount of information during your preflight weather briefing?

In the "about right" category, pilots responded that DUATs achieved this about one-half the time and FSS about three-quarters. DUAT briefings contain tremendous amounts of needless information. This is unavoidable with today's technology. A major part of the FSS briefing requires the briefer to summarize data, thus eliminating much superfluous information. Recall our discussion of background information. This allows the FSS briefer to eliminate strictly VFR information from an IFR briefing and vice versa. These are additional factors in many pilots' preference for FSS briefings.

How understandable do you find the DUAT/FSS briefing?

DUATs scored about 50 percent on this question, while the FSS reached 90 percent. Again, this is not surprising since the FSS briefer can translate and interpret information. Many briefers, with their training, experience, and knowledge of local terrain and weather features, pass on this expertise during their briefings.

AVweb, an aviation Web site, recently polled its subscribers about aviation weather. As interesting as the direct responses were some of the comments. The straw poll confirmed most pilots' attitudes as revealed in the survey conducted by the Mitre Corporation.

The vast majority of pilots who responded believe that weather collection and dissemination should be the responsibility of government agencies.

Not unexpectedly, half of those who responded would like weather in plain language. Interestingly, several comments supported the use of METAR and TAF codes. In one response, an airline dispatcher pointed out the difficulty of reviewing many METARs or TAFs when they were translated into plain language. Another complained about the large amount of text that plain language generates. One commented: "I never did understand what's so tough about learning the METAR/TAF codes." (I certainly concur with these observations.) Since DUAT vendors provide a decode function, is this not a moot point?

When asked about their primary source of weather information, a majority indicated a computer-based service. This should be no surprise since the poll was conducted on a Web site. However, one respondent commented that the DUAT, "was not user friendly." There were many comments that reflected the use of TV and the Internet for preliminary weather, DUATs for an initial briefing, then a call to the flight service station for a final check. This agrees with the Mitre survey, indicating that many pilots want the human interaction provided through the FSS.

Other comments ran the entire gamut from government "bashing" and venting, to misunderstanding, and finally to some interesting, positive suggestions.

One pilot complained about the lack of cloud tops and the blanket predictions of icing. Certainly better predictions of cloud tops and icing are a valid concern. We also know that many pilots do not understand the scope and purpose of weather products or how to interpret them. One commented that private weather sources were more accurate than the government. I find this difficult to believe, since most use the same products. There always seems to be a comment about forecasts being in error and being better in the past. Although I have no empirical data to support my conclusions, my experience contradicts this notion. Another wanted prog charts issued three times a day. Aren't they now issued four times a day? Several wanted more emphasis on PIREPs—a valid concern supported by most weather providers and users.

A number of comments concerned graphic products. This is somewhat of a hot issue at AWC. The Aviation Weather Center does offer graphic products (adds-awc-kc.noaa.gov). Certainly, graphic products are nice. But they have to make weather briefing and interpretation easier. Many private vendors produce beautiful graphic products. However, a flight decision cannot be based on them alone. A pilot or briefer has to go back to the textual data of METARs, TAFs, and the FAs.

A final comment from what would appear to be an airline pilot or dispatcher: This person pointed out the possible conflict of interest of private weather sources. That is, the vendor might skew the forecast to what the customer needed to dispatch a flight. Something to think about. This, however, is not to say that many private and airline forecasts are skewed. American Airlines, along with others, has been producing their own forecasts for years, without any apparent degradation in safety.

A FINAL COMMENT   AOPA President Phil Boyer recently gave testimony to a Congressional subcommittee. In his testimony he provided examples of general aviation safety programs jeopardized by the administration's budget. He specifically cited cuts in programs to improve the delivery of weather information to pilots. Among these was the failure of the FAA to modernize flight service station equipment. He emphasized the dramatic difference between FSS weather graphics and those that pilots can find on the Internet. Boyer added, "What's lacking on the Internet is the expert knowledge provided by FAA's highly trained FSS specialists."

THE NOTICE TO AIRMEN SYSTEM

Failure to check Notices to Airmen (NOTAMs) has lead many a pilot into an embarrassing and potentially hazardous situation. Increased use of DUATs and other commercially available briefing systems means that interpreting and understanding NOTAMs will take on a greater significance. Add to this recent changes in the FAA's NOTAM system with the introduction of ICAO abbreviations and date/time groups. These factors will challenge pilots to a much greater degree than in the past.

The Federal Aviation Administration advertises the status of components or hazards in the National Airspace System (NAS) through aeronautical charts, the *Airport/Facility Directory,* other publications, and the National Notice to Airmen System. Changes are normally published on charts or in the directory, or they appear in the *Notices to Airmen* publication. Published NOTAMs are sometimes referred to as *Class II.* Class II is the international term used to identify NOTAMs that appear in printed form for mail distribution. The need for current charts and publications cannot be overemphasized.

CASE STUDY   A pilot called flight services and requested a briefing from Bishop to Santa Cruz, California. The briefer explained that the airport was closed. "Oh, I must be using an old chart," said the pilot. Indeed, the airport had been closed for 2 years.

Aeronautical information not received in time for publication is distributed on the FAA's telecommunications systems. Information includes unanticipated or temporary changes and hazards when the duration is for a short period or until published. Unpublished NOTAMs are not necessarily given during abbreviated or outlook briefings, but they are routinely provided as part of an FSS standard briefing.

NOTAMs are issued for the commissioning and decommissioning of facilities, or restrictions to landing areas (runways and waterways), including closures, braking action, and problems due to snow, ice, slush, or water. Lighting aids that affect landing areas or are part of an instrument approach procedure, along with *pilot-controlled lighting* (PCL), receive NOTAM distribution, as well as other lighting aids (airport beacons, VASIs, wind-T lights, obstruction lights, etc.). NAVAID status, together with hours of operation for air traffic control facilities, controlled airspace, and services (EFAS, HIWAS, AWOS/ASOS, etc.) receive NOTAM distribution. Unpublished NOTAMs are divided into three groups:

- NOTAM (D)
- NOTAM (L)
- FDC NOTAMs

Flight service station controllers are responsible for classification, format, and transmission of NOTAM (D)s and NOTAM (L)s. The country is divided into flight plan areas. NOTAM responsibility for airports and NAVAIDs within the flight plan areas belongs to a designated FSS. The tied-in FSS that is responsible for a facility can be determined from the *Airport/Facility Directory*.

Like METARs, PIREPs, and TAFs, NOTAM codes were changed in 1996. However, for NOTAMs the change was postponed several times. (It appears the FAA's philosophy is: "We never have time to do it right; but, we always have time to do it over.") There were two major changes: the use of international abbreviations and a 10-digit date/time group.

The date/time group consists of year, month, day, and time (UTC). Times used on NOTAMs are now all UTC. The day begins at 0000Z and ends at 2359Z. For example, 9810291400 decodes as year 98 (1998), month 10 (October), day 29, and time 1400Z. A common abbreviation used with date/time groups is WEF. WEF translates as: "with effect from or effective from." Use care when determining effective times! These date/time groups can be very confusing. The absence of a date/time group means the condition is in effect and will continue *until further notice* (UFN). However, UFN is not transmitted in the NOTAM text. To indicate a condition is in effect and will exist until a specified time, the abbreviation TIL (until), followed by a year/date/time group, describes the effective period.

Runways are identified by magnetic bearing (12/30, 12, or 30). If the magnetic bearing has not been established, the runway is identified by the nearest eight points of the compass (NE-SW, northeast/southwest runway; N 200 N-S, north 200 ft of the north/south runway).

NOTAM (D)s begin with the character "!" which is an automatic data processing code. This is followed by the *accountability location* (LOCID)—for example, OAK for Oakland, California. Next comes the NOTAM number. NOTAM (D)s are numbered by month of issuance and numbered consecutively during the month. NOTAM number 03/005 was issued during March, and it is the fifth NOTAM issued for the LOCID in that month. Next comes the specific facility or location. This is the LOCID of the airport, facility, or navigational aid affected. The last two items are the condition reported and effective times, if required.

The following is an example of NOTAM (D) format:

!OAK 03/005 OAK VORTAC OTS WEF 9803021800-9803022200

This is Oakland, California, NOTAM 03/005. It was issued during the third month of the year ("03"/005) and was the fifth NOTAM issued for Oakland (03/"005"). The Oakland VORTAC is scheduled to be *out of service* (OTS) on March 2, 1998 ("980302"1800) at 1800Z until March 2, 1998, at 2200Z.

NOTAM (D)s

NOTAM (D)s contain information that might influence a pilot's decision to make a flight, or require alternate routes, approaches, or airports. They are considered "need-to-know" and are issued for certain landing area restrictions, lighting aids, special data, and air navigation aids that are part of the National Airspace System. NOTAM (D)s are issued for all public use airports listed in the *Airport/Facility Directory*.

NOTAM (D) information is given distant dissemination on the FAA's telecommunications system. NOTAM (D)s are prepared and transmitted using abbreviations contained in App. A, Abbreviations.

NOTAM (D)s can be divided into four major categories:

- Movement areas
- Lighting aids

- NAVAID/communications/services
- Special data

Movement areas refers to airports and seaplane bases. Airport management personnel are responsible for reporting area restrictions for NOTAM issuance. FAA flight service station controllers format and disseminate the information.

SFO 10L/28R CLSD WEF 9905170700-9905171500

San Francisco runway 10 Left/28 Right will be closed on 5/17/99 between 0700Z and 1500Z: UTC dates and times. The runway is scheduled to be closed between 11 p.m. on the 16th until 7 a.m. on the 17th, Pacific Standard Time. Pilots must use care converting UTC to local times. (Gotta love those year/date/time groups brought to you from the friendly folks at ICAO!)

SJC 12R/30L CLSD 0800-1400 DLY TIL 9905271400

This San Jose runway closure will occur daily (DLY) from 0800Z through 1400Z until 5/27/99 at 1400Z, which is daily from midnight local through 6 a.m., until the 27th at 6 a.m.

BUR 7/25 CLSD TKOF 12500/OVR

BUR 7/25 E 254 CLSD

Burbank runway 7/25 is closed for takeoffs to aircraft with gross weights of 12,500 lb and over, probably because, as the second NOTAM reveals, the eastern 254-ft section of the runway is closed.

NOTAM (D)s advertise the following conditions as they refer to snow, ice, and water:

- Runway friction measurements
- Runway braking action
- Snow
- Ice
- Slush
- Water conditions

Civil pilots now have access to objective runway friction measurements at some airports. The Greek letter mu (μ, pronounced "myôô") designates a friction value representing runway surface conditions. Values range from 1 to 100, where 0 is lowest friction value and 100 is the maximum. With snow or ice on the runway, a mu value of 40 or less is the level at which the aircraft braking performance starts to deteriorate and directional control begins to be less responsive. The lower the mu value, the less effective braking performance becomes and the more difficult direction control becomes.

Airport management will conduct friction measurements on runways covered with snow or ice. NOTAMs will be issued only when one or more values are below 40. Values will be reported for the first third, middle third, and last third of the runway. Pilots can expect to receive mu values from ATC or as a NOTAM (D). When friction measurements are reported for more than one runway, each will be advertised as a separate NOTAM, specifying the runway. For example:

DCA 36 MU 42/35/48

At Washington National Ronald Reagan Airport (DCA), the runway 36 mu values are 42, 35, and 48 for the first, middle, and last third of the runway, respectively.

No correlation has been established between mu values and the descriptive terms "fair," "poor," and "nil" used in braking action reports. Pilots should use mu information with other knowledge including type, weight, wind conditions, and previous experience.

When braking action is reported by airport management, it receives NOTAM (D) dissemination. Braking action reported by a pilot is distributed as a PIREP. In both cases, *braking action* (BA) is reported as FAIR, POOR, or NIL. The type of vehicle, when used, will not appear in the NOTAM.

There are no hard-and-fast definitions of reported braking action. *Fair* would tend to indicate that braking action was certainly not what could be expected on a dry runway, but the pilot was able to control the airplane. *Nil* represents the lack of any braking action. The pilot has essentially no braking control. And *poor* is somewhere in between. Any of these reports or a mu NOTAM should alert the pilot to braking problems at the airport. The bottom line: If you haven't had training or experience with snow- or ice-covered runways, avoid them!

Snow, ice, and water are self-explanatory. Slush is a soft, watery mixture of snow or ice on the ground that has been reduced by rain, above-freezing temperatures, or chemical treatment. When snow, ice, slush, and water conditions are reported, their depth is expressed in terms of *thin*—less than 1/2 in (THN), 1/2 in, and 1 in. Additional amounts are reported in 1-in increments. If the surface is not completely covered, it is reported as having *patches* (PTCHY). The absence of PTCHY means the entire landing area is covered.

Reports of *snow* (SN) on the runway may be modified by one of the following:

- LSR, loose snow
- PSR, packed or compacted snow
- WSR, wet snow (not slush)
- RUF FRZN SN, rough frozen snow

Conditions may be combined to better describe the surface. For example, THN LSR OVER 1 IN PSR translates to less than 1/2 in of loose snow, over 1 in of packed snow:

CLM 8/26 THN SN

In this example, the Port Angeles, Washington, runway 8/26 is reported as covered with thin snow. THN indicates the layer is less than 1/2 in deep.

Ice is reported using the abbreviation IR or SIR. SIR means packed or compacted snow and ice on the runway. Other descriptors may be used to better define conditions.

PAE PTCHY 1 IN RUF IR

There is patchy 1 in of rough ice on the runway at Paine, Washington.

Slush is reported using SLR. For example:

IAD 1L/19R 1/2 IN FRZN SLR

At Washington's Dulles International Airport, there is 1/2 in of frozen slush (may be described as RUF IR, but not wet snow).

Water on the runway is reported in a similar manner, using the abbreviation WTR. It may be reported as PTCHY but never as puddles:

EWB PTCHY THN WTR

At New Bedford, Massachusetts, there are patchy areas of thin—less than 1/2 in—water covering the airport surfaces.

Drifting or *drifted snow* (DRFT) describes one or more drifts. When drifts are variable in depth, the greatest depth is reported. When drifts are identified, pilots should consider that these conditions prevail throughout the airport surface—runways, taxiways, and ramps.

SFF 4 IN LSR 9 IN DRFT

In this example, 4 in of loose snow covers all airport surfaces at Spokane's Felts Field, with 9-in snow drifts.

Plowed (PLW) or *swept* (SWEPT) indicates that a portion of the surface has been cleared. (PLW will not be reported when the entire runway has been plowed.) It is either bare, or it has depth, coverage, and conditions different from the surrounding area, which appear in the NOTAM. When known, the surrounding areas will be specified as *remainder* (RMNDR):

MHT 17/35 SWEPT 100 WIDE RMNDR THN SIR

Runway 17/35 has been swept 100 ft wide. The remainder of the runway at Manchester, New Hampshire, is covered with thin snow and ice—less than 1/2 in.

Treated runways are also reported. A *sanded runway* (SA) means the entire runway has been sanded. If less than the published dimensions have been treated, it is indicated in the NOTAM. Since a *deicing liquid* (DEICED LIQUID) or *solid* (DEICED SOLID) may be operational, it is reported in the NOTAM:

PSM 16/34 PTCHY THN IR SA 100 WIDE DEICED LIQUID/SOLID 100 WIDE

At Pease International, Portsmouth, New Hampshire, runway 16/34 has patchy, thin—less than 1/2 in—ice on the runway. The center 100 ft has been sanded, and both liquid and solid deicers have been used.

Snowbanks (SNBNK) or *berm* (BERM) are assumed to be at the runway edge. Should the runway be *plowed* or *swept*, the SNBNK is at the edge of plowed or slept areas:

0B5 PTCHY IR 3FT SNBNK BA NIL

Turners Falls Airport, Montague, Massachusetts, reports patchy ice on the runway, 3-ft snowbanks, and the braking action is nil.

NOTAM (D)s are issued for airports certificated under 14 CFR Part 139. This refers to *aircraft rescue and fire fighting* (ARFF) equipment availability. The ARFF index for each certificated airport is published in the AFD. When ARFF equipment is inoperative or unavailable and replacement equipment is not available immediately, a NOTAM is issued advertising the fact and indicating the current status of ARFF equipment. ARFF status affects only air carrier and air taxi operations.

Lighting aids consist of runway lights, approach lighting systems, pilot-controlled lighting, and obstruction lights.

MIT 12/30 RY LGTS PCL CMSND/MED INTST CONT UNTIL 0800 OTRW KEY 122.8 FIVE TIMES WI 5 SEC MED INTST

At Minter Field, Shafter, California, *pilot-controlled lighting* (PCL) has been commissioned for runway 12/30. Lights are on medium intensity continuously until 0800Z, after which they must be activated by keying 122.8 five times within 5 s for medium intensity.

The FAA now advertises obstruction light outages as NOTAM (D)s when they meet the following criteria:

- Located within 5 statute miles (SM) of an airport and are 200 ft AGL or higher
- 500 ft AGL or higher, regardless of distance from a public use airport
- Location is within 500 ft either side of the centerline of a charted helicopter route

OAK TOWER 1281 (405AGL) 4.4 NE LGTS OTS TIL 9812250001

In this example there is an unlighted tower 1281 ft MSL (405 ft AGL) 4.4 nautical miles (NM) northeast of the airport. Both MSL and AGL heights are reported when known.

NAVAIDs, communications outlets, and ATC services that are part of the NAS receive NOTAM (D) dissemination. When a component of NAS is expected to be shut down for more than 1 h, 30 min of radar, a NOTAM (D) is issued. NOTAM (D)s are issued for communications outlets when it is the only transmitting frequency or when all multiple frequencies are out of service at the outlet. For example, if an FSS loses 122.2, MHz, but another frequency is available at the same outlet, a NOTAM (D) will not be issued.

SAC TAR 30 NMR MYV045005 OTS TIL 9903092359

The Sacramento terminal area surveillance radar (SAC TAR) within a 30-NM radius of the Marysville VOR 045 radial at 5 mi (30 NMR MYV045005) will be out of service until March 9, 1999, at 2359Z (9903092359). This alerts pilots that radar services to aircraft not equipped with transponders will be unavailable. This might be significant for a nontransponder aircraft if an approach existed that required radar to execute the procedure.

SWR TACAN AZM OTS

This NOTAM advertises the fact that the Squaw Valley (SWR) TACAN azimuth is out of service. This would normally affect only military aircraft. The VOR and DME functions of the facility are operating normally.

UKI LLZ DME 15 OTS

The Ukiah localizer DME for Runway 15 is out of service. Be careful: Only the DME, not the localizer, is out.

SBA ILS DME UNMTN 0700-1400 DLY

The Santa Barbara ILS DME is *not monitored* (UNMTN) between the hours of 0700Z and 1400Z daily. This does not mean the DME is off or out of service, only that if the facility fails, a NOTAM would not be issued immediately; a pilot report would probably be the first indication of an outage. Most NAVAIDs within NAS are monitored.

MQO VOR 290-090 UNUSBL BYD 7 BLO 7000

The Morro Bay VOR is unusable between the 290 and 090 radials beyond 7 nm below 7000 ft MSL. If an airway, approach segment, or fix is within the specified area, it could not be used for navigation.

When a *nondirectional radio beacon* (NDB) that serves as an *outer compass locator* (LO) is not working, the NOTAM is formatted as follows:

LVK REGIA NDB/ILS LO OTS

In this case the REGIA NDB that serves as the LO for the Livermore, California (LVK), ILS is out of service.

DAG VORTAC UNMON/VOICE OTS

The Daggett VORTAC is in an unmonitored status, but it is not out of service. Most likely the monitor line is down because voice communications through the VOR are also out of service. Weather advisory broadcasts will not be available.

PDT APP 1330-0630 DLY

Pendelton, Oregon, approach control services operate daily between 1330Z and 0630Z.

ONP CESA 1600-0300 DLY

The Newport, Oregon, class E surface area is effective daily between 1600Z and 0300Z daily. Loran-C and Global Positioning System NOTAMs are only available from the FSS on request. However, GPS NOTAM (D)s are disseminated for areas where GPS signals are unreliable. For example:

!GPS 02/065 ZOA GPS UNRELIABLE WITHIN A 305 NM RADIUS OF CHINA LAKE VOR (NID) AT FL400. AFFECTED AREA WILL DECREASE WITH ALTITUDE TO A 38 NM RADIUS AT THE SURFACE. IFR OPERATIONS BASED UPON GPS NAVIGATION SHOULD NOT BE PLANNED IN THE AFFECTED AREA DURING THE PERIODS INDICATED. THESE OPERATIONS INCLUDE DOMESTIC RNAV OR LONG-RANGE NAVIGATION REQUIRING GPS. THESE OPERATIONS ALSO INCLUDE GPS STANDALONE AND OVERLAY INSTRUMENT APPROACH OPERATIONS WEF 9902251900-9902252030

The FAA issues *special traffic management program advisory messages*—flow control— as NOTAM (D)s. For example:

JFK TMPA SEE ATCSCC MSG EFF 9906101900-9906102300

There is a traffic management program advisory in effect for JFK between 1900Z and 2300Z on June 20, 1999. These NOTAMs alert the pilot that flow control is in effect, without the details of the restriction. Pilots will need to check with an FSS for specific information contained in the ATC systems command center advisory.

Flight service stations normally, upon request, provide *military operations area* (MOA) and *military training route* (MTR) activity. AFSSs have the capability of retrieving MOA and MTR activity by MOA name and MTR route number; therefore, pilots requesting their status should provide the briefer with the MOA name or MTR route.

Special data NOTAMs consist of the commissioning, decommissioning, or outage of weather reporting services, restricted areas, aerial refuelings, and airshows, acrobatic flights, parachute jumping, and airspace reservations. As well as AWOS/ASOS, these NOTAMs include *low-level wind shear alter systems* (LLWASs), *runway visual ranges* (RVRs), and *terminal Doppler weather radar* (TDWR). Remember, like all NOTAMs if the information is already published in the *Airport/Facility Directory* or *Notices to Airmen,* it will not be distributed in a NOTAM (D).

Changes to *departure procedures* (DPs) and *standard terminal arrival routes* (STARs) receive NOTAM (D) distribution. These NOTAMs are issued by the U.S. NOTAM Office (USNOF), NOTAM accountability USD. For example:

!USD 10/017 RNO WAGGE ONE DEPARTURE MINIMUM CLIMB RATES:
RWY 16L: 610 PER NM TO 9000
RWY 16R: 490 PER NM TO 9700 WEF 9810142040
!USD 10/018 RNO RENO TWO DEPARTURE TAKE-OFF MINIMUMS:
RWY 16L: STANDARD WITH A MINIMUM CLIMB OF 610 PER NM
TO 7500
RWY 16R: STANDARD WITH A MINIMUM CLIMB OF 490 PER NM TO
9000
WEF 9810142040

NOTAM (L)s

The criteria for NOTAM (L)s have changed significantly over the years. NOTAM (L)s have always been a bastard operation. However, with the introduction of DUATs, many previously NOTAM (L) criteria items now receive NOTAM (D) distribution—for example, all public use airports and most tower light outages. Tower light outages that do not meet the criteria for NOTAM (D) are disseminated as a NOTAM (L)—that is, any obstruction 200 ft AGL or less and more than 5 SM from a public use airport.

BEWARE SCUD RUNNERS AND HELICOPTER PILOTS   Most of these obstructions are not charted. Pilots can normally not expect to receive light outages for obstructions 200 ft or less and more than 5 mi from a public use airport. Pilots who conduct such operations would be well advised to contact the local FSS for NOTAM (L) obstruction light outages.

Only distributed locally, NOTAM (L)s advertise conditions or hazards that do not meet the criteria for a NOTAM (D). Landing area information that does not restrict or preclude the use of the runway is issued as a NOTAM (L): cracks and soft edges, people and equipment on or adjacent to the runway, frost heaves in a runway, bird activity, and taxiways.

NOTAM (L)s also pertain to items such as taxiway lights, airport rotating beacons, runway end identifier lights, threshold lights, VASI, PAPI, ATIS, TWEB, and UNICOM. They also pertain to a single-frequency outage when there is more than one frequency available. For example, if a tower's ground control frequency is out of service, the information would be issued as a NOTAM (L) because the local control frequency is operating normally.

The FAA defines *local dissemination* as the area affected by the aid, service, or hazard being advertised—that is, within the issuing FSS and the appropriate towers or centers when necessary.

FDC NOTAMs

FDC NOTAMs contain regulatory information. They include conditions that fall into the following categories:

- Interim IFR flight procedures
- Temporary flight restrictions
- Flight restrictions in the proximity of the U.S. president and others
- 14 CFR part 139 certificated airport condition changes
- Snow conditions affecting glide slope operation
- Air defense emergencies

- Emergency flight rules
- Substitute airway routes
- Special data
- Charting corrections
- Laser light activity

Interim IFR flight procedures include airway structure changes, instrument approach procedure changes, and airspace changes in general. Temporary flight restrictions involve disaster areas, special events generating a high degree of interest, and hijacking. Presidential aircraft include the aircraft and the entourage of the president, the vice president, or other public figures designated by the White House.

FDC NOTAM numbers are assigned consecutively beginning with 0001 each year. The year of issuance and the serial number are separated by a slant (9/1323, issued 1999 number 1323). FDC NOTAMs of a temporary nature are indicated by FI/T, and permanent information, by FI/P. When an FDC NOTAM affects a specific location, the LOCID for that facility is used; airway changes will contain the appropriate ARTCC's LOCID.

FDC NOTAMs are transmitted over the FAA's telecommunications system. Automated FSSs will provide FDC NOTAMs for the route regardless of the distance. During a standard briefing, FDC NOTAMs that are pertinent, on hand, and not yet published are provided. When published in the *Notices to Airmen* publication, they are only available on request. This is extremely important to National Ocean Service (NOS) chart users because NOS charts are not updated as often as other commercially available charts. NOS users should routinely check the *Notices to Airmen* publication.

FDC 9/1769 SMO FI/T SANTA MONICA, SANTA MONICA, CA.
NDB-B ORIG PROC NA.

This FDC was issued in 1999 (FDC "9"/1769). It was the 1769th FDC issued during that year (FDC 9/"1769"), and it pertains to the Santa Monica airport (SMO). The information contained in the NOTAM is temporary (FI/T, flight information of a temporary nature). The SMO NDB-B original issuance approach procedure is not authorized.

FDC 4/4047 STS FI/T SONOMA COUNTY SANTA ROSA, CA.
ILS RWY 32 AMDT 15...VOR RWY 32 AMDT 18...VOR/DME RWY 14
AMDT 1...CHANGE NOTE TO READ: WHEN CDSA NOT IN EFFECT,
EXCEPT OPERATORS WITH APPROVED WEATHER REPORTING SERVICE,
USE TRAVIS AFB /SUU/ ALTIMETER SETTING AND INCREASE ALL
DH'S AND MDA'S BY 390 FEET.

This FDC changes a note on the three approaches listed to increase the decision heights and minimum descent altitudes by 390 ft when the *class D surface area* (CDSA) is not in effect, unless the pilot has access to an approved weather reporting service. When class D airspace is not in effect (the tower is closed), the altimeter setting is not available. Pilots without approved weather reporting service must use the Travis AFB (SUU) altimeter. The use of a remote altimeter setting requires an increase in minimums. (Automated weather observation systems solve this limitation.)

!FDC 9/0964 SJC FI/P SAN JOSE INTL SAN JOSE CA. CORRECT
U.S. TERMINAL PROC VOL 2 OF 2 DATED 28 JAN 99 PAGE 401. GPS
RWY 30L ORIG...PLAN VIEW: HOLDING PATTERN AT SUNNE SHOULD
READ: 123 DEG INBOUND AND 303 DEG OUTBOUND.

This is an example of an error in the terminal procedures publications. Recall the FAA never has time to do it right, but always has time to do it over!

!FDC 8/1667 ZOA FI/T AIRWAY ZLA ZOA V25 SAN MARCUS /RZS/ VORTAC, CA TO POZOE INT, CA MEA 9500. POZOE INT, CA TO PASO ROBLES /PRB/ VORTAC, CA MEA 7000.

!FDC 4/2590 ZOA FI/T AIRWAY ZOA, NV. V165 MUSTANG (FMG) VORTAC, NV TO PYRAM INT, NV MOCA 10000.

These are examples of airway changes. Note the LOCIDs are the appropriate ARTCC LOCIDs. At the present, computerized briefings are unable to distinguish route of flight, except through the entire ARTCC airspace. Therefore, many nonappropriate NOTAMs will appear. The FSS controller sifts through the mass of data to provide only that information that is relevant to the flight. Pilots using DUATs are required to accomplish this task on their own.

!FDC 9/1081 ZOA CA ..FLIGHT RESTRICTIONS SAN FRANCISCO, CA. FEBRUARY 25-26, 1999. PURSUANT TO TITLE 14 SECTION 91.141 OF THE CODE OF FEDERAL REGULATIONS, AIRCRAFT FLIGHT OPERATIONS ARE PROHIBITED WITHIN THE FOLLOWING AREA(s) UNLESS OTHERWISE AUTHORIZED BY ATC.
    1.5 NMR BLW 1500 FEET AGL OF 37.47.35N/122.26.40W OR THE SAU118005 FROM 1930 LOCAL UNTIL 2200 LOCAL 02/25/99. PRIVATE RESIDENCE.
    1.5 NMR BLW 1500 FEET AGL OF 37.47.30N/122.24.35W OR THE SAU113007 FROM 2145 LOCAL 02/25/99 UNTIL 1030 LOCAL 02/26/99. FAIRMONT HOTEL.
    1.5 NMR BLW 1500 FEET AGL OF 37.47.21N/122.24.21W OR THE SAU111007 FROM 1030 LOCAL UNTIL 1330 LOCAL 02/25/99. GRAND HYATT HOTEL.

Note that in this presidential NOTAM, times are local. This proves the observation that the only thing consistent with the FAA is inconsistency!

The *Notices to Airmen* Publication

Published every 28 days, *Notices to Airmen* is now available on the Internet at www.faa.gov/ntap. This Web site also contains airshow information. Special events are listed at www.faa.gov/ats/ata/. Once placed in the publication, FDC NOTAMs are removed from the telecommunication circuits. NOTAMs of a permanent nature are carried only until published on the appropriate aeronautical chart or in the *Airport/Facility Directory*. Therefore, pilots must check this document for current information. The publication is divided into four parts.
    Part 1 is divided into three sections. Section 1 contains airway NOTAMs, sorted alphabetically by ARTCC and descending FDC NOTAM numerical order. Section 2 provides airport, facility, and procedural NOTAMs. These include chart corrections, airports, facilities, procedural NOTAMs, and any other information required, listed alphabetically by state. Section 3 contains general FDC NOTAMs—those not listed under a specific LOCID. This information consists of flight advisories and restrictions.
    Part 2 contains revisions to 14 CFR 95 minimum enroute IFR altitudes and changeover points.

Part 3 incorporates international Notices to Airmen. This section includes significant international information and data that may affect a pilot's decision to enter or use areas of foreign or international airspace. Foreign country data are listed alphabetically, followed by international oceanic airspace notices and U.S. overland/oceanic notices.

Part 4, graphic notices, contains special notices too long for other sections that concern a wide or unspecified geographical area or items that do not meet other section criteria. Information in Part 4 varies widely, but it is included because of its impact on flight safety.

Using the NOTAM System

It's still the responsibility of the pilot to obtain pertinent NOTAM information for a flight. Here's the bottom line. Request a standard briefing. Pilots can expect to receive NOTAMs on the status of NAVAIDs, airway changes, and airspace restrictions unless there is a temporary NOTAM system outage. In that case, pilots will be advised of this fact, and they will have to check with FSSs enroute and at the destination to ensure receipt of current NOTAMs.

Pilots using DUATs or other commercially available systems may have to decode and translate NOTAMs. Even with systems that have a decode function, the pilot is still responsible for interpretation and application. Remember the DUAT disclaimer: "Nonassociated FDC NOTAMs are available. Do you request them?..." These would include temporary flight restrictions and airway changes. Commercial systems do not provide NOTAM (L)s, GPS RAIM NOTAMs, or information contained in the *Airport/Facility Directory* or the *Notices to Airmen* publication. The contents of these documents still remain the responsibility of the pilot. Should any doubt exist about the meaning or intent of a NOTAM, consult a flight service station for clarification.

The procedures discussed apply equally to VFR and IFR flights. Only by understanding the system can pilots ensure that they meet their regulatory obligation of obtaining all available information.

Checking NOTAMs is like going to the restroom before a flight. We know we should, but sometimes it's just not convenient. Both oversights can lead to a very uncomfortable flight!

Some pilots think NOTAMs are a bunch of bull; sometimes they are.

BELIEVE IT OR NOT!   !CNO 09/004 CNO ARPT UNSAFE LOOSE BULL

STRATEGIES FOR INTERPRETING AND FLYING THE WEATHER

CHAPTER 21

ALTIMETRY, AIRSPEED, AND AIRCRAFT PERFORMANCE

The two primary flight instruments directly affected by air density are the altimeter and airspeed indicator. Air density also affects aircraft performance, including powerplant output. A pilot, to safely and effectively use the aircraft, must understand how the atmosphere affects these instruments and their aircraft.

The following terms represent various altitude and airspeed references. Pilots must understand and be able to apply these terms to aircraft operations:

- Indicated altitude
- True altitude
- Pressure altitude
- Density altitude
- Indicated airspeed
- Calibrated airspeed
- True airspeed

Each of the preceding altitudes and airspeed terms will be defined and applied in the following sections.

Three flight instruments operate from air density, two have already been mentioned. In addition to the altimeter and airspeed indicator, the vertical speed indicator operates off of differences in air density. These instruments are connected to the aircraft's pitot-static system. Refer to Fig. 21-1. Ram air pressure is connected to the airspeed indicator. The airspeed indicator measures the differential pressure between the ram air and static air pressure from the static ports. The altimeter and vertical speed indicators are also vented to ambient, or static, air pressure through the static ports, usually located on the side of the aircraft.

To prevent loss of air pressure should the static ports become blocked, for example, iced over, many aircraft are equipped with an alternate static source. In an emergency the alternate static source vents the static line to the cabin. If the alternate source is vented inside the aircraft, where static pressure is usually lower than outside static pressure, selection of the alternate source may result in the following instrument indications:

- Airspeed reads greater than normal.
- Altimeter reads higher than normal.
- Vertical speed indicates a momentary climb.

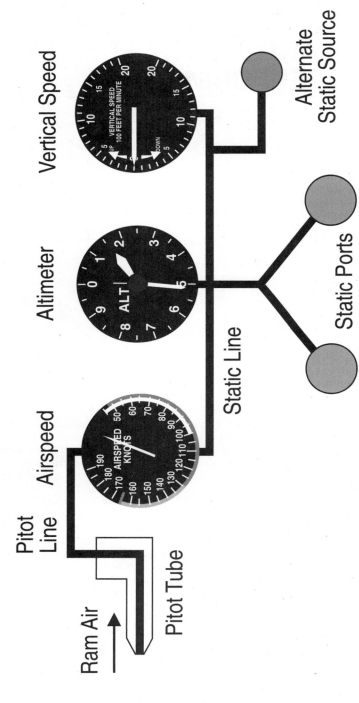

FIGURE 21-1 Pitot-static system. The airspeed indicator measures the differential pressure between the ram and static air pressure. The altimeter and vertical speed indicators are also vented to ambient or static air.

Airspeed and altitude read higher than normal because static pressure inside the aircraft is lower than the outside environment. Vertical speed momentarily indicates a climb because the instrument senses a momentary decrease in pressure, as if the aircraft were climbing into less dense air.

THE ALTIMETER

The altimeter measures the height of the aircraft. (This instrument's name was pronounced "al'te-mê'ter" by the British, but that pronunciation didn't appeal to Americans, so in the United States it's pronounced "al-tim'eter.") Similar to an aneroid barometer, the altimeter measures changes in pressure as the aircraft climbs or descends. Figure 21-2 shows the aneroid wafers. As pressure decreases, the wafers expand, and through mechanical linkage the change in altitude is reflected on the altimeter face. Conversely, when pressure increases, the wafers compress, and the altimeter indicates a descent. Without a means of adjustment, the altimeter indicates correct altitude only under standard atmospheric conditions of pressure and temperature.

Since atmospheric conditions vary continuously, the altimeter must be adjustable for a nonstandard environment. Most altimeters are equipped with an altimeter setting window, sometimes known as the *Kollsman window* (Fig. 21-2). (Paul Kollsman, a German-born aeronautical engineer, invented the method to correct the altimeter for nonstandard pressure in 1928. This was a major step in allowing pilots to fly solely by instruments.) The altimeter setting window allows the pilot to adjust the instrument for nonstandard pressure using the altimeter set knob (Fig. 21-2). Typically altimeters can be adjusted for pressures from 28.10 to 31.00 in Hg.

DEFINITION   The altimeter setting is a value determined for a point 10 ft above an airport (approximate cockpit height) that will correct an aircraft's altimeter to read airport elevation. It corrects the altimeter for both nonstandard pressure and temperature at the surface.

What happens if the pilot fails to adjust the altimeter enroute? A pilot departs Oklahoma City, Oklahoma (OKC), under higher-than-standard atmospheric conditions (30.05 in Hg) and selects that altimeter setting in the altimeter setting window (Fig. 21-3). The pilot climbs to an indicated altitude of 8500 ft, proceeds enroute to Denver (DEN), and maintains 8500 ft on the altimeter. Let's say that at Denver, Colorado, atmospheric pressure is 1 in lower than Oklahoma City (29.05 in Hg). As the aircraft proceeds west, pressure decreases. (In Fig. 21-3 this is represented by the lowering of the 850- and 700-mb constant pressure surfaces.) The aneroid wafers sense this decrease in pressure, expand, and indicate a climb. The pilot, maintaining 8500 ft, is in a continuous slight descent. Arriving over Denver, without setting the altimeter to the local altimeter setting, the aircraft's true altitude is only 7500 ft!

CASE STUDY   A final thought on altimeter settings. We were flying the Bonanza from Bakersfield, California, to Las Vegas, Nevada. A weather advisory was in effect for severe turbulence, and there were strong, gusty surface winds in the Las Vegas area. For some reason the *automatic terminal information service* (ATIS) at Las Vegas was unavailable. We contacted approach for clearance into class B airspace and reported our altitude as 8500 ft. Approach reported our *mode C* (altitude reporting transponder) readout as 8000 ft! The aircraft's mode C transponder reports pressure altitude. The air traffic control computer adjusts this reading, based on the altimeter setting, to display indicated altitude to the controller. We had not updated our altimeter setting since leaving Bakersfield. There was a 0.50-in difference at Las Vegas. The steep pressure gradient and lower pressure at Las Vegas were causing the strong winds and turbulence. From high to low, lookout below!

ALTIMETER

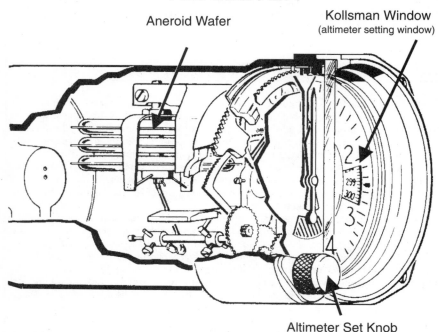

Aneroid Wafer

Kollsman Window
(altimeter setting window)

Altimeter Set Knob

FIGURE 21-2 Similar to an aneroid barometer, the altimeter measures changes in pressure as the aircraft climbs or descends.

Nonstandard temperature affects the altimeter, although not to the same degree as pressure. Since warm air is less dense than cold air, the aneroid will display a lower altitude in cold air than in warm. Indicated altitude corrected for temperature is often called *true altitude*. However, the strict definition of *true altitude* is the actual or exact altitude above mean sea level. For example, with a temperature 10° colder than standard and a pressure 0.5 in below standard (worst-case scenario), the aircraft would be approximately 500 ft below indicated altitude.

Figure 21-4 illustrates the effects of nonstandard temperature on indicated altitude. In an atmosphere where the decrease in temperature with altitude is standard, indicated altitude equals true altitude (center example Fig. 21-4). When the lapse rate is warmer than standard, indicated altitude is higher than true altitude; with the lapse rate colder than standard, indicated altitude is lower than true altitude (Fig. 21-4).

For most practical purposes we can ignore temperature-induced altimeter error. Exceptions would be flying under visual flight rules at low altitude—within 1000 ft of the surface—in extremely cold weather, especially at night or in low-visibility conditions where visual contact with terrain cannot be maintained. These factors explain the requirement for a local altimeter setting for instrument approach procedures, and they are one reason that, in the absence of a local altimeter setting, approaches are not authorized and result in higher landing minimums.

Pressure altitude is the altitude indicated on the altimeter when it is set to 29.92 in Hg or 1013.2 mb. (When the altimeter setting equals 29.92 in Hg, indicated altitude and pressure

EFFECTS OF CHANGING PRESSURE ON UNCORRECTED ALTIMETER

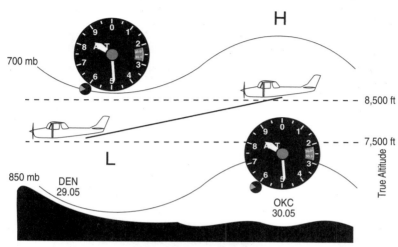

FIGURE 21-3   Failure to reset the altimeter can lead to serious altitude errors.

EFFECTS OF NONSTANDARD TEMPERATURE ON INDICATED ALTITUDE

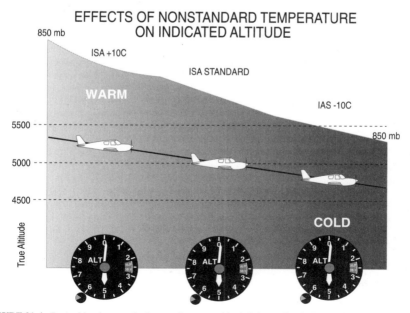

FIGURE 21-4   In a colder-than-standard atmosphere, true altitude is lower than indicated altitude.

INDICATED vs. PRESSURE ALTITUDE

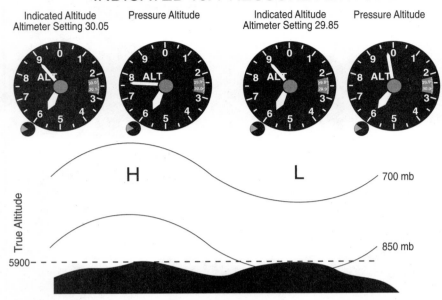

FIGURE 21-5   Most aircraft performance charts are based on pressure altitude.

altitude are the same—standard conditions.) Altitudes on the ISA chart (Fig. 1-3) are pressure altitudes. Pressure altitude is used when flying at or above 18,000 ft in the United States and in aircraft performance calculations to correct for nonstandard pressure.

As mentioned, most performance charts require the use of pressure altitude. In areas of high pressure, pressure altitude is lower than indicated altitude because the molecules of air are squeezed closer together. Conversely, in areas of low pressure, pressure altitude is higher than indicated altitude. This is illustrated in Fig. 21-5. In the example we have an airport elevation of 5900 ft. (Why 5900 ft? This is the elevation of the Truckee-Tahoe Airport in the California Sierra Nevada mountains. With the introduction of high temperature, it makes a good density altitude discussion.) Since atmospheric pressure changes about 1 in Hg per 1000 ft altitude, the difference between indicated altitude and pressure altitude can be determined from the difference between the airport's altimeter setting and 29.92.

Refer to Fig. 21-5. In the case of higher-than-standard pressure, the altimeter setting at Truckee is 30.05. The difference between the altimeter setting and standard is 0.13 in Hg, or 130 ft. (Note that each 0.10 of pressure equals 100 ft.) Pressure altitude is 5770 ft. In the lower-than-standard example, the difference is 0.07 in Hg, or 70 ft. This results in a pressure altitude of 5970 ft. The bottom line: In areas of higher-than-standard atmospheric pressure, pressure altitude is lower than standard—increasing aircraft performance; in areas of lower-than-standard atmospheric pressure, pressure altitude is higher than standard—decreasing aircraft performance.

In the United States, pilots operating below 18,000 ft MSL are required to fly with an altimeter setting of a station within 100 nautical miles of the aircraft. With the local altimeter setting in the window, the altimeter indicates altitude above mean sea level (MSL), indicated altitude. At and above 18,000 ft, pilots fly at pressure altitude—altimeter set to 29.92 in Hg, regardless of the local altimeter setting. Pressure altitude is used to eliminate station barometer errors and some altimeter instrument errors and to relieve air traffic controllers and pilots from having to issue and reset the altimeter at frequent intervals due to the high speeds typically flown at these altitudes.

FLIGHT LEVEL   When pressure altitude is being flown, altitudes are referred to as "flight level" (FL)—for example, "flight level one eight zero" (FL180 or a pressure altitude of 18,000 ft).

Under conditions of unusually low pressure, the lower flight levels (FL180 to FL200) may become unusable. This is because pressure altitudes conflict with the higher available indicated altitudes (17,000 ft). Extremely high pressures, above 31.00 in Hg, pose another problem. The altimeter cannot be set to the appropriate altimeter setting. Special procedures are then implemented. These procedures, along with lowest usable flight level, are contained in FARs and the *Aeronautical Information Manual* (AIM).

Pilots flying flight levels are actually flying a constant pressure surface. (Previously, we mentioned how the height of a constant pressure surface changes due to differences in temperature. Remember our axiom "when flying from hot to cold": Constant pressure surfaces are lower in cold air than in warm air.) The aircraft will climb or descend along with the height of the pressure surface. In the contiguous United States, terrain clearance is not a problem, even during periods of extremely low pressure and temperature. However, in other parts of the world the transition between indicated altitude and pressure altitude may be different. The following codes are used internationally to define the appropriate altimeter setting:

- QNH
- QNE
- QFE

QNH represents indicated altitude—the pilot is flying an altitude above mean sea level. QNE is pressure altitude. We may occasionally see QFE. QFE is the altimeter setting used so that the altimeter will read "zero" when the aircraft is on the ground.

CASE STUDY   While flying in England (low, flat terrain), we occasionally used QFE. I explained to an English pilot that in the States, we used indicated altitude or QNH. He thought that was kind of silly, until I explained that it would be rather hard—in fact, impossible—to crank 5000 ft out of the altimeter setting window for a landing at Denver, Colorado (elevation 5431 ft).

In a previous chapter we defined *density*. Air density is affected by atmospheric pressure, altitude, temperature, and to a minor degree humidity—moisture. *Density altitude* is pressure altitude corrected for nonstandard temperature. Density altitude affects aircraft performance, which is discussed later in this chapter.

AIRSPEED

Like the altimeter, air density affects the airspeed indicator. The airspeed indicator is a sensitive, differential pressure gauge. The instrument indicates the speed of the airplane through the air—not necessarily over the ground. The airspeed indicator measures the difference between the impact pressure of the air in the pitot tube and the static pressure of the ambient air surrounding the airplane, as shown in Fig. 21-1. A pointer on the face of the instrument registers the difference in pressure.

FACT   Not "pilot tube," the *pitot* tube is named after its eighteenth-century French hydraulic engineer inventor Henri Pitot.

Indicated airspeed (IAS) is the direct instrument reading obtained from the indicator, uncorrected for installation and instrument error or for variations in atmospheric density.

Calibrated airspeed (CAS) is indicated airspeed corrected for installation and instrument error. These errors are usually greatest at low airspeeds, with flaps deployed. Normally, the manufacturer provides a table to convert indicated to calibrated airspeed for different speed ranges and flap settings.

True airspeed (TAS) is calibrated airspeed corrected for atmospheric density—pressure and temperature. Because of the normal decrease in pressure and temperature with altitude, for a given indicated airspeed, true airspeed increases as altitude increases. The pilot normally uses a flight computer to calculate true airspeed based on pressure altitude, temperature, and calibrated airspeed.

AIRCRAFT PERFORMANCE

All aircraft have limitations, including turbojet aircraft, which often fly at the edge of their performance envelope. Nonstandard conditions can critically affect performance. The pilot's task is to determine aircraft performance based on actual or forecast conditions. This may require the information in Fig. 1-3, International Standard Atmosphere.

Certain performance charts require the pilot to determine temperature at altitude relative to ISA. For example, the forecast temperature over Reno at flight level 300 (30,000 ft) is $-42°C$. Figure 1-3 indicates that the standard temperature for that pressure altitude is $-45°C$. Certain flight computers can be used to determine ISA temperatures; these computers have a true altitude computation window. With the scales aligned (10 on the outer scale with 10 on the inner scale), standard temperature is read under pressure altitude. Under 30,000 ft, the pressure altitude, -45, appears. Therefore, the forecast temperature is ISA $+3$, or 3° warmer than standard.

STRATEGIES

Now let's put the theory into practical application. Aircraft operations can be divided into takeoff, climb, cruise, descent, and landing.

High temperatures at high-altitude airports produce high-density altitude. Atmospheric pressure and humidity are also factors; however, temperature and elevation are paramount. Surface temperature forecasts are not normally available, but maximum temperatures typically occur beginning by midmorning and continuing through late afternoon. Arrival and departure times must be planned based on aircraft performance.

Under certain conditions the pilot can get the airplane off the ground only to be trapped in ground effect. *Ground effect* is the temporary gain in lift during flight at altitudes of about the wing span of the airplane, due to compression of the air between the wing and the ground. If this is allowed to continue beyond the end of the runway, there is only one result—an airplane accident!

The following is a typical scenario: A pilot attempts to take off from a relatively short runway, during midday, with an aircraft at or above maximum gross takeoff weight. The aircraft accelerates and lifts off, and it may initially establish what the pilot perceives as a positive rate of climb. However, the aircraft then begins to settle, and with no runway remaining, it impacts the terrain. When this happens, pilots often blame downdrafts or wind shear when weather conditions are conducive to neither.

CASE STUDY   The pilot, along with three other pilots, was planning to fly to an aviation safety seminar in a Piper Arrow. The pilot set the flaps to 10°, and he applied full power to utilize a rolling takeoff. The pilot rotated near the end of the 3000-ft strip. The airplane climbed to about 40 ft before settling into trees. An investigation revealed that the airplane was about 175 pounds over maximum gross takeoff weight, the density altitude was 4500 ft, and the runway had a 0.5 percent upslope grade. The *pilot's operating handbook* (POH) recommended a 25° flap setting, with full power before brake release for an obstacle clearance takeoff.

The NTSB determined the probable cause as the pilot's inadequate planning, preparation, and takeoff technique.

There were about six factors that contributed to this incident, but the elimination of any one of the six might have prevented the accident. As we'll see in Chap. 27, Risk Assessment and Management, a pilot's fate often rests in his or her own hands.

Most flight computers provide for density altitude calculations. There are also graphs and charts that give this information, including Table B-1, Pressure Altitude, and Fig. B-1 on density altitude. Recall that Fig. 21-5 was designed to illustrate conditions at the Truckee-Tahoe Aairport, elevation 5900 ft. From the AWOS, we obtain the following data: temperature, 85°F; altimeter setting, 29.53 in Hg. What is the density altitude?

From Fig. B-1 we see that 85°F equals 29°C. From Table B-1, an altimeter setting of 29.53 in Hg results in a pressure altitude of 6300 ft (5900 + 400 = 6300). Now, from Fig. B-2 we can read off the density altitude as 9300 ft.

CASE STUDY   I have flown Cessna 150s out of Bryce Canyon, Utah (elev. 7586), and Grand Canyon, Arizona (elev. 6606), and a Cessna 210 out of Mammoth Lakes, California (elev. 7128)—density altitude 10,000 ft. There is no additional hazard in such operations as long as we calculate, and do not attempt to exceed, aircraft performance.

After calculating that the aircraft has sufficient performance for conditions—which we all know is required by regulations—it's a good idea to determine an abort point on the runway. If the aircraft is not airborne and climbing out of ground effect, this point should allow the pilot to come to a safe stop on the remainder of the runway. Remember, aircraft performance data are based on a brand new airframe and engine and perfect pilot technique.

CASE STUDY   The pilot of a Beech A36 Bonanza attempted to depart Las Vegas, New Mexico, from a 5000-ft strip, density altitude 8800 ft. After aborting two takeoffs, due to what the pilot reported as engine problems, the pilot tries a third time. The pilot tried to get the airplane airborne at the end of the runway but was unsuccessful. The airplane hit the ground and rolled off the runway. The airplane POH indicated that the takeoff distance should have been about 2000 ft. This is an example of a pilot's failure to determine the required takeoff distance, then abort the takeoff at midfield when sufficient speed was not attained.

Normally, a pilot of a carbureted, fixed-pitch propeller airplane will adjust the mixture, prior to takeoff, for maximum engine RPM. Some pilots run the engine to maximum power during the runup, then adjust the mixture. (Cessna recommends this procedure when density altitude reaches 3000 ft. However, always check the pilot operating handbook for the airplane you're flying for recommended procedures.) I prefer to adjust the mixture to a rough setting at runup RPM, then during the initial takeoff roll make final mixture adjustments. This procedure puts extra workload on the pilot during this critical phase but less stress on the airplane.

A pilot of a nonturbocharged, constant speed propeller airplane will use the procedure described above, except lean for maximum manifold pressure. The pilot must realize that manifold pressure will be reduced proportionally to density altitude. For example, with a density altitude of 6000 ft, the pilot could expect only 24 in of manifold pressure at full throttle.

An advantage of a turbocharged engine is the development of sea level power at altitude. The turbo Cessna 182 POH recommends leaning to smooth engine operation.

On airplanes equipped with a fuel flow indicator—for example, the Bonanza—the pilot can lean to density altitude takeoff fuel flow. This setting is a rough estimate, and smooth engine operation should always be maintained.

Multiengine pilots must consider that airport density altitude may exceed the inoperative engine ceiling. For example, an early-model nonturbocharged Cessna 310 has a single-engine ceiling of approximately 6800 ft, Piper Aztec 7500 ft, and Piper Seminole only 5000 ft. And, remember, these altitudes apply to standard conditions. Density altitude in the summer, especially in the west, can often exceed these values!

Pilots cannot allow themselves to be lured into a false sense of security because the airport is at a relatively low elevation.

CASE STUDY   The pilot of a Piper Saratoga, a turbocharged airplane, elected to depart from a 1913-ft grass strip in Texas, field elevation less than 1000 ft, air temperature 34°C. The density altitude was approximately 3000 ft. The grass was wet and from 4 to 6 in tall. The aircraft struck power lines that were about 30 ft tall. The aircraft was slightly over gross weight. Based on the POH, with a dry runway, the aircraft required 1700 ft to clear a 50-ft obstacle. For a wet, sod runway, the distance required would be approximately 2300 ft!

Most density altitude accidents have involved improper takeoff procedures. In nearly half, the pilot attempted takeoff in excess of maximum gross weight. About one-third exceeded the climb performance of the aircraft. All were preventable.

Climb and cruise performance are also affected by density altitude.

CASE STUDIES   We had filed an IFR flight plan from Lancaster's Fox Field in California's Mojave Desert to Ontario, California. I requested 7000 ft and planned to go through the San Fernando Valley because of lower minimum altitudes. The clearance came back, "Cleared via the Cajon two arrival; climb and maintain 11,000." The surface temperature was 30°C, and the Cessna 150 was not going to 11,000 that day! After negotiating with a rather perturbed ground controller, I received my requested routing. Pilots must know their aircraft's performance and not allow ATC, or anyone else for that matter, to push them into an untenable—in this case unobtainable—position.

Let's consider another example: The average temperature from Oakland to South Lake Tahoe at 13,500 ft is −04°C. Figure 1-3 indicates that the standard temperature for 13,500 ft is −10°C. The forecast temperature is 6° warmer, which is above standard. The air is less dense than standard, so aircraft performance will be less than performance charts advertise. Based on the above conditions, density altitude is about 15,000 ft!

My 1966 Cessna 150 had a book ceiling of 12,650 ft. We planned to traverse the 9943-ft Tioga Pass in California's Sierra Nevada range. The winds were out of the northeast at only 10 kn, resulting in a slight downdraft from the wind flowing up the east slopes and down the west slopes as we approached the pass. Temperature was slightly above standard, and in combination with the wind, the aircraft wouldn't climb out of 9,500 feet. We had to proceed north along the west slopes of the mountains to Ebbett's Pass at 8732 ft, where we were able to safely cross the mountains.

It is important to gain the required altitude prior to reaching the pass or crest of the mountains, and with sufficient room to make a comfortable course reversal, if required. One technique is to approach the ridge line at a 45° angle. If the ridge cannot be cleared, the pilot usually has to make only a 45° turn away from the mountains to reach lower terrain. How can we tell if we'll clear the crest? If we're above the crest, the terrain beyond will appear to be rising in relation to the crest of the mountains.

TABLE 21-1 Wind Chill Factors

Air temp., °C	Wind 10 kn	Wind 20 kn	Wind 30 kn	Wind 40 kn
20	15	13	12	12
10	1	−3	−5	−4
0	−9	−15	−19	−20
−10	−23	−32	−35	−37
−20	−35	−46	−50	−52
−30	−48	−60	−66	−69

Aircraft performance values with advertised service ceilings of 13,100 ft are based on standard conditions. Differences are usually not significant unless the pilot is operating at the limit of the aircraft's performance. Unfortunately, this occurs every year with pilots that attempt to cross the Sierra Nevada or Rocky Mountains in conditions well above standard. Some pilots can't understand why an aircraft with a service ceiling of 13,100 ft can't climb above 12,000 ft with an outside air temperature of 0°C. Density altitude at 12,000 ft and temperature 0°C is 13,100 ft! And, this is an ideal case. A runout engine, poor leaning technique, being over gross weight, and the possibility of turbulence and downdrafts would further decrease performance. Some have mused that you can walk across the Rockies and Sierra Nevada on the wreckages of Cessna 172s and Piper Cherokees. You can't fool Mother Nature; attempts can be fatal.

Descents are normally not a problem. But don't forget to richen the mixture, if required.

A pilot should use the same indicated airspeed values for high-density altitude approach and landing as those used at sea level. Why? Instrument indications are based on air density. But airplane speed over the ground (ground speed) will be higher. This is one reason for longer ground rolls at higher-density altitudes. Multiengine pilots, especially, should know density altitude prior to landing. If a go-around should be required, the pilot should know if the airplane is above its single-engine ceiling.

The bottom line: Calculate existing density altitude, and don't expect the aircraft to exceed its design limitations. Most flight computers have density altitude functions. Various tables and charts have been developed for this purpose. Appendix B, Graphs and Charts, contains a density altitude chart. There is no excuse for not obtaining and applying this information.

High and low temperatures affect aircraft operation and performance. Temperature also affects our operation and performance. It can get mighty uncomfortable on cold and windy days, and normally temperature decreases with altitude. The *wind chill factor* is the cooling effect of wind on temperature. Wind has the effect of lowering the apparent temperature sensed by our body. This is graphically depicted in Table 21-1, Wind Chill Factors.

If the air temperature is reported as 10°C with a wind of 25 kn, what is the wind chill factor? From Table 21-1 we see that the wind chill at 10° and 20 kn is −3 and 30 knots −5. Interpolating, we calculate the wind chill factor at −4°C.

CASE STUDY On most light, single-engine aircraft, cabin heat is obtained by routing outside air through a muffler shroud that surrounds the engine exhaust stacks. This raises the temperature of the air by about 20°C. To illustrate, consider a trip flown from Reno to Lovelock in Nevada. At 7500 ft the *outside air temperature* (OAT) was −18°C. The air entering the cabin was between 0° and 5°! Bring a wind breaker or jacket, and dress appropriately when conditions warrant.

CHAPTER 22
LOW CEILINGS AND VISIBILITIES

Low ceilings and visibilities are directly related to stability and vertical motion. We typically think of clouds and precipitation associated with this hazard. Precipitation in itself is not a hazard—after all, aircraft are Sanforized; they don't shrink. (Waterproof is another story. I've been rained and snowed on inside the cabin of more than one airplane. General Aviation waterproofing, especially with older aircraft, leaves something to be desired. But that's another story.) Heavy rain, snow, freezing rain, and hail present specific aviation hazards. Each will be reviewed. We have also included the hazards of sand and volcanic ash. The chapter concludes with strategies to avoid the dangers of low ceilings and visibilities. As we'll see, the IFR pilot is not immune to these hazards.

We now have enough background to apply the "theory" to actual flight situations. We will begin with low ceilings and visibilities, which are significant hazards, especially to the pilots restricted to visual flight rules. A VFR-only pilot needs a natural horizon or ground contact to maintain control of the aircraft. Even though the private pilot's curriculum includes basic, or emergency, instrument flying, the training is not sufficient for any length of time in instrument meteorological conditions. An all to frequent causal factor in accidents is: "Continued VFR flight into adverse weather." However, armed with knowledge of these phenomena, a pilot can avoid the hazard and nullify the risk.

The IFR pilot is also concerned with ceiling and visibility, especially in the takeoff, approach, and landing phases. Few general aviation aircraft are equipped with fully hands-off, auto landing systems. Therefore, low ceilings and visibilities in takeoff and landing phases are critical to IFR operations. No segment of aviation is immune.

CASE STUDY A U.S. Air Force T-43 (Boeing 737) crashed in low ceiling and visibility conditions in Bosnia, killing all aboard included Secretary of Commerce Brown.

Of those pilots involved in low ceiling accidents, more than half had instrument ratings. Also, more than half were fatal. In about one-third of the cases, there was no record of a weather briefing. A significant number of accidents occurred to IFR pilots descending below approach minimums.

In accidents involving fog, more than half were fatal. In more than two-thirds, the weather was reported as IMC and there was no record of a weather briefing. Again, IFR pilots were not immune. Descending below minimum and improper approach procedures were significant causal factors.

Like fog, rain and other low visibility accidents closely match.

In accidents involving snow, about one-third were fatal. However, rather than ceiling and visibility, the primary accident causal factor was loss of control on landing.

Although low ceilings and visibilities are often found together, there are situations in which only one or the other phenomenon exists. Therefore, we have three situations: low ceiling with good visibility, low visibility with no significant ceiling, and the worst case, both low ceiling and visibility.

A *ceiling* is defined as a *broken* (5/8 to 7/8 coverage) or *overcast* (8/8 coverage) layer. A ceiling restricts or precludes a VFR pilot from climbing or descending through the layer.

Ceiling heights are either measured or estimated. Reported ceilings may not be representative of the surrounding area. A sage axiom in aviation is that "weather reports may not be accurate, but they're official." This doesn't mean reports should be ignored; however, they should be viewed with caution, especially during marginal conditions or at night.

To review, there are two types of ceilings. One measures a distinct cloud base. This is the first broken or overcast layer aloft in a METAR report. The second is an indefinite ceiling. An indefinite ceiling is reported when an obscuring phenomenon (fog, snow, rain, smoke, etc.) hides the whole sky. It, in fact, is the vertical visibility into a surface base obscuration.

Normally, a pilot can expect ground contact while flying in areas with a reported partial obscuration. This is why they are not considered ceilings. Penetrating a partial obscuration VFR requires that the appropriate horizontal visibility be maintained for that airspace.

Visibilities aloft are most often reduced by rain, snow, dust, smoke, and haze. And, with reduced visibilities, usually less than 5 mi, the apparent visibility looking toward the sun can be almost nil! Pilot reports are the only source of slant range and inflight visibility.

Obstructions to vision, caused by fog, haze, dust, and smoke, are reported in weather observations when visibility is less than 7 mi. When these phenomena exist with visibilities 7 mi or greater, a remark might describe the condition. Pilot weather briefers sometimes use the term *unrestricted* to describe visibilities of 7 mi or greater. This sometimes causes confusion. A pilot told "visibility unrestricted" might respond, "What about the haze?" Thus, the phrase "visibility unrestricted" does not imply that smoke, haze, dust, or even fog are not present, just that visibility is 7 mi or greater.

Areas of low ceilings and visibilities are graphically depicted on weather depiction charts, advertised in the AIRMET Bulletins, reported in METAR reports, and forecast in TAFs.

FOG

Fog, a cloud with its base on the ground, occurs most frequently in coastal regions due to the large amount of water vapor available. However, fog can form anywhere. The rapidity with which fog can form makes it especially hazardous. It is not unusual for visibility to drop from more than 3 mi to less than half a mile in a few minutes.

Fog forms by any atmospheric process that does one or both of the following:

- Cools the air to its dew point
- Raises the dew point to the air temperature—adds water vapor

Fog, ground fog, and ice fog describe the same condition. *Ground fog,* normally less than 20 ft deep, reduces visibility horizontally rather than obscuring the sky. Usually localized, formed by radiational cooling, ground fog tends to dissipate rapidly once clearing begins. True *ice fog* forms in cold weather at temperatures around −30°C from radiational cooling and exhibits the same characteristics as radiation fog. (Recall from Chap. 8 that FZFG is reported in METARs any time the temperature is below 0°C.)

Radiation fog forms when air cools from contact with the ground and becomes saturated. This occurs at night and tends to be most dense, with lowest visibilities, around sunrise. Clear skies, light winds (less than 5 kn), high relative humidity, and stable air are favorable conditions for the formation of radiation fog. On calm, cool nights, high pressure traps low-level moisture. Look for fog over lakes and in river valleys. (Where do we like to build airports? Of course, in river valleys and next to large bodies of water!) We have already discussed how low water vapor content in the atmosphere increases terrestrial radiational and cooling; therefore, dry air aloft enhances the formation of radiation fog. Radiation fog shows up well on visual satellite imagery. Overcast skies, strong winds, low relative humidity, and unstable air prevent or retard the formation of radiation fog.

Radiation fog tends to be patchy and shallow, usually burning off by midmorning. It tends to form in valleys after moisture has been added at the surface from passing storms. As high pressure—clear, stable conditions—build into an area, circumstances are right for the formation of radiation fog. This condition can become persistent in California's central valley during winter and early spring. (In this area the low fog is known as "tule fog" with tops usually less than 3000 ft. *Tule* [tôô'lê] is a Spanish word for bulrushes, a marsh plant that grows during this season.) Figure 10-1 is a visual image that shows tule fog in California's central valley. Note how well the lateral extent of the fog shows up on the satellite picture. The satellite clearly indicates that the fog extends from the central valley through the Carquinez Straits into the San Francisco Bay.

Other areas conducive to the development of radiation fog are the valleys east the Cascade Range in the Pacific Northwest, Great Basin, Snake River Valley, and valleys of the Appalachian Mountains.

Radiation fog does not usually form until the second day after storm passage. This is due to atmospheric mixing following the front. A strong inversion develops that locks in moisture at lower levels, causing radiation fog to form. Zero-zero conditions over widespread areas can persist for days or even weeks, until the moisture evaporates, or another storm system moves through the area.

IFR pilots, normally, will have no difficulty operating in conditions caused by radiation fog, as long as they don't mind flying above zero-zero surface conditions. Landing minimums might not prevail until late morning or afternoon, if at all, especially in valleys with fog depths of 2000 to 3000 ft. Why does this occur? During the winter months, days are short; this reduces the length of time for burnout (evaporation) by the Sun. Longer nights provide longer periods for the fog to develop. The VFR pilot will be delayed until the condition dissipates. Some pilots routinely move their aircraft to mountain, or higher, elevation—that is, to airports above the fog layer during winter months.

Expect radiation fog the following morning when at dusk the skies are clear, the wind is light, and the temperature-dew point spread is 8°C or less. Since fog evaporates from the edges, radiation fog tends to dissipate from the mountains toward the center of the valley. Unfortunately, many airports are located at the lowest elevations in the center of the valley, where fog remains the longest.

Advection fog forms when moist air moves over colder ground or water. The air cooled from below becomes saturated. Unlike radiation fog, advection fog can form under an overcast. This is a persistent condition along the Pacific coast during the summer months. The prevailing onshore flow moves the layer into coastal sections and valleys. It is usually deepest and farthest inland at sunrise and retreats toward the ocean during the day. Figure 10-3 illustrates the effectiveness of visual satellite imagery for determining the extent of the layer—as long as higher cloud layers are not present. It is especially useful in areas without weather reporting stations. Winds of 5 to 15 kn tend to cause low ceilings rather than fog.

One feature of the coastal stratus is the eddying effect of coastal winds, in particular the *Catalina eddy*. This is a southeasterly current of air, along the immediate coast of southern California, flowing contrary to the main northwesterly flow. Its vortex is in the vicinity of Catalina Island. This southeasterly current disappears above the temperature inversion, which caps the layer of cool, moist air in which the eddy is embedded. It usually develops when a rather strong current of air flows southeastward over the ocean near the coast with falling pressures inland. Coastal stratus increases rapidly—often unforecast—and is carried

FIGURE 22-1   Upslope fog can be widespread, and it will persist as long as favorable conditions continue.

further inland with bases as well as tops higher than normal. A solid overcast well inland may persist for several days.

Upslope fog forms as air is forced upward, expands, and cools adiabatically. Moist air must be forced upslope, which requires a wind of 5 to 15 kn. This condition occurs during winter and spring in the midwest where the terrain rises steadily from the Gulf of Mexico to the Rockies. Figure 22-1 shows an area of extensive upslope fog that has developed over the Texas panhandle.

Upslope fog can be widespread and will persist as long as favorable conditions continue. During upslope fog conditions, the VFR pilot is pretty much out of luck. The IFR pilot might not be much better off. He or she might encounter IFR landing minimums, but the condition often exists over areas the size of several states. A legal IFR alternate might be beyond the aircraft's range. Here again, satellite imagery is useful in determining the extent of upslope fog, as long as higher cloud layers are not present.

Expect upslope fog and low clouds when wind blows up sloping terrain and the temperature-dew point spread is small enough that air will cool to saturation.

Rain-induced fog, also known as *frontal fog,* occurs when warm rain falls through cooler air, evaporates, and condenses, forming fog; it can be dense and will persist as long as the rain continues. Winds must generally be light. This condition is usually associated with weak cold fronts that are stationary, warm, or shallow. A potential for clear icing exists in areas of rain-induced fog. Satellite imagery is of no use in determining the extent of the fog because of the presence of higher cloud layers.

Expect rain-induced fog when the temperature-dew point spread is small and continuous rain or drizzle is falling in areas associated with stationary, warm, and weak cold fronts.

Steam fog develops as cold air moves over warm water. Evaporation from the water takes place and saturation occurs. Low-level turbulence develops as the warm water heats lower levels, creating a shallow layer of instability. Also known as *evaporation fog,* steam fog occurs in the autumn in cold climates over lakes, such as the Great Salt Lake in Utah and the Great Lakes.

Expect steam fog when wind is blowing from a cold land area over warm water.
Look for the development of fog when wind is blowing from a warm surface over a cold surface and the dew point is warmer than the colder surface. In general, be prepared for fog when the temperature-dew point spread is 3°C or less and decreasing. But remember, it usually takes more than just a close temperature-dew point spread for fog to develop. Expect little if any improvement in visibility when fog exists below overcast skies or when rain or drizzle is forecast to continue.

At times, when conditions are favorable for fog, a very low cloud layer will form. This is especially true over flat terrain when winds exceed about 15 kn. These foglike clouds, formed in stable air with smooth flying conditions, often exist together with fog.

HAZE AND SMOKE

Haze is caused by the suspension of extremely small, dry particles invisible to the eye but sufficiently numerous to reduce visibility. The term *haze,* when it is combined with smoke is often used to describe conditions in metropolitan areas—sometimes translated as hack and cough. Large anticyclones—high-pressure cells—can dominate the southeast United States, trapping haze and pollutants, especially in industrial areas. Above the haze layer, visibilities are unrestricted and temperatures cool, resulting in a much more comfortable flight.

Surface inversions form at night due to radiational cooling. As the surface cools, the lower layer of the atmosphere cools, while temperatures aloft remain unchanged. Inversions aloft often trap haze and smoke, sometimes reducing visibility aloft to less than 3 mi.

Strong inversions over cities or industrial areas can trap haze and smoke, reducing visibility to less than 1 mi. Landmarks are all but invisible. At altitude, dense haze often appears solid, like a cloud layer, in the distance; this occurs with high relative humidity. When the sun strikes the layer, light waves scatter, causing the layer to appear white. This accounts for apparent inconsistencies in surface observations. Reports might report clear or partially obscured skies with visibilities of 3 to 5 mi in haze and smoke. A pilot looking into the sun often has no forward visibility or natural horizon. Pilots caught in such situations, technically VFR, have become spatially disoriented and lost control of the aircraft.

Figure 22-2 illustrates a typical hazy day, under mostly clear skies, and a stable air mass. Notice how the terrain becomes obscure in the background and even appears as if it might be clouds. The light band in the middle of the picture is an area of relatively clear air above the surface-based haze layer and below another haze layer aloft.

Haze can obscure mountain ranges in the west, making pilotage navigation difficult. Under hazy conditions, it's all but impossible to distinguish one range from another. In the midwest and east, haze can obscure thunderstorms.

Haze and smoke are usually restricted to an area below 5000 ft, although layers can extend to above 10,000 ft. During the devastating September 1987 forest fires in California, smoke tops were reported as high as 19,000 ft, with visibilities aloft zero.

Recall from Chap. 8 the Redding, California, observation that reported an indefinite ceiling 3000 ft; the sky was completely hidden by smoke and haze. A pilot flying in this area can expect to maintain ground contact only within about 3000 ft of the surface. This situation could be extremely dangerous for the VFR pilot. Should the pilot climb above 3000, he or she would be in IFR conditions without ground contact, and most probably without the natural horizon. Pilots attempting to operate in similar conditions have lost aircraft control with fatal results.

Expect restrictions to visibility due to haze and smoke in areas of stable air and light winds. Visibilities may be less than 1 mi with strong, low inversions. Surface heating often breaks through the inversion, and along with increased winds, visibilities tend to improve during the afternoon. This is a typical case during the summer and fall in the Los Angeles

FIGURE 22-2   Haze trapped in inversions aloft can reduce visibility at altitude as well as on the surface.

Basin. Expect little if any improvement in visibility when haze or smoke exist under an overcast.

CASE STUDY   The following observation was taken at March AFB, California: KRIV...RMK CREPUSCULAR RAYS SW. *The Glossary of Meteorology* defines *crepuscular* rays as "literally, 'twilight rays'; alternating lighter and darker bands [rays and shadows] that appear to diverge in fan-like array from the sun's position at about twilight." Towering cumulus produce this effect, especially with haze in the lower atmosphere. This would seem a rather complicated way of saying HAZY TCU SW (hazy with towering cumulus southwest).

DUST, SAND, AND VOLCANIC ASH

Dust and *blowing dust,* a combination of fine dust or sand particles suspended in the air, can be raised to above 16,000 ft by the wind. Visibilities, surface and aloft, can be at or near zero. Because of its fine particles, dust can remain suspended for days once the wind subsides. Figure 22-3 shows blowing dust being raised from the surface in California's Mojave Desert by strong winds.

Blowing sand, made up of particles larger than dust, usually remain within a few hundred feet of the surface. It can also reduce visibility to near zero. But when the wind subsides particles fall back to the surface and visibility improves rapidly.

CASE STUDY   A pilot approaching Lovelock, Nevada, skeptical of a reported visibility of 2 mi in blowing sand, reported his flight visibility was 20 mi. Upon landing, however, he concurred, stating the tops of the blowing sand were at 200 ft above ground level (AGL).

FIGURE 22-3   Because of its fine particles, dust can remain suspended for days once the wind subsides.

When blowing sand reduces visibility to less than 3 mi, the condition is advertised in a SIGMET.

Expect reduced visibilities due to blowing dust and sand when strong winds are forecast and terrain surface is barren. This is especially true in deserts and the Great Plains.

Dust devils, reported as *dust/sand whirls* in the METAR code, are whirlwinds that form on clear, hot days with light winds. They have diameters of 10 to 50 ft and extend from the surface to several thousand feet. Wind speeds within the rotation vary from 25 to more than 75 kn. Dust devils are capable of substantial damage, but the majority are small.

CASE STUDY   We flew into a dust devil doing pattern work at Lancaster's Fox Field in California's Mojave Desert. The encounter was equivalent to light to moderate turbulence—it shook the Cessna 150, and the low pressure in the vortex caused both windows to pop open!

Volcanic ash became a significant weather phenomenon in the continental United States with the eruption of Mt. Saint Helens in 1980. A typical eruption can release 500 million tons of ash into the atmosphere. Since 1980 there have been a hundred or so eruptions worldwide, over a dozen in the United States—mostly Alaska. Volcanic ash is difficult to distinguish from cloud. Volcanic ash is seldom detected on ATC or aircraft radar. However, NEXRAD, the NWS's advanced Doppler radars, where available, should provide a new tracking capability for volcanic ash.

Volcanic ash consists of fine particles of rock powder, blown out from the volcano. The particles remain suspended in the atmosphere for long periods, extend well into the Flight Levels, and may drift thousands of miles. Ash power, up to one-eight inch in diameter can be very abrasive. Volcanic ash can be extremely destructive to aircraft leading edges, windscreens, and engines. Turbojet aircraft engines are especially susceptible.

CASE STUDY   Royal Dutch airlines flight KLM867, a Boeing 747, encountered an ash cloud from the eruption of Mt. Redoubt in 1989. The aircraft lost all four engines to flame out! ATC vectored the aircraft back to Anchorage. During the descent, the crew was able to restart two of the engines and make a safe emergency landing. KLM867 was one of several aircraft damaged by the eruption. Such encounters occur every year.

Ash on the airport is another significant hazard. Ash can be blown into the air, causing reduced visibilities and aircraft damage. Wet ash on the runway reduces braking action, and crews are trained to reduce or avoid the use of reverse thrust.

Volcanic ash is also a hazard to general aviation aircraft. Ash affects windscreens and leading edges. It blocks air filters and pitot systems. We have already discussed the hazard to turbojet engines—many of which are used on corporate twins and helicopters. And the hazards of reduced visibilities and slippery runways cannot be overlooked.

Like icing and thunderstorms, the key to volcanic ash hazard is avoidance. Do not fly into or climb though ash clouds. If you inadvertently enter an ash cloud, reverse course and descend. Finally, report the encounter to alert other pilots of the hazard, and assist officials to track the cloud.

Reports and forecasts of volcanic ash are available in METAR reports, SIGMETs, Center Weather Advisories, and the Volcanic Ash Forecast Transport and Dispersion Chart.

PRECIPITATION

Snow, drizzle, and rain are the most common forms of precipitation restricting visibility. Of these, snow is usually the most effective in reducing visibility. Heavy snow frequently reduces visibility to near zero. On the other hand, rain rarely reduces visibility to below 1 mi and has a tendency to wash haze and smoke out of the air. Cold rain will even remove fog from the air. Conversely, drizzle often accompanies fog, haze, and smoke, resulting in lower visibility than occurs in rain.

Figure 22-4 shows a downburst over the mountains west of Reno, Nevada. Notice how the precipitation obscures the terrain, another hazard associated with this phenomenon. The AIRMET Bulletin warns pilots about mountain obscuration in clouds and precipitation.

Rain or drizzle on the windscreen can reduce the pilot's visibility considerably below visibility outside the aircraft. Dry snow does not adhere to the windscreen and does not greatly affect visibility through the windscreen.

Blowing snow, like blowing sand, reduces visibilities near the surface when strong winds blow over freshly fallen snow. Visibility can be near zero close to the surface, with rapid clearing after the wind subsides. Expect reduced visibilities due to blowing snow when strong winds are forecast and terrain surface is snow covered.

Whiteout is an atmospheric optical phenomenon in which the pilot appears to be engulfed in a uniformly white glow. Neither shadows, horizon, nor clouds are discernible; sense of depth and orientation is lost. Whiteout occurs over an unbroken snow cover and beneath a uniformly overcast sky, when light from the sky is about equal to that from the snow surface. Blowing snow may be an additional cause.

To the VFR pilot, whiteout is disastrous. Snow-covered terrain, an overcast, and already reduced visibilities are a strong no-go indicator. At the very first sign, the pilot's only option is 180° turn to, what it is hoped, better conditions.

CASE STUDY   IFR pilots are not immune to whiteout. In one case, a pilot's first destination did not have an instrument approach. It was snowing, and the pilot reported "whiteout" conditions. The pilot diverted to an airport with an instrument approach.

FIGURE 22-4 Precipitation obscures the terrain—another hazard associated with this phenomenon.

Reported weather was visibility one-half variable between 1/4 and 1 mi in snow, indefinite ceiling 600 ft. Radio contact was lost after the second missed approach. Wreckage was located an hour later.

The following is strictly speculation. Because the crash occurred after the declaration to miss the approach, it appears the pilot decided to miss after the missed approach point. The transition between instrument and visual flight under these conditions is extremely difficult, especially with a single-pilot operation. The pilot may have acquired ground contact straight down, but slant range and apparent whiteout conditions would have precluded visual contact with the approach environment and airport.

Many commercial operations require two pilots just for this scenario. One pilot stays on the instruments, and the other looks for the airport. I had a similar incident while training an instrument pilot. We were flying at night in rain and fog, ceiling and visibility were at minimums. At the missed approach point, my student started the missed approach. At that point I caught a glimpse of the approach lights. We were certainly not in a position to land and continued the missed approach procedure.

TERRAIN OBSCUREMENT

We have already touched on the subject of terrain or mountain obscurement. Haze and smoke typically obscure distance landmarks, making visual navigation difficult. Precipitation, especially when heavy or accompanied by low visibilities, can obscure terrain. This was illustrated in Fig. 22-4. More typically, terrain is obscured by clouds.

To alert pilots to the potential dangers of cloud or precipitation obscurement, the NWS advertises these conditions through a weather advisory. The AIRMET Bulletin contains a section in AIRMET SIERRA addressing mountain obscuration. This forecast outlines areas where extensive or widespread obscurement is expected.

It might be a good idea to review the difference between IFR conditions—which are also advertised in AIRMET SIERRA—and mountain obscuration. Typically advisories for IFR conditions apply to flat, nonmountainous, terrain. For example, coastal valleys, California's central valley, the high plateau of the intermountain region of the west, and the Great Plains. Advisories for mountain obscurement apply to the mountains—coastal mountains, Cascades, Sierra Nevada, Rockies, Appalachian, and so on.

Mountain obscurement infers that VFR flight through valleys is usually possible, but visual flight over mountains or through mountain passes may not be possible. Figure 22-5 illustrates mountain obscurement. With good visibility, this hazard should not present a problem. Notice in Fig. 22-5 that VFR conditions are good in the valleys around the mountain, with the peak hidden in clouds. These clouds are sometimes classified as *cumulogranite*.

CASE STUDY   Returning from Oshkosh, we planned a leg from Pueblo, Colorado, to Farmington, New Mexico. Low pressure was affecting the weather, and an advisory for mountain obscurement was in effect. We successfully negotiated the Sangre de Cristo range between Pueblo and Alamosa, Colorado. VFR flight was no problem in the San Luis Valley. However, the San Juan range between Alamosa and Farmington were not to be conquered. We tried to top the clouds without success. "Plan B" was to fly south toward Taos where the terrain was lower. West of Taos a safe pass allowed us to proceed, uneventfully—except the increased flight time made that last cup of coffee at Pueblo most uncomfortable, to Farmington. We always had an out. If the pass was closed, both Taos and Alamosa could be used as alternates. (One more comment from a practical point of view: Throughout the flight we updated weather, our route of flight, and estimated time of arrival with flight service.)

FIGURE 22-5   Clouds associated with mountain obscurement are sometimes classified as *cumulogranite*.

Figure 22-4 illustrates how precipitation, especially when heavy, can dramatically reduce visibility and obscure terrain. The heavy rain is occurring. The mountains are clearly visible in the rain-free air but almost completely obscured in the background. Visibility outside the rain area is excellent. These rain showers should be avoided.

SPECIAL VFR

Basic VFR weather minimums within controlled airspace are designed to allow pilots to fly visually. This requires a visual horizon or contact with the ground and enough visibility to see and avoid terrain, obstructions, and other aircraft. Basic VFR weather minimums were developed in the 1930s when aircraft speeds averaged between 50 and 150 kn, and the minimums do not take into account the increased speeds of today's aircraft. Cloud clearance and visibility are, as the regulations state, minimums. "Minimum" does not necessarily equate to "safe."

Pilots have been known to become disoriented and lose control of the aircraft in visibilities as much as 5 mi. The pilot in command is still responsible for determining if the flight can be safely conducted, based on his or her experience and capabilities. I recommend low-time pilots obtain training from a competent instructor in operations under special VFR, and in actual special VFR weather conditions, should they wish to operate in this environment.

Aircraft can be safely flown visually in less than basic VFR conditions required for controlled airspace. *Special VFR* allows pilots to operate in weather with a visual horizon or contact with the ground and enough visibility to avoid terrain and obstructions. Under such conditions someone else, air traffic control, must ensure separation from other aircraft. Although ATC provides separation from other aircraft, it is still the pilot's responsibility to maintain terrain and obstruction clearance.

These operations must be conducted in accordance with special VFR weather minimums. Special VFR applies only to operations within low-density, surface-based class B, C, D, and E airspaces. Special VFR is prohibited within high-density class B airspace, which is indicated on aeronautical charts by the phrase "NO SVFR" in the airport data block. Weather minimums for an airplane operating under special VFR are clear of clouds and 1 statute mile visibility. To operate at night under the provisions of special VFR, the pilot and aircraft must be equipped and certified for IFR.

Prior to departure, or before entering less-than-basic VFR weather conditions, the pilot must obtain a clearance from the ATC facility with jurisdiction over the airspace. IFR operations have priority over special VFR.

CASE STUDY   A rather shaky voice called Van Nuys ground and requested taxi instructions. The visibility was less than basic VFR, and the controller asked what type of clearance the pilot would like. The pilot replied, "I don't have one of those," presumably referring to an instrument rating. The controller responded, "That's OK, I've got plenty."

Special VFR must be requested by the pilot. ATC is not allowed to suggest the procedure.

It might seem that special VFR is of little practical use. This is not the case, however, because special VFR has a specific and practical application. Special VFR is intended to allow a pilot to depart or enter surface-based controlled airspace when conditions are less than basic VFR but safe enough for contact flying.

This often occurs in metropolitan areas where surface visibilities are reduced to less than 3 mi but remain above 1 mi, in haze, smoke, and fog. Usually at a few thousand feet AGL, visibilities improve significantly. This procedure can also be used to allow a pilot to depart controlled airspace into uncontrolled airspace, with its reduced VFR requirements.

Pilots must be careful—special VFR can be a clearance to nowhere. Notice that the provisions of special VFR do not relieve the pilot from maintaining appropriate minimum safe altitudes as required by regulations.

CASE STUDIES   Fresno was reporting visibility 1 mi, ceiling 400 overcast. A pilot departed special VFR and flew 15 mi before tangling with some high tension wires.

Departing Long Beach, California, one morning with a surface visibility of 2 1/2 mi, we requested and received a special VFR clearance. We were cleared out of class D airspace to the north: "Climb and maintain VFR conditions. Report VFR on top or leaving class D airspace." As soon as we topped the haze, we had almost unlimited flight visibility, reported on top, and were cleared to leave the frequency. Pilots must report on top or leaving the surface-based controlled airspace. ATC is providing separation, so when the pilot fails to report, the airspace must be sterilized until the aircraft is located. This causes extensive, unnecessary delays to all.

Arrivals are conducted the same way. The pilot reports over the airport, or other prominent landmark, in VFR conditions above the visibility restriction, and contacts the control facility for a special VFR clearance.

CASE STUDIES   Special VFR might be more efficient than an IFR approach. Santa Barbara was reporting 10 mi visibility, ceiling 500 ft broken. Even though the ceiling was less than VFR, we could see the runway. We requested special VFR and made a straight-in approach, rather than a 15-mi round trip, which would have been required to execute the ILS approach. On another occasion our destination was Crescent City, California. The VOR was out of service, eliminating an IFR approach. Weather was, 1 mi visibility ceiling, 500 ft overcast. Finding a hole by the coastline, we requested a special VFR clearance through the Crescent City FSS. Because of the low ceiling and visibility, I slowed the Mooney to approach speed, about 100 kn. There was no sane reason to be blasting along in these conditions at 160 kn. We knew the tops were at 1500 ft; if the ceiling or visibility dropped, we could have climbed through the clouds to on top. ATC was providing separation, so there shouldn't have been any other aircraft in that airspace.

I was flying from Ontario to my home airport Whitman Airpark in the Los Angeles Basin one rainy afternoon. I had obtained clearance through the Burbank class C airspace when the controller advised visibility was 2 1/2 mi in rain and fog and requested my intentions. I requested and received a special VFR clearance out of class C airspace to the northwest. I reported leaving class C airspace and landed at my destination, which was in class G airspace.

Delays, extensive at times, should be anticipated. A pilot can never count on making it in special VFR; a solid VFR alternate should always be within reach.

STRATEGIES

Every year, pilots become lost, even lose control of the aircraft, flying in reduced visibilities. Often conditions can be improved by climbing to a higher altitude. Once above the layer, slant range visibility is usually greater with a distinct horizon preventing disorientation. The seemingly obvious assumption—the closer to the ground the better to see it—often isn't true.

CASE STUDY   The pilot departed a southern California airport during the evening for a flight to Monterey. Arriving in the Monterey area about midnight, the pilot found the airport overcast. The pilot landed on a highway south of the Reid-Hillview Airport in the San Francisco Bay area. There was no record of the pilot checking enroute for a weather update, which would have revealed the onset of coastal stratus! There is no rational reason for this accident to have occurred. The pilot landed safely and was applauded by some. Unfortunately for aviation, opponents of Reid-Hillview cited it as another reason to close the airport.

In the preceding example even the required fuel reserve may not have been adequate. A 45-min fuel reserve doesn't make any sense with the nearest suitable alternate 50 min away. What would have happened if the coastal stratus extended beyond the airplane's range? Most likely the accident would have been fatal! How could this incident have been prevented? Simple: Update weather enroute with flight watch or flight service.

When conditions are favorable for coastal stratus, VFR pilots should plan arrivals and departures during the afternoon hours. If this is not possible, moving the aircraft to an airport a few miles inland will often allow a morning departure.

CASE STUDIES   It's often hazy in the southern portion of California's San Joaquin Valley. A pilot flew for over an hour looking for the Porterville Airport. The pilot ran out of fuel and landed in a field. In spite of talking to UNICOM, this pilot could not locate the airport. A week or so later another pilot had a similar experience, under the same weather conditions. Unable to locate the Porterville Airport, this pilot called air traffic control. Specialists at the flight service station provided a *direction finder* (DF) steer to Bakersfield. The FAA has personnel and equipment ready and waiting. But the pilot does have to ask for assistance. Let ATC help before an incident becomes an accident.

This pilot was operating in the southeast United States, in an area of reduced visibility in haze and fog. The non-instrument-rated pilot had difficulty maintaining heading and altitude. Approaching the airport to land, the aircraft was seen in a descending right turn. The pilot crashed into trees and then hit the ground. The crash was fatal. Visibility in the vicinity of the crash site was estimated to be between 1 and 2 mi in fog. The pilot told several others of concern about flying in hazy conditions.

This pilot's first error was flying in conditions below weather minimums. The pilot was obviously uncomfortable about the flight. If that's the case, why depart in the first place? What can be so important as to endanger your life? If any doubt exists, don't go! Second, the pilot could have climbed to a higher altitude, above the haze and fog layer. This may have allowed the pilot to maintain aircraft control. Once on top, the pilot could have obtained the services of ATC to locate or assist in finding a suitable landing site.

With low ceilings and visibilities, the greatest hazard occurs when both prevail. We were on a flight from Shreveport, Louisiana, to Mineral Wells, Texas. Ceilings were below 1000 ft, but visibility was unlimited. In the sparsely populated Texas countryside, it was easy to maintain legal distance from objects on the ground and avoid towers and power lines. As we flew south of Dallas, the clouds lowered to the ground. In the non-instrument-equipped Cessna 150, there was only one course of action—reverse course and land. We were delayed another day before we could proceed west.

Pilots with even limited instrument capability—I'm speaking of aircraft and equipment—have the option of filing IFR. Always the best situation is to "get into the system" before reaching IFR conditions. Here are some examples.

When I first came into the FAA, I was assigned to the Lovelock, Nevada, FSS. I would routinely fly from Lovelock to Van Nuys, California. My Cessna 150, although equipped and certified for IFR, did not have the capability to fly IFR over the Sierra Nevada mountains. But, if the only weather was coastal stratus, I would file VFR to Palmdale and IFR from Palmdale to Van Nuys. It's always easier to pick up a profiled clearance before reaching congested terminal airspace.

On one of our trips to Oshkosh, the leg from Mason City, Iowa, to Madison, Wisconsin, was plagued with marginal VFR weather toward the destination. I elected, rather than to try to fly under the weather or attempt to pick up IFR in the Madison area, to file IFR from Mason City. The weather was clear, so we departed VFR and picked up the clearance with center on climbout. It was a good plan. With marginal weather in the Madison area, controllers were refusing popups and instructing VFR aircraft to remain clear of class C airspace. Pilots were advised to contact the next sector for traffic advisories. This is not a criticism of the controllers. There is only so much airspace, especially in terminal areas.

Plan A doesn't always work. After remaining overnight at North Platte, Nebraska, we planned to continue westbound to Cheyenne. A front had passed through the previous day, and the ground was moist. Where is the airport? Of course, next to the river where the fog and low clouds are usually the worst. The weather to the west was good, and I obtained a special VFR clearance. After takeoff, it became apparent it wasn't going to work. We had no option except to land, return to the airport office, and file an IFR flight plan. Twenty minutes later we were on our way. (When you have time to spare, go by air.)

An AIRMET was in effect for mountain obscuration in Idaho. During the briefing for the flight from Winnemucca, Nevada, to Idaho Falls, the briefer advised that "V-F-R flight was not recommend" (VNR). This was a little silly because our flight took us through the Snake River Valley, not the mountainous areas that were obscured. Upon opening our flight plan and, again, updating weather enroute, we were VNR'd. The valley was perfectly fine.

The next morning we prepared to fly from Idaho Falls to Billings, Montana. Weather at both ends was good, with no weather advisories for the route. What we found over Yellowstone National Park is illustrated in Fig. 22-6. As we approached the park, the valleys were fogged in, and there were several layers aloft. Most of the mountains were obscured. Because of the minimum enroute altitudes and freezing levels, IFR (on purpose) was not an option. We picked our way through, at times climbing to over 13,000 ft in the Cessna 172. We always had the option to reverse course and return to Idaho Falls or divert into West Yellowstone. We were lucky and on the north side of the park were able to descend below the clouds, verify with flight watch that Billings weather was good, and proceed on to our destination.

This scenario has the potential for disaster. Navigation was difficult because of sparse navigational aids and terrain obscurement. As is my practice, I had a dead-reckoning course planned and calculated. Throughout the flight I was evaluating options based on weather and fuel: proceed, return, or divert.

Oh, a side note: We were not VNR'd on the flight to Billings. There were no weather advisories. The point is that the existence, or lack, of an advisory does not preclude, or guarantee, a safe flight. Pilots must evaluate each flight separately, based on the reported and forecast weather, and their aircraft, training, and experience.

STRATEGY   Because of the stacking effect (discussed in Chap. 8), a cloud layer may appear solid. How can you tell if the layer is solid or has breaks? If there are breaks, the sun shows through in the distance. This can be seen in Fig. 22-6. Note the sun shining through to the surface in the distance. But remember there are no absolutes. Clouds may merge, and breaks may be too small to safely negotiate VFR.

Recall that we mentioned that no segment of aviation is immune from low ceiling and visibility accidents. The typical IFR scenario is descending below minimums. It has happened to the airline and military, as well as general aviation pilots. We mentioned the Air Force T-43 crash in Bosnia. Investigators discovered, among other things, that the aircraft was not properly equipped for the approach to be used and was not flown at proper airspeeds. It's foolhardy to attempt an approach without the appropriate equipment functioning normally. This includes appropriate and current charts.

FIGURE 22-6   Pilots must use caution when flying between cloud layers; they aren't called "sucker holes" for nothing!

Any time during the approach, if anything isn't normal, abandon the approach! Southwest Airlines recently lost a Boeing 737, which reportedly was high and fast, on its approach to runway 8 at Burbank, California. Runway 8 is relatively short for jets. A number of years ago I flew jump seat into Burbank on a Boeing 727. Believe me, there is no room for error.

The dos and don'ts of flying in low ceilings and visibilities:

- Do obtain a standard weather briefing.
- Do update weather enroute.
- Do get into the IFR system before encountering poor weather.
- Do ask for help whenever the situation becomes doubtful or uncertain.
- Do have options and a plan for each option.
- Don't let the briefer make the decision (VNR or no VNR).
- Don't fly below legal or personal minimums.
- Don't run out of options; land before you do.

> **CASE STUDY**   A Navion pilot requested and received a special VFR clearance from the Salinas, California, tower out of class D airspace. The pilot then reported clear of class D airspace. The airplane crashed 8 nm from the airport in rising terrain. At the time of the accident, the weather was reported: KSNS 151647Z 29010KT 5SM BR OVC004 15/13 A2996

Was this pilot legal? Weather minimums, yes. Minimum safe altitudes, no. Unfortunately, this pilot will not be able to represent himself at the NTSB hearing; he was fatally injured.

If used, these strategies can prevent low ceiling and visibility accidents. Taking everything into account, all are preventable. Almost one-third of the pilots involved in low ceiling and two-thirds involved in low visibility accidents had no record of a weather briefing. And, although there are no specific records, it's doubtful they obtained weather enroute.

When air is cooled to its dew point or moisture is added to raise the dew point, or both, fog can form. Temperature and dew point within 3°C is an indicator for the possible development of fog. FSS briefers normally provide temperature/dew point under these conditions. Some FAA aviation weather broadcasts use this criterion. The formation of fog, however, requires more than just a close temperature/dew point spread. However, when conditions are right, a low, dense fog layer can form over parts of the airport, undetected by either automated or manual observations—especially during the hours of darkness. Such conditions dictate additional vigilance and caution.

CHAPTER 23
TURBULENCE

Typically aviation accidents are not directly related to turbulence. However, turbulence is a contributing factor to spatial disorientation, loss of control, and structural damage. Therefore, a sound understanding of turbulence, where it is likely to occur, and strategies to reduce its effects are essential.

The intensity of turbulence is, to some degree, affected by aircraft type and flight configuration. United States Air Force studies have shown the following to generally increase the effects of turbulence:

- Decreased weight
- Decreased air density
- Decreased wing sweep angle
- Increased wing area
- Increased airspeed

Wind shear (WS) and especially *low-level wind shear* (LLWS) have a significant effect on aviation operations. Wind shear is any change in wind speed or direction, either vertically or horizontally, over a relatively short distance. LLWS is defined as wind shear that occurs within 2000 ft of the surface. LLWS is divided into two categories: convective and nonconvective. *Convective*, or *thunderstorm-related, LLWS* will be addressed in Chap. 25, along with our discussion of thunderstorms hazards.

WIND SHEAR AND TURBULENCE Pilots should realize there is a difference between wind shear and turbulence. Wind shear causes airspeed changes in one direction (plus or minus, but not both) with a sustained change in vertical speed. Turbulence, on the other hand, causes airspeed fluctuations (both plus and minus) with no appreciable, or only a momentary, change in vertical speed.

Nonconvective LLWS can be caused by:

- Fronts
- Low-level jet streams
- Terrain
- Valley winds
- Sea breezes
- Lee-side effect
- Inversions

A Santa Ana or similar foehn-like winds can produce severe turbulence. The occurrence, exact location, and intensity of turbulence and LLWS are difficult to predict.

TURBULENCE INTENSITIES

Classifications for the intensity of turbulence can be found in the *Aeronautical Information Manual* and *Aviation Weather Services*; however, I prefer the following:

Light

A turbulent condition during which your coffee is sloshed around but doesn't spill, unless the cup's too full. Unsecured objects remain at rest; passengers in the back seat are rocked to sleep.

Moderate

A turbulent condition during which even half-filled cups of coffee spill. Unsecured objects move about; passengers in the back seat are awakened by a definite strain against their seat belts.

Severe

A turbulent condition during which the coffee cup you left on the instrument panel whizzes by the passengers in the back seat. The aircraft might be momentarily out of control, but you don't let on. Those passengers not using their seat belts are peeling themselves off the cabin ceiling.

Extreme

Usually associated with rotor clouds in a strong mountain wave or a severe thunderstorm, extreme turbulence is a rarely encountered condition where the aircraft might be impossible to control. The turbulence can cause structural damage. Your passengers are becoming concerned by the beads of sweat on your brow, and your white knuckles, and the new frequency and new transponder code you have just selected—121.5 and 7700.

MECHANICAL TURBULENCE

An object placed in any moving air current impedes the flow, causing abrupt changes in wind direction. As the current closes in behind the object, eddy currents develop leeward of the obstruction. This turbulence is caused by the obstruction and not by any meteorological phenomena inherent in the air itself. Turbulence caused in this manner is termed *mechanical.*

Air flowing through mountainous terrain is forced upward on the windward side and spills downward over the leeward side, as illustrated in Fig. 3-5. The degree of turbulence induced by the mountains depends upon the shape and size of the mountains, the direction and speed of the wind, and the stability of the air. Downdrafts on the leeward side may be dangerous and can place an aircraft in an attitude from which it may not be able to recover.

As well as gustiness, surface winds in excess of 20 kn indicate moderate or greater mechanical turbulence, especially over rough terrain. Favorable conditions for turbulence exist just before, during, and after storm passage, especially when winds blow perpendicular to mountain ridges. For example, consider a wind from 230° at 45 gusting to 90 kn that occurred at Mammoth Lakes, California, just after storm system passage with winds perpendicular to the rugged California Sierra Nevada mountains.

Recall the Mammoth surface observation that resulted in a 35-kn crosswind component for the sustained speed, and 70 kn for the gusts! Obviously, this airport would not be suitable for landing. Most aircraft manuals specify a maximum demonstrated crosswind component. Every year pilots attempt to test these values—some pilots even succeed. Each pilot should know his or her limitations and that of their aircraft. As a flight instructor, I always give my students specific crosswind limitations, always with an alternate should they be exceeded. We'll revisit this issue in Chap. 27, in the section on personal minimums.

Strong mountain waves can generate moderate or greater mechanical turbulence. However, this is not always the case.

CASE STUDIES I encountered a mountain wave in California's Owens Valley while flying a Cessna 150. With cruise power and attitude the airplane rode the wave, at the rate of 500 ft/min, from 8500 ft to 13,500 ft, and back down again. The ride was absolutely smooth!

Notwithstanding the previous example, the November 1977 *Approach* magazine reported: "A Navy T-39 trainer was flying a low-level, high-speed navigational training route in mountainous terrain when it encountered severe turbulence. Gust acceleration loads were so high that aircraft design limits were exceeded, resulting in separation of the tail...." All aboard were killed. The article went on to say, "Mountain waves should never be taken lightly. In addition to the T-39 crash, mountain waves were identified in the crash of a C-118 and in extensive damage to a B-52....While this type of turbulence is obviously critical to traditional low fliers like helicopters, all aircraft are susceptible."

THERMAL TURBULENCE

Daytime heating causes rising air currents that produce *thermal turbulence*—also called *convective turbulence*. Thermal turbulence usually occurs within 7000 ft of the surface in stable or conditionally unstable air. This means that vertical movement requires an initiating force, in this case surface heating. In stable or extremely dry air, skies remain clear. Should air parcels reach the lifted condensation level, saturation occurs, and stratocumulus, or fair weather cumulus, clouds will form. These clouds are most often scattered and rarely become overcast. This is illustrated in Fig. 23-1. There are rising currents in the clouds and descending currents in the clear air. Flight below the clouds will be turbulent, and above the clouds, smooth. The air is stable since there is little vertical development.

Should the air be conditionally unstable—a parcel of air that becomes unstable on the condition that it is lifted to the *level of free convection* (LFC)—cumuliform clouds form, which can develop into air mass thunderstorms. Figure 23-2 illustrates this situation. Rather than the scattered stratocumulus or fair weather cumulus of Fig. 23-1, towering cumulus have developed. In this case, expect rain showers and thunderstorms to form.

Uneven surface heating affects the landing approach. Rocky terrain, plowed fields, and paved areas produce predominantly upward currents. These currents force the aircraft above the normal glide path, resulting in overshooting the touchdown point. Trees, rivers, lakes, and green fields produce predominantly downward currents. These currents allow the aircraft to fall below the normal glide path, causing it to undershoot the touchdown point. Often both types of currents are present, requiring the pilot to make several corrections on the final approach course.

FIGURE 23-1 Rising air currents in the clouds and descending currents in the clear result in turbulence below and smooth air above the clouds.

FIGURE 23-2 Should the air be conditionally unstable, cumuliform clouds may form, which can develop into air mass thunderstorms.

Although thermal turbulence rarely becomes severe, it can be extremely uncomfortable and annoying. Because thermal turbulence is caused by surface heating, it can usually be avoided by flying before midmorning or waiting until late afternoon. Otherwise, the only remedy is to climb above the turbulent layer, which might be marked by clouds. A word of caution to the VFR pilot: If you elect to fly above the clouds, be careful not to get caught on top. Should the air mass be conditionally unstable, clouds can build at an alarming rate and close up even faster.

CASE STUDIES Flying a new Cessna 150 from Kansas to California I found myself flying between Winslow, Arizona, and Needles, California, during the afternoon. Skies were clear, and winds aloft, light and variable. I had to climb to 12,500 ft to reach smooth, cool air. Descending into Needles, the turbulence was continuous light to moderate below about 11,000 ft, surface winds were calm.

On another occasion, I was flying one afternoon between Oklahoma City and Amarillo, Texas, in air that was conditionally unstable. As is my habit, I flew above the clouds in clear, smooth, cool air. The cumulus appeared to top out at about 9000 ft. So I thought I'd climb to a cruising altitude of 10,500. Something was strange. I was in what appeared to be level flight, but only indicating 60 kn in my Cessna 150. A scan of the instrument panel revealed the problem. I was not in level flight but instead in a climb attitude above a sloping cloud deck! Topping the clouds was impossible, and I was forced to descend and bounce the rest of the way to Amarillo at 4500 ft.

WIND SHEAR–INDUCED TURBULENCE

Wind shear–induced turbulence is caused by frontal systems, and is associated with clear air turbulence and thunderstorms. Because of their significance we will discuss them in subsequent sections. In this section we will review three types of wind shear–induced turbulence: inversion-induced, evaporative cooling, and Kelvin-Helmholtz (K-H) wind shear turbulence.

Inversion-induced wind shear turbulence develops along the boundary between cool air trapped near the surface and warm air aloft. The turbulence tends to be strongest in valleys during morning hours when temperature differences are greatest. Moderate or greater turbulence may be encountered penetrating the layer. After an initial outside air temperature rise during the climb through the haze, the air on top will be clear and cool. It is possible to have several haze or smoke layers trapped in inversions aloft.

CASE STUDIES In the Los Angeles Basin, a haze boundary often develops with a Santa Ana condition. Warm, dry desert air overruns haze trapped in cooler, moist marine air. Moderate or greater wind shear turbulence can be expected penetrating the transition zone. Takeoff and landing can be hazardous with the boundary close to the runway, often marked by a distinct transition between clear and smoggy air.

A similar condition occurs in the San Francisco Bay area. Strong offshore winds override cool marine air, causing an area of strong wind shear. At times the shear line reaches the surface. At the Oakland airport, surface winds for runway 27 can be northeast at 15 gusting to 20 kn and for runway 29 northwest at 5 kn. Needless to say, severe LLWS can develop.

Another type of wind shear, which like inversion-induced wind shear occurs in a vertical, rather than a horizontal, plane, is *evaporative cooling turbulence*. Evaporative cooling turbulence develops in areas of precipitation. This usually occurs in a dry environment with convective activity. Precipitation evaporates and cools the air, causing downdrafts. A pilot penetrating these areas will encounter wind shear turbulence, which can be severe. Turbulence can be avoided by circumnavigating areas of precipitation.

K-H turbulence is caused by wind shear. It develops when there is a proper balance of wind shear and stability. Stability is important; if the atmosphere is too unstable, mixing occurs and the waves do not develop. With large wind speed changes over a short distance in a stable atmosphere, the ingredients are right for K-H turbulence.

Under proper conditions an undulating wave forms. If the shear is strong enough, the crests overrun the troughs. With sufficient moisture a herring-bone-shaped cloud develops. Wavelengths are much shorter than in a mountain wave, usually less than 2 mi. The implications are severe wind shear and, like the mountain wave, strong updrafts on the windward side of the wave and strong downdrafts on the lee side.

There is evidence to show that K-H wind shear and turbulence have either caused or been a contributing factor in several aircraft accidents. Like wave rotor clouds, any time a pilot sees a cloud indicating an overturning of the air, such as K-H clouds, use caution. This is especially true in mountainous regions. K-H clouds are not a type reported by weather observers, although perhaps they should be.

Dust devils (PO in the METAR code) occur in hot, dry climates. Although not usually hazardous, an encounter can produce moderate turbulence due to wind shear.

CASE STUDY   We flew into a dust devil doing pattern work at Lancaster's Fox Field in California's Mojave Desert. The encounter was equivalent to light to moderate turbulence; it shook the Cessna 150, and the low pressure in the vortex caused both windows to pop open!

FRONTAL TURBULENCE

Frontal turbulence is caused by surface temperature differences exceeding 5°C within 50 mi of the front, and it usually occurs below 15,000 ft MSL. Since temperature is the determining factor, speed or type of front is not involved in the extent of frontal turbulence. Other types of turbulence, however, such as mechanical or wind shear, may also accompany a front. In fact, rapid changes in wind direction and speed below 3000 ft AGL within 200 mi of an advancing front may produce low-level wind shear.

The following procedures can be used to avoid or minimize frontal turbulence. To avoid the phenomenon, fly above the affected area or remain on the ground until frontal passage. Turbulence can be minimized by penetrating the front at a right angle, thus reducing exposure.

CLEAR AIR TURBULENCE

Turbulence encountered in clear air not associated with cumuliform clouds, usually above 15,000 ft and associated with wind shear is classified as *clear air turbulence* (CAT). Slight, rapid, and somewhat rhythmic bumpiness without appreciable changes in altitude or attitude defines *CHOP*. Since CHOP does not cause appreciable changes in altitude or attitude, it would not be severe.

CAT is very patchy and transitory in nature. The dimensions of these turbulent patches are quite variable, generally on the order of 2000 ft in depth, 20 mi in width, and 50 or more mi in length. The patches elongate in the direction of the wind. The dimensions of these areas are on the microscale, and the exact position of specific areas is difficult, if not impossible, to forecast.

CAT is greatest in areas of strongest shear. Moderate or greater CAT occurs most frequently during January and February. Jet stream winds are at lower altitudes, and they are more frequent and stronger during the winter than in summer.

Wind speeds and curvature of contours provide a clue to clear air turbulence. Often the curvature of the contours has more effect on the severity of the turbulence than does wind speed. Recall the discussion in Chap. 12. Areas of potential turbulence occur in the following:

- Sharp troughs
- In the neck of cutoff lows
- In a divergent flow

Turbulence in these areas can exist despite relatively low wind speeds.

A factor associated with the jet is wind shear turbulence. With an average depth of 3000 to 7000 ft, a change in altitude of a few thousand feet will often take the aircraft out of the worst turbulence and strongest winds. Maximum jet stream turbulence tends to occur above the jet core and just below the core on the north side. Additional areas of probable turbulence occur where the polar and subtropical jets merge or diverge.

WAKE TURBULENCE

Although wake turbulence from larger aircraft is a factor enroute, it is a more serious concern during takeoffs and landing. During the takeoff and landing phases, the aircraft is close to the ground and is moving at relatively slow speeds.

A wing producing lift generates a disturbance or wake caused by a pair of counterrotating vortices trailing the wing tips. Vortices occur because of the pressure differential above and below the wing at the tips. After completion of the rollup, the wake consists of two counterrotating cylindrical vortices. The strength of the vortex depends on the weight, speed, and shape of the wing. The greatest vortex strength occurs when the generating aircraft is in the following circumstances:

- Heavy
- Slow
- Clean—landing gear and flaps retracted

Refer to Fig. 23-3. Vortices develop as soon as the aircraft leaves the ground and cease when the aircraft touches down. The vortex circulation is outward, upward, and around the wing tip when viewed from either ahead or behind. Vortices remain spaced a bit less than a wingspan apart, drifting with the wind, at altitudes greater than a wingspan above the ground. Vortices from larger aircraft sink at a rate of several hundred feet per minute, slowing their descent and diminishing in strength with time and distance behind the generating aircraft. Atmospheric turbulence hastens breakup. When the vortices of larger aircraft sink close to the ground (100 to 200 ft), they tend to move laterally over the ground at 2 or 3 kn. A light wind of 1 to 5 kn could result in the upwind vortex remaining over the runway.

CASE STUDY The flight from Van Nuys to Long Beach was on a typical Los Angles Basin day. Above the haze there was clear, smooth air. I spotted a "three-holer" (Boeing 727) on approach to Los Angeles International. It was above my flight path, so I planned to cross about 3 mi behind. I penetrated the wake at a 90° angle. It grabbed the airplane and almost instantaneously dropped me about 50 ft. Because I was belted in, I was also yanked down. My recollection of the incident was seeing my notepad flying out of my pocket in front of my face and two ashtrays crossing in front of me dumping butts all over the place. I remember the first thing I did was check to see if the tail was still there; it was.

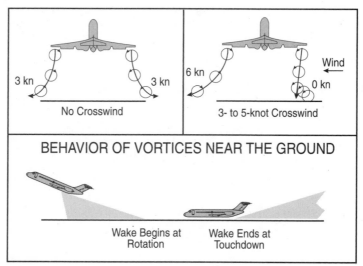

3 kn   3 kn

No Crosswind

6 kn   Wind

0 kn

3- to 5-knot Crosswind

BEHAVIOR OF VORTICES NEAR THE GROUND

Wake Begins at   Wake Ends at
Rotation   Touchdown

FIGURE 23-3   Vortices develop as soon as the aircraft leaves the ground and cease when the aircraft touches down.

IFR operations are not immune to the hazards of wake turbulence.

CASE STUDIES   In one fatal accident, ATC placed a Piper Navajo too close behind a Boeing 727. The Piper was less than 3 mi behind the 727 and above its glide path when the Piper was cleared for the approach. Apparently, the Piper was in a steep rate of descent, trying to intercept the glide slope, when it encountered the 727's wake and crashed.

A Twin Commander pilot following the 727 reported severe turbulence. Although the Commander was nearly 5 mi behind the jet, the NTSB believes the pilot most likely lost control due to slow speed and possible icing and wake turbulence.

Pilots can never allow ATC to put them in an untenable position. The Piper pilot should have initiated a missed approach rather than attempt to dive to the glide slope. Situation awareness was also a factor. The proximity to the jet should have been another clue that a missed approach was in order. Apparently the Commander pilot allowed the airspeed to decay well below normal approach speed. This coupled with possible ice and wake turbulence led to disaster.

Wake turbulence encounters with the Boeing 757 have received a lot of press lately. A Boeing 737 captain said a 757 wake rolled his aircraft 45° on approach to Salt Lake City. That must have been exciting. Keep in mind that wake turbulence from much smaller aircraft, including helicopters, can significantly affect small aircraft.

Because of the hazards of wake turbulence, controllers are required to apply specific minimum separation between aircraft of various weight categories. This will be in the form of time or distance. For example, a 2-min interval will be provided for a small airplane departing behind a heavy airplane, or 6 mi for a small airplane landing behind a heavy jet. Keep in mind that pilots have the option to request additional wake turbulence separation. Like icing and thunderstorms, avoidance is the key to handling the wake turbulence hazard.

STRATEGIES

By understanding the causes of turbulence and the forecast, a pilot can determine the most likely areas and plan accordingly. This includes reducing to turbulence air penetration speed, securing objects, and briefing passengers before entering areas of probable turbulence.

We know that the delineated area in the AIRMET Bulletin is usually larger than the actual affected area, and we must consider phenomena moving through the area during the forecast period. This explains some widely held misinterpretations. Turbulence will not necessarily occur at every location within an advisory area, during the entire forecast period. Forecasts for mechanical turbulence often cover wide areas. However, the greatest intensity will occur in the vicinity of mountains, leaving valleys and coastal areas relatively smooth. Often, the forecaster reflects this by stating: OCNL MOD TURBC VCNTY MTNS.

Evaluate *pilot weather reports* (PIREPs) within the context of METAR reports, forecasts, and other PIREPs. A single report of severe turbulence from a Beech Sundowner under clear skies and light winds should be viewed with skepticism. On the other hand, a report of severe turbulence from a Cessna 172 with conditions favorable for a mountain wave, and advisories in effect, should be taken very seriously. PIREPs that are not objective are worse than useless. Not only do they give a false impression to other pilots but forecasters must take them as fact and issue advisories, which undermine forecast credibility.

Recall that a recommended strategy is to approach the mountain range at a 45° angle. If the aircraft cannot safely clear the crest, only a 45° turn is required to reach lower terrain. This brings up the question: How does a pilot know that the aircraft will clear a mountain ridge? If the terrain behind the mountain crest is getting bigger, the aircraft will clear the ridge; if terrain is disappearing behind the crest, abort the attempt; the aircraft will not clear the ridge.

To avoid the worst mechanical turbulence, remain at least 5000 ft above mountain crests. In the west, most light aircraft simply don't have that performance. This leaves three options: Select a course with lower terrain, wait it out, or take a chance on getting your fillings knocked loose and maybe losing the airplane. Passenger comfort and safety should be the priority consideration.

Knowledge is the key to avoiding strong or gusty surface wind accidents. Three areas of knowledge are required. First, know your own limitations. Second, know the performance and limitations of your aircraft. Third, know the surface wind conditions.

What are your personal minimums? In our Air Force Aero Club, student pilots were limited to 10 kn of surface wind, with no more than a 5-kn crosswind component. As an instructor, I impose specific limits based on the individual student's training and experience. For solo cross-country flights, my students have always had an alternate in case adverse winds develop. When instructing in the Mojave Desert, we necessarily have to train in higher winds and crosswinds because of the typically strong winds in that area. It makes no sense to attempt a takeoff or landing with a 25-kn crosswind when you've trained to a maximum 15-kn crosswind. Again, more about this in Chap. 27, Risk Assessment and Management.

What is the capability of your aircraft? I have a friend with thousands of hours of military experience. He owns a Kitfox. His personal limits are 15 kn of wind and no more than 10 kn of crosswind component. These limits are based more on the aircraft than his experience.

How can we determine wind conditions? More and more airports have either towers or automated weather reporting systems. If not, check with a nearby field or observe smoke or trees on the ground. Note wind drift in the pattern and plan accordingly. Finally, if anything isn't right, go around! On approach to Ontario, California, during a Santa Ana wind, a DC-8 made a go-around at about 100 ft when it was caught by a strong gust. During the takeoff roll abort, if everything isn't going right, the slower the aircraft, the easier it is to control.

Pilots can avoid the worst effects of thermal turbulence by planning flights above the altitudes or avoiding the time of day when turbulence is strongest. Avoid midday through late afternoon, if possible.

Evaporative cooling turbulence develops in the vicinity of virga; precipitation evaporates and cools the air, causing downdrafts. A pilot penetrating these areas will encounter wind shear turbulence, which can be severe. Flying through steady precipitation is generally smooth, but showers tend to be turbulent. Avoiding showers will not only result in improved forward visibility but a smoother flight.

A primary reason for high-altitude flying is to avoid mechanical, frontal, and mountain wave turbulence; however, the flight levels have their own problems—wind shear or clear air turbulence. When pilots encounter problems, flight watch can help find a smooth altitude or alternate route. If the pilot elects to change altitude, an update of actual or forecast winds aloft is often a necessity.

Although there are no hard-and-fast rules as to exactly where turbulence will develop or its intensity, Table 23-1 provides an overview of turbulence locations, probability, and intensity.

If reports or forecasts indicate turbulence, a pilot can minimize the hazard when he or she encounters it. Turbulence imposes gust loads that appear to be almost instantaneous. Gust loads increase with the speed of the aircraft and gust velocity. The pilot's first task should be to secure loose objects.

If you encounter or expect light or moderate turbulence, avoid flight in the caution range—the yellow arc on the airspeed indicator. If you encounter or expect severe turbulence, reduce to maneuvering speed. Maneuvering speed is not printed on the airspeed indicator, although it's usually placarded in the vicinity of the instrument. Because the gust load factor decreases as wing loading increases, maneuvering speed increases with the aircraft's gross weight. Therefore, it's often necessary to determine maneuvering speed based on gross weight.

In moderate or greater turbulence, fly attitude rather than altitude. Disengage the autopilot altitude hold if it's in use. The aircraft should already be slowed to turbulent air penetration speed. Don't chase airspeed or altitude. It's like riding a horse; go with it rather than fighting it. The object is to avoid imposing additional abrupt maneuvering loads. For the most part, ignore altitude unless terrain clearance becomes a problem; VFR, try to avoid IFR cardinal altitudes (4000, 5000, 6000, etc.), or opposite direction VFR altitudes; IFR, inform air traffic control of the problem and any altitude deviations required. ATC increases vertical separation during severe conditions.

A wake turbulence encounter can result in one or two jolts of varying severity or it can be catastrophic. The probability of induced roll increases when the encountering aircraft's heading is generally aligned with the flight path of the generating aircraft. The key to wake turbulence avoidance is knowledge and defensive flying. Avoid the area below and behind the generating aircraft, especially at low altitude where even a momentary encounter could be hazardous. Pilots should be particularly alert in calm wind conditions and situations in which the vortices could do one of the following:

- Remain in the touchdown area.
- Drift from aircraft operating on a nearby runway.
- Sink into the takeoff or landing path from a crossing runway.
- Sink into the traffic pattern from other airport operations.
- Sink into the flight path of VFR aircraft.

TABLE 23-1 Locations and Conditions for Probable Turbulence

Light turbulence

---

1. In hilly and mountainous areas even with light winds
2. In and near small cumulus clouds
3. In clear air convective currents over heated surfaces
4. With weak wind shear in the vicinity of the following:
   a. Troughs aloft
   b. Low-pressure areas aloft
   c. Jet streams
   d. The tropopause
5. Within 5000 ft of terrain:
   a. When winds are near 15 kn
   b. Where the air is colder than the underlying surface

Moderate turbulence

---

1. In mountainous areas with a wind of 25–50 kn perpendicular to the ridge
   a. From the surface to 5000 ft above the tropopause:
      (1) Within 5000 ft of the ridge
      (2) At the base of relatively stable air below the tropopause
      (3) Within the tropopause
2. In and near dissipating thunderstorms
3. In and near towering cumulus
4. Within 5000 ft of the surface:
   a. When surface winds exceed 25 kn
   b. In areas of strong surface heating
   c. At the boundary of strong inversions
5. In fronts aloft
6. Where:
   a. Vertical wind shear exceeds 6 kn per 1000 ft
   b. Horizontal wind shear exceeds 18 kn per 150 mi

Severe turbulence

---

1. In mountainous areas when wind exceeds 50 kn perpendicular to the ridge
   a. Within 5000 ft of the ridge
   b. At and below the ridge in rotor clouds or rotor action
   c. At the tropopause
   d. At the base of a stable layer below the tropopause
   e. Extending outward on the lee of the ridge for 50 to 150 mi
2. In and near growing and mature thunderstorms
3. In towering cumulus clouds
4. Fifty to 100 mi on the cold side of the center of the jet stream, in troughs aloft, and in lows aloft when:
   a. Vertical wind shear exceeds 6 kn per 1000 ft
   b. Horizontal wind shear exceeds 40 kn per 150 mi

Extreme turbulence

---

1. In mountain waves, in and below the level of well-developed rotor clouds, sometimes extending to the ground
2. In growing severe thunderstorms, especially squall lines, with:
   a. Hailstones three-quarters of an inch or greater
   b. Heavy intensity radar echoes
   c. Continuous lightning

The following vortex avoidance procedures are recommended:

- *Departing behind a larger aircraft.*   Note the larger aircraft's rotation point, and rotate prior to that point, continue to climb above the larger aircraft's climb path until turning clear of its wake. Avoid subsequent headings that will cross below or behind a larger aircraft.

- *Intersection takeoffs on the same runway.*   Be alert to adjacent larger aircraft operations, particularly upwind of your runway. If you receive an intersection takeoff clearance, avoid subsequent headings that will cross below a larger aircraft's path.

- *Departing or landing after a larger aircraft executing a low-approach, missed-approach, or touch-and-go landing.*   Because vortices settle and move laterally near the ground, the vortex hazard may exist along the runway and in your flight path after a larger aircraft has executed one of these maneuvers, particular with a light quartering wind. You should ensure that an interval of at least 2 min has elapsed before your takeoff or landing.

- *Landing behind a larger aircraft on the same runway.*   Stay at or above the larger aircraft's final approach flight path, note its touchdown point, and land beyond that point.

- *Landing behind a larger aircraft when a parallel runway is closer than 2500 ft.* Consider possible drift to your runway. Stay at or above the larger aircraft's final approach flight path and note its touchdown point.

- *Landing behind a larger aircraft on a crossing runway.*   Cross above the larger aircraft's flight path.

- *Landing behind a departing larger aircraft on the same runway.*   Note the larger aircraft's rotation point and land well before its rotation point.

- *Landing behind a departing larger aircraft on a crossing runway.*   Note the larger aircraft's rotation point. If you are past the intersection, continue the approach and land prior to the intersection. If the larger aircraft rotates prior to the intersection, avoid flight below the larger aircraft's flight path. Abandon the approach unless a landing is ensured well before reaching the intersection.

- *Enroute.*   Avoid flight below and behind a large aircraft's flight path. If a larger aircraft is observed above on the same track (meeting or overtaking), adjust your position laterally, preferably upwind.

CHAPTER 24
ICING

Structural icing affects aircraft in three ways: flight and engine instruments, induction systems, and aircraft structures. An erroneous assumption is that icing affects only the IFR pilot. This is not true. Icing affects the VFR pilot as well. In fact, as we shall see, based on total numbers, VFR pilots are involved in more icing-related accidents than IFR pilots. Therefore, our discussion applies equally to both types of operations. The last item in this section is runway ice. Snow- or ice-covered runways can have a significant effect on taxi, takeoff, and landing operations. Pilots that fly in these conditions must be prepared for these situations. And, again, runway hazards apply to both VFR and IFR operations.

The chapter concludes with strategies to avoid or counter the affects of icing. For most light aircraft, avoidance is the only solution. Once flight instruments have failed, the engine has faltered due to carburetor ice, or the airfoils are incapacitated by ice, the pilot has few, if any, options. Through personal and accident scenarios, and a sound understanding of the subject, we will learn to preclude or avoid hazardous icing situations.

Icing affects airframes, engines, propellers and rotors, and aircraft flight instruments and radios. Supercooled water drops produce airframe icing—visible, liquid water with temperatures at or below 0°C. When these droplets hit an aircraft, they freeze. Icing can also occur with water droplets slightly warmer than 0° if the airframe temperature is at or below zero—such as on an aircraft descending from an area of cold air. Aerodynamic cooling can also lower the temperature of an airfoil to 0°C even though the ambient temperature is a few degrees warmer. When the temperature reaches −40°C or less, it is generally too cold for supercooled droplets and airframe ice to form.

Ice on an airfoil disrupts the smooth flow of air; it decreases lift and increases drag. Lift may be decreased by as much as 50 percent and drag increased by as much as 35 percent. Ice increases the weight of the aircraft and may affect the engine, reducing power output or in extreme cases causing engine failure.

Ice forms on propellers and in jet engine inlets. Even a small amount of ice, if not evenly distributed, can cause stress on the piston engine mounts and propeller. When the propeller sheds ice, a momentary increase in vibration and stress occurs. This can be very exciting as chunks of ice hit the fuselage. Finally, ice affects aircraft intakes and carburetors.

Ice also affects radio communications by reducing antenna efficiency or causing antennas to break off.

Ice, especially on aircraft without ice protection equipment, can obstruct the pilot's view. Needless to say, it's extremely dangerous attempting to land with an iced-over windscreen. Then add the fact that ice can disrupt the function of control surfaces, reduce the effectiveness of brakes, and interfere with landing gear operation.

To sum up, ice adversely affects an aircraft in the following ways:

- Increased drag
- Loss of lift
- Increased weight

- Reduced or no power
- Interference with control surfaces
- Reduce effectiveness of brakes
- Interference with landing gear operation
- Increased vibration and structural stress
- Reduced or precluded forward vision
- Loss of, or false, instrument readings
- Loss of, or reduced, radio navigation and communications

Icing can form as slowly as 1/2 in per hour or as rapidly as 1 in per minute! Icing potential exists any time visible moisture exists—clouds or precipitation—at temperatures of 0°C or less. This contradicts the notion that icing can occur only in clouds. We mentioned that freezing rain or freezing drizzle can be the most serious icing hazard that the potential for icing exists in wet snow conditions. Both phenomena can affect the VFR pilot. The greatest icing potential occurs between the freezing level and −10°C to −15°C, or within a layer approximately 5000 and 7500 ft deep. Icing has been encountered in convective clouds at altitudes of 30,000 to 40,000 ft in temperatures less than −40°C. During the winter of 1996 to 1997, central California was struck by a series of storms called the "pineapple express." These storms begin in the latitudes of Hawaii and bring moist, unstable air to California. Icing potential is enhanced by the upslope from the Sierra Nevada mountains. On several occasions aircraft were reporting moderate to severe mixed icing up to flight level 260. The temperature at that altitude was −26°C.

Icing is as difficult to forecast as it is hazardous. Forecasters must determine which areas contain enough moisture to form clouds, which cloud areas will most likely contain supercooled droplets during the forecast period, and the freezing level. Needless to say, this is not an easy task. Pilots should consider these as forecasts of icing potential. They alert the pilot to the need to consider the possibility of icing in clouds and precipitation within the areas and altitudes specified.

ICING TYPES AND INTENSITIES

Icing is classified by its formation and appearance. Pilots should use the following terms when reporting icing:

- Rime icing
- Clear icing
- Mixed icing

Rime Ice

Rime ice is a milky, opaque, and granular deposit with a rough surface. It normally forms when small supercooled water droplets instantaneously freeze upon impact with the aircraft. Figure 24-1 shows a wing with rime ice. It is most frequently encountered in stratiform clouds at temperatures between 0° and −20°C. This icing condition is usually widespread due to the character of stratiform clouds. Rime ice is relatively easy to remove with ice protection equipment.

FIGURE 24-1 Rime ice is a milky, opaque, and granular deposit with a rough surface. It forms when small supercooled water droplets instantaneously freeze upon impact with the aircraft. (*Photo courtesy of NASA.*)

Clear Ice

Clear ice is glossy and forms when large supercooled water droplets flow over the aircraft's surface after impact and then freeze into a smooth sheet of solid ice. It is most frequently encountered in cumuliform clouds or freezing precipitation. Clear ice is illustrated in Fig. 24-2. Brief, but severe accumulations occur at temperatures between 0° and −10°C, with reduced intensities at lower temperatures, and in cumulonimbus clouds down to as low as −25°C. Clear ice is usually not as widespread as rime ice because it occurs in cumuliform clouds, but its intensity tends to be more severe. Clear ice is more difficult to remove than rime ice.

Rime Ice and Clear Ice (Mixed Icing)

Mixed ice is a hard, rough, irregular, whitish conglomerate formed when supercooled water droplets vary in size or are mixed with snow, ice pellets, or small hail. Deposits become blunt with rough bulges building out against the airflow. Of the three types of ice, mixed ice is the most difficult to remove.

A condition favorable for rapid accumulation of clear ice is freezing rain below a frontal surface. Icing can also become severe in cumulonimbus clouds along a surface cold front or above a warm front. Icing is also more probable and more severe in mountainous regions than over flat terrain. Mountain ranges cause rapid upward vertical current on their windward side, which supports large water drops. The movement of frontal systems across mountain ranges often combines frontal lift with upslope effect, creating extremely severe icing zones, with the most severe icing taking place above the crest and to the windward side of the ridges.

FIGURE 24-2   Clear ice is glossy and forms when large supercooled water droplets flow over the aircraft's surface after impact and freeze. (*Photo courtesy of NASA.*)

Like turbulence, pilots have a tendency to overestimate icing intensity, especially when the pilots are new or low-time.

CASE STUDY   A recently rated instrument pilot, after experiencing his second encounter with icing in a Cessna 172, reported the intensity as severe. The encounter lasted about 30 min; the pilot was unable to maintain altitude and was forced to descend. This description, however, is only of moderate intensity.

Icing intensity has been classified for reporting purposes in the *Aeronautical Information Manual* and *Aviation Weather Services*. However, like turbulence, I prefer more descriptive definitions.

Trace

Ice becomes perceptible, and the rate of accumulation is slightly greater than the rate of sublimation. It is not hazardous even though ice protection equipment is not utilized, unless encountered for more than 1 h. Your spouse admires how pretty it looks on the wing; ATC has just instructed you to climb. You advise them icing is probable and request descent. The controller calmly replies that in that case, "you can declare an emergency or land." Shortly, you're handed off to the next controller. You inquire about a lower altitude, and the controller responds, "Is that Terry up there?" (A friend at Los Angeles Center.) A lower altitude is approved in about 15 min.

Light

The rate of accumulation can create a problem if the flight continues for more than 1 h. Occasional use of ice protection equipment removes or prevents accumulation. Ice should not present a problem if the ice protection equipment is used. Your student hasn't noticed the ice yet; your pilot friend in the back seat is hoping he has enough life insurance; you're negotiating with ATC for a lower altitude, which they can approve in 15 mi. This will take only about 8 min, but each minute seems like 10.

Moderate

The rate of accumulation, even for short periods, becomes potentially hazardous, and the use of ice protection equipment or flight course diversion becomes necessary. On his second encounter with ice, a friend and his passengers, in an aircraft without ice protection equipment, survived moderate icing only because the terrain was lower than the freezing level.

Severe

The rate of accumulation is such that ice protection equipment fails to reduce or control the hazard. Immediate diversion is necessary. This is a situation in which the person in the left seat very rapidly ceases being the pilot and becomes a passenger; the wing is an ice cube. Certain pilots will report icing intensity as heavy. This is a misnomer—all ice is heavy! (It seems that the maximum intensity reported for icing was "heavy" until 1968. That's almost before my time!)

The term *ice protection equipment* in the *icing intensity* definitions refers to aircraft and equipment certified for flight in known icing conditions. Although many aircraft have limited ice protection equipment (pitot heat, prop anti-ice, alternate static source, etc.), it should never be construed as adequate to allow for flight in icing. The purpose of the equipment is to aid emergency measures only, should icing be inadvertently encountered.

This brings up the question: What is *known icing?* You won't find it in FAR Part 1, *Definitions and Abbreviations,* or the pilot/controller glossary. Icing is difficult to forecast and transitory in nature. Do we want a forecast of icing to forbid flight? Would a report of light icing above 8500 ft, with bases 8000 and tops 9000, preclude flight for aircraft not certified for flight in known icing? What if the terrain were at 7800 and tops 15,000? Do we really want a hard answer? If some in the FAA had their way, the definition would be, "any time there is visible moisture and a temperature of +5°C or less"! Should this or a similar proposal ever be adopted, it would certainly mark the end to many useful icing PIREPs. If icing is reported or forecast and we fall out of the sky and survive, or require emergency or special handling from ATC, we're a candidate for a violation.

FACT A National Transportation Safety Board (NTSB) decision in 1993 held a pilot in violation of Federal Aviation Regulations. The board found that a pilot cannot pick and choose between forecasts and PIREPs. A forecast for icing is sufficient to warrant the violation.

More often than not, however, ATC is so busy and so happy to get us out of their hair we'll never hear another word. No one should interpret this as meaning that I, the FAA, or ATC condone such actions. Icing for aircraft not certified for flight in icing conditions is to be avoided! At present, the decision as to whether the flight can be made safely rests solely with the pilot, which is where it will stay until we prove we're not worthy of the responsibility.

PIREPs are the only source of reported icing. Forecasts for icing are contained in the AIRMET Bulletins and SIGMETs.

SUPERCOOLED LARGE DROPLET ICING

As a result of several accidents due to icing, the Aviation Weather Center (AWC) of the National Weather Service has revised its icing advisories—AIRMETs and SIGMETs—to imply hazards due to the presence of *supercooled large droplets* (SLDs).

SLDs are the size of freezing drizzle or freezing rain droplets, and they are much larger than cloud droplets. A forecast for SLDs implies rapid accumulation of mixed or clear ice, possibly forming aft of the aircraft's ice protected areas. This is illustrated in Fig. 24-3.

To alert pilots of potential SLD occurrences aloft, the AWC will issue advisories containing the terms MIXED OR CLEAR ICING IN CLOUDS OR PRECIPITATION (MXD/CLR ICICIP) or CLEAR ICING IN PRECIPITATION (CLR ICGIP), which means precipitation-size drops aloft. Even though the rate of accumulation is only moderate, the presence of SLDs poses a significant hazard, even to aircraft with ice protection equipment.

To indicate areas of SLD, there may be an AIRMET within an AIRMET to highlight the threat. For example, an AIRMET for moderate rime icing below 14,000 ft may cover a relatively large area. A second AIRMET wholly within the first AIRMET's coverage may forecast moderate mixed or clear icing in clouds and precipitation below 10,000 ft. This alerts pilots to the SLD threat within the second area below an altitude of 10,000 ft, with a potential of cloud-size droplets between 10,000 and 14,000 ft.

It appears the most likely areas for SLD occurrence are the following:

FIGURE 24-3 Supercooled large droplets can produce ice well aft of the aircraft's ice protected areas. (*Photo courtesy of NASA.*)

- 25 to 300 mi ahead of a warm front
- 30 mi either side of an occluded front
- 25 to 130 mi ahead of a Pacific cold front
- 25 to 130 mi behind an arctic front

INSTRUMENT ICING

Ice can affect aircraft instruments. (Recall our discussion of the pitot-static system in Chap. 21 and Fig. 21-1.) Iced over pitot-static instruments can cause false readings or render the instruments useless. Clogging of the pitot tube by ice affects the airspeed indicator only. Many aircraft are equipped with an alternate static source, vented inside the cabin, for emergency use. Static pressure inside the cabin is usually lower than outside static pressure. Therefore, the altimeter reads higher than normal; indicated airspeed is greater than normal; and the vertical speed indicator shows a momentary climb, then operates normally. Instrument icing has caused jet air carrier as well as general aviation accidents.

CASE STUDIES A Boeing 727 was lost due to an iced over pitot tube. As static pressure decreased during the climb, the airspeed indicator showed speed increasing. The autopilot attempted to hold airspeed by increasing pitch, resulting in a stall. A Boeing 737 crashed because of an iced-over engine power sensor—the airplane was simply not developing takeoff power, even though the instrument indicated it was.

STRUCTURAL ICING

Structural icing accidents accounted only for about 40 percent of total accidents involving icing. The majority of icing accidents are attributed to carburetor or induction system icing, with fewer than 10 percent involving icy runways. Both induction and runway icing will be addressed later in this chapter. Most structural icing accidents occurred when the pilot continued flight into known icing, severe weather, or deteriorating weather conditions. A lesser amount occurred on approach or landing in icing conditions or with ice accumulation.

As well as wing and fuselage icing, another hazard of structural icing is *tailplane* or *empennage stall*. A tailplane stall occurs, like the wing, when the critical angle of attack is exceeded. Since the horizontal stabilizer counters the natural nose-down tendency caused by the center of lift of the main wing, the airplane will react by pitching down, sometimes uncontrollably, when the tailplane stalls. Application of flaps can aggravate or initiate the stall. A pilot should use caution when applying flaps during an approach if there is the possibility of icing on the tailplane.

Perhaps the most important characteristic of a tailplane stall is the relatively high airspeed at the onset and, if it occurs, the suddenness and magnitude of the nose-down pitch. A stall is more likely to occur when the flaps are approaching the fully extended position, after nose-down pitch and airspeed changes following flaps extension, or during flight through gusty winds.

Another type of icing that must not be ignored is frost. As you will recall, condensation is the change of state from water vapor to liquid water. When moist air comes in contact with a cool surface, then cools to its dew point, dew appears. If the dew point temperature is below freezing, the water vapor *sublimates*—changes from a vapor to a solid—producing frost.

The effects of frost may be more subtle than the effects of ice. Although the airfoil's aerodynamic contour remains relatively unchanged, considerable roughness, resulting in

FIGURE 24-4   The solution to the frost hazard is to hangar the aircraft and arrange for a deicing service, or plan a departure later in the day when the sun has melted the ice.

increased drag, occurs. Under no circumstances should a takeoff be attempted with frost on the aircraft. A heavy coating of frost can cause a 5 to 10 percent increase in stall speed. Just as there is no such thing as "a little pregnant," there is no such thing as "a little frost." Frost is illustrated in Fig. 24-4. Note in the foreground that the sun has melted the ice, but considerable ice remains on the wing.

> **CASE STUDY**   We made a trip from Mammoth Lakes to Livermore, California, in late June. As is my habit, I planned an early morning departure, assuming density altitude would be the most significant factor. Arriving at the airport around dawn, to my surprise, I found frost on the wings! I pointed the airplane into the sun and in about 15 min we wiped the melting frost from the ship. This just goes to prove you have to be ready for, and aware of, everything.

INDUCTION/CARBURETOR ICING

As mentioned earlier, the biggest cause of icing accidents are attributed to induction or carburetor icing. Most involved the lack, or improper use, of carburetor heat.

> **CASE STUDY**   On a flight from Van Nuys, California, to San Francisco in a Cessna 172, we encountered light icing after an ATC instruction to climb. I periodically applied carburetor heat. Something unusual occurred. With the carburetor heat on, the engine ran fine: off, the engine faltered. On the ramp at San Francisco, we parked next to a Navion that had also flown from Los Angeles but who had flown at a higher altitude and encountered more ice. Sure enough, in the Navion's air filter was a large chunk of ice. I realized that the carburetor heat in the 172 was functioning as an alternate air source. I'm sure this seems ridiculously obvious; it didn't at the time, which illustrates the hazards of learning by experience.

The induction system includes the air filter, ducting, and fuel metering device. Induction system icing consists of any ice accumulation that blocks any component of the system.

Air filter icing occurs when flying in areas of visible moisture with temperatures at or below freezing. For VFR pilots, air filter icing should occur only in areas of freezing precipitation or wet snow.

Induction system icing takes place any time structural icing occurs. A symptom of air filter icing is a more-or-less gradual decrease in power. Should air filter icing occur, apply carburetor (carb) heat or alternate air. The application of carb heat or alternate air bypasses the air filter. Leave carb heat or alternate air on until above-freezing temperatures melt the ice. The use of carb heat or alternate air results in unfiltered air entering the induction system. Except for operational checks, avoid engaging either control on the ground.

In addition to air intake icing, normally aspirated engines can develop ice in the carburetor throat. Refer to Fig. 24-5. The vaporization of fuel, along with the adiabatic expansion of air as it passes through the fuel discharge nozzle, venturi, throttle valve, and passages to the engine, causes sudden and significant cooling. If the air temperature drops below the dew point, water vapor in the air condenses into water droplets. Therefore, water can form in the carburetor in cloudless skies. This cooling can reduce the temperature in the carburetor to below freezing, and with sufficient moisture present, ice will form. Known as *carburetor icing,* ice can form with outside air temperatures as high as 32°C. The formation of carburetor ice restricts engine power and may result in complete engine stoppage.

As air accelerates through the carburetor and fuel evaporates, temperatures can be lowered as much as 34°C. Whether ice will develop depends on the velocity of the fuel-air mixture, outside air temperature, humidity, and carburetor system. Conditions most favorable for carburetor ice are outside temperatures between −7° and 21°C, high relative humidity, and low-power settings.

Carburetor heat preheats the air before it reaches the carburetor. Carburetor heat is usually adequate to prevent icing, but it may not always clear ice that has already formed. Pilots should monitor engine performance for the first signs of icing, especially during favorable conditions. When ice is detected or suspected, immediately apply full carburetor heat. Leave heat on until all ice has been removed.

Using full carburetor heat will initially cause an additional loss of power and engine roughness. The added loss of power and roughness result from the richer mixture due to warmer, less dense air and melting ice passing through the engine. You must resist the natural urge to remove carb heat. Leave full heat on until an increase in rpm or manifold pressure and smooth engine operation resume.

Under severe conditions it may be necessary to leave carburetor heat on for an extended period. When carb heat must be left on, relean the mixture for maximum rpm and smoothest operation. A cruise power setting of 75 percent or less with any amount of heat will not damage the engine.

CASE STUDY On a flight from Page, Arizona, to Las Vegas, Nevada, in our Cessna 150, conditions for carburetor ice were ideal. The temperature was about 15°C in rain showers. At the first indication of carb ice, heat was applied. It cleared up the ice, but as soon as it was removed, the rpm began dropping. The only solution was to leave carb heat on and live with the hundred or so loss in engine rpm, even after releaning the mixture.

Carbureted engines are more susceptible to icing during reduced-power operation. Some aircraft and engine manufacturers recommend the use of carburetor heat during all power reductions, others only when ice is suspected. Pilots should know and follow the aircraft manufacturer's recommendations. If full power is required, such as for a go-around, full carburetor heat and full power might cause early detonation or engine damage. It will certainly prevent the engine from developing full power, which might be critical in low-

CARBURETOR

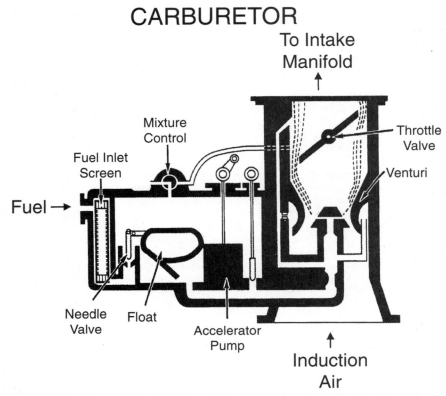

FIGURE 24-5   The vaporization of fuel, along with the adiabatic expansion of air, passing through the carburetor causes sudden and significant cooling.

power aircraft at high-density altitudes. Again, know and follow the manufacturer's recommendations.

> **CASE STUDY**   Pay attention! I had remained overnight in Amarillo, Texas, because of a line of thunderstorms that approached from the west. The Cessna 150 was parked into the wind when torrential rains moved through the area. The next morning was clear, with abundant surface moisture, temperature was about 15°C, and nearly 100 percent relative humidity.

My first clue of trouble was the increased throttle setting required to obtain idle rpm. Engine runup also took more throttle than usual. I suspected carburetor ice and a water-saturated air filter because of the conditions. I had a 13,000-ft runway.

Full throttle gave me only about 2200 rpm. The increased ground run to rotation speed—about 7000 ft—should have been another clue. I was off the ground, with no runway remaining, and 200 ft of altitude when the engine started losing rpm. I applied carburetor heat, and the engine was running very rough producing about 1700 rpm.

There was a tremendous psychological urge to reduce carb heat and get that rpm back. I was preparing to crash straight ahead, but the engine was still producing power and I decided to make a 180° turn and land on a taxiway. Then I informed a surprised tower controller of what happened. Remember, a pilot's first job is to fly the airplane.

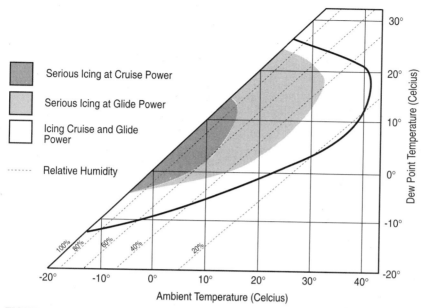

FIGURE 24-6   This chart graphically depicts carburetor icing potential; but like structural icing, carburetor icing is complex and there are few hard-and-fast rules.

Twenty minutes of running the engine and one aborted takeoff later, I launched into the air. The engine again performed normally above the shallow, moist layer. It was a perfect example of having the clues and ignoring them. I was extremely fortunate.

Some airplanes have a carburetor air temperature gauge. This gauge measures the temperature in the carburetor. The yellow arc indicates temperatures at which icing is most likely to occur. When ice is suspected or detected, the pilot can apply enough heat to raise the temperature out of the danger zone. Do not use partial carburetor heat without a carburetor air temperature gauge. Applying partial heat or leaving it on for an insufficient time might aggravate the situation.

PIREPs are the pilot's only source of reported induction system icing: iced-over air intake systems and carburetor ice. Both result in loss of engine power and, left unchecked, complete power failure. The solution is to use the alternate air source on fuel-injected engines or carburetor heat on normally aspirated—carbureted—engines. Induction system icing takes place any time structural icing occurs. A symptom is a gradual loss of power.

Figure 24-6 shows a composite carburetor icing probability chart. This chart was derived from FAA and Transport Canada carburetor icing probability charts. Like structural icing, carburetor icing is complex, and there are few hard-and-fast rules. The probability chart graphically depicts carburetor icing potential previously discussed.

STRATEGIES

During the preflight weather briefings, pay particular attention to reports or forecasts of icing. As well as the AIRMET Bulletins, SIGMETs, and PIREPs, look for the following precipitation types.

Ice pellets do not bring about the formation of structural ice, except when mixed with supercooled water. Frequently, ice pellets or ice pellet showers indicate areas of freezing rain above. Since snow grains are already frozen, they typically do not present an icing hazard.

Freezing rain (FZRA), *freezing drizzle* (FZDZ), and *ice pellets* (PLs) alert pilots to significant icing conditions and to the fact that there is warmer air somewhere above the station.

Ice crystals do not generally result in the formation of structural icing. Ice crystals can accumulate in inlets and ducts and dislodge as a mass that can create choking and possible engine damage. A mixed condition where both ice crystals and supercooled droplets exist probably constitutes the worst condition for engine and intake icing.

Dry snow does not lead to the formation of aircraft structural ice since the particles are dry and do not adhere to aircraft surfaces, except for heated engine inlets. However, wet snow—snow that contains a great deal of liquid water—produces structural icing.

If the briefing indicates even a remote possibility of encountering ice, the pilot should accomplish several tasks. Ensure the pitot heat works with a very light touch during preflight. (Be careful, it gets hot!) Check ice protection equipment for proper operation. Remember that a heated pitot is an anti-icing device, to be turned on before encountering ice; deicing equipment usually requires ice buildup before activation; improper operation can actually increase ice buildup and prevent its removal! Check alternate air or carburetor heat for an alternate source of air should the air filter ice over. It is also necessary to preflight all aircraft deice and anti-icing systems.

CASE STUDIES The pilot of a twin-engine airplane was enroute when the airplane encountered icing conditions. When the pilot activated the deice boots, the right wind deice boot failed to function. As the airplane slowed for landing, the asymmetrical ice buildup caused instability that the pilot was unable to control. The airplane crashed on landing seriously injuring the occupants. There was no mention of the pilot's checking the boots prior to departure.

One of our local pilots at Fresno had a similar occurrence in a Piper Aerostar. During the first storm of the season, upon activating the deicing boots, the pilot noted that one inflated, the other did not. Our pilot was visibly shaken by the experience. The bottom line: Check all aircraft equipment prior to departure.

When dealing with icing always have a way out.

CASE STUDY As a new instrument instructor, I took an instrument student on a flight from Van Nuys to Lancaster's Fox Field. The freezing level was forecast to be at 6000 ft. The minimum altitude for the route was 7000. Cloud bases were at 6500, well above terrain. Sure enough, we picked up trace to light rime icing. We had neglected to turn on the pitot heat, and a reverse cone of rime ice grew from the pitot tube. I, matter-of-factly, pointed this out to my student, and we turned on the heat, which immediately corrected the problem. On descent into Fox, the ice made a deafening noise as it broke off the tail surfaces. I do not recommend this procedure!

The objective in dealing with ice is to minimize exposure. With temperatures 0°C or less, avoid flying in clouds or precipitation. Should you encounter ice, immediately notify ATC and initiate a plan of action. The first consideration, if the aircraft has sufficient performance, might be to climb to colder air or above the clouds based upon PIREPs and the weather briefing. Ice will slowly sublimate—change from a solid directly into a gas—when on top. Or descend to warmer air based upon the actual freezing level on climbout. You did note the freezing level on climbout? This requires a careful check of *minimum enroute altitudes* (MEAs). Finally, if you have to, turn around; presumably you came from an ice-free area. The point is, do something!

CASE STUDY   A nonturbocharged Baron, without ice protection equipment, departed Reno, Nevada, for southern California. Moderate icing and severe turbulence were forecast. The pilot elected to fly a direct course along the crest of the Sierra Nevada mountains, the route along which the most intense icing and turbulence could be expected. The aircraft iced up, resulting in a fatal accident.

The pilot had no way out because the MEA was the aircraft service ceiling. The terrain was well above the freezing level, and the pilot failed to reverse course at the first sign of ice. What other options were available? The pilot could have crossed the mountains near Sacramento, minimizing exposure to ice, and once over the Sierras, it was all downhill. The pilot could have flown toward Las Vegas where the weather was considerably better, or simply waited for better weather conditions.

When the Baron became ice covered, the pilot had no option but to ride it to the crash site. Attempting flight under these conditions and with this type of equipment was quite literally suicide.

CASE STUDY   A Bonanza pilot departed the San Francisco Bay area on a flight to Los Angeles. Icing above 7000 ft was forecast and reported. The pilot elected to fly at 11,000 ft. The pilot's last words were, "I've iced up and stalled." The crash occurred in the San Joaquin Valley where the elevation was near sea level. Minimum altitudes in the vicinity of the crash were well below the freezing level. The pilot simply did nothing until aircraft control was lost.

An advantage of turbocharged and pressurized aircraft is the ability to fly high, above the weather. Icing is normally not a significant factor in the flight levels, except around convective activity or in the summer when temperatures can range between 0° and $-10°C$ at these altitudes. Just because it's summer doesn't necessarily mean there is no icing potential.

CASE STUDY   The following illustrates the decision-making process with forecast icing. The aircraft was a turbo Mooney on a flight from Bakersfield to Hayward, California. The synopsis indicated moisture and a stable air mass. Bases along the route were reported around 5000 and tops 9000 to 11,000 ft, the freezing level 7000. Except for the coastal mountains, terrain along the route was close to sea level—terrain, as we have mentioned, is a very important factor. The flight was planned at 12,000 ft because tops were relatively low and the aircraft had the performance to quickly climb through the potential icing layer. During the climb, trace to light icing was encountered. Once on top, the ice sublimated quickly. There were some buildups above 12,000 ft. Deviations to avoid these clouds were obtained from ATC. By circumnavigating the buildups, icing and turbulence were avoided. You can put money on the assumption that there was ice in those clouds.

Should this be attempted in an airplane with lesser performance? Absolutely not! The bases and tops were known quantities. The airplane had the performance to quickly climb on top. Had that not been possible, the pilot had the option to return—cloud bases were over 4,000 ft above terrain and well below the freezing level. Was the icing forecast correct? Yes. Had emergency assistance been required, the pilot would have been a candidate for a violation. I do not intend to imply that this procedure is recommended or endorsed. The decision rests solely with the pilot, based on his or her training and experience, and the capability of the aircraft.

If a descent through an icing layer is required, remember the objective is to minimize exposure. Under such circumstances, negotiate with ATC to obtain a continuous descent. Avoid, if possible, level flight in clouds. ATC is usually very responsive to such requests.

A popular notion in some aviation publications is that a pilot's mere mention of ice will receive emergency-like handling. Icing might be an emergency, but remember the controller's job is to separate aircraft within a finite amount of airspace. ATC might have to assign a higher altitude, but it cannot, and should not, be expected to fly the aircraft or assume the responsibility of the pilot in command. To paraphrase: An accurate report of actual icing conditions is worth a thousand forecasts.

In freezing precipitation, aircraft without a heated pitot and alternate static source, especially in IFR conditions, would be in serious trouble. Another significant factor, especially for aircraft without ice protection equipment, is that accumulated ice could be carried all the way to the ground, making landing extremely hazardous. It cannot be overemphasized that this hazard can affect VFR as well as IFR operations. Should this phenomenon be encountered in aircraft without ice protection equipment, virtually the only option, and certainly the safest, is to fly into warmer air and land. Pilots who fly into ice, with aircraft not certified for flight in icing conditions, must have the right stuff. Because in every sense of the word, they become test pilots.

Continually updating the weather picture is the key to managing a flight, especially in aircraft with limited or no ice protection equipment. Icing can be significant during descent, especially when the destination temperature is at or below freezing. Flight watch can provide information on tops, temperatures aloft, reported and forecast icing, and current surface conditions.

On airplanes equipped with a float type of carburetor, the use of carb heat is normally recommended during reduced- (out of the airspeed indicator's green arc) or closed-throttle operations. Since the heat is generated by the running engine, during extended periods of low throttle operation, it may be desirable to advance the throttle periodically to ensure proper engine operation. On airplanes with a pressure carburetor—much less susceptible to icing—carb heat may be recommended only if icing conditions exist.

Remove carburetor heat prior to a go-around or balked landing. With carburetor heat, loss of power may be critical at low altitudes and low airspeeds, and engine damage may occur with full heat on a go-around at takeoff power on some powerplants.

Low temperatures cannot be ignored. Snow, snow melt, freezing rain, and frost produce structural ice that can be difficult to remove from a parked aircraft. Even a thin layer of frost can severely affect performance, and it must be removed before takeoff, according to federal regulations.

If an aircraft is parked in an area of blowing snow, special attention should be given to openings in the aircraft where snow can enter, freeze solid, and obstruct operations. These openings should be free of snow and ice before flight. Some of these area are the following:

- Pitot tubes
- Static system sensing ports
- Wheel wells
- Heater intakes
- Carburetor intakes
- Tail wheel areas
- Control surfaces and cables

CASE STUDIES   Some years ago four of us flew a Cessna 172 into the South Lake Tahoe Airport in November. After our stay, about three-quarters of an inch of ice had formed on the airplane. Being naive at the time about such conditions, I assumed it would blow off during the takeoff roll—silly me.

An experienced tower controller, however, suggested we clean the ice. It had to be scraped off with a plastic scraper! Later I calculated we had between 300 and 400 lb of ice on the airplane. I had no experience with this condition at the time. I shutter to think what would have happened during a takeoff in a low-performance airplane, 300 lb over gross, at a high-altitude airport, with three-quarters of an inch of ice on the airfoils. This is a perfect example of where the test almost came before the lesson. This was the case with a Cessna Caravan pilot. The aircraft was parked outside where freezing rain fell. The next day the pilot removed about 80 percent of the snow covering the wings, which left a coarse layer of ice about three-sixteenths of an inch thick. The aircraft crashed after takeoff. The probable cause was determined to be the pilot's failure to remove ice from the airframe prior to takeoff.

The solution to these problems is to hangar the aircraft, arrange for a deicing service, or plan the departure later in the day when the sun has melted the ice. During taxi, avoid areas of standing water or slush. Slush thrown into wheel wells, wheel pants, and control surfaces can freeze, resulting in locked controls and frozen landing gear or brakes. With a descent through an icing layer, and surface temperatures close to or below freezing, a pilot must be prepared for an approach and landing with airframe icing, possibly to an ice- or snow-covered runway.

If snow, ice, or slush are on the runway, aircraft control might be difficult, especially in high winds, with reduced braking efficiency, resulting in a longer-than-normal ground roll. Without windscreen deice—that's *deice,* not defrost—a pilot could be faced (pardon the pun) with zero forward visibility during the landing. Regulations require the pilot to consider "runway lengths at airports of intended use." Pilots operating in this environment must consider these factors when selecting destination and alternate airports, or even the advisability of making the flight.

During the winter, both VFR and IFR pilots may have to contend with a snow-or ice-covered runway. Airports may be closed due to snow or ice accumulation or for snow removal. Be sure to check Notices to Airmen (NOTAMs) for airport closures, runway conditions, and braking action reports. It's also a good idea to check enroute with an FSS for updates on destination and alternate runway conditions.

Recall from Chap. 20 that civil pilots have access to objective runway friction measurements at some airports. The Greek letter mu (μ) designates a friction value representing runway surface conditions. The lower the mu value, the less effective braking performance becomes and more difficult direction control becomes. Pilots can expect to receive mu values from ATC or a Notice to Airmen. Also recall that no correlation has been established between mu values and the descriptive terms "good," "fair," "poor," and "nil" used in braking action reports. Use mu information with other knowledge including aircraft type and weight, wind conditions, and your previous experience. If you haven't had training or experience with snow- or ice-covered runways, avoid them!

Arriving pilots during instrument weather conditions may have to descend through an icing layer all the way to the ground. On landing, a pilot may have an iced-over windscreen and adverse flight characteristics, including tailplane icing. When the pilot lowers the flaps for landing, the change in angle of attack might cause the tailplane to stall! This is why some advise the use of no flaps when ice is present. Higher stall speed due to ice may require a higher approach airspeed. With ice, a no-flaps, higher-than-normal airspeed, a longer-than-normal ground roll, possibly on an ice-covered runway, may be required.

Recall our discussion of flight and instrument icing problems. The solution to an iced-over pitot or engine instrument probe is crosscheck. Crosschecking all flight or engine instruments will reveal the discrepancy. With the problem diagnosed, a plan can be developed to work around the situation.

CASE STUDY The investigation of a Cessna 152 accident revealed that the crankcase breather line was plugged with ice, and the oil had been forced out through the engine's nose seal. The probable cause was determined to be the pilot's inadequate preflight that failed to detect the ice-clogged breather.

As with turbulence, there are no hard-and-fast rules as to exactly where ice will develop. I had planned to include a table providing an overview of icing locations, probability, and intensity—similar to the table in Chap. 23 on turbulence. However, after consulting with NWS icing experts, it seems that such generalities are just not possible. This highlights the hazards associated with icing.

Winter can be one of the best flying seasons, with its cool temperatures and usually excellent flying weather between storms. But, like mountain flying in the summer, winter flying has its own unique problems and solutions.

FACT   No aircraft is tested or certified for flight in severe icing conditions, especially areas of supercooled large droplets.

CHAPTER 25
THUNDERSTORMS

Thunderstorms contain just about every weather hazard known to aviation. The first section of this chapter discusses most hazards associated with thunderstorms. Since convective *low-level wind shear* (LLWS) is such a serious hazard, we have dedicated a separate section to this phenomenon. Convective, or thunderstorm, LLWS is caused by a microburst. Microbursts are inherently random and difficult to forecast. The hazards of wind shear microbursts affect general aviation as well as air carrier operations. In the convective low-level wind shear portion of the chapter, we will discuss the causes, recognition, and avoidance of this phenomenon.

CASE STUDY Just when you think you've seen all the hazards associated with thunderstorms, something different comes up. Consider, for example, U.S. Air Force Lt. Col. William Rankin. He was forced to bail out of his jet at 47,000 ft over Virginia—you guessed it!—into a raging thunderstorm. As he fell through the storm, he plummeted to 10,000 ft before opening his chute. For the next 40 min the colonel rode the storm's updrafts and downdrafts, with lightning bolts flashing all around. He finally wound up in some trees about 65 mi from where he bailed. His adventure gives a whole new meaning to "sport parachute jumping."

In the final section of the chapter, strategies, we will evaluate accident statistics. With a knowledge of thunderstorm phenomena, we can develop techniques to avoid their hazards. The section concludes with some dos and don'ts of thunderstorm avoidance and recommended strategies should a thunderstorm be inadvertently penetrated. However, it cannot be overemphasized that the key to avoiding thunderstorm hazards is to avoid thunderstorms!

THUNDERSTORM HAZARDS

Thunderstorm hazards consist of turbulence, icing, precipitation—including hail, lightning, tornadoes, gusty surface winds—including low-level wind shear, effects on the altimeter, and low ceilings and visibilities. Under certain conditions, these storms can produce high-density altitude. Thunderstorm hazards are illustrated in Fig. 25-1.

It should be no surprise that thunderstorms have the potential to produce severe to extreme turbulence. Vertical motion is the structural basis of the cell. In thunderstorms, the widths of updrafts and downdrafts may vary from a few feet to several thousand feet. These drafts affect the aircraft's altitude as it flies through the thunderstorm. It is virtually impossible to hold altitude. Altitude changes of several thousand feet are not unusual. This is illustrated by the close proximity of the updrafts and downdrafts in Fig. 25-1.

Downdrafts continue below the base of the cloud with significant speed to within 300 to 400 ft of the ground. These drafts constitute a significant hazard to flight beneath the thunderstorm, which is often in heavy rain and poor visibility. The most severe turbulence

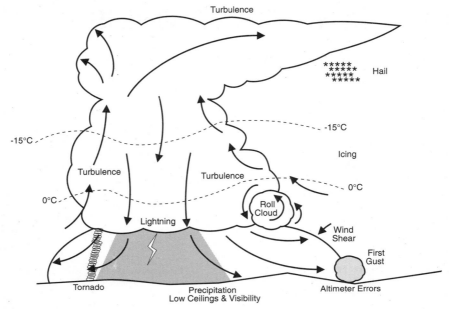

FIGURE 25-1   Thunderstorms can produce every kind of aviation weather hazard.

is encountered most frequently near the freezing level, but it can also occur from the ground to above the cloud tops. Significant turbulence can also occur in clear air well away from the cell itself.

Icing is another principal hazard in thunderstorms. Expect icing in all storms at elevations above the freezing level. Although thunderstorm clouds are usually limited in diameter, flying in them for even a short time can result in severe icing. The most severe icing can be expected between the freezing level and −15°C, as illustrated in Fig. 25-1.

Thunderstorms contain great amounts of liquid moisture, even though rain is not falling. The amount of liquid water decreases above the freezing level where snow becomes more predominant. Snow mixed with supercooled water exists everywhere above the freezing level, creating an icing hazard.

Hail can be one of the worst hazards of thunderstorm flying. Great amounts of hail and the largest stones generally are found in the larger and taller storms, but many thunderstorms have no hail associated with them. In general, large hail occurs in severe thunderstorms. Frequently hail is carried aloft and tossed out the top or side of the cloud by updrafts and thus may be encountered in clear air several miles from the cloud. Hail frequently exists in thunderstorms even though it is not reported at the ground. Flight beneath the anvil should be avoided because of the hail hazard, as illustrated in Fig. 25-1.

The largest hailstones and greatest frequency occur in a mature thunderstorm cell, usually at altitudes between 10,000 and 30,000 ft. Hail usually falls in streaks or swaths beneath the thunderstorm, covering an area about 1/2 mi long and 5 mi wide. However, encounters can occur in clear air outside the storm cell. Hail typically occurs in the rain area within the cloud, under the anvil or other overhanging cloud, and up to 4 mi from the cloud. Hail can also include other forms of frozen precipitation with differing origins, such as snow. Thunderstorms that are characterized by strong updrafts, large liquid water content, large cloud-drop size, and great vertical height are favorable to hail formation. The violent updrafts keep hailstones suspended for several up and down cycles. Each cycle adds a layer to the hailstone until it can no longer be suspended in the cloud.

Hail can cause severe damage to objects on the ground as well as aircraft in flight. Blunted leading edges, cracked windscreens, and frayed nerves are common results of a hail encounter. A number of multiple turbine engine power loss and instability occurrences, forced landings, and accidents have been attributed to operating airplanes in extreme rain or hail. Investigations have revealed that rain or hail concentrations can be amplified significantly through the turbine engine core at high flight speed and low engine power conditions. Rain or hail may degrade compressor stability, combustion flameout margin, and fuel control rundown margin. Ingestion of extreme quantities of rain or hail through the engine core may ultimately produce engine problems, including surging, power loss, and flameout. Pilots must be familiar with this phenomenon and comply with the manufacturer's recommendations. Like most thunderstorm hazards, avoidance may be the only safe alternative.

Lightning experienced in a thunderstorm can cause temporary blindness so that the pilot may temporarily lose control of the aircraft because he or she cannot read the instruments. Lightning damage to navigational and electronic equipment also can create a hazard. Small punctures may result in aircraft skin from direct lightning strikes. Lightning is found throughout the thunderstorm cloud, but it is most frequent and severe from the freezing level up to $-10°C$. The aircraft, like cars on the ground, typically insulates the passengers from lightning hazards. The only known incident of lightning downing a jetliner occurred on December 8, 1963. Over Elkton, Maryland, lightning struck the aircraft, exploding three of its fuel tanks. Eighty-one people perished. Although lightning strikes an airplane approximately every 3000 h, significant damage is the exception, rather than the rule. Another exception occurred in May 1996 when lightning struck a Beech King Air. The result: a cabin fire. Lightning strikes can burn wire, magnetize airframes, destroy composite structures, fuse control surfaces, cause turbojet compressor stall and flameout, and, although very rare, ignite fuel tanks. The answer: Avoid thunderstorms!

CASE STUDY An Air Force commander ordered a military KC135 to penetrate an intensity level 4 thunderstorm to refuel an SR-71. The pilot, who was also a meteorologist, had always wondered about the diameter of a lightning bolt. You guessed it! A lightning strike punched a 4-in-diameter hole in the radome. (Allegedly, the commander who ordered the thunderstorm penetration was soon looking for a new position.)

Lightning is classified by its origin and destination. Lightning can be from cloud to ground, in cloud, cloud to cloud, or cloud to air. And astronauts have observed lightning coming out of the tops of clouds—cloud to space? Figure 25-2 shows lightning that would be classified as cloud to ground.

Another electrical phenomenon associated with thunderstorms is *corona discharge,* colloquially known as *St. Elmo's fire.* St. Elmo's fire becomes visible as bluish static electric streaks dancing across the windscreen. Aircraft flying through or in the vicinity of thunderstorms often develop corona discharge streamers from antennas and propellers, and even from the entire fuselage and wing structure. It produces the so-called precipitation static. Precipitation static, however, usually affects only low-frequency radio communications and navigation.

Tornadoes are produced by severe thunderstorms. They are whirlpools or air, cloud, and debris that range in diameter from 100 ft to a half-mile. Pressure is extremely low in the center of the small concentrated vortex. Tornado winds probably reach 200 to 300 kn, although based on damage patterns, winds of over 400 kn are indicated. Tornadoes appear as funnel-shaped clouds from the base of thunderstorms and usually move 25 to 50 kn. Their paths range from a few miles to probably less than 50 mi; although squall lines and frontal systems can produce a series of tornadoes that can cover hundreds of miles. Their exact path is erratic and unpredictable.

FIGURE 25-2   The solution to the lightning hazards is to avoid thunderstorms.

The late Professor Ted Fujita, a University of Chicago meteorologist, developed a scale to estimate peak winds inside a tornado based on damage. Table 25-1 contains the six-category Fujita scale.

Technically, to be called "tornadoes" they must touch the ground. When they occur over water, they are known as *waterspouts*. When the characteristic whirling clouds extend downward from the parent cloud but do not reach the surface, they are called *funnel clouds*. Families of tornadoes or funnels have been observed from appendages of the main cloud extending several miles outward from the area of lightning and precipitation. A tornado vortex extends a great distance into the parent cloud, and pilots may encounter extreme turbulence in the imbedded vortex that is not visible. Frequently cumulonimbus mammatus clouds occur in connection with violent thunderstorms and tornadoes.

An average of approximately 200 tornadoes per year occur in the United States; about 90 percent develop ahead of cold fronts. Tornadoes are not a common factor for the west coast states or the intermountain region of Idaho, Montana, Nevada, Utah, and Arizona and the western portions of Wyoming, Colorado, and New Mexico. Tornadoes occur most frequently in the Great Plains states east of the Rocky Mountains. However, they have occurred in every state and Canada. Tornadoes occur with isolated thunderstorms at times, but more frequently with cold fronts or squall line thunderstorms. Reports or forecasts of tornadoes are indications that atmospheric conditions in the area are favorable for extreme turbulence.

The most severe thunderstorms develop in a wind shear environment—that is, wind speed must increase with height. At the same time, the wind direction veers (a clockwise change in wind direction). Typically, surface wind blows from the southeast at between 15 and 20 kn; at 5000 ft wind is from the south at between 30 and 35 kn; at 15,000 ft wind is from the southwest at about 50 kn. This situation allows the storms to have separate updrafts and downdrafts—steady-state thunderstorms. Low-level air feeds the storm from

TABLE 25-1 Fujita Scale of Tornado Intensity

F0	Up to 63 kn
F1	64–97 kn
F2	98–136 kn
F3	137–179 kn
F4	180–226 kn
F5	227–277 kn

the southeast. This air is lifted in an updraft and exits the storm at high levels toward the east. Air at midlevels approaches the storm from the southwest. It passes around the updraft, gets caught in a downdraft on the north side of the storm, and exits on the back side as part of the low-level outflow. Pilots can expect to see discussions of this wind shear environment in the outlook portions of convective SIGMETs and convective outlooks.

Figure 25-3 illustrates the typical synoptic situation for tornadoes in the central United States. A low-level jet of moist air feeds the systems from the south. The upper-level tropopause jet acts like a vacuum sucking up the moist air. The dashed-line box shows the usual location where tornadoes develop.

Forecasts for tornadoes are contained in convective SIGMETs, alert weather watches, severe weather watch bulletins, and the convective outlooks.

Fair-weather waterspouts form over warm and shallow coastal waters. They are much smaller and less intense than the average tornado and tend to form through convergence in unstable air beneath developing cumulus. They tend to be stronger than their dust devil cousins because of the energy available, through the release of latent heat, from the ocean surface.

Refer to Fig. 25-1. Gusty and variable surface winds are associated with thunderstorms. Usually the first gust, or gust front, precedes arrival of the roll cloud and onset of rain as the thunderstorm approaches. Frequently it stirs up dust and debris as it plows along, announcing the thunderstorm's approach. The strength of the first gust frequently is the strongest observed at the surface during a thunderstorm. It may approach at 100 kn in extreme cases. The roll cloud is not always present but is found most frequently on the leading edge of fast-moving thunderstorms. It represents an extremely turbulent area.

A *gustnado* describes a funnel cloud that develops along the gust front; the funnel cloud is not a tornado. It is believed that the gustnado receives its initial rotation from the shift in wind directions across the gust front. Cold, dense air behind the gust front lifting the warm air ahead imparts a rotating motion in the wind shear zone.

Pilots should avoid takeoff or landing when a thunderstorm is within 10 to 20 mi of the airport. This is the region of the strongest and most variable winds. Caution must also be exercised following thunderstorm passage. A strong, gusty outflow boundary can follow the storm. Along with these winds are downbursts and microbursts that produce severe low-level wind shear.

Pressure usually falls rapidly with the approach of a thunderstorm. Then it rises sharply with the onset of the gust front and arrival of the cold downdraft and heavy rain. Pressure then normally falls back as the storm moves on. This cycle of pressure change may occur within 15 min. Height indicated on a pressure altimeter during the storm may be in error by over 100 ft.

Heavy rain brings lowering ceilings and visibilities. With severe storms, ceilings and visibilities can be at or near zero. Figure 25-4 shows a thunderstorm downburst in Idaho's Snake River Valley.

We have already mentioned the thunderstorm phenomenon of a *heat burst*. Carolyn Kloth, forecaster at the NWS's Aviation Weather Center, has proposed that a heat burst may be a another thunderstorm hazard for aviation—high-density altitude. Heat bursts increase turbulence and wind shear, develop in radar echo-free air, and affect the pressure altimeter. They typically occur in the dissipating stage of nighttime thunderstorms.

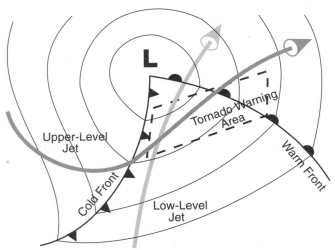

FIGURE 25-3   This is the typical synoptic situation for tornadoes in the central United States.

FIGURE 25-4   With severe storms, ceilings and visibilities can be at or near zero.

CONVECTIVE LOW-LEVEL WIND SHEAR

Wind shear, a rapid change in wind direction or speed, has always been around. Convective activity produces *severe wind shear,* which is defined as a rapid change in wind direction or velocity, causing airspeed changes of greater than 15 kn or vertical speed changes of greater than 500 ft/min. The microburst produces the most severe wind shear threat.

Rain-cooled air within a thunderstorm produces a concentrated rain or virga shaft less than 1/2 mi in diameter, which forms a downdraft. The downdraft or downburst has a very sharp edge and forms a ring vortex upon contact with the ground, and it spreads out, causing gust fronts that are particularly hazardous to aircraft during takeoff, approach, and landing. Reaching the ground, the burst continues as an expanding outflow.

A microburst consists of a small-scale, severe, storm downburst less than 2 1/2 mi across. This flow can be 180° from the prevailing wind, with an average peak intensity of about 45 kn. Microburst winds intensify for about 5 min after ground contact and typically dissipate about 10 to 20 min later. Microburst wind speed differences of almost 100 kn have been measured. On August 1, 1983, at Andrews Air Force Base, indicated differences near 200 kn were observed. Some microburst events are beyond the handling capability of any aircraft and pilot. Although normally midafternoon, midsummer events, microbursts can occur any time, in any season.

Avoidance is the best defense against a microburst encounter. When the possibility of microbursts exists, the pilot must continually check all clues. Therefore, pilots must learn to recognize situations favorable to this phenomenon.

Figure 25-5 illustrates the effects of a microburst on a landing aircraft—this is the same scenario that claimed the Delta flight. The aircraft is established on the glide path, and it may even have the runway in sight. The outflow from the microburst causes increasing

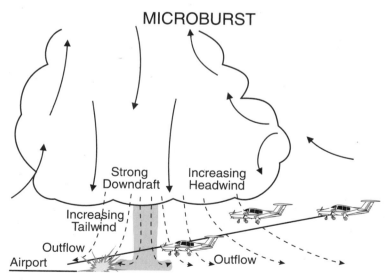

FIGURE 25-5   As is true of most thunderstorm hazards, avoidance is the best defense against a microburst encounter.

headwind. This has the effect of causing the aircraft to climb. Typically the pilot increases the rate of descent. Just as the rate of descent increases, the aircraft flies into a strong downdraft. This increases the rate of descent. The pilot attempts to climb back to the proper glide path. This may result in reduced airspeed just as the aircraft enters the areas of increasing tailwind. If the aircraft is close to the ground, recovery may not be possible. The gray area in Fig. 25-5 represents the microburst. Note the characteristic upward curl as the wind spreads out upon contact with the ground.

The scenario is similar for aircraft taking off. The aircraft encounters increased headwind during the takeoff roll. At rotation or just after liftoff, the aircraft encounters the strong downdraft. This is immediately followed by an increasing tailwind. Again, if the aircraft is close to the ground, recovery may not be possible.

CASE STUDY   American Airlines flight 1420 crashed while attempting to land at Little Rock's National Airport. The crash occurred just before midnight on June 1, 1999. On approach a microburst hit the airport with a wind gust to 76 kn.

KLIT 020458Z 29010G76KT 210V030 1/2SM +TSGSRA...

Note that in addition to the gust, wind was variable between 210° and 030° to 180°—a classic indication of a microburst.

The FAA, along with a group of aviation specialists, has developed AC 00-54, *Pilot Windshear Guide.* Although primarily for the airlines, much of the information can be applied to general aviation.

Wind Shear Recognition

The following discussion is based on AC 00-54, *Pilot Windshear Guide,* and the Department of Commerce publication *Microbursts: A Handbook for Visual Identification.* The latter publication is for sale by the Superintendent of Documents. It contains an indepth, technical explanation of the phenomenon along with numerous color photographs depicting microburst activity. It should be part of every pilot's library.

Microbursts can develop any time convective activity, such as thunderstorms, rain showers, or virga occur, associated with both heavy and light precipitation. Approximately 5 percent of all thunderstorms produce microbursts. And more than one microburst can occur with the same weather system. Therefore, pilots must be alert for additional microbursts—if one has already been encountered or reported—and prepare for turbulence and shear as subsequent microbursts interact. Microbursts are characterized by precipitation or dust curls carried back up toward the cloud base, horizontal bulging near the surface in a rain shaft forming a foot-shaped prominence, an increase in wind speed as the microburst expands over the ground, and abrupt wind gusts.

Microbursts can occur in extremely dry, as well as wet, environments. The lack of low clouds does not guarantee the absence of shear. Microbursts can develop below clouds with bases as high as 15,000 ft. As virga or light rain falls, intense cooling causes the cold air to plunge, resulting in a dry microburst. Evaporative cooling turbulence, associated with this phenomena, has already been discussed. Anvils of large dry-line thunderstorms can produce high-level virga and result in dry microbursts. High-based thunderstorms with heavy rain should be of particular concern. This was the type that produced intense wind shear in the 1985 Delta accident in Dallas/Fort Worth.

Embedded microbursts are produced by heavy rain from low-based clouds in a wet environment. A wet microburst might first appear as a darkened mass of rain within a light rain shaft. As the microburst moves out along the surface, a characteristic upward curl appears.

The potential for wind shear and microbursts exists whenever convective activity occurs. Pilots should review forecasts (FAs, TAFs, WSTs, and AWWs) for thunderstorms—thunderstorms imply low-level wind shear—or the inclusion of LLWS. Check surface reports and PIREPs for wind shear clues: thunderstorms, rain showers, gusty winds, or blowing dust. Dry microbursts are more difficult to recognize. Check surface reports for convective activity (VCRA, CBs, or VIRGA) and low relative humidity (15° to 30°C temperature/dew point spread).

The Low-Level Wind Shear Alert System (LLWAS) has been installed at 110 airports in the United States. The system detects differences between wind speed around the airport and a reference center-field station. Differences trigger an alert. Sensors are not necessarily associated with specific runways; therefore, descriptions of remote sites are based on the eight points of the compass: "Center field wind three one zero at one five. North boundary wind zero niner zero at three five."

The FAA's integrated wind shear detection plan consists of the following programs:

- Integrated Terminal Weather System (ITWS)
- Terminal Doppler Weather Radar (TDWR)
- Weather System Processor (WSP)
- Low-Level Wind Shear Alert System (LLWAS)

This equipment is used to detect wind shear and microbursts and to alert controllers, who then pass the information onto the pilot. ITWS, TDWR, and WSP process radar data; LLWAS uses a series of wind instruments to detect the presence of hazardous wind shear and microbursts in the vicinity of the airport. Locations are contained in the *Aeronautical Information Manual* and the *Airport/Facility Directory*.

Like most equipment, these systems have limitations. It is important for pilots to understand that TDWR does **not** do the following:

- Warn of wind shear outside of the arrival and departure ends of the runway
- Detect nonconvective wind shear
- Detect gusty or cross-wind conditions
- Detect turbulence

Pilots can expect wind shear and microburst alerts from these agencies to appear in the following format:

RUNWAY 27 ARRIVAL, MICROBURST ALERT, 35 KNOT LOSS 2 MILES FINAL, THRESHOLD WIND 250 AT 20.

On the final approach corridor to runway 27, at approximately 2 mi from the runway, the crew can expect a 35 kn loss of airspeed. The pilot can then anticipate a wind shear encounter and apply escape procedures. The surface wind at the approach end of runway 27 is 250° at 20 kn. The pilot can expect to exit the event before reaching the runway threshold.

RUNWAY 27 ARRIVAL, WIND SHEAR ALERT, 20 KNOTS LOSS 3 MILES FINAL, THRESHOLD WIND 200 AT 15.

At approximately 3 mi on final, the crew can expect to encounter a wind shear event, with 20 kn loss of airspace and possible turbulence. With threshold wind of 200° at 15 kn, the crew can expect to exit the shear zone before reaching the runway.

The lack of an LLWAS alert does not necessarily indicate the absence of wind shear. LLWAS also has limitations. Magnitude of the shear might be underestimated. Surface obstructions can disrupt or limit the airflow near the sensor, and due to the location of sensors, microburst development might go undetected, especially in the early stages. Sensors are located at the surface; therefore, microburst development that has not yet reached the surface will be undetected, and because coverage exists only near the runways, microbursts on approach will not be observed. Even with these limitations, LLWAS can provide useful information about winds near the airport.

Airborne radar returns of heavy precipitation indicate the possibility of microbursts. Although potentially hazardous, dry microbursts might produce only weak radar returns. Strong wind shear might occur as far as 15 mi from storm echoes. Radar echoes can be misleading by themselves, and it might require a Doppler radar to spot the danger of a dry microburst. The southwest edge of an intense storm can appear weak both visually and on radar; however, this area is known to spawn tornadoes and severe wind shear. Convective weather approaching an airport, the downwind side, tends to be more hazardous than activity moving away.

No quantitative means exist for determining the presence or intensity of microburst wind shear. Pilots must exercise extreme caution when determining a course of action. Microburst wind shear probability guidelines have been developed by the FAA, and they apply to operations within 3 mi of the airport, along the intended flight path and below 1000 ft AGL. Probabilities are cumulative; therefore, when more than one point exists, probability increases.

The following indicate a high probability of wind shear with the presence of convective weather near the intended flight path:

1. Localized strong winds reported, or observed blowing dust, rings of dust or tornadolike features

2. Visual or radar indications of heavy precipitation

3. PIREPs of airspeed changes 15 kn or greater

4. LLWAS alert or wind velocity change of 20 kn or greater

A pilot must give critical attention to these observations. A decision to avoid, divert, or delay is wise.

The following indicate a medium probability of wind shear with the presence of convective weather near the intended flight path:

1. Rain showers, lightning, virga, or moderate or greater turbulence reported or indicated on radar

2. A temperature/dew point spread of 15° to 30°C

3. PIREPs of airspeed changes of less than 15 kn

4. LLWAS alert or wind velocity change of less than 20 kn

A pilot should consider avoiding these conditions. Precautions are indicated.

The FAA states: "Pilots are...urged to exercise caution when determining a course of action." Probability guidelines "should not replace sound judgment in making avoidance decisions." In aviation weather, there are no guarantees. The lack of high- or medium-probability indicators in no way promises the absence of wind shear when convective weather is present or forecast. Avoidance is the best precaution.

Review the following Colorado Springs (COS) METAR and UUA for wind shear probability indicators:

METAR KCOS 202150Z 36012G24KT 45SM -TSRA BKN090 OVC250 23/08 A3019 RMK TS OVH MOV E OCNL LTGCG SW E

COS UUA /OV COS/TM 2156/...TP PA60/RM AIRSPEED +- 40 KTS, THOUGHT I WAS IN THE TWILIGHT ZONE

A thunderstorm with rain showers, a 15° temperature/dew point spread, strong, gusty surface winds, and lightning are being reported. Add the PIREP, and there are two high- and two moderate-probability indicators. This is an example of the dry environment. A pilot must be watchful for visual microburst and LLWS clues.

This METAR indicates a wet microburst environment:

METAR KPHL 032250Z 36018G24KT 1SM +TSRA BR VV006...RMK FQT LTGICCG

A thunderstorm with heavy precipitation, strong gusty winds, and lightning is being reported. There is certainly a high probability of wind shear and microbursts.

Takeoff, Approach, and Landing Precautions

Select the longest suitable runway for takeoff. Determine at what point the takeoff can be aborted with enough runway to stop the aircraft. Certain manufacturers provide tables for takeoff calculations; otherwise, the pilot will have to base this distance on landing roll tables and experience. Use the recommended flap setting, if available, for gusty wind or turbulent conditions. Use maximum-rated takeoff power. This reduces takeoff roll and overrun exposure. Consider increased airspeed at rotation to perhaps improve the ability of the airplane to negotiate wind shear or turbulence after liftoff. Do not use a speed reference flight director. Be alert for airspeed fluctuations that might be the first signs of wind shear. Should shear be encountered with sufficient runway remaining, abort the takeoff. This decision, however, can be made only by the pilot, based on his or her training and experience. After takeoff, use maximum rated power and rate of climb to achieve a safe altitude, at least 1000 ft AGL.

Select the longest suitable runway to land. Consider a recommended approach configuration with a higher-than-normal approach speed. Turbulent air penetration or maneuvering speed should be considered. Establish a stabilized approach at least 1000 ft AGL with configuration, power, and trim set to follow the glideslope without additional changes. Any deviation from glideslope or airspeed change will indicate shear. The autopilot, except for autoflight systems, should be disengaged, with the pilot closely monitoring vertical speed, altimeter, and glideslope displacement. Ground speed and airspeed comparisons can provide additional information for wind shear recognition. Increased approach speed, while providing an extra margin for safety, will require longer-than-normal landing distance.

Wind Shear Recovery Technique

Wind shear recovery technique has not yet been developed for small aircraft. The following wind shear recovery technique, developed for airline aircraft, has been adapted from AC 00-54. It is, however, logical and applicable to most wind shear encounters in practically any aircraft.

Wind shear recognition is crucial to making a timely recovery decision. Encounters occur infrequently with only a few seconds to initiate a successful recovery. The objective is to keep the airplane flying as long as possible in hope of exiting the shear. The first priority must be to maintain airplane control. The following guidelines were developed for the airlines; exact criteria cannot be established. Whenever these parameters are exceeded, recovery and/or abandoning the takeoff or approach should be strongly considered. It must be emphasized that it is the responsibility of the pilot to assess the situation and use sound judgment in determining the safest course of action. It might be necessary to initiate recovery before any of these parameters are reached:

1. Plus or minus 15 kn indicated airspeed
2. Plus or minus 500 ft/min vertical speed
3. Plus or minus one dot glideslope displacement
4. Unusual throttle position for a significant period of time

If any condition is encountered, aggressively apply maximum rated power. Avoid engine overboost unless it is required to avoid ground contact. While on approach, do not attempt to land. Establish maximum rate of climb airspeed. As with any turbulent condition, pitch up in a smooth, steady manner. Should ground contact be imminent, pitch up to best-angle-of-climb airspeed, being careful not to stall the airplane. Controlled contact with the ground is preferable to an uncontrolled encounter. When airplane safety has been ensured, adjust power to maintain specified limits. When the airplane is climbing and ground contact is no longer an immediate concern, cautiously reduce pitch to desired airspeed.

The key to the thunderstorm and LLWS hazard is avoidance. A superior pilot uses superior knowledge to avoid having to use superior skill. At the first sign of severe shear, reject the takeoff or abandon the approach. It is easier to explain an aborted takeoff or missed approach to passengers rather than explain an accident to the FAA and insurance company—assuming you're still around to do so.

Avoidance is the operative word with thunderstorms, microbursts, and wind shear. A pilot's proper application of many resources—training, experience, visual references, cockpit instruments, weather reports and forecasts—make avoidance possible.

STRATEGIES

Accident statistics show the majority of thunderstorm accidents occur to non-instrument-rated, low-time private pilots; over half resulted in fatalities. Ironically, most received a preflight weather briefing. Most occurred when pilots initiated IFR flight into adverse weather, attempted VFR flight into deteriorating weather, or attempted to fly in or around thunderstorms. Some occurred when flight was continued into areas of embedded thunderstorms. Others resulted in loss of control due to high, gusty winds or crosswinds.

The violent nature of thunderstorms causes gust fronts, strong updrafts and downdrafts, and wind shear in clear air adjacent to the storm out to 20 mi with severe storms and squall lines. Precipitation, which is detected by radar, generally occurs in the downdraft, while updrafts remain relatively precipitation free. Clear air or lack of radar echoes does not guarantee a smooth flight in the vicinity of thunderstorms.

Our first defense is a complete preflight weather briefing. Recall from Chap. 4 our group of hot-air balloon pilots downed in California's Napa Valley by thunderstorms. These pilots typically call the FSS and request an abbreviated briefing of specific weather reports and winds aloft forecasts. Since they rarely ask for the synopsis or area forecast, and often scattered thunderstorms are not covered in weather advisories, they would not receive information on convective activity. A complete, or standard briefing, just prior to departure cannot be overstressed.

Storm detection equipment is airborne weather radar or lightning detection equipment. Both radar and lightning detectors have limitations. Pilots using this equipment to avoid thunderstorms must understand their operation and limitations. Just reading the manual is certainly not enough to prepare a pilot to translate the complex symbology presented into reliable information. Pilots should obtain training courses with appropriate instructors and simulators to properly use this equipment.

More and more general aviation aircraft are equipped with airborne weather radar and lightning detection equipment. However, these systems are plagued by low power, attenuation, and limited range. A pilot might pick his or her way through a convective area only to find additional activity beyond. The following PIREP illustrates just such an occurrence:

MSY UUA /OV NEW 150020/TM 2015/FLDURD/TP C550/TB SEV/RM OCCURRED IN AREA WHERE ACFT RADAR DID NOT INDICATE PCPN. BOTH CREW INJURED.

This incident occurred over New Orleans in thunderstorm weather. Both crew members of a Cessna Citation were injured when the aircraft encountered severe turbulence in an area where their airborne weather radar indicated no precipitation.

Recall from Chap. 11 the limitations of airborne weather radars and lightning detection equipment. Airborne weather radars are low power, generally with a wavelength of 3 cm. Precipitation attenuation, which is directly related to wavelength and power, can be a significant factor. With lightning detection equipment, the absence of dots or lighted bands does not necessarily mean no thunderstorms. Precipitation intensity levels of 3 and occasionally 4 (moderate to heavy) would be indicated on radar without activating the lightning detection system. A clear display indicates only the absence of electrical discharges. This is why many authorities recommend a combination of radar and lightning detection systems as the best thunderstorm avoidance method. Keep in mind that these are avoidance, not penetration, devices. Thunderstorms imply severe or greater turbulence, and neither radar nor lightning detection systems, at present, directly detect turbulence.

One solution to the limitations of airborne weather radar and lightning detection systems is flight watch. With real-time NWS NEXRAD Doppler weather radar, flight watch has the latest information. Well before engaging any convective activity, a pilot should consult flight watch to determine the extent of the system and its movement, intensity, and intensity trends. Armed with this information, the pilot can determine whether to attempt to penetrate the system or select a suitable alternate. ATC prefers issuing alternate clearances to handling emergencies in congested airspace and severe weather.

CASE STUDY A Bonanza pilot approached an area of thunderstorms in California's central valley. The pilot received the latest weather radar and satellite information, as well as PIREPs and surface observations from flight watch. The pilot safely traversed the area with minimum diversion or delay.

That is not a very exciting story, but that's the purpose of flight watch—to assist pilots in conducting uneventful flights. Flight watch has been around for more than 20 years. In spite of this, its function, and the best way to use this important service, is not known to, or understood by, many pilots.

CASE STUDIES We were trying to get from Springfield, Illinois, to Oklahoma City. Arriving in Saint Louis, the ground-based weather radar showed cell after cell along our route to the southwest. We decided to wait it out until the next day. However, about six that evening, the front passed and the weather cleared. We decided to continue on to Kansas City, which was also behind the front. As darkness fell, the southern horizon was ablaze with continuous lightning.

Checking weather from Kansas City to Oklahoma City indicated clearing, but there was still some thunderstorm activity. After departure we again saw lightning on the horizon. That night it just wasn't meant to be. We returned to Kansas City and spent the night. The next day was bright and clear. I was attending the FAA's Air Route Traffic Control School at the time. I missed half a day. It would have sure been embarrassing to have been involved in an aircraft accident!

Returning from Oshkosh, thunderstorms were forecast in the Colorado area. Sure enough, as we approached Pueblo, there were several cells. A check with flight watch indicated Pueblo was in the clear and Colorado Springs would make a suitable alternate, should Pueblo's weather deteriorate. By keeping visual contact with the storms, we circumnavigated to the north around the heavy rain, lightning, and buildups, and we made an uneventful landing. There really is no reason to get caught in a cell.

A notion held by some pilots is that the absence of a weather advisory means there is no significant weather. The lack of a weather advisory does **not** guarantee the absence of hazardous weather. An unfortunate pilot learned this lesson the hard way.

CASE STUDY   The synopsis described a moist unstable air mass. Thunderstorms were not forecast for the time of flight, but they were expected to develop; thunderstorms, however, were already being reported along the route. The pilot, without storm detection equipment, encountered extreme turbulence inadvertently entering a cell. The pilot, with three passengers, had filed an IFR flight plan based on the fact that there were no advisories. After the encounter, the pilot could not understand why a precaution or advisory regarding that system had not been provided. There were no advisories in effect because, at the time of the briefing, none were warranted. The pilot had the clues—moist unstable air; thunderstorms already reported—but he put complete trust in a forecast that included no precautions or advisories.

The existence of an advisory, or lack thereof, does not relieve the pilot from using good judgment and applying personal limitations. Like all pilots, I have had on occasion to park my turbo Cessna 150 and take one of American's Boeing 727s. These instances lend credence to the axiom: "When you have time to spare, go by air; more time yet, take a jet." When you don't have the equipment or qualifications to handle the weather, don't go! More about risk and decision making in Chap. 27, Risk Assessment and Management.

The preceding examples illustrate several decisions. One resulted in a canceled flight, another a routine flight—although the route was changed—and the last was almost fatal. My intent is not to brag about my skills or to criticize another individual. Instead, I hope to show the decision-making process, which should always be based on available information and a knowledge of the weather and of personal and aircraft limitations.

All too often, FSS briefers hear pilots flying aircraft without storm detection equipment say, "Thunderstorms, ah; well I'd better go IFR." Not for me, thanks. I want to be clear of clouds where I can see and avoid convective activity. Pilots who fly in conditions favorable for thunderstorms without storm detection equipment, and the knowledge to use it, sooner or later will end up—more likely upside down—in a thunderstorm cell.

The following are some dos and don'ts of thunderstorm avoidance:

Do avoid by at least 20 mi any thunderstorm identified as severe or giving an intense radar echo. This is especially true under the anvil of a large cumulonimbus.

Do clear the top of a severe thunderstorm by at least 1000 ft for each 10 kn of wind speed at the cloud top.

Do regard as severe any thunderstorm with tops 35,000 ft or higher.

Don't land or take off in the face of an approaching thunderstorm.

Don't attempt to fly under a thunderstorm, even if you can see through to the other side.

Don't try to circumnavigate thunderstorms covering more than half of the area, even with storm detection equipment.

Don't attempt to enter areas of embedded thunderstorms without storm detection equipment.

If thunderstorm penetration cannot be avoided, the following steps are recommended before entering the storm:

- Tighten seat belts and shoulder harnesses; secure all loose objects.
- Plan a course through the storm in the minimum time.
- To avoid the most critical icing, establish a penetration altitude below the freezing level or above −15°C.
- Turn on pitot heat and carburetor or inlet heat.
- Establish power setting for turbulence penetration airspeed.
- Turn up cockpit lights to highest intensity to lessen danger of temporary blindness from lightning.
- Disengage autopilot altitude and speed hold.
- Tilt airborne radar antenna up and down occasionally. Tilting may help detect a hail shaft or a growing thunderstorm cell.

The following are some dos and don'ts during thunderstorm penetration:

- Do keep your eyes on the instruments. Looking outside increases the danger of lightning blindness.
- Do maintain a constant attitude; let the aircraft ride with the turbulence. Maneuvers to maintain altitude increase gust loading.
- Don't change power settings.
- Don't turn back once in the thunderstorm. A straight course through the storm most likely will get you out of the hazards most quickly. Turning increases gust loading.

Three final words regarding thunderstorms: Avoid, avoid, avoid!

When it comes to thunderstorms, microbursts, and wind shear, a pilot's proper application of many resources—training, experience, visual references, cockpit instruments, weather reports, and weather forecasts—make avoidance possible.

A popular aviation saying goes: "Aviation in itself is not inherently dangerous. But to an even greater degree than the sea, it is terribly unforgiving of any carelessness, incapacity, or neglect."

CHAPTER 26
STRATEGIES FOR WEATHER SYSTEMS

Thus far we've discussed aviation weather theory and strategies for handling specific weather hazards. The discussion has been directed, generally, at local weather events, such as high-density altitude, and local areas of low ceiling and visibility, turbulence, icing, and thunderstorms. These phenomena are often produced by major weather systems, and individual hazards may not be specifically addressed here. In this chapter our focus is on weather systems. That is, large areas of organized weather such as frontal systems, troughs, and low-pressure areas. These phenomena can be divided into two distinct events: frontal weather systems and nonfrontal weather systems. We have touched on these subjects throughout the book. Here we will put our knowledge of these events into practical application of avoidance or penetration.

A pilot's ability to handle weather systems depends on a number of factors:

- The pilot's capability, experience, and currency
- The aircraft's capability and equipment
- The weather system to be negotiated

For example, a VFR-only pilot, or non-current-instrument pilot, is precluded from flying in less the VFR weather conditions; most single-engine aircraft without ice protection and storm detection equipment should not attempt to penetrate icing conditions or convective weather. On the other hand, weak weather systems may be safely negotiated by the most novice pilot. Larger aircraft, with appropriate equipment, have the capability to handle or avoid most serious weather events. In Chap. 27 we will specifically address the issues of risk assessment and management, as well as personal minimums.

Like any weather situation, the key to hazardous weather avoidance is knowledge. This knowledge consists of a thorough understanding of basic weather phenomena, a complete weather briefing just prior to departure, and frequent weather updates enroute.

FRONTAL WEATHER SYSTEMS

Basically pilots have two options when dealing with frontal systems: penetrate or avoid. If a pilot elects to penetrate, there are several additional options. These options are directly related to pilot and aircraft capabilities. A pilot may elect to fly over the front, penetrate at high or low altitude, or attempt to fly under the front. The pilot may want to fly through the front at a 90° angle to reduce exposure, or accept the consequences and fly parallel to the frontal band.

There are really only two ways to avoid or reduce exposure to frontal turbulence: Fly above the front or penetrate the front as close to perpendicular as possible. Penetrating a front at a 90° angle may work for a cold or stationary front, but it may not be satisfactory for a warm or occluded front because of the extensive areas of frontal weather associated with these fronts.

Icing problems exist mostly with spring and autumn fronts. Winter fronts, where clouds and precipitation are already frozen, do not present an icing hazard. In the summer, high freezing levels are common. However, in aviation weather, there are virtually no absolutes. A winter front in the southern tier or states may produce serious icing; in the summer, icing conditions may exist into the lower flight levels. The flight decision is often governed by terrain. Freezing levels are a paramount concern in flight planning. In the mountains and high plateaus, the freezing level is often at the surface.

Usually the solution to the icing problem is to fly high or low. Fly above the area of potential icing or below the freezing level. This might not be possible in aircraft with marginal performance, especially in the western United States with its high minimum enroute altitudes. Another option is to avoid visible moisture—precipitation and clouds. That is, remain clear of clouds and areas of precipitation when temperatures are less than 0°C.

Thunderstorms with all their ominous hazards are to be avoided. For aircraft with limited performance and without storm detection equipment, the only sane solution is to remain on the ground if storms cannot be visually circumnavigated.

NOTE   In the following sections, we will explore frontal penetration. The fact that the cold front illustration shows unstable conditions (cumuliform clouds) and the warm front illustration shows stable conditions (stratiform clouds) does not imply that cold fronts always produce convective activity or that warm fronts do not produce thunderstorms. Recall our discussion of fronts in Chap. 7.

Cold Fronts

Below is the Chicago area forecast synopsis discussed in Chap. 15:

```
CHIS FA 300940
SYNOPSIS VALID UNTIL 310400
AT 10Z CDFNT FROM LS SWWD THRU NWRN IA INTO NWRN KS THEN
WWD THRU CO. HI PRES OVR OH VLY AND MT. THE CDFNT WL CONT
EWD AND BY 00Z WL EXTEND FROM LWR MI SWWD INTO SRN KS AS HI
PRES BLDS OVR DKTS. MRNG FOG/ST OVR ERN GRTLKS WL IPV BY
16Z. AFTN/EVE TSTMS MOST ACTV ALG FNT FROM IL NEWD THRU
MI...WILLIAMS...
```

The CHIS synopsis has been plotted in Fig. 26-1.

First let's consider a VFR flight from Green Bay, Wisconsin (GRB), to Denver, Colorado (DEN). The pilot has a number of options. First, the pilot might be well advised to delay the flight until after frontal passage. Based on the synopsis, this should occur between about 1300Z and 1400Z. Recall that with a cold front, after frontal passage the weather typically improves rapidly. In addition to better weather, as high pressure builds into the area, the pilot could expect better tail winds in the anticyclonic circulation.

CASE STUDY   We had remained overnight in Huntington, West Virginia. The next day we planned to fly to Cincinnati, Ohio, for some radio repairs. Our only option was VFR. A relatively weak front was forecast to move through the area. At the time there was a weather service office on the field. I asked if the front had passed. Unfortunately, they were unable to give me a positive answer. I called the tower to coordinate light signals out of the airport, filed a VFR flight plan with NO TRANSMITTER in remarks, and requested the tower forward our departure time to flight service.

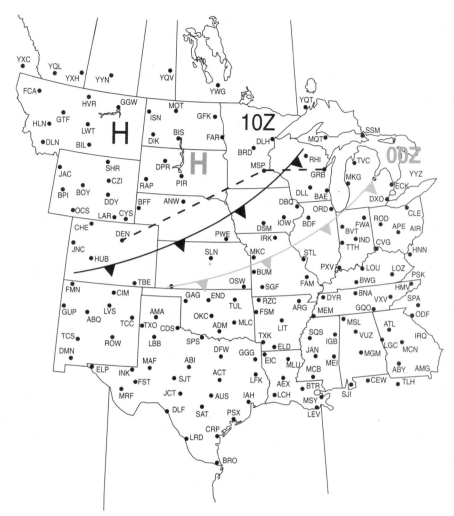

FIGURE 26-1   Frontal weather can be minimized by penetrating the front as close to a 90° angle as possible.

As it turned out, we did go IFR—I followed rivers, the Ohio to be exact. It was pretty dicey for the first 15 mi as we flew low over the river. This can be potentially dangerous at low altitudes, in poor visibilities, with all the cables and catenaries crossing the river. Then the weather improved dramatically as we exited the frontal boundary. Overflying Lunken Field in Cincinnati, I heard the tower call, "Aircraft over the field, if you're Cessna Five Two Yankee, rock your wings." Flight service had passed on our radio trouble, call sign, and color to the tower.

With today's technology, we could use the satellite picture and radar to determine the extent of the front cloud band and confirm frontal passage. Even if these products are not directly available at the airport through DUATs or other commercial sources, a call to flight service will give us the answer.

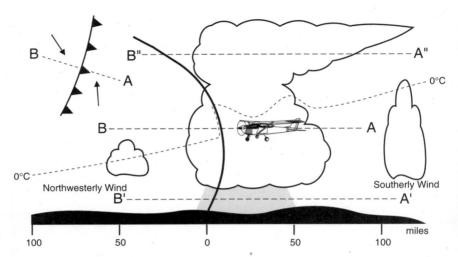

FIGURE 26-2   Cold front. Frontal icing can be avoided by flying below the freezing level or avoiding areas of clouds and precipitation.

An IFR flight might be planned direct to Minneapolis, Minnesota (MSP), to penetrate the front at a right angle and minimize exposure to its weather. When flying through a front, a pilot should be prepared for the changes in flight conditions from one air mass to the other. This is sometimes quite abrupt. Abrupt changes indicate a narrow frontal zone. At other times, the changes are very gradual, indicating a broad, weak, or diffuse frontal zone. Pilots should anticipate changes in cloud cover, temperature, moisture, wind, and pressure when penetrating a frontal zone. (These changes were discussed in Chap. 6.)

Here again, the pilot could use the satellite picture and radar to determine the extent and intensity of the front. Along with appropriate forecasts, the pilot could determine if convective activity was present or forecast to make the go–no-go decision.

The pilot might expect the conditions illustrated in Fig. 26-2. The pilot will be penetrating the cold front from the warm sector (A) to the cold sector (B). Our pilot has decided to penetrate the front below the freezing level in the warm sector. The pilot could expect no icing problem until crossing the frontal boundary, as indicated by the 0° isotherm in Fig. 26-2. However, icing could be expected after penetrating the frontal surface aloft.

How would the pilot recognize frontal passage? Frontal passage aloft typically would occur at some point beyond the location of the front on the surface. It would be indicated by a change from a left crosswind to a right crosswind, as illustrated by the wind arrows in Fig. 26-2, and a decrease in outside air temperature. Pressure would decrease until reaching the frontal boundary, then increase. The pressure change may be significant and the pilot would be well advised to obtain frequent altimeter setting updates. With a relatively steep frontal surface, the pilot could expect rapidly improving conditions after penetrating the frontal zone. This could be confirmed enroute with flight watch.

Upon frontal slope penetration, the pilot would have the three icing options mentioned in Chap. 24: Climb to colder air above, descend to warmer air below, or reverse course to warmer air behind.

As previously mentioned, a pilot might elect to fly low (A' to B'). In the example, this would certainly eliminate the icing problem. The pilot may have to contend with low ceilings and visibility and precipitation. The pilot could expect a relatively narrow area of low ceilings and visibility. With strong cold fronts, ceilings and visibilities are generally good except in areas of heavy showers. Fog is unlikely because of gustiness and strong winds, but with the result of increased turbulence. With an active front, such as the one we encountered on our flight from St. Louis to Kansas City, the only option was to wait on the ground. However, with care a weak front may be negotiated. If the pilot elects to fly high (A'' to B''), again any serious icing should not be encountered.

Essentially, the opposite scenario would be true for a pilot penetrating from the cold sector to the warm sector. Upon frontal slope penetration, somewhere prior to the location of the front on the surface, the pilot would experience a change from a left crosswind component to a right crosswind—the same as a flight from A to B. But the pilot would also experience an increase in temperature, and he or she could expect a considerable amount of poor weather before exiting the frontal system.

CASE STUDIES   Our route from Tulsa, Oklahoma, to Springfield, Illinois, was dominated by a weak cold front. Weather reports and forecasts indicated relatively clear conditions on both ends of the route and tops to about 6000 ft. Beneath the clouds, conditions were turbulent, with marginal ceilings and visibilities. The decision was clear: With an instrument-rated and current pilot, and an aircraft with instrument capability, VFR over the top was acceptable. We filed VFR, and the flight was completed without incident.

Should flight on top be tried by a non-instrument-rated pilot? You have to evaluate the risk. An engine failure could be disastrous. Navigational error or instrument malfunction could result in getting caught on top. If this should occur, it cannot be overemphasized that the pilot should obtain assistance from ATC as soon as possible.

Depending on the type of front and associated weather, a pilot may elect to penetrate the front.

CASE STUDY   A stable Pacific cold front was along our route from Van Nuys to San Francisco. We were flying a Cessna 172. Thunderstorms were neither reported nor forecast. The freezing level was expected to be around 9000 ft in the south and 6000 ft in the north. We flight-planned along the coastal route, where minimum altitudes were 6000 ft or lower. Under such circumstances flight over the coastal mountains into California's central valley, with minimum altitudes around 10,000 ft, was not an option.

A pilot may wish to avoid frontal weather by either flying around the front or waiting for frontal passage. I have already related our decision on the flight from Saint Louis to Kansas City with an active front producing thunderstorms. As well as the weather, flight decisions must be based on the capability of the pilot and aircraft. In that case frontal thunderstorms made the flight decision easy—No go!

Warm Fronts

Now let's review a flight penetrating a warm front from the cool sector to the warm sector, as illustrated in Fig. 26-3. The pilot has elected to fly well above the freezing level, thus minimizing the risk of icing. However, entering the frontal inversion, the pilot would expect to encounter freezing rain or drizzle—a very hazardous icing zone. Here the pilot has a fourth icing option—climb to warmer air above. After penetrating the frontal surface, icing would no longer be a problem. The pilot could expect to penetrate the frontal slope well in advance of the surface front location. Frontal slope penetration would be indicated by a wind shift from southeast to southwest and an increase in outside air temperature. Like a cold front, pressure would decrease until reaching the frontal boundary, then rise. The pressure change would not be as dramatic as a cold front, but the pilot would still be well

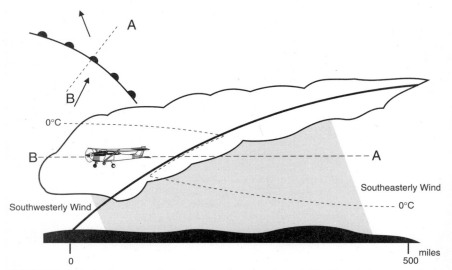

FIGURE 26-3   Penetrating a warm front, pilots must be prepared to contend with extensive areas of clouds and precipitation.

advised to obtain frequent altimeter setting updates. After frontal penetration the pilot could expect extensive clouds and precipitation before exiting the frontal weather. With an unstable air mass, embedded thunderstorms would be an additional hazard. Along with the forecast, current weather radar could be used to determine convective activity. Should thunderstorms be reported or forecast, access to real-time weather (storm detection equipment) is a must. Because cells are typically embedded, visual avoidance is normally not an option.

Again, the opposite scenario would be true for a pilot penetrating from the warm sector to the cool sector. In our example, the pilot would be far below the freezing level until encountering the frontal slope, located well ahead of the front's surface position. Frontal penetration would be indicated by an increased right crosswind component and a drop in temperature. The pilot may encounter freezing precipitation, with an extensive area of clouds to negotiate, before exiting the frontal weather. Suitable alternates, especially for an aircraft with a limited range, may be another problem.

With warm fronts, VFR pilots will have to contend with widespread area of low ceilings and visibilities and the possibility of freezing precipitation. Pilots must use caution under these circumstances and have a suitable alternate airport close at hand during the entire flight. This will require careful navigation and frequent weather updates.

Occluded and Stationary Fronts

As previously mentioned, weather and flight conditions associated with an occluded front will have characteristics of both cold and warm fronts. Therefore, the pilot must be prepared to contend with the hazards of both types of fronts.

Flight through stationary fronts is generally smooth, except when cumuliform clouds are present. Low clouds, ceilings, and visibilities may persist in the cold sector. Don't be fooled by stationary fronts. They can be either completely devoid of weather or provide the initial lifting mechanism to trigger thunderstorms.

CASE STUDY The previous night's weather news told of a stationary front between Albuquerque, New Mexico, and Amarillo, Texas. The preflight briefing the next morning confirmed its location. The front was completely devoid of all but cirrus clouds. Frontal penetration was marked by an area of light turbulence and a wind shift.

Enroute Considerations

The preflight briefing should provide the location and expected movement of frontal systems along our proposed route of flight. However, sometimes these buggers have minds of their own—that's the weather not the forecasters.

By observing winds aloft and weather conditions, especially cloud types, we can often get a clue that the forecast isn't going according to plan. If we're IFR, our observations of wind direction and speed and outside air temperature can provide the same clues.

We should check with flight watch for updated weather reports and forecast. In this way we can track the accuracy of the forecast, and—especially at the first signs of unforecast conditions—we can consult flight watch for changing conditions.

Clues to unforecast frontal movements are clouds and/or wind and pressure changes either before or after the time they were forecast to have occurred. For example, if we're expecting a front to move beyond our destination but current conditions are not improving as rapidly as forecast, the front has slowed, and we had better find an alternate. The same would be true if we're attempting to beat a front to our destination. If conditions are deteriorating more rapidly than forecast, the front has accelerated—look out! IFR alternate minimums and fuel reserves take these factors into account. VFR pilots, with the exception of having minimal fuel reserves, are pretty much on their own. It makes no sense to plunge into marginal or deteriorating conditions without adequate fuel reserve and suitable alternates. And the excuse "I didn't know conditions were that bad" is just that, an excuse!

We have our own eyes and the resources of ATC, especially the FSS, and in particular flight watch. There is really only one reason for getting caught in adverse weather—poor planning and failure to update weather enroute. Chapter 27 provides specific examples.

NONFRONTAL WEATHER SYSTEMS

VFR pilots must be careful dealing with surface and upper-level lows and troughs, especially in the absence of a front. Often weather conditions will be VFR to marginal VFR at the surface. In fact, pilots are often able to operate VFR beneath ceilings with relatively good visibility, except in the vicinity of showers. Here's the rub. In mountainous areas there is almost always mountain obscuration. This prevents VFR operations in these areas. Pilots have to be especially careful not to be caught in box canyons. IFR pilots will typically have to contend with low freezing levels, turbulence, and thunderstorms—often embedded. To safely operate in these areas aircraft must have sufficient performance and ice protection and storm detection equipment. If you don't, don't go!

CASE STUDY We had planned to fly the "turbo" Cessna 150—that's an attempt at a little humor—from Los Angeles to San Antonio, Texas, for the AOPA convention. We made it as far as El Paso. A large upper-level low was located over central Texas, bringing low ceilings and visibility, icing, and embedded thunderstorms. And this was October! The flight decision was a "no brainer." No go. This was a situation where we had to park the Cessna 150 and take one of American's 727s.

As the previous case study shows, upper-level weather systems can bring the weather down to the ground. This situation often occurs over the intermountain region and plains states. Since these systems can be slow moving, the weather may be unflyable for days, even by the airlines.

Upper-level weather systems tend to develop clouds and weather in bands. Typically aviation forecasts cannot take this phenomenon into account. Therefore, the forecaster will cover the area with a "broad-brush" approach. Consider, for example, the following forecast:

SRN CA
CSTL SXNS...BKN/SCT020-030 BKN080 LYRD TOPS FL200 WDLY SCT
-SHRA ISOLD -TSRA CB TOPS FL300.
INTR SXNS...SCT/BKN080-100 BKN120 TOPS FL200 ISOLD -SHRA.

A pilot seeing a clear area may assume the weather has passed. However, the weather can then deteriorate with the approach of the next band. With closed lows and hurricanes, expect weather in bands. The solution is the satellite picture. The satellite will often confirm or refute the existence and location of weather bands. A satellite loop can provide some indication of the movement of the bands. A pilot may then be able to conduct a flight in relatively good weather between the bands. Here again, keeping a close watch on the weather, frequent updates enroute, and suitable alternates are a must.

Another factor associated with upper-level lows and troughs is an unstable air mass. The low or trough may move through an area bringing clear skies behind. However, with surface heating, abundant moisture, and an unstable air mass, guess what comes next? You got it—thunderstorms! Usually the forecast has this pretty well in hand. Unfortunately, the scattered nature of this event makes the exact location of convective activity difficult to predict.

Refer to Fig. 26-4. This satellite picture shows a major cold front along the Pacific coast. Behind the front, in the unstable air mass, rain showers and thunderstorms have developed. Notice the sharp trailing edge of the frontal band. After frontal passage, a pilot may be suckered into the assumption that all significant weather has passed. However, the satellite picture and associated radar reports will confirm the unstable air mass and the existence of convective weather in the cold air behind the front. Circumnavigation should be possible, but if a cell develops over your destination, you had better have an alternate.

Note the cloud band through west, central Texas in Fig. 26-4. This is a line of severe thunderstorms. Typically convective SIGMETs, along with severe thunderstorm and tornado warnings will be in effect. However, warning areas will usually depict much larger areas than those shown by the satellite or radar. That's because they must cover the expected movement of the storms. The satellite, along with current radar observations, will help the pilot get a much better sense of exactly where the weather is occurring.

As another example, recall the discussion in Chap. 12, and Fig. 12-12, which depicted an upper-level low. Even without a surface front, there are extensive areas of IFR ceilings and visibility, and snow covers large parts of the Ohio Valley, Appalachians, and Mid-Atlantic and New England states.

Sources of severe weather information associated with upper-level weather systems are *convective SIGMETs* (WSTs), *alert weather watches* (AWWs), and the *convective outlooks* (ACs). These products contain information on the atmosphere that relate to severe weather. By now we should be able to understand and interpret most or all of the forecaster's discussion. These products provide additional insight into the overall weather picture.

A couple of words of caution: FSS briefers will not normally provide the outlook portion of WSTs—which contain the discussions—except on request. (Ironically, some in the

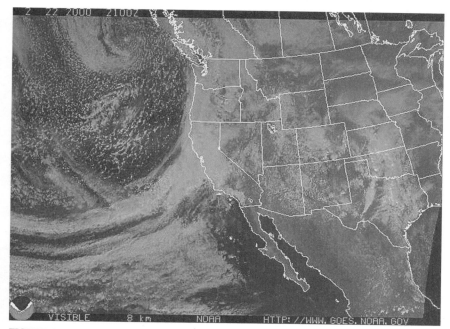

FIGURE 26-4 After frontal passage, a pilot may be suckered into the assumption that all significant weather has passed.

FAA have proposed elimination of the outlook. To a competent FSS briefer and weather savvy pilot, the WST outlook contains much useful information.) Remember, the convective outlook is just that, an outlook. The AC can never be used as a substitute for appropriate aviation weather forecasts (convective SIGMETs, area forecasts, and terminal aerodrome forecasts).

The jet stream, with its hazards of turbulence and strong winds, is often associated with upper-level weather systems. We have already discussed their significance and strategies to avoid the hazards associated with them, along with vorticity. The mention of these phenomena in forecasts or their appearance on charts should alert us to the fact that vertical motion is, or will be, occurring. Enroute, pilots should watch for unforecast wind speeds or shifts, or temperature changes. These can signal a change in the weather pattern, which should cue the pilot to obtain additional information, such as current weather reports and revised forecasts. Of course, a PIREP will also alert forecasters, briefers, and other pilots to a potentially hazardous, unforecast change in the weather.

A hurricane is a nonfrontal weather system that contains just about every possible hazard and covers large areas. Like thunderstorms, the key to handling the hurricane hazard is to avoid it. Recall that thunderstorm tops associated with tropical cyclones frequently exceed 50,000 ft. Wind in a typical hurricane is strongest at low levels, decreasing with altitude. However, winds in excess of 100 kn at 18,000 ft are not uncommon. Severe to extreme turbulence and severe low-level wind shear are to be expected. Turbulence increases in intensity in spiral rain bands and becomes most violent in the wall cloud surrounding the eye. And severe icing can be expected above the freezing level, which may be in the lower flight levels. Altimeter errors, due to the extreme low pressure, may be as much as 2000 ft.

Flying to the right of the storm, a pilot can take advantage of a tailwind; to the left, the pilot will encounter the strongest headwinds, increasing fuel consumption and prolonging the risk.

CASE STUDY   The crew of the record-setting *Voyager* round-the-world flight in 1986 used this technique. In the western Pacific, off the Philippine Islands, they altered their flight plan to take advantage of the tailwinds north of a tropical storm. However, they had to be careful because the airplane was not constructed to withstand any significant amount of turbulence.

CHAPTER 27
RISK ASSESSMENT AND MANAGEMENT

With our knowledge of the three-dimensional atmosphere, we should be able to put together the complete picture. By understanding why some weather systems are benign and others severe, and how they are modified, we can now integrate this knowledge with the preflight weather briefing and updates enroute, to allow us to make intelligent, safe weather decisions. But that is only half of the safety equation. We have reviewed numerous accidents and weather scenarios. Most frequently an accident occurred because the pilot failed to obtain complete and accurate information, attempted to exceed aircraft performance, or simply continued flight beyond his or her capability or below safe minimums. Virtually all were preventable! Now it's time to introduce the human factor.

There are a number of items, although not necessarily weather related, that the pilot must consider. These include terrain, time of day, alternates if the planned flight cannot be completed, and the pilot's physiological and psychological condition.

Throughout the book we've related terrain to specific aviation weather hazards and flight situations. As well as weather conditions, pilots must include aircraft performance in their decisions. *Aircraft performance and equipment* refer to density altitude, service ceiling, availability of supplemental oxygen, and ice protection and thunderstorm avoidance equipment. If the aircraft does not have the performance or equipment, don't go! Other factors would be alternate landing sites and time of day. For example, risks are increased flying over mountainous terrain at night in low-performance, single-engine aircraft.

The pilot's physiological and psychological condition must also be considered. Proper rest and good mental condition are paramount. It makes no sense for a tired, hungover pilot to attempt to take on weather, or any flying for that matter. Other factors include the pilot's training and experience. Pilots need to set their own personal minimums based on these and other factors. It has been said that a new private pilot certificate or instrument rating is a ticket to learn.

As observed by the U.S. Supreme Court: "Safe does not mean risk free." We know about certification and operating rules. We have a complete picture of the weather. We know our aircraft and its equipment, and ourselves and our passengers. How do we decide if a particular flight is safe? How do we assess the risk and manage that risk? We'll specifically address these questions in this chapter.

In the final section we'll take a look at some actual flight scenarios. Like all the weather reports and forecasts contained in this book, they were obtained during and represent an actual weather event.

FLIGHT PLANNING STRATEGIES

Flight planning strategies begin with sound assessment of personal minimums. With personal minimums established, we can move on to flight preparation. This is followed by evaluating the weather. Finally, we'll consider areas that are all too often overlooked: the pilots and passengers.

Unlike being a "little pregnant," there is some middle ground when it comes to the weather. For example, we can plan the flight in stages, landing short of our ultimate destination. We can take a look at the weather. However, there are two caveats to this option. First, we must know when to abandon the plan. When it's not meant to be, it's not meant to be! We must know when to call it a day. Second, we must have an alternate plan or two (Plan B, Plan C, Plan *N*). More about this later in the chapter.

Personal Minimums

As previously mentioned, a new certificate or rating should be considered a "ticket to learn." In the introduction we discussed the fact the student pilots are guided through their initial training under the direct supervision of a flight instructor. Military pilots are shepherded by more experienced pilots. And, ironically, the airlines have an extensive dispatch system to ensure the safety of their flights. However, after initial certification there is no such safety umbrella for the general aviation pilot. This fact can be directly related to the accident record. In a very real sense, we hold our fate in our own hands.

When we talk about "personal minimums," there are a number of factors to consider:

- Training
- Experience
- Currency
- Aircraft
- Weather
- Time of day
- Physical condition
- Psychological condition

As our level of training and experience increases, we may wish to consider different minimums. As a flight instructor, I tailored student minimums to their training and experience. For example, I had a student flying out of Lancaster's Fox Field, in California's Mojave Desert. We trained in strong, gusty surface winds. When the student was proficient I would increase the minimums. Some pilots obtain an instrument rating without ever having flown in the clouds. Do they have the experience to operate in actual instrument conditions? A prudent pilot would have another qualified, experienced pilot or flight instructor along until they became familiar with flight in clouds.

Currency with the type of operation is another personal minimum factor. Here again, *legal* does not necessarily mean *safe*. If we've been recently qualified to fly at night, we would certainly want to gain experience before tackling weather close to either VFR or IFR minimums during this time of day.

Table 27-1 contains suggested minimums based on pilot qualification and type of operation. [If the FAA put this together, it would be called a "matrix." So what's the difference between a *table* and a *matrix?* The cost, of course. The FAA would grant a contract to a university or corporation worth tens of thousands of dollars to come up with such a matrix. Your tax dollars at work(?).]

Refer to Table 27-1. The left column describes pilot qualification [*student* (STU); *private* (PVT); *commercial* (COM); *instrument* (INST)]. Two additional categories are *dual flight instruction VFR* (DUAL V) and *IFR* (DUAL I). Across the top are operational categories. Operational categories are subdivided into *day* and *night,* along with winds-aloft

TABLE 27-1 Personal Minimums

Pilot	Cross-country			Surface wind			Local		Pattern	
	Day	Night	Winds aloft	Crosswind	Sustained	Gusts	Day	Night	Day	Night
STU	7500/7	NA	25 KT	7 KT	15 KT	NONE	5000/5	NA	2000/3	NA
PVT	7500/5	7500/5	25 KT	10 KT	20 KT	5 KT	4000/5	4000/5	2000/3	2000/3
COM	7500/3	7500/3	35 KT	POH	25 KT	10 KT	4000/3	4000/3	2000/3	2000/3
DUAL V	FAR	4000/5	35 KT	POH	PD	PD	FAR	4000/3	FAR	2000/3
DUAL I	FAR	800/2	35 KT	POH	PD	PD	FAR	4000/3	FAR	1500/3
INST	500/1*	1000/2								

*Or FAA published takeoff and IFR departure minimums, including climb gradients, whichever is greater.

limits for cross-country and surface-wind limitations. Note that student night solo is *not authorized* (NA), even though it is permitted by regulations with specific training and an instructor's endorsement. Ceiling and visibility are given in feet and statute miles (7500/7, ceiling 7500 ft and visibility 7 statute miles). For dual flights, basic 14 CFR 91 limitations are authorized (FAR), with sustained surface winds and gusts left to the discretion of the *instructor pilot* (PD). For commercial pilots and dual flights, maximum demonstrated crosswind components as listed in the *pilot's operating handbook* (POH) are allowed.

Before we continue, let's concede that the minimums described in Table 27-1 are not hard and fast. Because of their training and experience, instructors may wish to raise or lower a student's solo limitations. A private pilot with years of experience and thousands of hours in a specific make and model of aircraft may be competent to exercise limits in the commercial or dual categories. We know that the maximum demonstrated crosswind component is not an absolute limit. Pilots should consider the personal minimums in Table 27-1 as a beginning point. Also note that the relatively high local and cross-country ceilings are due to the high terrain in the western United States. In the midwest and along the east coast, in the absence of mountainous terrain, these would normally be lower. Our U.S. Air Force Aero Club in England had ceiling minimums of 2500 ft because of the flat terrain. This was certainly sufficient for operations in eastern England.

Notice in Table 27-1 that night minimums are typically higher than day. This is a direct reflection of the additional hazards of night flight. Instrument minimums are also typically higher than those specified in the regulations. These minimums were developed for single-engine and light twin-engine airplanes. Why? Even though a single-engine pilot, operating under 14 CFR 91, is not directly prohibited from making a zero-zero takeoff, it isn't safe. Consider that in the event of an engine failure, the pilot is still required to be able to "make an emergency landing without undue hazard to persons or property on the surface." Also, recall that "no person may operate an aircraft in a careless of reckless manner so as to endanger the life or property of another."

Table 27-1 does not address personal fuel minimums. The following fuel reserves are recommended and should be strongly considered:

- Minimum 1 h reserve for all flights
- Minimum 2 h of fuel for pattern or local flights
- Full fuel for cross-country and IFR flights

On certain airplanes, filling the tanks to the tabs may be satisfactory to accommodate additional passengers and baggage. Most four-place aircraft are designed to accommodate four passengers with a partial fuel load, or fewer passengers and baggage with a full fuel load, but not both! Pilot's must understand the limitations of their aircraft.

Preparation

In order to apply a weather briefing to a flight, we must have done our homework. What is the terrain like along the route? What are the minimum altitudes? Are there suitable alternates? What if Plan A does not pan out? Therefore, it's incumbent upon the pilot—for every briefing—to study the terrain, routes, and possible alternates for the proposed flight. For example, recall my experience with Los Angeles Center. I had flight planned the flight below the freezing level, but due to traffic I had to climb into icing conditions. The objective is to have an out. If there are no "outs," the flight is a definite no go!

To apply a weather briefing to a flight, we must consider the aircraft we're planning to fly. Is the aircraft ready for cold or hot weather operations? Does the aircraft have ice protection or storm detection equipment? How about the pilot and passengers? Human factors are often overlooked and take on additional significance during cold or hot weather, especially with potential icing or thunderstorm conditions.

Evaluating the Weather

Now we can apply our knowledge of weather reports and forecasts, and the weather briefing, to flight situations along with our personal minimums for ceiling, visibility, wind, and fuel. As far as ceiling and visibility go, personal minimums apply. For turbulence, pilot and passenger comfort and tolerance are paramount. With icing, the goal is to minimize exposure. If thunderstorms are forecast, our only logical course is to avoid them.

Pilots planning a flight below the freezing level can normally not expect to receive an icing advisory during an FSS preflight briefing, because icing will not affect their proposed flight. Some briefers fail to understand and consider this and issue the advisory even through it is not a factor. This practice undermines the credibility of both the forecast and the briefing. Pilots planning flights and briefed for low altitudes should keep this point in mind in the event that they should elect, or be instructed by ATC, to climb to a higher altitude.

This is another reason to request temperatures for your planned cruising altitude. From this information you can determine how close you are to the freezing level. By comparing forecast temperature with observed temperature, you can get a sense of forecast accuracy. A large discrepancy may indicate a busted forecast. Pilots might well consider the advisability of accepting the clearance without additional information. This point also applies to rerouting. Should the pilot or ATC reroute the aircraft, weather advisories that were not pertinent during the briefing may then apply.

Based on the weather briefing and a complete picture of conditions, plan the route of flight to avoid potentially hazardous areas. Penetrate frontal areas at as close to a 90° angle as possible. Circumnavigate hazardous areas. Land short and wait until the hazard no longer exists. Delay the flight until the hazard has passed.

In the case of icing, our first choice might be to climb above clouds into clear air. This assumes we know where the tops are and our aircraft has the capability to climb through an icing layer to this altitude. Climb to air that is too cold for icing. This, again, assumes the previous caveats. Or descend below the freezing level. This requires a knowledge of the freezing level, terrain, and minimum altitudes. We can avoid the icing hazard by remaining clear of clouds and precipitation when flying above the freezing level. ATC always seems to have a way of destroying "the best-laid plans." Like the Scouts, "Be prepared." Finally, if necessary, turn around—presumably we came from an ice-free area.

CASE STUDY   Among the many pilots that I have had the pleasure of serving at the Oakland, California, FSS was a particular local air taxi pilot. He routinely flew from Oakland, through the Sacramento Valley, to northern California. He provided all the necessary background information and always requested a standard briefing. Then, since it's not part of a standard briefing, at the conclusion, he would always ask for the closest area of clear conditions. What an excellent idea! This was one of the most prepared pilots I know. Should he have engine, navigation, or electrical problems, he knew the closest location of clear weather. We should all add this technique to our personal briefing requirements.

An advantage of turbocharged and pressurized aircraft is that these planes can fly high, above the weather. Turbulence can usually be avoided. Convective activity is more easily circumnavigable. Icing is normally not a significant factor in the flight levels, except around convective activity or in the summer. Just because it's summer doesn't necessarily mean there is no icing potential.

CASE STUDY The pilot was flying a Mooney from Little Rock, Arkansas, to Charleston, West Virginia. The pilot's ultimate destination was Massachusetts. The briefing included weather advisories for turbulence and light to moderate rime icing below 12,000 ft in clouds and precipitation. Scattered to broken clouds were forecast at 2000 to 3000 ft. Charleston: 26009G21KT 7SM -SN BKN027 OVC040 01/M04.... The briefer editorialized about conditions "not looking too bad." The pilot proceeded on top for the western portion of the flight.

The Mooney was on frequency when the center received a PIREP departing Charleston reporting moderate rime ice. As the flight approached Charleston, the ATIS reported: 26010KT 7SM -SG OVC013 01/M01 A3000. Charleston approach control received numerous reports for light to moderate mixed and rime icing.

The pilot was cleared to descend and received vectors for the approach. On descent, the pilot lost control. The airplane was reported to have rocked from side to side and crashed.

The pilot received the essential information: icing below 12,000 ft. The temperature at Charleston was right at the freezing point, with light snow reported. Any ice would be carried all the way to ground. The pilot was flying from good weather to bad.

STRATEGY For new private and instrument pilots, it's usually best to begin challenging weather by going from poor conditions to good. After you've gained experience, and with a healthy respect for the weather, consider flying from good conditions to poor—as long as you have an alternate plan.

Upslope, caused by the rising terrain of the Appalachian Mountains, would have enhanced any icing present. PIREPs confirmed the forecast. The pilot had the opportunity to land short or reverse course, but he exercised neither of these options. If the pilot had landed short, there was a good possibility of overflying the Appalachians. The pilot could then land closer to the east coast, in an area not affected by upslope and with higher ground temperatures. Although we'll never know, it appears this pilot was locked into one, poorly conceived plan. Never let the briefer make the go–no go decision. The briefer is a resource.

For example, let's consider the "VFR flight is not recommended" statement. It leaves a great deal of leeway for the briefer; some use this statement more than others. The inclusion of this statement should not necessarily be interpreted as an automatic cancellation, nor its absence as a go-for-it day. Notice that VNR applies to sky condition and visibility only. Few FSS briefers understand the provisions of special VFR. Hazardous phenomena, such as turbulence, icing, winds, and thunderstorms, of themselves, do not warrant the issuance of this statement. It is important to remember that this is a recommendation.

It's been my experience that VFR flight is possible between 50 and 60 percent of the time that this statement has been issued for flights that I planned VFR. Remember that this is based on my training and experience. This doesn't mean it should be ignored but that we must take a careful look at the complete picture.

Surface wind reports and forecasts are an important part of weather evaluation for any flight, but especially to and from areas with surface snow, ice, or slush. It's essential to calculate crosswind or tailwind components. Significant crosswinds or tailwinds may result in a no-go decision. Recall the limits in Table 27-1, Personal Minimums.

Aircraft and Engine Considerations

If you're based in a warm climate, you may not be familiar with the aircraft manufacturer's recommendation for winterizing the aircraft. Most mechanical equipment, including aircraft and components, have specific design temperatures. Know and follow the manufacturer's recommendation.

Some manufacturers recommend engine and oil covers, and baffles. When baffles are installed, a cylinder head temperature gauge is recommended, particularly if wide temperature differences are expected. Engine oil is extremely important in low temperatures. Be sure the proper weight oil is used in low temperatures. Pay particular attention to the crankcase breather in cold weather. A number of engine failures have resulted from frozen crankcase breather lines, which caused pressure to build up, sometimes blowing off the oil fill cap or rupturing a case seal, causing loss of oil. Water vapor, a byproduct of combustion, can condense and freeze. Special care is recommended during the preflight to assure that the breather system is ice free.

CASE STUDY The investigation of a Cessna 152 accident revealed that the crankcase breather line was plugged with ice and that the oil had been forced out through the engine's nose seal. The probable cause was determined to be the pilot's inadequate preflight that failed to detect the ice-clogged breather.

Many aircraft are equipped with cabin heater shrouds, which enclose the muffler or portions of the exhaust systems. It is imperative that a thorough inspection of the heating system be made to eliminate the possibility of carbon monoxide entering the cabin. Each year accident investigations have revealed that carbon monoxide has been a probable cause in accidents that have occurred in cold weather operations. Many pilots employ a small placard that indicates the presence of carbon monoxide.

A dead battery can be very annoying, especially in cold weather. With a dead battery, you may have one of three options: Charge the battery, use an external power source, or handprop the airplane. Assuming there's a maintenance shop on the field, you may wish to change the battery. Unfortunately, this takes considerable time. Many airport operators have *auxiliary power units* (APUs). If the aircraft is equipped with an auxiliary power receptacle, this may be the easiest and quickest solution. Be sure to follow the manufacturer's recommendations. If the airplane has a generator, or an alternator with some power left in the battery, handpropping is another alternative.

Handpropping is an extremely dangerous procedure. Handprop an airplane only when absolutely necessary, and only after taking proper precautions. Never handprop unless a qualified person, thoroughly familiar with the operation of all controls, is seated in the airplane with the brakes set. Leave the wheels chocked and at least the tail tiedown secured. The ground should be firm and free of debris. Loose gravel or a slippery surface (ice and snow) might cause the person to slip or fall into the propeller. If you think you may need to handprop an airplane, obtain instruction from a qualified flight instructor or mechanic.

Pilot and Passenger Considerations

As well as the aircraft, summer and winter flights must consider the pilot and passengers. Are we properly dressed for the environment? In addition to personal comfort, we must be psychologically and physiologically prepared to "commit aviation."

For the present, consider proper apparel. Dress for flying. This applies equally to pilot and passengers. If you can inspect the airplane and not get dirty, you haven't done a thorough job! Wear slacks. Shoes should be flat-soled for safety and to ensure proper flight control operation. Avoid loose-fitting clothes or jewelry that could get caught on sharp edges around the airplane.

It can get mighty uncomfortable on cold and windy days, and normally temperature decreases with altitude. Cabin heat is usually obtained by routing outside air through a muffler shroud that surrounds the engine exhaust stacks, on most light, single-engine aircraft. This raises the temperature of the air by about 20°C. To illustrate, consider a trip from Reno to Lovelock in Nevada. At 7500 ft the *outside air temperature* (OAT) was −18°C. The air entering the cabin was between 0° and 5°! Bring a wind breaker or jacket when conditions warrant.

Are we flying over sparsely populated or mountainous terrain? Then we should carry waterproof jackets, long pants, boots, and gloves. Do we have proper survival gear for a forced landing? Our survival might depend on being properly equipped for a emergency landing, possibly in below-freezing conditions, and being able to survive until rescued.

CASE STUDY   Years ago an airplane crashed in California's rugged Sierra Nevada mountains. This incident was subsequently made into the TV movie *I Alone Survived*. The accident perfectly illustrates the hazards of flying over wilderness without proper clothing or survival equipment. There were two fatalities. The flight was from the San Francisco Bay area to Death Valley. Both relatively warm areas. The crash occurred on the crest of the Sierra Nevada mountains, at an elevation above 11,000 ft. Neither the pilot nor the passengers had survival gear or proper clothing. The sole survivor had to walk many miles out of the mountains.

We in the southwest have to be especially aware of the hazards associated with a crash landing. We typically fly from warm, populated coastal areas and valleys to freezing, snow-covered wilderness areas. General aviation pilots aren't the only ones seduced by our climate.

CASE STUDY   The Navy has a major training facility at Lemoore, California. They routinely fly to their gunnery and bombing ranges, across the Sierras to northern Nevada. Occasionally, they get lax and wear their flight suits over their skivvies. To keep this practice from getting out of hand, they occasionally helicopter the pilots to a place at about the 8000-ft elevation for an overnighter, with only what they're wearing and their survival gear.

It's relatively inexpensive to put together a first-aid/survival kit and carry a nonbreakable jug of water. This, along with proper clothing for the terrain to be flown, should be adequate. According to Murphy's law, the only time you'll ever need this equipment is when you didn't bring it!

Everyday illnesses can seriously degrade pilot performance. Illness can produce distracting symptoms that impair judgment, memory, alertness, and the ability to make calculations. Even if symptoms appear to be under control with medication, the medication itself often impairs performance.

Did you know a visit to your dentist, and ingesting the accompanying pain reliever, may seriously impair your performance? Cough suppressants behave in the same way. The safest rule is not to fly while suffering from any illness. If you have questions about a particular malady, contact your aviation medical examiner for advice.

Minimum time between alcohol consumption and flying is contained in the regulations. However, as is often the case, *minimum* does not necessarily equate to *safe*. Research indicates that as little as 1 ounce of liquor, one bottle of beer, or 4 ounces of wine can impair our flying skills. Alcohol consumed in these drinks is detectable in our blood and on the breath for at least 3 hours. Alcohol also renders us much more susceptible to disorientation and hypoxia.

Day-to-day living experiences also affect our flying ability and safety. How? Well, get up at "oh-dark-thirty," go to work, work a full day—trying to get out as much work as possible—drive to the airport and begin the preflight inspection. Think about it. We just drove to and from work; freeway, traffic, someone just cut you off—they're probably on their way to the airport to go flying—and to say the least, we're perturbed. Now, continue with the flight.

Fatigue has, believe it or not, set in. Fatigue can be described as either *acute* (short term, gone after a good night's sleep or perhaps a nap) or *chronic* (long term, those all-nighters preparing for final exams or partying). This is just one of many everyday living occurrences that cause fatigue. Fatigue is the tiredness felt after physical or mental strain, including muscular effort, immobility, heavy mental workload, strong emotional pressure, monotony, and lack of sleep. Fatigue can be minimized with proper rest and sleep, regular exercise, and proper nutrition (M&Ms, Coke, coffee, and donuts don't count).

Other health factors related to safety are stress and emotional well-being. Stress from daily activities, like work or home management, are typically not relieved by flying. Stress and fatigue can be an extremely hazardous combination.

Emotions, upset by events like a serious argument, death, separation or divorce, loss of a job, and financial problems, also affect our ability to fly safely. If you experience an emotionally upsetting event, you should not fly until you have given yourself adequate time to recover.

Our involvement in all health aspects of flying continues until the day we hang up our certificate. Pilots are prohibited from flying with any known medical condition that does not meet the standards of their medical certification.

The FAA has developed a—somewhat hokey, but nonetheless useful—physical and mental checklist. We're safe to fly by not being impaired by the following:

- Illness
- Medication
- Stress
- Alcohol
- Fatigue
- Emotion

I'M SAFE! (I told you it was hokey.) If you're not at your peak, seriously consider canceling and rescheduling the flight. Please believe me, you can really mess up your day—in more ways than one—if you don't.

Any discussion of fitness for flight would not be complete without a word about *hazardous attitudes*. Volumes have been written on the subject; you're encouraged to do more research into the matter, to know as much about it as possible. What is it? Research shows that most preventable accidents have one common factor, or as HAL9000 would put it "human error." All too often, we are our own worst enemy.

There are five identified attitudes that adversely affect our ability to make sound decisions. The first is called *macho*. This is an attitude by which an individual thinks he or she must continually demonstrate that he or she is better than others. This usually results in unsafe actions. Although typically associated with men, women can be just as susceptible.

Closely associated with macho is *antiauthority*. People with this attitude believe that the rules don't apply to them. This attitude manifests itself as, Don't tell me what I can do!

As our experience grows through training and experience, we gain respect for aviation, the nature of flight, and our own mental attitude. We hope that the mental attitude we develop will be positive. But we must be careful not to slip into the third of negative attitudes: *invulnerability*. Some people feel that bad things happen only to the others. I hope we all can see how a combination of macho, antiauthority, and invulnerability are a road to disaster!

The last two negative attitudes are *impulsiveness* and *resignation*. Impulsive individuals feel they must do something, anything, immediately! A prime example is when an engine on a multiengine airplane fails and pilots respond by becoming so impulsive that they shut down the good engine or stall the airplane. It has happened to air carriers, as well as to general aviation. They don't take the time to evaluate the situation or to consider options and risks before taking action. Finally, people who just give up demonstrate the attitude of *resignation*. It's their fate or bad luck. An example was the pilot, while flying in IFR conditions lost the attitude and heading indicators, who advised ATC: "I've lost the gyros, we're going in." Good thing the astronauts and mission control on *Apollo 13* weren't afflicted with this attitude!

The first step in correcting a hazardous attitude is to recognize the behavior:

- Macho
- Invulnerability
- Antiauthority
- Impulsiveness
- Resignation

Armed with this knowledge, you and your instructor should be able to identify any negative attitudes and change them!

EVALUATING THE RISK

In the area of risk, it's often helpful to look at statistics. This is not the end of assessment and management, but the beginning. Of all accidents, almost all involved low-time pilots. Using myself as an example, I gained quite a lot of training and experience, without being exposed or trained for operations in a hazardous environment. Another significant accident area is the training environment itself.

CASE STUDY The flight instructor stated that he had given the student a simulated engine-out emergency. The student had completed the emergency and had initiated a climb from the low approach. The engine failed to develop power. A 1700-ft-long airstrip was selected for the "real thing." The airplane went off the end of the runway and collided with the ground. After the accident, the engine started and operated normally. No mechanical problems were noted during the wreckage examination. A review of weather data disclosed that conditions were favorable for the formation of carburetor ice. The flight instructor reported using the carburetor heat, but during a subsequent conversation said the engine was not cleared for more than 2 min during the descent for the simulated emergency.

This has ominous implications. Our cadre of flight instructors have not been properly trained and are, therefore, unable to properly train students! All these accidents are preventable.

Accident prevention is part of the National Aeronautics and Space Administration's commitment to "aeronautics." To this end, they have developed scenarios of precursors to aviation accidents. A *precursor* is a factor that precedes and indicates or suggests that an incident or accident will occur.

Refer to Fig. 27-1. Each "wheel" represents one precursor. It might be physical incapacity, poor judgment, aircraft deficiency, failure of the ATC system, the weather, or another factor that in itself would not create an incident or accident but when combined with other factors leads to disaster.

CASE STUDY Seven-year-old Jessica Dubroff accompanied her father (a passenger) and the pilot in command in an attempt to set a so-called transcontinental record involving 6660 mi of flying in 8 consecutive days. (I say "so-called record" because this was nothing more than a publicity stunt. It reminds me of telling friends that my son soloed at age 3 months. He was the sole occupant of the airplane as we pulled it over to the wash rack.) The first leg of the trip, about 8 h of flying, had been completed the previous day, which began and ended with considerable media attention.

On the second day they participated in media interviews, preflight, and then loaded the airplane. The pilot in command received a weather briefing that included weather advisories for icing, turbulence, and IFR conditions due to a cold front moving through the area.

NASA

NASA's ACCIDENT PRECURSOR SCENARIO

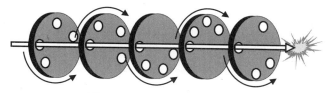

Alignment = Incident or Accident

FIGURE 27-1 Precursors might be physical incapacity, poor judgment, aircraft deficiency, failure of the ATC system, or the weather.

The airplane was taxied in rain for takeoff. While taxiing, the pilot acknowledged receiving information that the wind was from 280° at 20 gusting to 30 kn. A departing Cessna 414 pilot reported moderate low-level wind shear of plus and minus 15 kn. The airplane departed toward a nearby thunderstorm and began a gradual turn to an easterly heading.

Witnesses described the airplane's climb rate and speed as slow, and they observed the airplane enter a roll and descent that was consistent with a stall. Density altitude at the airport was 6670 ft. The airplane's gross weight was calculated to be 84 lb over the maximum limit at the time of impact.

The probable cause was the pilot's improper decision to take off into deteriorating weather conditions. This included turbulence, gusty winds, an advancing thunderstorm, and possible carburetor and structural icing. The airplane was over gross weight. Density altitude was higher than that to which the pilot was accustomed. The result was a stall caused by failure of the pilot to maintain airspeed.

As in virtually all the previous examples, most accidents can be attributed to a series of relatively insignificant factors that, when taken together, cause an accident. Let's review the Dubroff accident in this context.

They were on a tight schedule. Publicity events had been scheduled in advance. The original takeoff time was delayed to allow Jessica additional sleep. The pilot was fatigued from the previous day's flight and obtained little rest during the night. The weather was marginal at best. The pilot had to obtain a special VFR clearance for departure. Who was really flying the airplane? The pilot in command was seated in the right seat of the Cessna Cardinal. Now add high-density altitude, an airplane over gross weight, and a mindset that they must go.

The first precursor was the need to keep a time schedule—sometimes referred to as "get-home-itis." Precursor number 2 was pilot fatigue. The next precursor was a high-density-altitude takeoff with an airplane over gross weight. The fourth precursor was the weather, with its low ceilings and visibility, gusty winds, wind shear, turbulence, icing, and thunderstorms. (You could count each of these weather factors as an individual precursor.) A fifth precursor was the pilot's attempt, under these very adverse conditions, to try to maintain control of the airplane. We will never know what exactly happened, but airplane control was lost. The deck was certainly stacked against them.

Like most accidents, I think we can see how breaking any one individual link could have prevented this accident. The first link, the time schedule: A friend, and excellent pilot, has the philosophy that there is never a reason that you absolutely have to be anywhere. In contrast, the Cessna Cardinal's pilot's mindset appeared to be, "we're going no matter what."

The second link was fatigue. It was reported that Jessica had slept most of the first leg. As we've discussed, pilot fatigue is a significant factor in the deterioration of both mental and physical skills. This certainly may have clouded the pilot's go–no-go decision, and failure to calculate gross weight and density altitude.

The weather was terrible. If the weather had been clear and calm, the pilot might have gotten away with fatigue, overloading the airplane, and his lack of experience with high-density altitude.

Now add the pressure of flying from the right seat, with a novice student in the left, in less than VFR conditions. Even a slight, momentary distraction under similar conditions can have serious consequences. It's reasonable to conclude that the pilot experienced sensory overload during climbout. All these factors together aligned the precursors, resulting in a fatal accident.

So how do we assess and manage risk? We apply *aeronautical decision making*—the ability to obtain all available, relevant information, evaluate alternate courses of action, then analyze and evaluate their risks, and determine the likely results. First, evaluate all the factors for a particular flight and decide if the risk is worth the mission. Our goal is to prevent the precursors from aligning. That's easy for me to say. This can be extremely simple or extremely complex. There are three elements in risk assessment and management: planning, aircraft, and pilot. *Planning* is the homework part of the flight, which we've previously mentioned. We study terrain, altitude requirements, and the environment. The environment includes the weather, our personal minimums, and alternatives. Now we evaluate the *aircraft*. Does it have the performance and equipment for the mission? If the answer is yes, we preflight the aircraft and determine whether it is airworthy. Assuming the *pilot* is "fit for flight," we're ready to go. Simple, huh.

Making the decision can be as simple as it would be for my friend John and his Kit Fox looking at an afternoon flight in the traffic pattern; or as complex as it would be for one of NASA's shuttle missions. To help I've developed the Risk Assessment and Management Decision Tree in Fig. 27-2.

Let's start with John's decision. *Planning:* Airport elevation 397 ft, runway 25L 2699 ft; pattern altitude 1400 ft; the environment, clear, cool, winds calm, alternate runway 25R. *Aircraft:* Performance of the Kit Fox OK; airplane equipped for flight in class D airspace. *Pilot:* Fit for flight. *Decision:* Go!

Don't worry. We're not going to evaluate a space shuttle mission. Instead, let's take an actual flight. We were flying from Oklahoma City to Palm Springs for the 1998 AOPA convention. The weather was good through Tucumcari, New Mexico, but it deteriorated between Tucumcari and Albuquerque. Passing Tucumcari, I checked with flight watch and received the bad news. The weather ahead was IFR to MVFR. The plan was to fly direct, via the Anton Chico and Otto VORs. With the deteriorating weather ahead, I decided to go IFR—I Follow Roads. With few landmarks and low ceilings and visibilities, the safest option was to follow I-40. The terrain and clouds merged about 20 mi west of Santa Rosa, New Mexico. It was afternoon, and we had been flying for about 4 h. With night approaching, poor weather, and fatigue a factor, the only viable option was to return and land at Santa Rosa.

Hal Marx (USMC retired), the Santa Rosa airport manager, fueled our airplane and gave us a lift into town, where we remained overnight. The next day wasn't any better and we spent another night.

We had been trying to get to Albuquerque for 2 days without success. The following morning wasn't much better but forecast to improve.

Planning: Santa Rita has a field elevation of 4782 ft. Along I-40 the high plateau of eastern New Mexico rises to over 7000 ft, with the pass through the Sandia Mountains at about the same elevation. Terrain is slightly lower to the north and south but still over 6000 ft. Because of the mountains, IFR MEAs vary from about 10,000 to 12,000 ft. Minimum alti-

RISK ASSESSMENT AND MANAGEMENT DECISION TREE

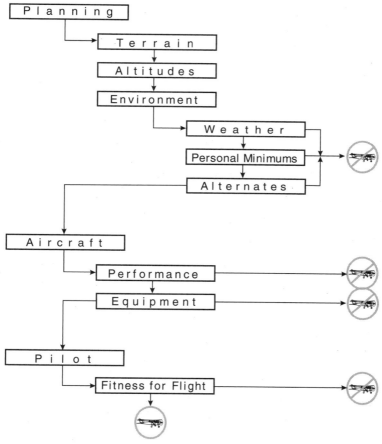

FIGURE 27-2   Risk assessment and management can be as simple as they would be for a flight around the airport traffic pattern or as complex as they would be for a space shuttle mission.

tudes would range from 6500 to 8500 VFR or 10,000 to 12,000 IFR. Even though we were flying a Cessna 172, we still had the option of going IFR or VFR.

Environment: Upslope due to rising terrain was, and continued to be, the culprit. MVFR to IFR ceilings, generally good visibility, high tops, freezing level at about 10,000 ft, conditions forecast to slowly improve during the day. When evaluating risk: Flying toward or in improving weather is better than flying toward or in deteriorating conditions.

With my training and experience, I have different personal minimums depending on the environment. I also have confidence in my ability to make the decision to turn around. As

John Hyde puts it, "Cowardice is the better part of valor." (Undoubtedly, an axiom he learned from his Army aviator days.)

How familiar are we with the aircraft? If we've just checked out in a high-performance aircraft, especially without previous experience, are we ready to fly it in minimum weather? Probably not. We would certainly want to consider the personal minimums in Table 27-1, or even raise them, especially for IFR.

Low ceilings and visibilities, even when technically legal, are often an unacceptable risk. Depending on training and experience, low ceilings with good visibilities may be acceptable.

Time of day is another factor to consider. There is no question that flying at night introduces additional challenges and risk.

Physical and psychological conditions have already been discussed. If we're not fit, we shouldn't go. Here's a good example of the application of personal minimums.

CASE STUDY One of our local flight standards operations inspectors had a flight in a Mooney 252 from Hayward to Ukiah in California. There was a low stratus layer over the San Francisco Bay. This individual had thousands of hours as a Navy P3 pilot. Even though this individual was qualified and current, he was not comfortable conducting this IFR operation. I volunteered to fly with him, and we had an uneventful flight.

The following is the decision-making process using personal minimums from Table 27-1, and the Risk Assessment and Management Decision Tree in Fig. 27-2. Recall that we were stuck in Santa Rosa, New Mexico, for 2 days. The weather wasn't much better the following day, but forecast to improve.

IFR flight: High minimum altitudes, low freezing level. I did not have approach charts for Albuquerque. The airplane would be at the limit of its performance envelope. The airplane was equipped for IFR operations, except that it was not certified for flight in icing. We would be at the MEA in probable icing conditions, unable to climb, over mountainous terrain. What alternates were available? None! Risk high. *Decision:* No go.

VFR flight: Plan A–climb to VFR on top and fly to Albuquerque and descend through broken clouds; forecast anyway. Plan B–fly under the clouds and land at Albuquerque. Plan C–fly south, along the railroad to Albuquerque. (For some reason, railroad engineers always select the lowest terrain.) Plan D–return to Santa Rosa. Risk, yes, but plenty of options. For me this was a "go-take-a-look" situation. Why? The area was sparsely populated, and there was good visibility and good weather at the departure airport. On the negative side, I was not familiar with the area, and lack of familiarity has led many pilots to disaster. It was daylight. A night flight, either IFR or VFR under these conditions, would have resulted in a no-go decision.

Airplane performance and equipment were go for the VFR plan. The pilot was fit for flight. *Decision:* Go.

Risk assessment and management do not stop with a go decision. We must reevaluate conditions throughout the operation, from preflight inspection to determining that a particular airport is suitable for landing. Should the airplane be unairworthy—this includes equipment—for the flight, the decision is no go. If conditions at the destination (wind, weather, surface conditions, etc.) change, we may have to divert. If we don't have an alternate plan, the risk is too high, resulting in a no-go decision.

With the preflight complete and 4 1/2 h of fuel, we departed and opened our VFR flight plan to Albuquerque. (A VFR flight plan, especially under these conditions, is part of risk management.) Ceilings were low, but visibility was excellent. It soon became apparent that Plan A, over the clouds, was not going to work. This was confirmed through a conversation with Albuquerque Radio advising that their weather had not improved. *Plan A:* No go.

Plan B–fly under the clouds. Approaching Clines Corners, terrain rises to about 7000 ft. The clouds went right down to the ground! When as pilots do we say no and call it a day?

I teach, or maybe it's preach, that the first time the thought "Should I really be here" or "Maybe I should turn around" occurs to you, you should see a red flag to take positive action at that moment! Don't push the weather, your aircraft, or yourself. Turn around and wait it out. We initiated a 180° turn. We would have been flying from poor weather to worse weather. Risk too high. *Decision:* No go.

At this point I had resigned myself to returning to Santa Rosa—Plan D. However, my wife said, "What about Plan C?" An increased risk accompanied Plan C. There were only a couple of dirt strips with high elevations and short runways for alternates. The terrain was lower, ceilings were low, but visibility remained excellent. For navigation we had the "iron compass" (railroad). I called Albuquerque Radio and changed our route and ETA. As is my practice, I made position reports and updated weather with flight service—another part of risk management. We always had the option of returning to Santa Rosa should the weather deteriorate. Albuquerque did not improve, and we landed short at Alexander, New Mexico. With the weather now improving from the west, the flight continued uneventfully on to Palm Springs.

PUTTING IT ALL TOGETHER

Let's apply our knowledge of weather to some actual flight situations. The following weather event occurred on December 21, 1998. We'll begin with a flight from Goodland to Wichita, Kansas.

Recall the Risk Assessment and Management Decision Tree (Fig. 27-2). Our first step is planning. A review of aeronautical charts, and the *Airport/Facility Directory* (A/FD) or *Terminal Procedures Publication* (TPP), reveals the following:

Goodland: Elevation 3656 ft; RWY 12-30 5419 ft; RWY 17-35 1800 ft (turf); RWY 5-32 3501 ft.

Wichita: Elevation 1332 ft; magnetic variation 7°E; RWY 1L-19R 10,300 ft; RWY 1R-19L 7302 ft; RWY 14/32 6301 ft.

Terrain: Relatively flat, sparsely populated prairie—except for the cities, sloping from about 3500 ft in western Kansas to about 1500 ft around Wichita.

Altitudes: VFR, 500 to 1500 ft AGL and up; IFR, minimum altitudes 5000 to 3000 ft.

Environment: Daylight, midday flight, departing around 18Z.

We have the capability of flying VFR or IFR and can select from a number of airplanes. We can take a Piper Archer or an ice-protected Commander 114. The choice will depend on the weather.

A pictorial view of the weather is often helpful in developing the complete picture. For this we need access to graphic products. With FAA and NWS consolidation, a visit to an aviation weather facility is often not practical. But with DUATs and the Internet, these products are becoming more accessible. Obtaining graphic products prior to the standard briefing provides a general picture of the weather. Like checking terrain, altitudes, and airport information, a preliminary look at the weather provides a helpful background for the preflight briefing.

Figure 27-3 contains a morning weather depiction and radar summary chart for December 21, 1998. An arctic cold front extends southeast of our route. The general weather along the route: departure VFR, enroute MVFR, and destination IFR.

Freezing drizzle in southeast Kansas is a red flag for SLDs. Station models show freezing drizzle behind the front from the Texas panhandle to Missouri. This location coincides with the location of SLDs 25 to 130 mi behind an arctic front, exactly where we would

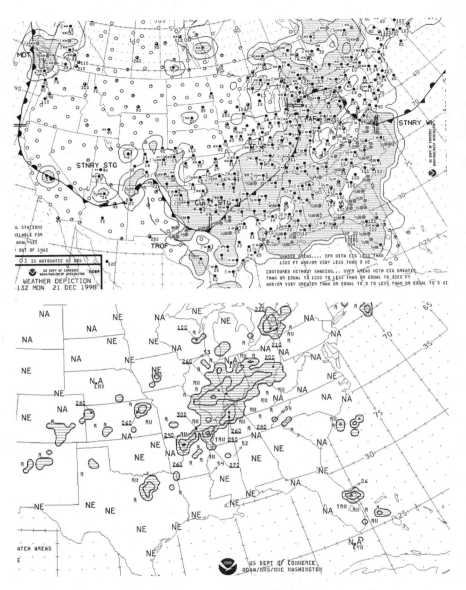

FIGURE 27-3 A pictorial view of the weather is often helpful in developing the complete picture.

expect this phenomenon. The arctic cold front is undercutting and lifting warmer air to the south. This feature is often referred to as *overrunning warm air aloft*. Precipitation falls as liquid into colder air below, then freezes on contact with a surface that is below freezing.

With our knowledge of weather patterns, we would expect improvement from the north during the day—we'll verify this with the area forecast and TAFs. If we're looking for alternates, north appears to be the best bet. Upslope continues over the plains of Colorado, with snow and IFR conditions—not favorable for an alternate. Widespread areas of freezing drizzle and IFR conditions continue south and east of our route. Even though we may find legal alternates in these areas, they might not be the best choice. *Legal* does not necessarily mean *safe*.

The radar chart is encouraging. It depicts scattered light to moderate precipitation, in the form of rain or rain showers. Rain indicates a stable air mass—typically less serious icing. Tops of precipitation in 6000- to 8000-ft range. With relatively stable air, we would expect cloud tops to be within several thousand feet of radar tops, in the 9000- to 12,000-ft range. Since icing tends to be more serious in cloud tops, based on this information, we would want to avoid the 9000- to 12,000-ft altitude range. No convective activity is reported for the route. What about precipitation in the vicinity of Wichita? The radar chart shows the symbol NA. For some reason the data are *not available*. With *no echoes* (NE) reported in Oklahoma and the weather depiction showing light, steady precipitation, we could reasonably expect the conditions shown in western Kansas to extend into the Wichita area. Note that the front is most active from northern Arkansas through the Ohio Valley.

Satellite imagery in Fig. 27-4 confirms our interpretation of the weather depiction and radar summary charts: widespread clouds, but relatively low tops, with no convective activity along the route. What about the clouds over Nebraska? The weather depiction shows mostly clear. These are most likely automated observations, even though not depicted as such, which report maximum cloud heights of 12,000 ft. The visible image shows relatively thin clouds, IR image cold tops. This is most likely a cirroform layer.

With access to the Internet, we can check the experimental icing products.

Experimental Neural Network Icing Products have been developed by the Experimental Forecast Facility (EFF) at the NWS's Aviation Weather Center (www.awc-kc.noaa.gov). They consist of an initial analysis and graphical forecasts out to about 12 h for various levels in the atmosphere.

The following are Neural Network Icing Product output values:

- 0: No icing
- 1: No icing to light icing
- 2: Light icing
- 3: Light to moderate icing
- 4: Moderate icing
- 5: Moderate to severe icing
- 6: Severe icing

Although useful in a general sense, this product is not a substitute for current weather advisories. Icing intensity and coverage normally correlates well with weather advisories. The model, however, cannot take all of the factors available to the forecaster into consideration. Future refinements of the model may be able to consider these factors.

Another experimental product has been developed by the National Center for Atmospheric Research (NCAR), Research Applications Program, Boulder, Colorado (www.ucar.edu/wx.html).

The Stovepipe icing algorithm uses information available from surface observations and three-dimensional gridded fields of temperature, relative humidity, and *geopotential height*—a measure of atmospheric energy—to create a three-dimensional diagnostic of icing conditions. Since new surface observations are available every hour, the algorithm is run hourly.

The algorithm is based on research that has shown that nearly all pilot reports of icing occur in regions of precipitation or overcast cloud conditions. Research has also shown that

IR SATELLITE IMAGE

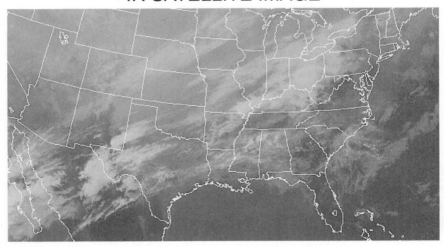

VISIBLE SATELLITE IMAGE

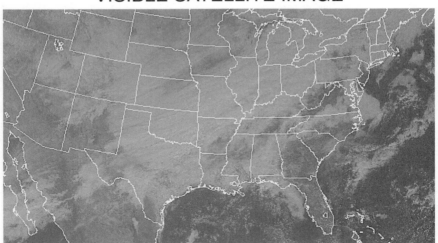

FIGURE 27-4 These satellite images indicate widespread clouds, but relatively low tops, with no convective activity along the route from Goodland to Wichita.

an unusually high number of moderate or greater intensity PIREPs that reported mixed or clear icing occurred in these areas. When these conditions are observed at the surface, precipitation-size supercooled large drops will exist through some depth above the surface.

This information has been employed by the Stovepipe algorithm, which looks for surface observations of freezing drizzle, freezing rain, and ice pellets within 100 km of each *rapid update cycle* (RUC) model grid point. If those conditions are found, the algorithm searches for a narrow range of temperature and relative humidity associated with the moderate or greater intensity PIREPs. These locations are where SLD conditions are likely to exist. If these conditions are not found, the algorithm checks for any other precipitation or overcast sky conditions. If those conditions are found, the algorithm applies a slightly different temperature and relative humidity range associated with PIREP occurrences of less serious icing.

The algorithm improves greatly upon those models that use temperature and relative humidity blindly since the Stovepipe model will diagnose icing in only those locations where at least overcast conditions exist—an advantage over the Neural Network. Basic temperature/dew point algorithms use thresholds of relative humidity that sometimes extend well below 70 percent, and can predict icing in the absence of clouds.

There are five plots given for the observation-based Stovepipe algorithm, as follows:

1. *Surface projection.* This plot shows all the locations where icing has been diagnosed at some level in the model. If red is shown, then SLD was diagnosed at some level, and general icing was probably found at other levels. If blue is shown, then general icing was diagnosed at some level, but SLD was not diagnosed at any level in the model sounding. The locations of all pilot reports of icing are superimposed on this plot so that a real-time evaluation of the algorithm's performance can be done at a glance. A description of the plotted PIREPs is given in the lower left corner of the plot. Real-time verification of the icing algorithm is given in the lower right portion of the plot, including what portion of the icing PIREPs was correctly diagnosed by each portion of the algorithm and one measure of how efficient the algorithm was for that hour.

2. *SLD icing bases.* Here, the altitudes of the base of the SLD layer at each model grid point is color coded by range. The ranges are indicated at the bottom left portion of the plot. The altitude of all PIREPs of moderate or greater intensity icing are given in thousands of feet. These PIREPs are fairly good indicators of aircraft encounters with SLDs.

3. *SLD icing tops.* Here, the altitudes of the top of the SLD layer at each model grid point is color coded by range. The ranges are indicated at the bottom left portion of the plot. The altitude of all moderate or greater intensity icing that was mixed or clear according to PIREPs is given in thousands of feet. These PIREPs are a fairly good indicator of aircraft encounters with SLDs.

4. *General icing bases.* Here, the altitudes of the base of the general icing layer at each model grid point is color coded by range. The ranges are indicated at the bottom left portion of the plot. The altitude of all PIREPs of moderate or greater intensity icing is given in thousands of feet.

5. *General icing tops.* Here, the altitudes of the top of the SLD layer at each model grid point is color coded by range. The ranges are indicated at the bottom left portion of the plot. The altitudes of all PIREPs of moderate or greater intensity icing are given in thousands of feet.

The Neural Network Icing product, surface to 6000-ft composite, is shown in Fig. 27-5. Moderate to severe icing is projected for northern Texas, most of Oklahoma, through southeastern Kansas, Missouri, and Illinois. These conditions are expected along and behind the cold front, where we would expect them. Light to moderate icing is forecast for western Kansas. Again, what we would expect in the colder air to the north.

Figure 27-6 contains the 6000- and 9000-ft Stovepipe icing potential projections. At the 6000-ft level, moderate potential exists over western Kansas and a high potential over southeastern Kansas, and in the same areas depicted by the Neural Network model. The 9000-ft level predicts moderate potential over Kansas, with isolated areas of high potential in the southeast, with the most serious icing at lower levels. Again, what we would expect.

From Fig. 27-7, icing tops for Kansas are forecast in the 15,000- to 20,000-ft range. That's pretty high, but remember that the model cannot take all factors into consideration. Also from Fig. 27-7, SLD tops are expected to be in the 5000- to 10,000-ft range. (I know

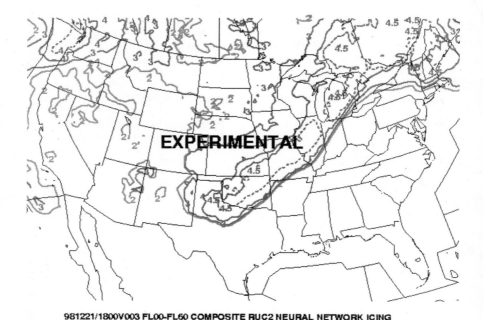

981221/1800V003 FL00-FL60 COMPOSITE RUC2 NEURAL NETWORK ICING

FIGURE 27-5 The Neural Network Icing Model predicts light to moderate icing for western Kansas, becoming more severe around Wichita.

you can't see the color—you'll just have to take my word.) We can conclude from these models that the best escape route would, again, be to the north.

With the preceding background, we're ready for the briefing from either FSS or DUATs. Each will contain essentially the same information. We'll call the FSS or log onto the computer, provide the necessary background information, and request a standard briefing. The DUATs briefing will appear in the following sequence.

The first product displayed is the area forecast:

```
CHIC FA 211045
SYNOPSIS AND VFR CLDS/WX
SYNOPSIS VALID UNTIL 220500
CLDS/WX VALID UNTIL 212300...OTLK VALID 212300-220500
ND SD NE KS MN IA MO WI LM LS MI LH IL IN KY


SYNOPSIS...10Z LOW PRES NRN LM WITH CDFNT TVC-SBN-ARG. HI PRES NERN SD.
BY 05Z CDFNT OVR XTRM ERN KY. HI PRES SERN KS. LK EFFECT SHSN/BLSN OVR
GRTLKS THRU 05Z.


KS
SERN...CIG OVC010-020 TOP 120. OCNL VIS 3-5SM -FZDZ BR. 15-17Z FZDZ BECMG SN.
21Z AGL SCT-BKN020 CIG OVC040 TOP 080. OTLK...VFR.
WRN...CIG BKN010-020 TOP 120. VIS 3-5SM -SN. 18Z AGL SCT-BKN020 CIG BKN040.
OCNL -SN. 22Z AGL SCT-BKN040 BKN100 TOP 150. OTLK...VFR.
CNTRL/NERN...CIG BKN-SCT010-015 OVC030 TOP 100. OCNL VIS 3-5SM -SN. 20Z AGL
SCT025 CIG BKN040 TOP 100. OTLK...VFR.
```

9000-FOOT ICING POTENTIAL

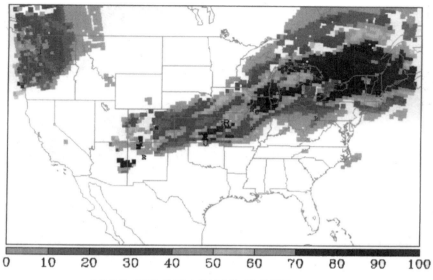

6000-FOOT ICING POTENTIAL

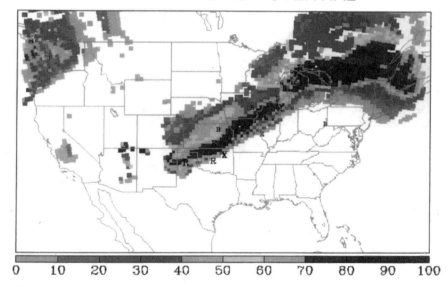

FIGURE 27-6 Not surprisingly, the Stovepipe Model predictions for icing potential agree with the Neural Network Model, and they are consistent with the weather advisories.

ICING LAYER TOPS

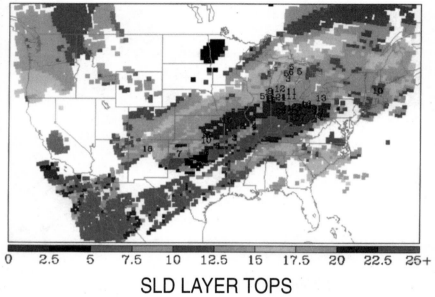

0   2.5   5   7.5   10   12.5   15   17.5   20   22.5   25+

SLD LAYER TOPS

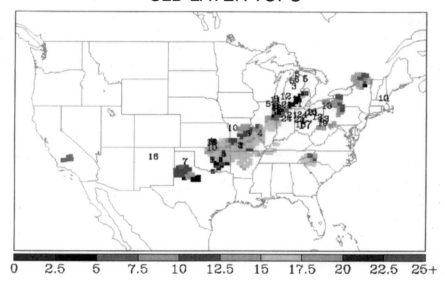

0   2.5   5   7.5   10   12.5   15   17.5   20   22.5   25+

FIGURE 27-7 Remember that the icing models cannot take all factors into consideration; they are not a substitute for current weather advisories.

The synopsis discloses a cold front east of our route in southeastern Missouri, moving eastward, high pressure over northeastern South Dakota. The synopsis confirms our analysis of the weather depiction and radar summary charts. The front is not mentioned in Oklahoma and Texas because it is relatively weak and diffuse and areas are not within the Chicago FA coverage. The front is, however, strongest in the Ohio Valley and Arkansas, even producing some thunderstorms. The front is forecast to move eastward during the period. High pressure is expected to move into southeastern Kansas. This confirms our expectation of better weather to the north.

The route forecast indicates for western Kansas scattered to broken clouds 2000 ft AGL, ceilings 4000 broken, tops 12,000 MSL, occasional light snow until mid to late afternoon; central Kansas 1000 to 1500 scattered to broken, 3000 overcast AGL, tops 10,000, visibility occasionally 3 to 5 mi in light snow; southeastern Kansas 1000 to 2000 AGL, tops 12,000 MSL, visibilities occasionally 3 to 5 mi in moderate freezing drizzle becoming snow by 17Z. Conditions improving during the afternoon. Note how the forecast very closely agrees with our analysis of the experimental icing products and satellite images. The forecast confirms our analysis that in general we'll want to avoid altitudes near the cloud tops, the 9000- to 12,000-ft range, unless we can remain clear of clouds.

Weather advisories are depicted in Fig. 27-8. AIRMET ZULU pertains to the second two-thirds of the flight; occasional moderate mixed or rime icing in clouds and precipitation below 18,000 ft, freezing level at the surface. This is consistent with the general icing tops predicted in the Stovepipe model, icing layer tops (Fig. 27-7). Why no icing in northwest Kansas? The air is too cold. Cloud droplets and precipitation will be frozen. The AIRMET does, however, imply possible SLDs (MXD ICGICIP) within its area of coverage.

The AIRMET refers us to SIGMET PAPA. The SIGMET, represented by the gray shaded area in Fig. 27-8, warns of occasional severe rime or mixed icing in clouds and precipitation below 10,000 ft—a definite indicator of SLDs. Again, this corresponds to the Stovepipe model SLD layer tops (Fig. 27-7). However, the SIGMET states that this condition will be ending after noon (CONDS ENDG AFT 18Z).

Current weather:

METAR KGLD 211653Z 35005KT 7SM OVC040 M22/M27 A3034

METAR KICT 211655Z 360020KT 2 1/2SM -SN OVC016 M13/M15 A3027

Current weather at Goodland is well above the lower limit of the weather depiction/FA VFR category. Surface temperatures are very cold, and wind is not much of a factor. We'll have to make sure there is no snow or ice on the airplane before takeoff. Some type of preheat should be used. Wichita is below FAR basic VFR, with light snow falling, winds out of the north at 20 kn, which could present a problem landing.

Pilot reports:

DDC UA /OV DDC/TM 1818/FLUNKN/TP SW4/RM BA FAIR-POOR/NEG
ICE DURGD

TOP UA /OV KFOE/TM 1732/FL140/TP FA20/TA 0/IC LGT-MOD MXD
025-140/RM FM ZKC

ICT UA /OV ICT/TM 1809/FLUNKN/TP BE9L/SK OVCUNKN-TOP095/IC
TRACE/RM IC DURGC

ICT UA /OV KICT 15SE/TM 1659/FLUNKN/TP DC9/SK OVCUNKN-
TOP092/TA 00 AT 092/IC MDT RIME 050/RM DURGC

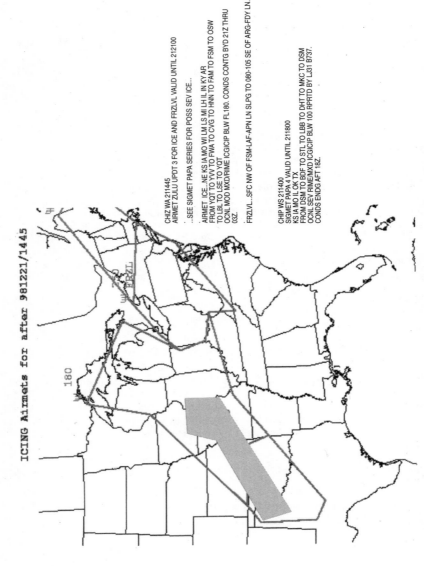

FIGURE 27-8 The gray shade graphically depicts the area described in Chicago SIGMET PAPA 4.

At Dodge City (DDC), southeast of Goodland—in western Kansas—a metroliner reports braking action fair to poor, no ice during descent. This implies a cloud layer, but what are the tops? Unfortunately, this is an example of an incomplete pilot report. Dodge City is just west of the AIRMET area. At Topeka (TOP), northeast Kansas, a Falcon jet reports light to moderate mixed icing from 2500 to 14,000 ft, well within the AIRMET area. This report was made through the Kansas City Center's CWSU (ZKC), but again cloud tops are missing.

The two Wichita pilot reports are extremely helpful. They report cloud tops at 9200 and 9500 ft. The King Air reports a trace of icing during climb; the DC9 experienced light to moderate rime ice at 5000 ft during climb. These reports confirm that the most serious icing has moved to the southeast.

Terminal aerodrome forecasts:

```
TAF KGLD 211722Z 211818 33010KT 3SM -SN BKN015 OVC140 TEMPO
1820 P6SM SCT015 OVC140
FM2100 34008KT P6SM FEW015 SCT140 SCT250
FM0100 32006KT P6SM FEW250


TAF KICT 211722Z 211818 36015G25KT 3SM -SN BR OVC025 TEMPO
1823 1SM -SN BR OVC015
FM2300 35015G25KT P6SM OVC025 TEMPO 2301 SCT015 BKN090
FM0100 33008KT P6SM SKC
```

The Goodland TAF reveals improving conditions throughout the day. (Note that the departure TAF is not normally part of an FSS briefing—except on request.) Goodland would be a good retreat plan should we run into problems. The Wichita TAF also indicates improvement, but with much lower conditions and slower improvement than expected at Goodland. IFR conditions are expected to continue through late afternoon, with northerly, gusty surface winds. Frozen precipitation is predicted through late afternoon. Decreasing frozen precipitation is a positive indicator of lessening serious icing potential. The light snow will have some effect on surface conditions. This needs to be monitored throughout the flight. The weather is expected to clear during the evening.

Wind and temperatures aloft forecast:

	3000	6000	9000	12000
GLD		3310	3418-25	3523-31
ICT	3115	3018-15	3921-21	2825-27

Winds are generally northwesterly averaging 15 to 25 kn, increasing with altitude. They are within our personal minimums (Table 27-1). Temperatures aloft are very cold. Above 9000 ft, the air is too cold for serious icing in stratiform clouds. This confirms our interpretation of no icing in western Kansas—downward vertical motion, cold temperatures, droplets in the form of ice crystals.

Notices to Airmen:

```
!GLD 12/005 17-35 CLSD
!GLD 12/006 5-32 CLSD
!GLD 12/007 12-30 1 IN LSR
!ICT 12/020 14-32 CLSD
!ICT 12/026 1L-19R 1/2 IN SIR
!ICT 12/027 1R-19L 1/2 IN SIR
```

The shorter runways are closed at both Goodland and Wichita. This is typical with a storm system moving through the area. Primary runways are cleared first, then air carrier taxiways and ramps. We'll use extra caution taxiing because taxiways and ramps, especially for general aviation, are typically not cleared as rapidly as those used for air carrier operations.

Let's analyze an IFR flight in an ice-protected airplane.

Takeoff. Runway 12-30 at Goodland has 1 in of loose snow on the runway. Since its length is 5000 ft, we'll select a marker halfway down the runway, at which, should we not be airborne, out of ground effect, with a positive rate of climb, we'll abort the takeoff. The Wichita forecast is well above our personal minimums (Table 27-1). Decision: Go.

Enroute. We have two options: Climb to on top or select an IFR altitude below about 9000 ft. We want to stay out of the tops as much as possible. Decision: Go. We'll check enroute for updates on weather conditions with flight watch. Should we wish to update surface conditions at Wichita, we'll have to contact an FSS. If conditions should deteriorate, we will return to Goodland. Goodland is not only a legal, but an acceptable alternate based on current conditions and the forecast. Why consider current conditions? If there were 8 in of snow and ice on the runways, it wouldn't make any sense to designate Goodland as an alternate, even if the forecast were acceptable.

Landing. We'll plan to stay at our cruising altitude until we can make a continuous descent for landing. Runways at Wichita have 1/2 in of snow and ice. Taking into account that TAF surface winds are true and runways numbered magnetic, from Fig. 8-2 showing crosswind components we calculate a crosswind of approximately 5 to 10 kn. This is acceptable—well within our personal minimums—but we'll keep a close eye on conditions prior to and during landing. We'll request braking action reports and closely monitor surface winds during descent and request a wind check prior to touchdown. Decision: Go.

How about an IFR flight in an airplane without ice protection equipment?

Takeoff. We apply the same analysis as to an ice-protected airplane. Decision: Go.

Enroute. Flying in clouds does not appear to be a viable option. We could fly on top, but we will still have to descend in icing conditions. Any ice would be carried all the way to the ground. Once in flight, even if we attempted to return, we could not reasonably expect to lose all of the ice prior to landing. Even though the weather is within our personal minimums, the lack of ice protection equipment or a positive exit route from any possible ice makes this decision a no-brainer. Decision: No go.

A no-go decision would also be a given if thunderstorms were forecast and the airplane were not equipped with storm detection equipment. Unless we had positive assurance we could see and circumnavigate any convective activity.

Now let's look at a VFR flight in an airplane without ice protection equipment.

Takeoff. Again, we employ the previous risk assessment and management technique, and our personal minimums. Since we're flying over flat terrain, we could logically decrease our ceiling minimums to 2500 ft. Decision: Go.

Enroute. We'll have to stay in VFR conditions. This requires a low-altitude flight below clouds. Ceilings and visibilities are low; however, from the forecast any precipitation should be frozen, in the form of snow.

Personal minimums, and training and experience play important parts of this decision.

Certainly this would be no place for a student or low-time, inexperienced pilot. (However, it might be an opportunity for some takeoff and landing practice on a snow-covered runway, with a qualified instructor.) For a VFR cross-country flight to Wichita, for most pilots, decision: No go.

An experienced pilot might elect to begin the flight. Here is the dreaded "Let's go take a look." There are no inherent additional risks to consider as long as we continue to apply sound risk assessment and management to the flight. In addition to those previously mentioned, consider flying a major highway, such as an interstate. After a snowfall, remember that the landscape will no longer look like the sectional chart. Many landmarks will most likely be covered with snow. Interstate highways are usually cleared of snow first, and as a last resort could be used as an emergency landing area. This reduces risk, but it does eliminate a direct flight to Wichita, but reduces risk.

Do we have a current sectional chart? These are certainly not conditions for flying with a WAC or outdated sectional. Carefully check for towers or other obstacles enroute. Ensure that there are suitable alternate landing fields. Here again, a nondirect route will reduce risk by providing additional suitable alternate landing areas. During the weather briefing, remember to check NOTAMs for all of these airports. Decision: Go.

Enroute risk assessment and management become strong players. We are flying from an area of relatively good weather to poor, but improving, conditions. Should we encounter freezing precipitation, our only option is an immediate retreat. We always should have an alternate in mind, should it become necessary. If the precipitation is snow, don't attempt to penetrate the area unless you can see the other side. Closely monitor the engine for possible carburetor or induction system ice. Check with flight watch or flight service enroute for updates, and provide pilot reports.

Consider the following:

CASE STUDY   The Cessna 172 pilot received a preflight weather briefing that included marginal VFR conditions and reported icing in clouds near the route of flight. The pilot pulled the carburetor heat control to the on position, and descended to about 500 ft AGL to maintain visual contact with the ground. About 1/8 in of ice had formed on the airplane, and the pilot reversed course in an attempt to locate an airport. The flight controls felt "sluggish." The pilot selected a field, configured the airplane, and made a precautionary landing. During the landing, the nose gear sank into the muddy field, then collapsed, and the airplane nosed over.

The NTSB determined the probable cause to be the pilot's continued flight into adverse weather. They sighted as factors the low ceiling, icing conditions, airframe ice, and the muddy field—an alignment of at least four precursors (Fig. 27-1). Unfortunately, this is an example of a situation in which the pilot failed to retreat before entering adverse weather conditions. I know it's easy for me to say, but the goal is to not enter these conditions in the first place!

Landing.   Current conditions and forecasts indicate VFR or special VFR conditions will prevail for Wichita for our arrival time. (Wichita is in class C airspace; certain class B airspace prohibits fixed-wing Special VFR.) Before 2300Z, personal minimums would eliminate all but flight instructor-qualified pilots. The predominate conditions should be VFR. Here again is no place for a student or low-time, inexperienced pilot, or a pilot without adequate aeronautical charts. Do we have an alternate, should Wichita fall below special VFR? There are several suitable alternate airports north and west of Wichita. We would not want to proceed beyond these points without positive assurance of landing at Wichita. If we can meet all of these requirements, then the decision is go.

Thus far we've examined several scenarios for a flight from Goodland to Wichita. What about flying to Wichita from the south, say, Oklahoma City? If this were a morning flight, we would be flying into the teeth of the SIGMET area. If we have ice protection equipment, we might consider a high flight, above the clouds and SLD tops. This would still require a descent into possible severe conditions, for which our aircraft is neither tested nor certified. Therefore, even with ice protection equipment the best, and safest, decision would be to wait for improving weather, which should only be a few hours away. For alternates, we would, again, look north, to the area of improving weather. Without ice protection for an IFR or a flight in VFR conditions, the decision: No go. Our best bet would be to wait until the weather system passed.

Flight into Arkansas and the Ohio Valley presents three additional hazards: widespread IFR conditions, high tops, and thunderstorms. With widespread IFR, a suitable alternate may be beyond the range of most small aircraft. Radar shows precipitation tops well above 20,000 ft. Getting on top may not be possible. The AIRMET, Neural Network, and Stovepipe models predict significant icing to 18,000 ft.

Flight to the west, for example, Denver, would have a different set of hazards. The area is under the influence of upslope—IFR conditions, with snow, and relatively low tops. Aircraft icing is not a serious factor because of the cold temperature. But low ceilings and visibilities, and snow-covered runways, continue to persist and will continue as long as upslope remains a factor. Whiteout would certainly be a factor in these areas. VFR flight would certainly be out. IFR flight would have to be contained with conditions close to, or below, minimums over a relatively widespread area. However, suitable alternates exist to the northeast and east, well within the range of most aircraft.

All pilots should understand the relationship between personal minimums, the fact that most accidents are a combination of events, and the existing techniques for applying risk assessment and management. These techniques will not prevent all accidents, but if they are conscientiously followed, they will result in a healthy dent in the accident record.

APPENDIX A
ABBREVIATIONS

Many abbreviations are used on aviation weather reports and forecasts to save space on telecommunication circuits, in computer equipment, and on charts. The abbreviations will normally be used for any derivative of the root word. However, to prevent confusion, variations are sometimes made by adding the following letters to the contraction of the root word.

able	BL
al	L
ally, erly, ly	LY
ance, ence	NC
ary, ery, ory	RY
der	DR
ed, ied	D
ening	NG
er, ier, or	R
ern	RN
ically	CLY
iest, est	ST
iness, ness	NS
ing	G
ity	TY
ive	V
ment	MT
ous	US
s, es, ies	S
tion, ation	N
ward	WD

The following abbreviations are normally used on aviation weather reports, PIREPs, forecasts, charts, and notices to airmen. The remainder of this appendix is divided into three sections: weather, notices to airmen, and international. To avoid duplication and save space, redundant abbreviations have been eliminated. Therefore, it may be necessary to check more than one section to decode a particular abbreviation.

WEATHER

ABNDT	Abundant
ABNML	Abnormal
ABT	About
ABV	Above
AC	Convective outlook; altocumulus
ACC	Altocumulus castellanus
ACCUM	Accumulate
ACFT	Aircraft
ACLT	Accelerate
ACLTD	Accelerated
ACLTG	Accelerating
ACLTS	Accelerates
ACPY	Accompany
ACRS	Across
ACSL	Altocumulus standing lenticularus
ACTV	Active
ACTVTY	Activity
ACYC	Anticyclone
ADJ	Adjacent
ADL	Additional
ADQT	Adequate
ADQTLY	Adequately
ADRNDCK	Adirondack
ADVCT	Advect
ADVN	Advance
ADVNG	Advancing
ADVY	Advisory
ADVYS	Advisories
AFCT	Affect
AFDK	After dark
AFOS	Automated field operations system
AFT	After
AFTN	Afternoon
AGL	Above ground level
AGN	Again
AGRD	Agreed
AGRMT	Agreement
AHD	Ahead

AK	Alaska
AL	Alabama
ALF	Aloft
ALG	Along
ALGHNY	Allegheny
ALQDS	All quadrants
ALSTG	Altimeter setting
ALTA	Alberta
ALTHO	Although
ALTM	Altimeter
ALUTN	Aleutian
AMD	Amend
AMDD	Amended
AMDG	Amending
AMDT	Amendment
AMP	Amplify
AMPG	Amplifying
AMPLTD	Amplitude
AMS	Air mass
AMT	Amount
ANLYS	Analysis
ANS	Answer
AOA	At or above
AOB	At or below
AP	Anomolous propagation
APCH	Approach
APLCN	Appalachian
APLCNS	Appalachians
APPR	Appear
APRNT	Apparent
APRX	Approximate
AR	Arkansas
ARND	Around
ARPT	Airport
ASAP	As soon as possible
ASL	Above sea level
ASMD	As amended
ASSOCD	Associated
ATLC	Atlantic
ATTM	At this time

ATTN	Attention
AVBL	Available
AVG	Average
AVN	Aviation model
AWT	Awaiting
AZ	Arizona
AZM	Azimuth
BACLIN	Baroclinic
BAJA	Baja California
BATROP	Barotropic
BC	British Columbia
BCH	Beach
BCKG	Backing
BCM	Become
BDA	Bermuda
BDRY	Boundary
BFDK	Before dark
BFR	Before
BGN	Begin
BHND	Behind
BINOVC	Breaks in overcast
BKN	Broken
BLD	Build
BLDG	Building
BLDS	Builds
BLDUP	Buildup
BLKHLS	Black Hills
BLKT	Blanket
BLKTG	Blanketing
BLKTS	Blankets
BLO	Below
BLZD	Blizzard
BND	Bound
BNDRY	Boundary
BNDRYS	Boundaries
BNTH	Beneath
BOOTHEEL	Bootheel
BR	Branch
BRG	Branching
BRS	Branches

BRF	Brief
BRK	Break
BRKG	Breaking
BRKHIC	Breaks in higher clouds
BRKS	Breaks
BRKSHR	Berkshire
BRM	Barometer
BTN	Between
BYD	Beyond
C	Celsius
CA	California
CAA	Cold air advection
CARIB	Caribbean
CASCDS	Cascades
CAVOK	Ceiling and visibility OK
CAVU	Ceiling and visibility unlimited
CB	Cumulonimbus
CC	Cirrocumulus
CCLDS	Clear of clouds
CCLKWS	Counterclockwise
CCSL	Standing lenticular cirrocumulus
CDFNT	Cold front
CDFNTL	Cold frontal
CFP	Cold front passage
CG	Cloud to ground
CHC	Chance
CHG	Change
CHSPK	Chesapeake
CI	Cirrus
CIG	Ceiling
CLD	Cloud
CLDNS	Cloudiness
CLKWS	Clockwise
CLR	Clear
CLRS	Clears
CMPLX	Complex
CNCL	Cancel
CNDN	Canadian
CNTR	Center
CNTRD	Centered

CNTRL	Central
CNTRLN	Centerline
CNTY	County
CNTYS	Counties
CNVG	Converge
CNVGG	Converging
CNVGNC	Convergence
CNVTN	Convection
CNVTV	Convective
CNVTVLY	Convectively
CNFDC	Confidence
CO	Colorado
COMPAR	Compare
COND	Conditions
CONT	Continue
CONTD	Continued
CONTLY	Continually
CONTG	Continuing
CONTRAILS	Condensation trails
CONTS	Continues
CONTDVD	Continental divide
CONUS	Continental United States
COORD	Coordinate
COR	Correction
CPBL	Capable
CRC	Circle
CRCLC	Circulate
CRCLN	Circulation
CRNR	Corner
CRS	Course
CS	Cirrostratus
CSDR	Consider
CST	Coast
CSTL	Coastal
CT	Connecticut
CTGY	Category
CTSKLS	Catskills
CU	Cumulus
CUFRA	Cumulus fractus
CVR	Cover

CYC	Cyclonic
CYCLGN	Cyclogenesis
DABRK	Daybreak
DALGT	Daylight
DBL	Double
DC	District of Columbia
DCR	Decrease
DE	Delaware
DEG	Degree
DELMARVA	Delaware-Maryland-Virginia
DFCLT	Difficult
DFNT	Definite
DFRS	Differs
DFUS	Diffuse
DGNL	Diagonal
DGNLLY	Diagonally
DIGG	Digging
DIR	Direction
DISC	Discontinue
DISRE	Disregard
DKTS	Dakotas
DLA	Delay
DLT	Delete
DLY	Daily
DMG	Damage
DMNT	Dominant
DMSH	Diminish
DNDFTS	Downdrafts
DNS	Dense
DNSLP	Downslope
DNSTRM	Downstream
DNWND	Downwind
DP	Deep
DPND	Deepened
DPNG	Deepening
DPNS	Deepens
DPTH	Depth
DRFT	Drift
DRZL	Drizzle
DSCNT	Descent

DSIPT	Dissipate
DSND	Descend
DSNT	Distant
DSTBLZ	Destabilize
DSTC	Distance
DTRT	Deteriorate
DURG	During
DURN	Duration
DVLP	Develop
DVRG	Diverge
DVV	Downward vertical velocity
DWNDFTS	Downdrafts
DWPNT	Dewpoint
DX	Duplex
E	East
EBND	Eastbound
EFCT	Effect
ELNGT	Elongate
ELSW	Elsewhere
EMBDD	Embedded
EMERG	Emergency
ENCTR	Encounter
ENDG	Ending
ENE	East-northeast
ENELY	East-northeasterly
ENERN	East-northeastern
ENEWD	East-northeastward
ENHNC	Enhance
ENTR	Entire
ERN	Eastern
ERY	Early
ERYR	Earlier
ESE	East-southeast
ESELY	East-southeasterly
ESERN	East-southeastern
ESEWD	East-southeastward
ESNTL	Essential
ESTAB	Establish
ESTS	Estimates
ETA	Eta model

ETC	Et cetera
ETIM	Elapsed time
EVE	Evening
EWD	Eastward
EXCLV	Exclusive
EXCP	Except
EXPC	Expect
EXTD	Extend
EXTN	Extension
EXTRAP	Extrapolate
EXTRM	Extreme
EXTSV	Extensive
F	Fahrenheit
FA	Aviation area forecast
FAH	Fahrenheit
FAM	Familiar
FCST	Forecast
FCSTD	Forecasted
FCSTG	Forecasting
FIG	Figure
FILG	Filling
FIRAV	First available
FL	Florida
FLG	Falling
FLRY	Flurry
FLRYS	Flurries
FLT	Flight
FLW	Follow
FLWG	Following
FM	From
FMT	Format
FNCTN	Function
FNT	Front
FNTGNS	Frontogenesis
FNTL	Frontal
FNTLYS	Frontolysis
FORNN	Forenoon
FPM	Feet per minute
FQT	Frequent
FRM	Form

FROPA	Frontal passage
FROSFC	Frontal surface
FRST	Frost
FRWF	Forecast wind factor
FRZ	Freeze
FRZG	Freezing
FRZN	Frozen
FT	Feet
FTHR	Further
FVRBL	Favorable
FWD	Forward
FYI	For your information
G	Gust
GA	Georgia
GEN	General
GEO	Geographic
GEOREF	Geographical reference
GICG	Glaze icing
GLFALSK	Gulf of Alaska
GLFCAL	Gulf of California
GLFMEX	Gulf of Mexico
GLFSTLAWR	Gulf of St. Lawrence
GND	Ground
GNDFG	Ground fog
GRAD	Gradient
GRDL	Gradual
GRDLY	Gradually
GRT	Great
GRTLKS	Great Lakes
GSTS	Gusts
GSTY	Gusty
GV	Ground visibility
HAZ	Hazard
HCVIS	High clouds visible
HDFRZ	Hard freeze
HDSVLY	Hudson Valley
HDWND	Head wind
HGT	Height
HI	High
HIER	Higher

HIFOR	High-level forecast
HLF	Half
HLSTO	Hailstones
HLTP	Hilltop
HLYR	Haze layer aloft
HND	Hundred
HR	Hour
HRS	Hours
HRZN	Horizon
HTG	Heating
HURCN	Hurricane
HUREP	Hurricane report
HV	Have
HVY	Heavy
HVYR	Heavier
HVYST	Heaviest
HWVR	However
HWY	Highway
IA	Iowa
IC	Ice
ICG	Icing
ICGIC	Icing in clouds
ICGIP	Icing in precipitation
ID	Idaho
IL	Illinois
IMDT	Immediate
IMPL	Impulse
IMPT	Important
INCL	Include
INCR	Increase
INDC	Indicate
INDEF	Indefinite
INFO	Information
INLD	Inland
INSTBY	Instability
INTCNTL	Intercontinental
INTL	International
INTMD	Intermediate
INTMT	Intermittent
INTMTLY	Intermittently

INTR	Interior
INTRMTRGN	Intermountain region
INTS	Intense
INTSFCN	Intensification
INTSFY	Intensify
INTSTY	Intensity
INTVL	Interval
INVRN	Inversion
INVOF	In vicinity of
IOVC	In overcast
IPV	Improve
IPVG	Improving
IR	Infrared
ISOL	Isolate
ISOLD	Isolated
JCTN	Junction
JTSTR	Jet stream
KFRST	Killing frost
KLYR	Smoke layer aloft
KOCTY	Smoke over city
KS	Kansas
KT	Knots
KY	Kentucky
LA	Louisiana
LABRDR	Labrador
LAT	Latitude
LCL	Local
LCTD	Located
LCTMP	Little change in temperature
LCTN	Location
LEVEL	Level
LFM	Limited fine mesh model
LFTG	Lifting
LGRNG	Long range
LGT	Light
LGTR	Lighter
LGWV	Long wave
LI	Lifted index
LK	Lake
LKLY	Likely

LKS	Lakes
LLJ	Low-level jet
LLWAS	Low-level wind shear alert system
LLWS	Low-level wind shear
LMTD	Limited
LN	Line
LN	Lines
LO	Low
LONG	Longitude
LONGL	Longitudnal
LRG	Large
LST	Local standard time
LTD	Limited
LTG	Lightning
LTGCC	Lightning cloud to cloud
LTGCCCG	Lightning cloud to cloud, cloud to ground
LTGCG	Lightning cloud to ground
LTGCW	Lightning cloud to water
LTGIC	Lightning in cloud
LTL	Little
LTLCG	Little change
LTR	Later
LTST	Latest
LV	Leaving
LVL	Level
LVLS	Levels
LWR	Lower
LYR	Layer
MA	Massachusetts
MAN	Manitoba
MAX	Maximum
MB	Millibars
MCD	Mesoscale discussion
MD	Maryland
MDFY	Modify
MDFYD	Modified
MDFYG	Modifying
MDL	Model
MDT	Moderate
ME	Maine

MED	Medium
MEGG	Merging
MESO	Mesoscale
MET	Meteorological
METRO	Metropolitan
MEX	Mexico
MHKVLY	Mohawk Valley
MI	Michigan
MID	Middle
MIDN	Midnight
MIL	Military
MIN	Minimum
MISG	Missing
MLTLVL	Melting level
MN	Minnesota
MNLND	Mainland
MNLY	Mainly
MO	Missouri
MOGR	Moderate or greater
MOV	Move
MPH	Miles per hour
MRGL	Marginal
MRNG	Morning
MRTM	Maritime
MS	Mississippi
MSG	Message
MSL	Mean sea level
MST	Most
MSTR	Moisture
MT	Montana
MTN	Mountain
MULT	Multiple
MULTILVL	Multilevel
MXD	Mixed
N	North
NAB	Not above
NAT	North Atlantic
NATL	National
NAV	Navigation
NB	New Brunswick

NBND	Northbound
NBRHD	Neighborhood
NC	North Carolina
NCWX	No change in weather
ND	North Dakota
NE	Northeast
NEB	Nebraska
NEC	Necessary
NEG	Negative
NELY	Northeasterly
NERN	Northeastern
NEWD	Northeastward
NEW ENG	New England
NFLD	Newfoundland
NGM	Nested grid model
NGT	Night
NH	New Hampshire
NIL	None
NJ	New Jersey
NL	No layers
NLT	Not later than
NLY	Northerly
NM	New Mexico
NMBR	Number
NMC	National Meteorological Center
NML	Normal
NMRS	Numerous
NNE	North-northeast
NNELY	North-northeasterly
NNERN	North-northeastern
NNEWD	North-northeastward
NNNN	End of message
NNW	North-northwest
NNWLY	North-northwesterly
NNWRN	North-northwestern
NNWWD	North-northwestward
NOAA	National Oceanic and Atmospheric Administration
NOPAC	Northern Pacific
NPRS	Nonpersistent
NR	Near

NRN	Northern
NRW	Narrow
NS	Nova Scotia
NTFY	Notify
NV	Nevada
NVA	Negative vorticity advection
NW	Northwest
NWD	Northward
NWLY	Northwesterly
NWRN	Northwestern
NWS	National Weather Service
NXT	Next
NY	New York
OAT	Outside air temperature
OBND	Outbound
OBS	Observation
OBSC	Obscure
OCFNT	Occluded front
OCLD	Occlude
OCLN	Occlusion
OCNL	Occasional
OCR	Occur
OFC	Office
OFP	Occluded frontal passage
OFSHR	Offshore
OH	Ohio
OK	Oklahoma
OMTNS	Over mountains
ONSHR	Onshore
OR	Oregon
ORGPHC	Orographic
ORIG	Original
OSV	Ocean station vessel
OTLK	Outlook
OTP	On top
OTR	Other
OTRW	Otherwise
OUTFLO	Outflow
OVC	Overcast
OVNGT	Overnight

OVR	Over
OVRN	Overrun
OVRNG	Overrunning
OVTK	Overtake
OVTKG	Overtaking
OVTKS	Overtakes
PA	Pennsylvania
PAC	Pacific
PBL	Planetary boundary layer
PCPN	Precipitation
PD	Period
PDMT	Predominant
PEN	Peninsula
PERM	Permanent
PGTSND	Puget Sound
PHYS	Physical
PIBAL	Pilot balloon observation
PIBALS	Pilot balloon reports
PIREP	Pilot weather report
PLNS	Plains
PLS	Please
PLTO	Plateau
PM	Post meridian
PNHDL	Panhandle
POS	Positive
POSLY	Positively
PPINE	PPI no echoes
PPSN	Present position
PRBL	Probable
PRBLTY	Probability
PRECD	Precede
PRES	Pressure
PRESFR	Pressure falling rapidly
PRESRR	Pressure rising rapidly
PRIM	Primary
PRIN	Principal
PRIND	Present indications are
PRJMP	Pressure jump
PROC	Procedure
PROD	Produce

PROG	Forecast
PROGD	Forecasted
PROGS	Forecasts
PRSNT	Present
PRSNTLY	Presently
PRST	Persist
PRSTNC	Persistence
PRSTNT	Persistent
PRVD	Provide
PS	Plus
PSBL	Possible
PSG	Passage
PSN	Position
PTCHY	Patchy
PTLY	Partly
PTNL	Potential
PTNS	Portions
PUGET	Puget Sound
PVA	Positive vorticity advection
PVL	Prevail
PVLT	Prevalent
PWR	Power
QN	Question
QPFERD	NMC excessive rainfall discussion
QPFHSD	NMC heavy snow discussion
QPFSPD	NMC special precipitation discussion
QSTNRY	Quasistationary
QUAD	Quadrant
QUE	Quebec
RADAT	Radiosonde observation data
RAOB	Radiosonde observation
RAOBS	Radiosonde observations
RCH	Reach
RCKY	Rocky
RCKYS	Rockies
RCMD	Recommend
RCRD	Record
RCV	Receive
RDC	Reduce
RDGG	Ridging

RDR	Radar
RDVLP	Redevelop
RE	Regard
RECON	Reconnaissance
REF	Reference
REPL	Replace
REQ	Request
RES	Reserve
RESP	Response
RESTR	Restrict
RGL	Regional model
RGT	Right
RGD	Ragged
RGLR	Regular
RGN	Region
RH	Relative humidity
RI	Rhode Island
RIOGD	Rio Grande
RLBL	Reliable
RLTV	Relative
RLTVLY	Relatively
RMN	Remain
RNFL	Rainfall
ROT	Rotate
RPD	Rapid
RPLC	Replace
RPLCD	Replaced
RPRT	Report
RPT	Repeat
RQR	Require
RS	Receiver station
RSG	Rising
RSN	Reason
RSTR	Restrict
RSTRG	Restricting
RSTRS	Restricts
RTRN	Return
RUF	Rough
RUFLY	Roughly
RVS	Revise

S	South
SASK	Saskatchewan
SATFY	Satisfactory
SBND	Southbound
SBSD	Subside
SBSDD	Subsided
SBSDNC	Subsidence
SBSDS	Subsides
SC	South Carolina
SCND	Second
SCNDRY	Secondary
SCSL	Standing lenticular stratocumulus
SCT	Scatter
SCTR	Sector
SD	South Dakota
SE	Southeast
SEC	Second
SELY	Southeasterly
SEPN	Separation
SEQ	Sequence
SERN	Southeastern
SEWD	Southeastward
SFC	Surface
SFERICS	Atmospherics
SGFNT	Significant
SHFT	Shift
SHLD	Shield
SHLW	Shallow
SHRT	Short
SHRTWV	Short wave
SHUD	Should
SHWR	Shower
SIERNEV	Sierra Nevada
SIG	Signature
SIGMET	Significant meteorological information
SIMUL	Simultaneous
SKC	Sky clear
SKED	Schedule
SLD	Solid
SLGT	Slight

SLO	Slow
SLP	Slope
SLT	Sleet
SLY	Southerly
SM	Statute mile
SMK	Smoke
SML	Small
SMRY	Summary
SMS	Synchronus meteorological satellite
SMTH	Smooth
SMTM	Sometime
SMWHT	Somewhat
SNBNK	Snow bank
SND	Sand
SNFLK	Snowflake
SNGL	Single
SNOINCR	Snow increase
SNOINCRG	Snow increasing
SNST	Sunset
SNW	Snow
SNWFL	Snowfall
SOP	Standard operating procedure
SPCLY	Especially
SPD	Speed
SPDS	Speeds
SPENES	Satellite precipitation estimate statement
SPKL	Sprinkle
SPKLS	Sprinkles
SPLNS	Southern Plains
SPRD	Spread
SPRL	Spiral
SQAL	Squall
SQLN	Squall line
SR	Sunrise
SRN	Southern
SRND	Surround
SRNDG	Surrounding
SS	Sunset
SSE	South-southeast
SSELY	South-southeasterly

SSERN	South-southeastern
SSEWD	South-southeastward
SSW	South-southwest
SSWLY	South-southwesterly
SSWRN	South-southwestern
SSWWD	South-southwestward
ST	Stratus
STAGN	Stagnation
STBL	Stable
STBLTY	Stability
STD	Standard
STDY	Steady
STFR	Stratus fractus
STFRM	Stratiform
STG	Strong
STLT	Satellite
STM	Storm
STN	Station
STNRY	Stationary
SUB	Substitute
SUBTRPCL	Subtropical
SUF	Sufficient
SUFLY	Sufficiently
SUG	Suggest
SUP	Supply
SUPG	Supplying
SUPR	Superior
SUPS	Supplies
SUPSD	Supersede
SVG	Serving
SEV	Severe
SVRL	Several
SW	Southwest
SWD	Southward
SWLG	Swelling
SWLY	Southwesterly
SWOMCD	SELS mesoscale discussion
SWRN	Southwestern
SWWD	Southwestward
SX	Stability index

SXN	Section
SYNOP	Synoptic
SYNS	Synopsis
SYS	System
TCNTL	Transcontinental
TCU	Towering cumulus
TDA	Today
TEMP	Temperature
THD	Thunderhead
THDR	Thunder
THK	Thick
THKNG	Thickening
THKNS	Thickness
THKST	Thickest
THN	Thin
THR	Threshold
THRFTR	Thereafter
THRU	Through
THRUT	Throughout
THSD	Thousand
THTN	Threaten
TIL	Until
TMPRY	Temporary
TMW	Tomorrow
TN	Tennessee
TNDCY	Tendency
TNGT	Tonight
TNTV	Tentative
TOPS	Tops
TOVC	Top of overcast
TPG	Topping
TRBL	Trouble
TRIB	Tributary
TRKG	Tracking
TRML	Terminal
TRNSP	Transport
TROF	Trough
TROFS	Troughs
TROP	Tropopause
TRPCD	Tropical continental

TRPCL	Tropical
TRRN	Terrain
TRSN	Transition
TRW	Thunderstorm
TRW+	Thunderstorm with heavy rain shower
TSFR	Transfer
TSHWR	Thundershower
TSHWRS	Thundershowers
TSNT	Transient
TSQLS	Thundersqualls
TSTM	Thunderstorm
TSTMS	Thunderstorms
TURBC	Turbulence
TURBT	Turbulent
TWD	Toward
TWI	Twilight
TWRG	Towering
TX	Texas
UDDF	Updrafts and downdrafts
UN	Unable
UNAVBL	Unavailable
UNEC	Unnecessary
UNKN	Unknown
UNL	Unlimited
UNRELBL	Unreliable
UNRSTD	Unrestricted
UNSATFY	Unsatisfactory
UNSBL	Unseasonable
UNSTBL	Unstable
UNSTDY	Unsteady
UNSTL	Unsettle
UNUSBL	Unusable
UPDFTS	Updrafts
UPR	Upper
UPSLP	Upslope
UPSTRM	Upstream
URG	Urgent
USBL	Usable
UT	Utah
UVV	Upward vertical velocity

UVVS	Upward vertical velocities
UWNDS	Upper winds
VA	Virginia
VARN	Variation
VCNTY	Vicinity
VCOT	VFR conditions on top
VCTR	Vector
VDUC	VAS Data Utilization Center (NSSFC)
VFY	Verify
VLCTY	Velocity
VLNT	Violent
VLY	Valley
VMC	Visual meteorological conditions
VOL	Volume
VORT	Vorticity
VR	Veer
VRBL	Variable
VRG	Veering
VRISL	Vancouver Island, British Columbia
VRS	Veers
VRT MOTN	Vertical motion
VRY	Very
VSB	Visible
VSBY	Visibility
VSBYDR	Visibility decreasing rapidly
VSBYIR	Visibility increasing rapidly
VT	Vermont
VV	Vertical velocity
W	West
WA	Washington
WAA	Warm air advection
WBND	Westbound
WDLY	Widely
WDSPRD	Widespread
WEA	Weather
WFO	Weather forecast office
WFOS	Weather forecast offices
WFP	Warm front passage
WI	Wisconsin
WIBIS	Will be issued

WINT	Winter
WK	Weak
WKDAY	Weekday
WKEND	Weekend
WKNG	Weakening
WKNS	Weakens
WKR	Weaker
WKN	Weaken
WL	Will
WLY	Westerly
WND	Wind
WNDS	Winds
WNW	West-northwest
WNWLY	West-northwesterly
WNWRN	West-northwestern
WNWWD	West-northwestward
WO	Without
WPLTO	Western Plateau
WRM	Warm
WRMR	Warmer
WRMFNT	Warm front
WRMFNTL	Warm frontal
WRN	Western
WRNG	Warning
WRNGS	Warnings
WRS	Worse
WSHFT	Wind shift
WSHFTS	Wind shifts
WSFO	Weather Service Forecast Office
WSFOS	Weather Service Forecast Offices
WSO	Weather Service Office
WSOS	Weather Service Offices
WSTCH	Wasatch Range
WSW	West-southwest
WSWLY	West-southwesterly
WSWRN	West-southwestern
WSWWD	West-southwestward
WTR	Water
WTRS	Waters
WTSPT	Waterspout

WUD	Would
WV	West Virginia
WVS	Waves
WW	Severe weather watch
WWAMKC	Status report
WWD	Westward
WWS	Severe weather watches
WX	Weather
WY	Wyoming
XCP	Except
XPC	Expect
XPLOS	Explosive
XTND	Extend
XTRM	Extreme
YDA	Yesterday
YKN	Yukon
YLSTN	Yellowstone
ZN	Zone
ZNS	Zones

NOTICES TO AIRMEN (NOTAMs)

ABN	Airport beacon (ICAO)
ABV	Above (ICAO)
ACC	Area control center (ARTCC) (ICAO)
ACCUM	Accumulate (FAA)
ACFT	Aircraft (ICAO)
ACR	Air carrier (FAA)
ACT	Active (ICAO)
ADZD	Advised (ICAO)
AFD	Airport facility directory (FAA)
AGL	Above ground level (ICAO)
ALS	Approach lighting system (ICAO)
ALT	Altitude (ICAO)
ALTM	Altimeter (FAA)
ALTN	Alternate (ICAO)
ALTNLY	Alternately (FAA)
ALSTG	Altimeter setting (FAA)
AMDT	Amendment (ICAO)

AMGR	Airport manager (FAA)
AMOS	Automatic Meteorological Observing System (FAA)
AP	Airport (ICAO)
APCH	Approach (ICAO)
APLGT	Airport lighting (ICAO)
APP	Approach control (ICAO)
ARFF	Aircraft rescue and fire fighting (FAA)
ARR	Arrive, arrival (ICAO)
ASOS	Automatic surface observing system (FAA)
ASPH	Asphalt (ICAO)
ATC	Air traffic control (ICAO)
ATIS	Automatic terminal information service (ICAO)
AUTH	Authority (ICAO)
AUTOB	Automatic weather reporting system (FAA)
AVBL	Available (ICAO)
AWOS	Automatic weather observing/reporting system (FAA)
AWY	Airway (ICAO)
AZM	Azimuth (ICAO)
BA FAIR	Braking action fair (ICAO)
BA NIL	Braking action nil (ICAO)
BA POOR	Braking action poor (ICAO)
BC	Back course (FAA)
BCN	Beacon (ICAO)
BERM	Snowbank(s) containing earth/gravel (FAA)
BLW	Below (ICAO)
BND	Bound (FAA)
BRG	Bearing (ICAO)
BYD	Beyond (FAA)
CAAS	Class A airspace (FAA)
CAT	Category (ICAO)
CBAS	Class B airspace (FAA)
CBSA	Class B surface area (FAA)
CCAS	Class C airspace (FAA)
CCLKWS	Counterclockwise (FAA)
CCSA	Class C surface area (FAA)
CD	Clearance delivery (FAA)
CDAS	Class D airspace (FAA)
CDSA	Class D surface area (FAA)
CEAS	Class E airspace (FAA)
CESA	Class E surface area (FAA)

CFR	Code of Federal Regulations (FAA)
CGAS	Class G airspace (FAA)
CHG	Change or modification (ICAO)
CIG	Ceiling (FAA)
CK	Check (ICAO)
CL	Centerline (ICAO)
CLKWS	Clockwise (FAA)
CLR	Clearance, clear(s), cleared to (ICAO)
CLSD	Closed (ICAO)
CMB	Climb (ICAO)
CMSND	Commissioned (FAA)
CNL	Cancel (ICAO)
COM	Communications (ICAO)
CONC	Concrete (ICAO)
CPD	Coupled (FAA)
CRS	Course (FAA)
CTC	Contact (ICAO)
CTL	Control (ICAO)
DALGT	Daylight (FAA)
DCMSN	Decommission (FAA)
DCMSND	Decommissioned (FAA)
DCT	Direct (ICAO)
DEGS	Degrees (ICAO)
DEP	Depart, departure (ICAO)
DEPPROC	Departure procedure (FAA)
DH	Decision height (ICAO)
DISABLD	Disabled (FAA)
DIST	Distance (ICAO)
DLA	Delay or delayed (ICAO)
DLT	Delete (FAA)
DLY	Daily (FAA)
DME	Distance measuring equipment (ICAO)
DMSTN	Demonstration (FAA)
DP	Dewpoint temperature (ICAO)
DRFT	Snowbank(s) caused by wind action (FAA)
DSPLCD	Displaced (FAA)
E	East (ICAO)
EB	Eastbound (ICAO)
EFAS	Enroute flight advisory service (FAA)
ELEV	Elevation (ICAO)

ENG	Engine (ICAO)
ENRT	Enroute (ICAO)
ENTR	Entire (FAA)
EXC	Except (ICAO)
FAC	Facility or facilities (ICAO)
FAF	Final approach fix (ICAO)
FAN MKR	Fan marker (ICAO)
FDC	Flight data center (FAA)
FI/P	Flight inspection permanent (FAA)
FI/T	Flight inspection temporary (FAA)
FM	From (ICAO)
FNA	Final approach (ICAO)
FPM	Feet per minute (ICAO)
FREQ	Frequency (ICAO)
FRH	Fly runway heading (FAA)
FRZN	Frozen (FAA)
FSS	Automated/flight service station (ICAO)
FT	Foot, feet (ICAO)
GC	Ground control (FAA)
GCA	Ground control approach (ICAO)
GOVT	Government (FAA)
GP	Glide path (ICAO)
GPS	Global Positioning System (FAA)
GRVL	Gravel (ICAO)
HAA	Height above airport (FAA)
HAT	Height above touchdown (FAA)
HDG	Heading (ICAO)
HEL	Helicopter (ICAO)
HELI	Heliport (FAA)
HIRL	High-intensity runway lights (FAA)
HIWAS	Hazardous inflight weather advisory service (FAA)
HLDG	Holding (ICAO)
HOL	Holiday (ICAO)
HR	Hour (ICAO)
IAF	Initial approach fix (ICAO)
IAP	Instrument approach procedure (FAA)
ID	Identification (ICAO)
IDENT	Identify, identifier, identification (ICAO)
IF	Intermediate fix (ICAO)
ILS	Instrument landing system (ICAO)

IM	Inner marker (ICAO)
IMC	Instrument meteorological conditions (ICAO)
IN	Inch, inches (ICAO)
INBD	Inbound (ICAO)
INDEFLY	Indefinitely (FAA)
INFO	Information (ICAO)
INOP	Inoperative (ICAO)
INSTR	Instrument (FAA)
INT	Intersection (ICAO)
INTL	International (ICAO)
INTST	Intensity (ICAO)
IR	Ice on runway(s) (ICAO)
KT	Knots (ICAO)
L	Left (ICAO)
LAA	Local airport advisory (FAA)
LAT	Latitude (ICAO)
LAWRS	Limited Aviation Weather Reporting Station (FAA)
LB	Pound(s) (FAA)
LC	Local control (FAA)
LOC	Local, locally, location (ICAO)
LCTD	Located (FAA)
LDA	Localizer type directional aid (FAA)
LDG	Landing (ICAO)
LGT	Light or lighting (ICAO)
LGTD	Lighted (FAA)
LIRL	Low-intensity runway lights (FAA)
LLWAS	Low-level wind shear alert system (FAA)
LLZ	Localizer (ICAO)
LM	Compass locator at ILS middle marker (ICAO)
LO	Compass locator at ILS outer marker (ICAO)
LONG	Longitude (ICAO)
LRN	Long-range navigation (FAA)
LSR	Loose snow on runway(s) (FAA)
LT	Left turn (FAA)
MAG	Magnetic (ICAO)
MAINT	Maintain, maintenance (ICAO)
MALS	Medium-intensity approach light system (FAA)
MALSF	Medium-intensity approach light system with sequenced flashing indicator lights (FAA)
MALSR	Medium-intensity approach light system with runway alignment (FAA)

MAPT	Missed approach point (ICAO)
MCA	Minimum crossing altitude (ICAO)
MDA	Minimum descent altitude (ICAO)
MEA	Minimum enroute altitude (ICAO)
MED	Medium (FAA)
MIN	Minute(s) (ICAO)
MIRL	Medium-intensity runway lights (FAA)
MLS	Microwave landing system (ICAO)
MM	Middle marker (ICAO)
MNM	Minimum (ICAO)
MOC	Minimum obstruction clearance (ICAO)
MNT	Monitor, monitoring, or monitored (ICAO)
MRA	Minimum reception altitude (ICAO)
MSA	Minimum safe altitude, minimum sector altitude (ICAO)
MSAW	Minimum safe altitude warning (FAA)
MSG	Message (FAA)
MSL	Mean sea level (ICAO)
MU	Mu meters (FAA)
MUD	Mud (FAA)
MUNI	Municipal (FAA)
N	North (ICAO)
NA	Not authorized (FAA)
NAV	Navigation (ICAO)
NB	Northbound (ICAO)
NDB	Nondirectional radio beacon (ICAO)
NE	Northeast (ICAO)
NGT	Night (ICAO)
NM	Nautical mile(s) (ICAO)
NMR	Nautical mile radius (FAA)
NONSTD	Nonstandard (FAA)
NOPT	No procedure turn required (FAA)
NR	Number (ICAO)
NTAP	Notice to airmen publication (FAA)
NW	Northwest (ICAO)
OBSC	Obscured, obscure, or obscuring (ICAO)
OBST	Obstruction, obstacle (ICAO)
OM	Outer marker (ICAO)
OPR	Operate, operator, or operative (ICAO)
OPS	Operation(s) (ICAO)
ORIG	Original (FAA)

OTS	Out of service (FAA)
OVR	Over (FAA)
PAEW	Personnel and equipment working (FAA)
PAPI	Precision approach path indicator (ICAO)
PAR	Precision approach radar (ICAO)
PARL	Parallel (ICAO)
PAT	Pattern (FAA)
PAX	Passenger(s) (ICAO)
PCL	Pilot controlled lighting (FAA)
PERM	Permanent (ICAO)
PJE	Parachute jumping exercise (ICAO)
PLA	Practice low approach (ICAO)
PLW	Plow, plowed (FAA)
PN	Prior notice required (ICAO)
PPR	Prior permission required (ICAO)
PRN	Psuedo-random noise (FAA)
PROC	Procedure (ICAO)
PROP	Propeller (FAA)
PSR	Packed snow on runway(s) (FAA)
PTCHY	Patchy (FAA)
PTN	Procedure turn (ICAO)
PVT	Private (FAA)
RAIL	Runway alignment indicator lights (FAA)
RAMOS	Remote automatic meteorological observing system (FAA)
RCAG	Remote communication air/ground facility (FAA)
RCL	Runway centerline (ICAO)
RCLL	Runway centerline lights (ICAO)
RCO	Remote communication outlet (FAA)
REC	Receive or receiver (ICAO)
RELCTD	Relocated (FAA)
RENL	Runway end lights (ICAO)
REP	Report (ICAO)
RLLS	Runway lead-in light system (ICAO)
RMK	Remark(s) (ICAO)
RMNDR	Remainder (FAA)
RNAV	Area navigation (ICAO)
RPLC	Replace (ICAO)
RQRD	Required (FAA)
RRL	Runway remaining lights (FAA)
RSR	Enroute surveillance radar (ICAO)

RSVN	Reservation (FAA)
RT	Right turn (FAA)
RTE	Route (ICAO)
RTR	Remote transmitter/receiver (FAA)
RTS	Return to service (ICAO)
RUF	Rough (FAA)
RVR	Runway visual range (ICAO)
RVRM	Runway visual range midpoint (FAA)
RVRR	Runway visual range rollout (FAA)
RVRT	Runway visual range touchdown (FAA)
RWY	Runway (ICAO)
S	South (ICAO)
SA	Sand, sanded (ICAO)
SAWRS	Supplementary aviation weather reporting station (FAA)
SB	Southbound (ICAO)
SDF	Simplified directional facility (FAA)
SE	Southeast (ICAO)
SFL	Sequenced flashing lights (FAA)
SID	Standard instrument departure (ICAO)
SIMUL	Simultaneous or simultaneously (ICAO)
SIR	Packed or compacted snow and ice on runway(s) (FAA)
SKED	Scheduled or schedule (ICAO)
SLR	Slush on runway(s) (FAA)
SN	Snow (ICAO)
SNBNK	Snowbank(s) caused by plowing [windrow(s)] (FAA)
SNGL	Single (FAA)
SPD	Speed (FAA)
SSALF	Simplified short approach lighting with sequence flashers (FAA)
SSALR	Simplified short approach lighting with runway alignment indicator lights (FAA)
SSALS	Simplified short approach lighting system (FAA)
SSR	Secondary surveillance radar (ICAO)
STA	Straight-in approach (ICAO)
STAR	Standard terminal arrival (ICAO)
SVC	Service (ICAO)
SVN	Satellite vehicle number (FAA)
SW	Southwest (ICAO)
SWEPT	Swept or broom(ed) (FAA)
T	Temperature (ICAO)
TACAN	Tactical air navigational aid (azimuth and DME) (ICAO)

TAR	Terminal area surveillance radar (ICAO)
TDZ	Touchdown zone (ICAO)
TDZLGT	Touchdown zone lights (ICAO)
TEMPO	Temporary or temporarily (ICAO)
TFC	Traffic (ICAO)
TFR	Temporary flight restriction (FAA)
TGL	Touch-and-go landings (ICAO)
THN	Thin (FAA)
THR	Threshold (ICAO)
THRU	Through (ICAO)
TIL	Until (ICAO)
TKOF	Takeoff (ICAO)
TRML	Terminal (FAA)
TRNG	Training (FAA)
TRSN	Transition (FAA)
TRSN	Transient (FAA)
TWR	Airport control tower (ICAO)
TWY	Taxiway (ICAO)
UNAVBL	Unavailable (FAA)
UNLGTD	Unlighted (FAA)
UNMKD	Unmarked (FAA)
UNMNT	Unmonitored (FAA)
UNREL	Unreliable (ICAO)
UNUSBL	Unusable (FAA)
VASI	Visual approach slope indicator system (ICAO)
VDP	Visual descent point (FAA)
VICE	By way of (FAA)
VIS	Visibility (FAA)
VMC	Visual meteorological conditions (ICAO)
VOL	Volume (FAA)
VOR	VHF omnidirectional radio range (ICAO)
VORTAC	VOR and TACAN (colocated) (ICAO)
W	West (ICAO)
WB	Westbound (ICAO)
WEF	With effect from or effective from (ICAO)
WI	Within (ICAO)
WKDAYS	Monday through Friday (FAA)
WKEND	Saturday and Sunday (FAA)
WND	Wind (FAA)
WPT	Waypoint (ICAO)

WSR   Wet snow on runway(s) (FAA)
WTR   Water on runway(s) (FAA)
WX   Weather (ICAO)

INTERNATIONAL

AAL   Above aerodrome level
ABT   About
ABV   Above
ACC   Area control center or area control
ACCID   Notification of an aircraft accident
ACK   Acknowledge
ACPT   Accept or accepted
ACT   Active or activated or activity
AD   Aerodrome
ADA   Advisory area
ADDN   Addition or additional
ADR   Advisory route
ADVS   Advisory service
ADZ   Advise
AFT   After ... (time or place)
AGL   Above ground level
AGN   Again
ALT   Altitude
AMSL   Above mean sea level
AP   Airport
APR   April
APSG   After passing
ARFOR   Area forecast
ARR   Arrive or arrival
AS   Altostratus
ASC   Ascent to or ascending to
ATP...   At... (time or place)
AUG   August
AVG   Average
B   Blue
BA   Braking action
BASE   Cloud base
BCFG   Fog patches

BDRY	Boundary
BECMG	Becoming
BFR	Before
BKN	Broken
BL...	Blowing
BLO	Below clouds
BLW	Below ...
BR	Mist
BRKG	Braking
BTL	Between layers
C	Center (runway identification) or degrees Celsius
CAT	Category or clear air turbulence
CCA	Corrected meteorological message
CI	Cirrus
CLA	Clear type of ice formation
CM	Centimeter
CNL	Cancel or canceled
COM	Communications
CONS	Continuous
COOR	Coordinate or coordination
COR	Correct or correction or corrected
COT	At the coast
COV	Cover or covered or covering
CUF	Cumuliform
CW	Continuous wave
DEC	December
DEG	Degrees
DEP	Depart or departure
DES	Descend or descending to
DEV	Deviation or deviating
DIF	Diffuse
DIST	Distance
DIV	Divert or diverting
DOM	Domestic
DP	Dewpoint temperature
DR	Low drifting
DRG	During
DS	Duststorm
DTG	Date-time group
DTRT	Deteriorate or deterioring

DU	Dust
DUC	Dense upper cloud
DUR	Duration
DZ	Drizzle
EB	Eastbound
ELEV	Elevation
EMBD	Embedded in a layer
ENE	East north east
ENRT	Enroute
ER	Here…or herewith
EST	Estimate or estimated or estimate
ETA	Estimated time of arrival
ETD	Estimated time of departure
EV	Every
EXC	Except
EXER	Exercises or exercising or to exercise
EXP	Expect or expected or expecting
FAX	Facsimile transmission
FBL	Light (e.g., FBL RA = light rain)
FC	Funnel cloud (tornado or water spout)
FEB	February
FG	Fog
FIC	Flight information center
FIR	Flight information region
FIS	Flight information service
FL	Flight level
FLD	Field
FLG	Flashing
FLT	Flight
FLUC	Fluctuating or fluctuation or fluctuated
FLW	Follow(s) or following
FLY	Fly or flying
FM	From
FPM	Feet per minute
FREQ	Frequency
FRI	Friday
FST	First
FU	Smoke
FZ	Freezing
FZDZ	Freezing drizzle

FZFG	Freezing fog
FZRA	Freezing rain
G	Green
GEN	General
GR	Hail
GRID	Processed meteorological data in the form of grid point values
GS	Small hail and/or snow pellets
HDG	Heading
HEL	Helicopter
HGT	Height or height above
HJ	Sunrise to sunset
HN	Sunset to sunrise
HOL	Holiday
HPA	Hectapascal
HURCN	Hurricane
HVY	Heavy
HYR	Higher
HZ	Haze
IAO	In and out of clouds
IC	Diamond dust (very small ice crystals in suspension)
ICE	Icing
ID	Identifier or identify
IFR	Instrument flight rules
IMC	Instrument meteorological conditions
IMPR	Improve or improving
IMT	Immediate or immediately
INBD	Inbound
INC	In cloud
INFO	Information
INOP	Inoperative
INP	If not possible
INPR	In progress
INSTL	Install or installed or installation
INTL	International
INTRP	Interrupt or interruption or interrupted
INSTF	Intensify or intensifying
INTST	Intensity
IR	Ice on runway
ISA	International standard atmosphere
JAN	January

JTST	Jet stream
JUL	July
JUN	June
KG	Kilograms
KM	Kilometers
KMH	Kilometers per hour
KPA	Kilopascal
KW	Kilowatts
LAM	Logical acknowledgment (message type designator)
LAN	Inland
LAT	Latitude
LDG	Landing
LEN	Length
LGT	Light or lighting
LGTD	Lighted
LMT	Local mean time
LOC	Local or locally or location or located
LONG	Longitude
LSQ	Line squall
LTD	Limited
LV	Light and variable (relating to wind)
M	Mach number (followed by figures)
M	Meters (preceded by figures)
MAP	Aeronautical maps and charts
MAR	March
MAX	Maximum
MAY	May
MCW	Modulated continuous wave
MET	Meteorological or meteorology
METAR	Aviation routine weather report
MIFG	Shallow fog
MIL	Military
MIN	Minutes
MNM	Minimum
MNT	Monitor or monitoring or monitored
MOD	Moderate
MON	Above mountains
MON	Monday
MOV	Move or moving or movement
MS	Minus

MSG	Message
MSL	Mean sea level
MT	Mountain
MTU	Metric units
MTW	Mountain waves
MWO	Meteorological watch office
MX	Mixed type of ice formation
NAT	North Atlantic
NAV	Navigation
NB	Northbound
NBFR	Not before
NC	No change
NE	North-east
NEB	North-eastbound
NEG	No or negative or permission not granted or that is not correct
NGT	Night
NIL	None, or I have nothing to send to you
NM	Nautical miles
NML	Normal
NOSIG	No significant change
NOTAM	A notice to airmen
NOV	November
NR	Number
NS	Nimbostratus
NSC	Nil significant cloud
NSW	Nil significant weather
NWB	North-westbound
NXT	Next
OAC	Oceanic area control center
OBS	Observe or observed or observation
OBSC	Obscure or obscured or obscuring
OBST	Obstacle
OCNL	Occasional or occasionally
OCT	October
OHD	Overhead
OPA	Opaque, white type of ice formation
OPMET	Operational meteorological (information)
OPN	Open or opening or opened
OPR	Operator or operate or operative or operating or operational
OPS	Operations

OTLK	Outlook
OTP	On top
OUBD	Outbound
PARL	Parallel
PCD	Proceed or proceeding
PL	Ice pellets
PLVL	Present level
PO	Dust devils
POSS	Possible
PRI	Primary
PROB	Probability
PROC	Procedure
PS	Plus
PSG	Passing
PSN	Position
PWR	Power
QBI	Compulsory IFR flight
QFE	Atmospheric pressure at aerodrome elevation
QUAD	Quadrant
R	Red or right or restricted area
RA	Rain
RAFC	Regional area forecast center
RAG	Ragged
RCH	Reach or reaching
RE	Recent
REP	Report or reporting or reporting point
REQ	Request or requested
RL	Report leaving
RMK	Remark
ROBEX	Regional OPMET bulletin exchange (scheme)
ROC	Rate of climb
ROD	Rate of descent
ROFOR	Route forecast
RPLC	Replace or replaced
RR	Report reaching
RRA	Delayed meteorological message
RTD	Delayed
RTE	Route
RTN	Return or returned or returning
RVR	Runway visual range

RWY	Runway
SA	Sand
SAP	As soon as possible
SARPS	Standards and Recommended Practices (ICAO)
SAT	Saturday
SB	Southbound
SEB	Southeast bound
SEC	Seconds
SEP	September
SEV	Severe
SFC	Surface
SG	Snow grains
SH	Showers
SIGMET	Information concerning enroute weather phenomena that may affect the safety of aircraft operations
SIGWX	Significant weather
SIMUL	Simultaneous or simultaneously
SKC	Sky clear
SKED	Schedule or scheduled
SLW	Slow
SN	Snow
SPECI	Aviation selected special weather report
SPECIAL	Special meteorological report
SPOT	Spot wind
SQ	Squall
SR	Sunrise
SRY	Secondary
SS	Sandstorm or sunset
STD	Standard
STF	Stratiform
STN	Station
STNR	Stationary
SUN	Sunday
SWB	Southwest bound
T	Temperature
TAF	Aerodrome forecast
TAS	True airspeed
TC	Tropical cyclone
TDO	Tornado
TEMPO	Temporary or temporarily

TEND	Trend forecast
THR	Threshold
THRU	Through
THU	Thursday
TIL	Until
TIP	Until past ... (place)
TL...	Till
TO	To... (place)
TOP	Cloud top
TR	Track
TROP	Tropopause
TS	Thunderstorm
TT	Teletypewriter
TUE	Tuesday
TURB	Turbulence
TYP	Type of aircraft
TYPH	Typhoon
UAB	Until advised by ...
UFN	Until further notice
UIR	Upper flight information region
UNA	Unable
UNL	Unlimited
UNREL	Unreliable
U/S	Unserviceable
UTC	Coordinated Universal Time
VA	Volcanic ash
VAL	In valleys
VC	Vicinity of the aerodrome
VCY	Vicinity
VER	Vertical
VFR	Visual flight rules
VHF	Very high frequency (30 to 300 MHz)
VIP	Very important person
VIS	Visibility
VMC	Visual meteorological conditions
VOLMET	Meteorological information for aircraft in flight
VOR	VHF omnidirectional radio range
VRB	Variable
VSA	By visual reference to the ground
W	White

WAC	World Aeronautical Chart (ICAO) 1:1,000,000
WAFC	World area forecast center
WB	Westbound
WDI	Wind direction indicator
WDSPR	Widespread
WED	Wednesday
WI	Within
WID	Width
WIE	With immediate effect or effective immediately
WILCO	Will comply
WINTEM	Forecast upper wind and temperature
WIP	Work in progress
WKN	Weaken or weakening
WO	Without
WRNG	Warning
WS	Wind shear
WTSPT	Waterspout
WX	Weather
X	Cross
XNG	Crossing
XS	Atmospherics
Y	Yellow
YR	Your

APPENDIX B
GRAPHS AND CHARTS

CELSIUS TO FAHRENHEIT AND DENSITY ALTITUDE

This section contains a Celsius to Fahrenheit conversion chart and approximate density altitude calculation chart and graph. Below is an example of an approximate density altitude calculation.

Recall from Chap. 1 that density altitude is based on field elevation corrected for nonstandard pressure and temperature. Given the following data, calculate approximate density altitude.

Field elevation: 4000 ft

Altimeter setting: 29.73 in Hg

Temperature: 75°F

Since our density altitude graph requires temperature in degrees Celsius and pressure altitude, these will be our first conversions.

From the temperature conversion chart, Fig. B-1, we determine that 75°F equals 24°C.

The pressure altitude chart, Table B-1, allows us to correct for nonstandard pressure. Note that the correction factor for 29.92 in Hg (standard pressure) is zero. Pressure higher than standard results in a pressure altitude lower than field elevation; pressure lower than standard causes pressure altitude to be higher than field elevation. Therefore, add positive

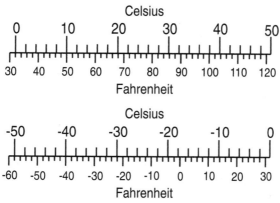

FIGURE B-1 Temperature conversion chart.

TABLE B-1 Pressure Altitude

ALSTG	COR	ALSTG	COR	ALSTG	COR	ALSTG	COR
29.10–29.14	+800	29.50–29.54	+400	29.90–29.94	0	30.30–30.34	−400
29.15–29.19	+750	29.55–29.59	+350	29.95–29.99	−50	30.35–30.39	−450
29.20–29.24	+700	29.60–29.64	+300	30.00–30.04	−100	30.40–30.44	−500
29.25–29.29	+650	29.65–29.69	+250	30.05–30.09	−150	30.45–30.49	−550
29.30–29.34	+600	29.70–29.74	+200	30.10–30.14	−200	30.50–30.54	−600
29.35–29.39	+550	29.75–29.79	+150	30.15–30.19	−250	30.55–30.59	−650
29.40–29.49	+500	29.80–29.84	+100	30.20–30.24	−300	30.60–30.64	−700
29.45–29.49	+450	29.85–29.89	+50	30.25–30.29	−350	30.65–30.70	−750

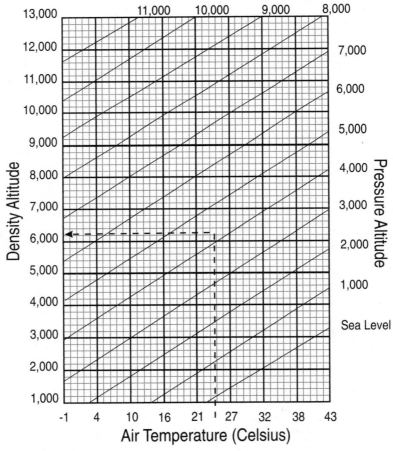

FIGURE B-2 Density altitude.

values to field elevation and subtract negative values.

From the pressure altitude chart, an altimeter setting of 29.73 raises pressure altitude by 200 ft. Therefore, field elevation 4000 ft plus 200 ft correction for pressure results in a pressure altitude of 4200 ft. The results thus far:

Pressure altitude: 4200 ft

Temperature: 24°C

Refer to the density altitude graph, Fig. B-2.

1. Locate the temperature (24°C) on the bottom, horizontal scale.
2. Track up the graph to the intersection of the 4200-ft pressure altitude (diagonal) line, right vertical scale. (This is illustrated by the vertical dashed line on the density altitude graph.)
3. Track left to the density altitude scale, left vertical scale; read approximate density altitude 6200 ft. (This is illustrated by the horizontal dashed line on the density altitude graph.)

CLOUD CHARACTERISTICS

See Table B-2.

WEATHER ADVISORY PLOTTING CHART/LOCATIONS

See Fig. B-3.

The following decodes the locations shown in this figure.

ABI	Abilene, TX
ABQ	Albuquerque, NM
ABR	Aberdeen, SD
ABY	Albany, GA
ACK	Nantucket, MA
ACT	Waco, TX
ADM	Ardmore, OK
AEX	Alexandria, LA
AIR	Bellaire, OH
AKO	Akron, OH
ALB	Albany, NY
ALS	Alamosa, CO
AMA	Amarillo, TX
AMG	Alma, GA
ANW	Ainsworth, NE
APE	Appleton, OH
ARG	Walnut Ridge, AR

TABLE B-2 Cloud Characteristics

Name	Category	Bases, ft	Height	Stability	Precipitation	Extent/LWC
Cirrus (curly)	Cirroform	ABV 20,000	High	Stable	None	Ice crystals
Cirrostratus	Cirroform	ABV 20,000	High	Stable	None	Ice crystals
Cirrocumulus	Cirroform	ABV 20,000	High	Unstable	None	Ice crystals
Altostratus (alto—high)	Stratiform	6500–20,000	Middle	Stable	RA SN	Widespread, solid and liquid water
Altocumulus	Cumuliform	6500–20,000	Middle	Unstable	VIRGA	Limited, solid and liquid water
Stratus (spread out)	Stratiform	SFC–6500	Low	Stable	DZ	Widespread, high LWC
Stratocumulus	Stratiform	SFC–6500	Low	Slightly unstable	DZ – RA	Limited, high LWC
Cumulus (heaped up)	Cumuliform	Near SFC	Vertical development	Unstable	SHRA GS	Limited, high LWC
Nimbostratus (nimbus—rain)	Stratiform	SFC–6500	Low	Stable	– RA RA	Widespread, high LWC
Cumulonimbus	Cumuliform	Near SFC	Vertical development	Unstable	+TSRA +RA GR	Limited, high LWC

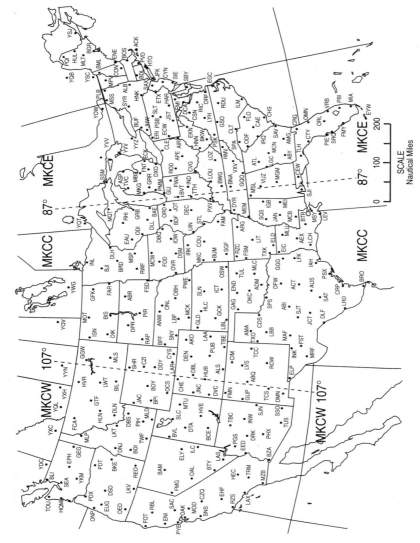

FIGURE B-3 Weather advisory plotting chart.

ASP	Oscoda, MI
ATL	Atlanta, GA
AUS	Austin, TX
BAE	Milwaukee, WI
BAM	Battle Mountain, NV
BCE	Bryce Canyon, UT
BDF	Bradford, IL
BDL	Windsor Locks, CT
BFF	Scottsbluff, NE
BGR	Bangor, ME
BIL	Billings, MT
BIS	Bismarck, ND
BJI	Bemidji, MN
BKE	Baker, OR
BKW	Beckley, WV
BLI	Bellingham, WA
BNA	Nashville, TN
BOI	Boise, ID
BOS	Boston, MA
BOY	Boysen Resv., WY
BPI	Big Piney, WY
BRD	Brainerd, MN
BRO	Brownsville, TX
BTR	Baton Rouge, LA
BTY	Beatty, NV
BUF	Buffalo, NY
BUM	Butler, MO
BVL	Booneville, UT
BVT	Lafayette, IN
BWG	Bowling Green, KY
BZA	Yuma, AZ
CAE	Columbia, SC
CDS	Childress, TX
CEW	Crestview, FL
CHE	Hayden, CO
CHS	Charleston, SC
CIM	Cimarron, NM
CLE	Cleveland, OH
CLT	Charlotte, NC
CON	Concord, NH

COU	Columbia, MO
CRG	Jacksonville, FL
CRP	Corpus Christi, TX
CSN	Cassanova, VA
CTY	Cross City, FL
CVG	Covington, KY
CYN	Coyle, NJ
CYS	Cheyenne, WY
CZI	Crazy Woman, WY
CZQ	Fresno, CA
DBL	Eagle, CO
DBQ	Dubuque, IA
DBS	Dubois, ID
DCA	Washington, DC
DDY	Casper, WY
DEC	Decatur, IL
DEN	Denver, CO
DFW	Dallas-Ft. Worth, TX
DIK	Dickinsin, ND
DLF	Laughlin AFB, TX
DLH	Duluth, MN
DLL	Dells, WI
DLN	Dillon, MT
DMN	Deming, NM
DNJ	McCall, ID
DPR	Dupree, SD
DRK	Prescott, AZ
DSD	Redmond, WA
DSM	Des Moines, IA
DTA	Delta, UT
DVC	Dove Creek, CO
DXO	Detroit, MI
DYR	Dyersburg, TN
EAU	Eau Claire, WI
ECG	Elizabeth City, NC
ECK	Peck, MI
EED	Needles, CA
EHF	Bakersfield, CA
EIC	Shreveport, LA
EKN	Elkins, WV

ELD	El Dorado, AR
ELP	El Paso, TX
ELY	Ely, NV
EMI	Westminster, MD
END	Vance AFB, OK
ENE	Kennebunk, ME
ENI	Ukiah, CA
EPH	Ephrata, WA
ERI	Erie, PA
ETX	East Texas, PA
EUG	Eugene, OR
EWC	Ellwood City, PA
EYW	Key West, FL
FAM	Farmington, MO
FAR	Fargo, ND
FCA	Kalispell, MT
FLO	Florence, SC
FMG	Reno, NV
FMN	Farmington, NM
FMY	Ft. Meyers, FL
FNT	Flint, MI
FOD	Ft. Dodge, IA
FOT	Fortuna, CA
FSD	Sioux Falls, SD
FSM	Ft. Smith, AR
FST	Ft. Stockton, TX
FWA	Ft. Wayne, IN
GAG	Gage, OK
GCK	Garden City, KS
GEG	Spokane, WA
GFK	Grand Forks, ND
GGG	Longview, TX
GGW	Glasgow, MT
GIJ	Niles, MI
GLD	Goodland, KS
GQO	Chattanooga, TN
GRB	Green Bay, WI
GRR	Grand Rapids, MI
GSO	Greensboro, NC
GTF	Great Falls, MT

HAR	Harrisburg, PA
HBU	Gunnison, CO
HEC	Hector, CA
HLC	Hill City, KS
HLN	Helena, MT
HMV	Holston Mountain, TN
HNK	Hancock, NY
HNN	Henderson, WV
HQM	Hoquiam, WA
HTO	East Hampton, NY
HUL	Houlton, ME
HVE	Hanksville, UT
HVR	Havre, MT
IAH	Houston International, TX
ICT	Wichita, KS
IGB	Bigbee, MS
ILC	Wilson Creek, NV
ILM	Wilmington, NC
IND	Indianapolis, IN
INK	Wink, TX
INL	International Falls, MN
INW	Winslow, AZ
IOW	Iowa City, IA
IRK	Kirksville, MO
IRQ	Colliers, SC
ISN	Williston, ND
JAC	Jackson, WY
JAN	Jackson, MS
JCT	Junction, TX
JFK	New York/Kennedy, NY
JHW	Jamestown, NY
JNC	Grand Junction, CO
JOT	Joliet, IL
JST	Johnstown, PA
LAA	Lamar, CO
LAR	Laramie, WY
LAS	Las Vegas, NV
LAX	Los Angeles Intl., CA
LBB	Lubbock International, TX
LBF	North Platte, NE

LBL	Liberal, KS
LCH	Lake Charles, LA
LEV	Grand Isle, LA
LFK	Lufkin, TX
LGC	La Grange, GA
LIT	Little Rock, AR
LKT	Salmon, ID
LKV	Lakeview, OR
LOU	Louisville, KY
LOZ	London, KY
LRD	Laredo, TX
LVS	Las Vegas, NM
LWT	Lewistown, MT
LYH	Lynchburg, VA
MAF	Midland, TX
MBS	Saginaw, MI
MCB	McComb, MS
MCK	McCook, NE
MCN	Macon, GA
MCW	Mason City, IA
MEI	Meridian, MS
MEM	Memphis, TN
MGM	Montgomery, AL
MIA	Miami, FL
MKC	Kansas City, MO
MKG	Muskegon, MI
MLC	McCalester, OK
MLD	Malad City, ID
MLP	Mullan Pass, ID
MLS	Miles City, MT
MLT	Millinocket, ME
MLU	Monroe, LA
MOD	Modesto, CA
MOT	Minot, ND
MPV	Montpelier, VT
MQT	Marquette, MI
MRF	Marfa, TX
MSL	Muscle Shoals, AL
MSP	Minneapolis, MN
MSS	Massena, NY

MSY	New Orleans, LA
MTU	Myton, UT
MZB	Mission Bay, CA
OAK	Oakland, CA
OAL	Coaldale, NV
OBH	Wolbach, NE
OCS	Rock Springs, WY
ODF	Toccoa, GA
ODI	Nodine, MN
OED	Medford, OR
OKC	Oklahoma City, OK
OMN	Ormond Beach, FL
ONL	O'Neil, NE
ONP	Newport, OR
ORD	O'Hare International, IL
ORF	Norfolk, VA
ORL	Orlando, FL
OSW	Oswego, KS
OVR	Omaha, NE
PBI	West Palm Beach, FL
PDT	Pendleton, OR
PDX	Portland, OR
PGS	Peach Springs, AZ
PHX	Phoenix, AZ
PIE	Saint Petersburg, FL
PIH	Pocatello, ID
PIR	Pierre, SD
PLB	Plattsburgh, NY
PMM	Pullman, MI
PQI	Presque Isle, ME
PSB	Phillipsburg, PA
PSK	Dublin, VA
PSX	Palacios, TX
PUB	Pueblo, CO
PVD	Providence, RI
PWE	Pawnee City, NE
PXV	Pocket City, IN
PYE	Point Reyes, CA
RAP	Rapid City, SD
RBL	Red Bluff, CA

RDU	Raleigh-Durham, NC
REO	Rome, OR
RHI	Rhinelander, WI
RIC	Richmond, VA
ROD	Rosewood, OH
ROW	Roswell, NM
RWF	Redwood Falls, MN
RZC	Razorback, AR
RZS	Santa Barbara, CA
SAC	Sacramento, CA
SAT	San Antonio, TX
SAV	Savannah, GA
SAX	Sparta, NJ
SBY	Salisbury, MD
SEA	Seattle, WA
SGF	Springfield, MO
SHR	Sheridan, WY
SIE	Sea Isle, NJ
SJI	Semmnes, AL
SJN	St. Johns, AZ
SJT	San Angelo, TX
SLC	Salt Lake City, UT
SLN	Salina, KS
SLT	Slate Run, PA
SNS	Salinas, CA
SNY	Sidney, NE
SPA	Spartanburg, SC
SPS	Wichita Falls, TX
SQS	Sidon, MS
SRQ	Sarasota, FL
SSM	Sault Ste Marie, MI
SSO	San Simon, AZ
STL	St. Louis, MO
SYR	Syracuse, NY
TBC	Tuba City, AZ
TBE	Tobe, CO
TCC	Tucumcari, NM
TCS	Truth or Consequences, NM
TLH	Tallahassee, FL
TOU	Neah Bay, WA

TRM	Thermal, CA
TTH	Terre Haute, IN
TUL	Tulsa, OK
TUS	Tucson, AZ
TVC	Traverse City, MI
TWF	Twin Falls, ID
TXK	Texarkana, AR
TXO	Texico, TX
UIN	Quincy, IL
VRB	Vero Beach, FL
VUZ	Vulcan, AL
VXV	Knoxville, TN
YDC	Princeton, BC
YKM	Yakima, WA
YOW	Ottawa, ON
YQB	Quebec, QB
YQL	Lethbridge, AB
YQT	Thunder Bay, ON
YQV	Yorkton, SK
YSC	Sherbrooke, QB
YSJ	St. John, NB
YVV	Wiarton, ON
YWG	Winnipeg, MB
YXC	Cranbrook, BC
YXH	Medicine Hat, AB
YYN	Swift Current, SA
YYZ	Toronto, ON

TWEB ROUTES

See Fig. B-4.

NWS WEATHER RADAR CHART/LOCATIONS

See Fig. B-5.
The following decodes the locations shown in this figure.

Alabama

- BMX: Birmingham/Alabaster (WSR-88D/WFO)
- MOB: Mobile Regional Airport, Mobile (WSR-88D/WFO)

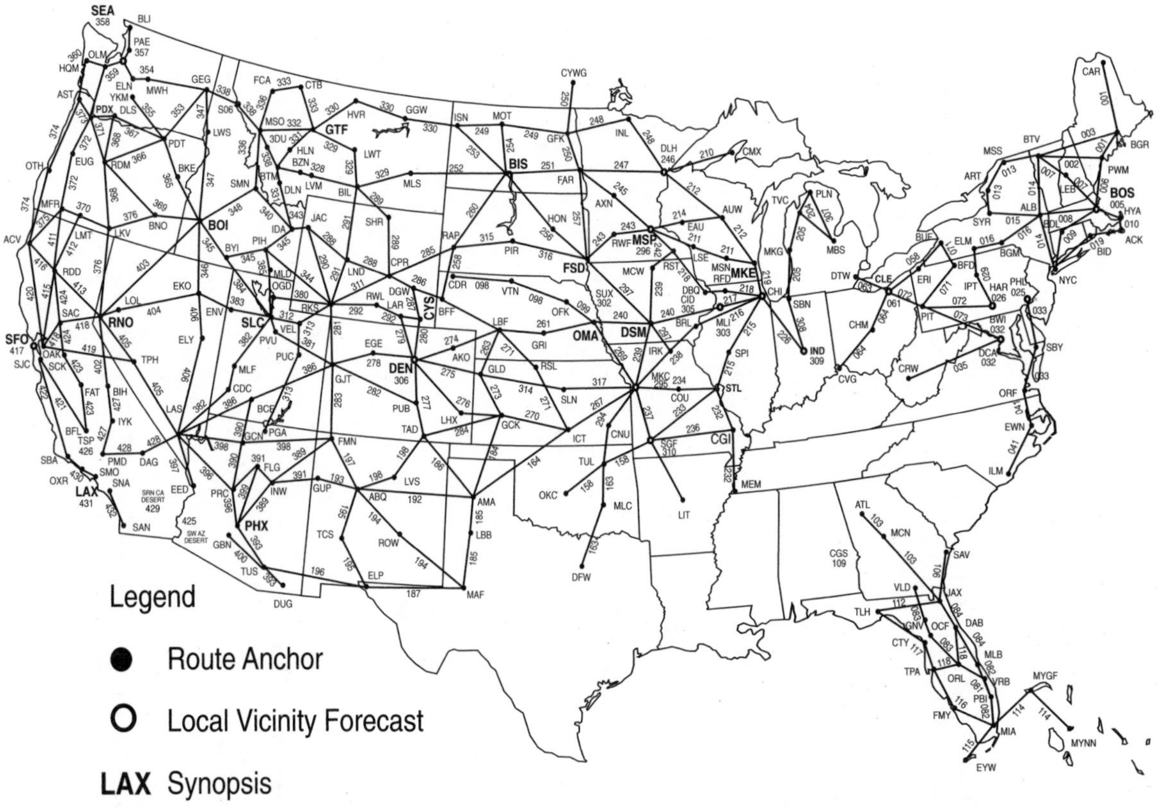

Legend

● Route Anchor

○ Local Vicinity Forecast

LAX Synopsis

FIGURE B-4 TWEB route forecasts.

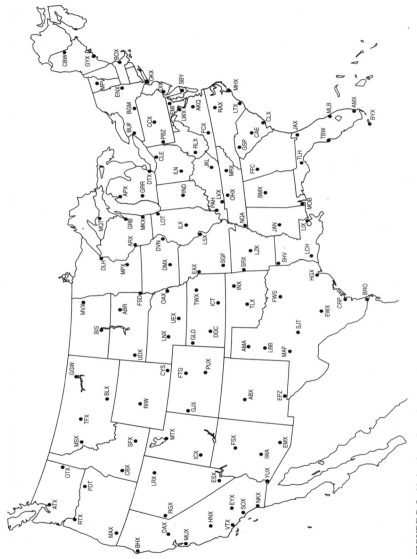

FIGURE B-5 National weather radar network.

Arkansas

- LZK: Little Rock (WSR-88D/WFO)
- SRX: Ft. Smith (WSR-88D)

Arizona

- EMX: Tucson (WSR-88D)
- FSX: Flagstaff (WSR-88D)
- IWA: Williams Gateway Airport, Phoenix (WSR-88D)
- YUX: Yuma (WSR-88D)

California

- BHX: Eureka (WSR-88D)
- DAX: Davis (WSR-88D)
- EYX: Edwards AFB (WSR-88D)
- HNX: San Joaquin/Hanford (WSR-88D/WFO)
- MUX: San Francisco (WSR-88D)
- NKX: Miramar Naval Air Station, San Diego (WSR-88D)
- SOX: Santa Ana Mountains (WSR-88D)
- VTX: Los Angeles (WSR-88D)

Colorado

- FTG: Front Range Airport, Denver (WSR-88D)
- GJX: Grand Junction (WSR-88D)
- PUX: Pueblo (WSR-88D)

Florida

- AMX: Richmond Heights (Miami) (WSR-88D/WFO)
- BYX: Key West (WSR-88D)
- JAX: Jacksonville International Airport (WSR-88D/WSO)
- MLB: Melbourne International Airport (WSR-88D/WSO)
- TBW: Tampa Bay-Ruskin (WSR-88D/WSO)
- TLH: Tallahassee Regional Airport (WSR-88D/WSO)

Idaho

- CBX: Boise (WSR-88D)
- SFX: Pocatello (WSR-88D)

Georgia

- FFC: Peachtree City/Falcon Field Airport (Atlanta) (WSR-88D/WSO)

Illinois

- ILX: Central Illinois, Lincoln (WSR-88D/WFO)
- LOT: Lewis University Airport, Chicago/Romeoville (WSR-88D/WFO)

Indiana

- IND: Indianapolis International Airport, Indianapolis (WSR-88D/WFO)
- IWX: North Webster (WSR-88D)

Iowa

- DMX: Des Moines/Johnston (WSR-88D/WFO)
- DVN: Davenport Municipal Airport, Davenport (WSR-88D/WFO)

Kansas

- DDC: Dodge City Regional Airport, Dodge City (WSR-88D/WFO)
- GLD: Renner Field (Goodland Municipal) Airport, Goodland (WSR-88D/WFO)

- ICT: Wichita Mid-Continent Airport, Wichita (WSR-88D/WFO)
- TWX: Topeka (WSR-88D)

Kentucky

- JKL: Julian Carroll Airport, Jackson (WSR-88D/WFO)
- LVX: Fort Knox (WSR-88D)
- PAH: Barkley Regional Airport, Paducah (WSR-88D/WFO)

Louisiana

- LCH: Lake Charles Regional Airport, Lake Charles (WSR-88D/WFO)
- LIX: New Orleans/Slidell (WSR-88D/WFO)
- SHV: Shreveport Regional Airport, Shreveport (WSR-88D/WFO)

Maine

- CBW: Houlton/Hodgton (WSR-88D)
- GYX: Portland/Gray (WSR-88D/WSO)

Massachusetts

- BOX: Taunton (Boston) (WSR-88D/WSO)

Michigan

- APX: Green Township, Alpena County (WSR-88D/WFO)
- DTX: Detroit/White Lake (WSR-88D/WFO)
- GRR: Kent County International Airport, Grand Rapids (WSR-88D/WFO)
- MQT: Marquette County Airport, Marquette (WSR-88D/WFO)

Minnesota

- DLH: Duluth International Airport, Duluth WSR-88D/WFO
- MPX: Minneapolis/Chanhassen (WSR-88D/WFO)

Mississippi

- JAN: Jackson International Airport, Jackson (WSR-88D/WFO)

Missouri

- EAX: Kansas City/Pleasant Hill (WSR-88D/WFO)
- LSX: St. Louis/Weldon Spring (WSR-88D/WFO)
- SGF: Springfield Regional Airport, Springfield (WSR-88D/WFO)

Montana

- BLX: Billings (WSR-88D)
- GGW: Glasgow International Airport, Glasgow (WSR-88D/WFO)
- MSX: Missoula (WSR-88D)
- TFX: Great Falls (WSR-88D/WFO)

Nebraska

- LNX: North Platte (WSR-88D)
- OAX: Omaha/Valley (WSR-88D/WFO)
- UEX: Blue Hill (Hastings) (WSR-88D/WFO)

Nevada

- ESX: Las Vegas (WSR-88D)
- LRX: Elko (WSR-88D)
- RGX: Reno (WSR-88D)

New Jersey

- DIX: Wrightstown (Philadelphia) (WSR-88D)

New Mexico

- ABX: Albuquerque (WSR-88D)
- EPZ: Santa Teresa (El Paso) (WSR-88D)

New York

- BGM: Binghamton (WSR-88D/WSO)
- BUF: Buffalo (WSR-88D/WSO)
- ENX: State University of New York/East Berne (Albany) (WSR-88D/WSO)
- OKX: Brookhaven National Laboratory (Long Island) (WSR-88D/WSO)

North Carolina

- LTX: Shallotte (Wilmington) (WSR-88D)
- MHX: Morehead City/Newport (WSR-88D/WSO)
- RAX: Raleigh/Durham (WSR-88D)

North Dakota

- BIS: Bismarck Municipal Airport, Bismarck (WSR-88D/WFO)
- MVX: Grand Forks (WSR-88D)

Ohio

- CLE: Cleveland-Hopkins International Airport (WSR-88D/WSO)
- ILN: Wilmington Airborne Airpark Airport (WSR-88D/WSO)

Oklahoma

- INX: Inola (Tulsa) (WSR-88D)
- TLX: Twin Lakes/Midwest City (Oklahoma City) (WSR-88D)

Oregon

- MAX: Medford (WSR-88D)
- PDT: Eastern Oregon Regional at Pendleton Airport, Pendleton (WSR-88D/WFO)
- RTX: Portland (WSR-88D)

Pennsylvania

- CCX: Centre County (State College) (WSR-88D)
- PBZ: Corapolis (Pittsburgh) (WSR-88D/WSO)

South Carolina

- CAE: Columbia (WSR-88D/WSO)
- CLX: Charleston (WSR-88D)
- GSP: Greenville/Spartenburg (WSR-88D/WSO)

South Dakota

- ABR: Aberdeen Regional Airport, Aberdeen (WSR-88D/WFO)
- FSD: Joe Foss Field Airport, Sioux Falls (WSR-88D/WFO)
- UDX: Rapid City (WSR-88D)

Tennessee

- MRX: Knoxille/Morristown (WSR-88D/WFO)
- NQA: Millington Municipal Airport, Millington (Memphis) (WSR-88D)
- OHX: Old Hickory (Nashville) (WSR-88D/WFO)

Texas

- AMA: Amarillo International Airport, Amarillo (WSR-88D/WFO)
- BRO: Brownsville/South Padre Island International Airport, Brownsville (WSR-88D/WFO)

- CRP: Corpus Christi International Airport, Corpus Christi (WSR-88D/WFO)
- EWX: San Antonio/New Braunfels (WSR-88D/WFO)
- FWS: Fort Worth Spinks Airport, Fort Worth (WSR-88D)
- HGX: Dickinson (Houston) (WSR-88D/WFO)
- LBB: Lubbock International Airport, Lubbock (WSR-88D/WFO)
- MAF: Midland International Airport, Midland (WSR-88D/WFO)
- SJT: Mathis Field Airport, San Angelo (WSR-88D/WFO)

Utah

- ICX: Cedar City (WSR-88D)
- MTX: Salt Lake City (WSR-88D)

Virginia

- AKQ: Wakefield (Richmond) (WSR-88D/WSO)
- FCX: Roanoke (WSR-88D)
- LWX: Sterling (Washington, D.C.) (WSR-88D/WSO)

Washington

- ATX: Everett (WSR-88D)
- OTX: Spokane (WSR-88D)

West Virginia

- RLX: Charleston (WSR-88D/WSO)

Wisconsin

- ARX: LaCrosse Ridge (WSR-88D/WFO)
- GRB: Austin Straubel International Airport, Green Bay (WSR-88D/WFO)
- MKX: Milwaukee/Sullivan Township (WSR-88D/WFO)

Wyoming

- CYS: Cheyenne Airport, Cheyenne (WSR-88D/WFO)
- RIW: Riverton Regional Airport, Riverton (WSR-88D/WFO)

AREA FORECAST AREAS OF COVERAGE, STATES

See Fig. B-6 on p. B-20.

AREA FORECAST DESIGNATORS (NW, SW, C, NE, SE)

See Figs. B-7 to B-11 on pp. B-21 to B-25.

ALASKA AND HAWAII AREA FORECAST DESIGNATORS

See Fig. B-12 on p. B-26.

ATLANTIC, CARIBBEAN, AND GULF OF MEXICO AREA FORECAST AREAS OF COVERAGE

See Fig. B-13 on p. B-27.

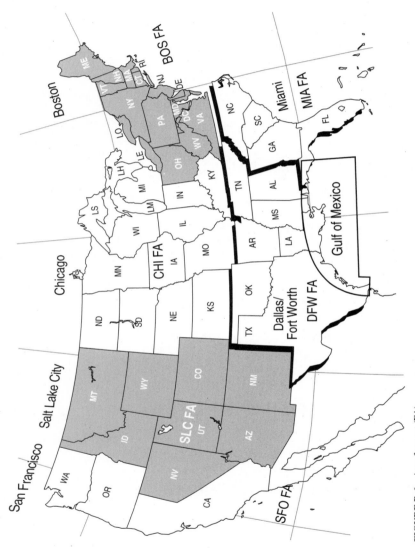

FIGURE B-6 Area forecast (FA).

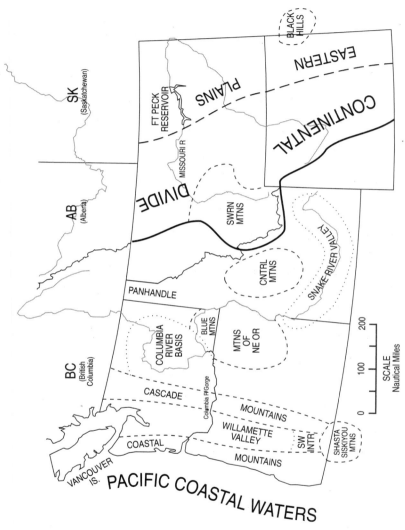

FIGURE B-7 Common geographical designators (northwest).

FIGURE B-8 Common geographical designators (southwest).

FIGURE B-9   Common geographical designators (central).

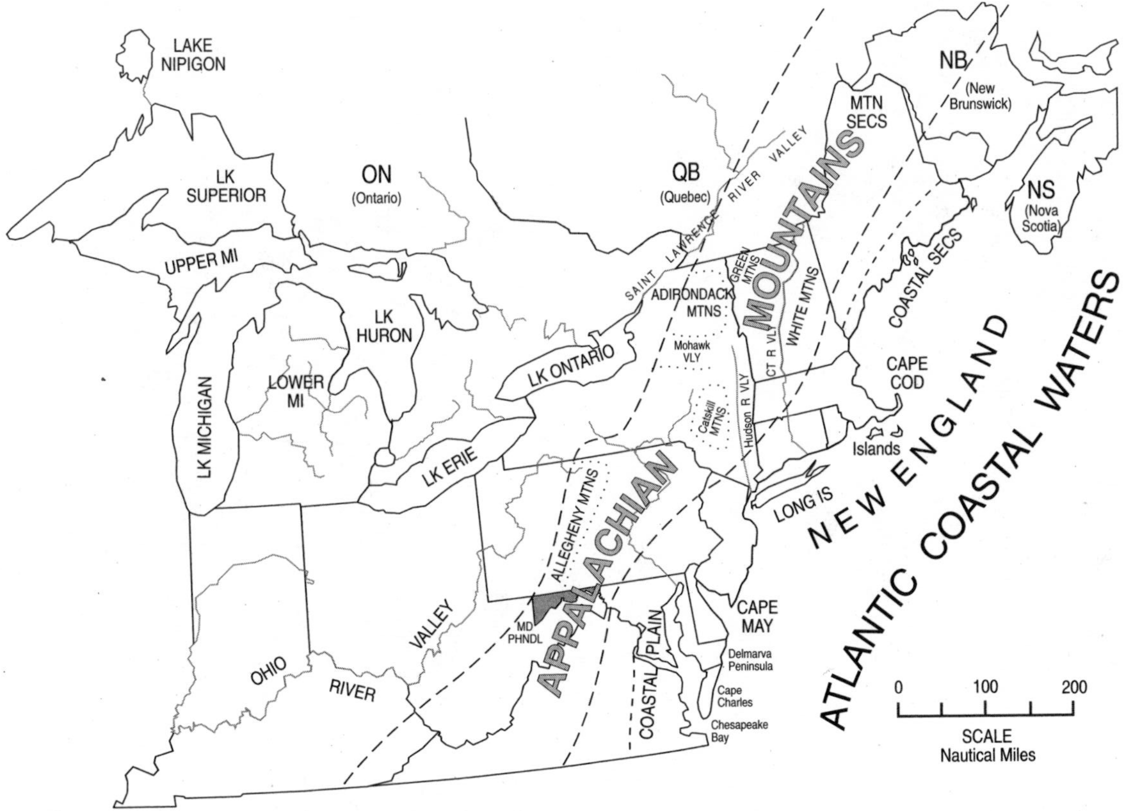

FIGURE B-10 Common geographical designators (northeast).

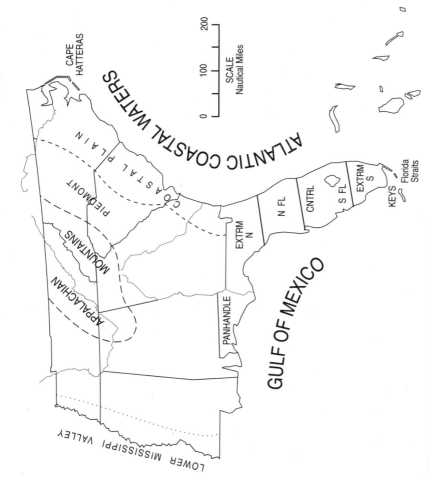

FIGURE B-11 Common geographical designators (southeast).

CAPE HATTERAS

ATLANTIC COASTAL WATERS

COASTAL PLAIN

PIEDMONT

MOUNTAINS

APPALACHIAN

EXTRM N

PANHANDLE

N FL

CNTRL

S FL

EXTRM S

KEYS

Florida Straits

GULF OF MEXICO

LOWER MISSISSIPPI VALLEY

SCALE
Nautical Miles

0   100   200

B-25

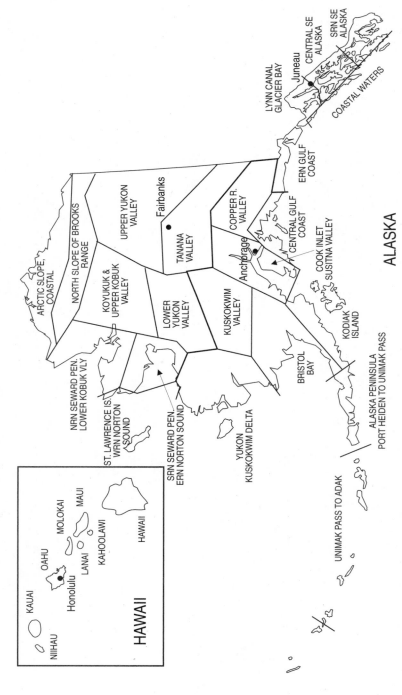

FIGURE B-12 Common geographical designators (Alaska and Hawaii).

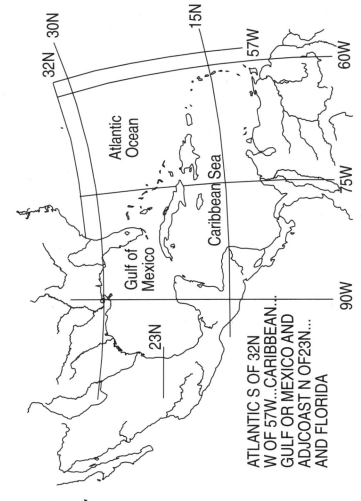

32N 30N

15N

57W

60W

Atlantic
Ocean

75W

Caribbean Sea

Gulf of
Mexico

90W

23N

ATLANTIC S OF 32N
W OF 57W...CARIBBEAN...
GULF OR MEXICO AND
ADJCOAST N OF 23N...
AND FLORIDA

FIGURE B-13 Common geographical designators (Atlantic, Caribbean, and Gulf of Mexico).

GLOSSARY

absolute instability   A condition of the atmosphere in which vertical displacement is spontaneous, whether saturated or unsaturated.

absolute stability   A condition of the atmosphere in which vertical displacement is resisted whether a parcel is saturated or unsaturated.

accretion   The deposit of ice on aircraft surfaces in flight as a result of the tendency of cloud droplets to remain in a liquid state at temperatures below freezing.

adiabatic process   A thermodynamic change of state with no transfer of heat or mass across the boundaries, during which compression always results in heating and expansion always results in cooling.

advection   The process of moving an atmospheric property from one location to another. *Advection* usually refers to the horizontal movement of properties (temperature, moisture, vorticity, etc.).

air mass   A widespread body of air, whose homogeneous properties were established while that air was over a particular region of the Earth's surface and that undergoes specific modifications while moving away from its source region.

AIRMET Bulletin   An inflight weather advisory program intended to provide advance notice of potentially hazardous weather to small aircraft.

air-to-air visibility   The visibility aloft between two aircraft or the aircraft and clouds.

air-to-ground visibility   See *slant range visibility*.

altimeter setting   A value of atmospheric pressure used to correct a pressure altimeter for nonstandard pressure.

anticyclonic   Having a clockwise rotation in the northern hemisphere, associated with the circulation around an anticyclone (high-pressure area).

anti-icing   Prevention of ice from forming on aircraft surfaces.

arc cloud   See *arc line*.

arc line   An arc-shaped line of convective clouds often observed in satellite imagery moving away from a dissipating thunderstorm area.

atmospheric phenomena   As reported on the METAR, weather that is occurring at the station and any obstructions to vision. Obstructions to vision are reported only when the prevailing visibility is less than 7 mi.

atmospheric property   A characteristic trait or peculiarity of the atmosphere such as temperature, pressure, moisture, density, and stability.

augmented   When referring to a surface weather observation, means someone is physically at the site monitoring the equipment and has overall responsibility for the observation.

AUTO   When used in METARs and SPECIs, indicates the report comes from an automated weather observation station that is not augmented.

automated observation   An observation derived without human intervention.

automated surface observing system (ASOS)   A computerized system similar to AWOS but developed jointly by the FAA, NWS, and the Department of Defense; in addition to standard weather elements, the system encodes climatological data at the end of the report.

automated weather observing system (AWOS)   A computerized system that measures sky conditions, visibilities, precipitation, temperatures, dew points, wind, and altimeter settings. It has a voice synthesizer to report minute-by-minute observations over radio frequencies, telephone lines, or local displays.

automatic terminal information service (ATIS)   A recorded service provided at tower-controlled airports to provide the pilot with weather, traffic, and takeoff and landing information.

backscatter device   An instrument used to measure surface visibility at an automated weather observing station.

backup   See *augmentation.*

baroclinic   A state of the atmosphere in which isotherms—lines of equal temperature—cross contours, temperature and pressure gradients are steep, and temperature advection takes place. A baroclinic atmosphere enhances the formation and strengthens the intensity of storms. It is characterized by an upper-level wave one-quarter wavelength behind the surface front.

barotropic   An absence of, or the opposite of, baroclinic. Theoretically, an entirely barotropic atmosphere would yield constant pressure charts with no height or temperature gradients or vertical motion.

berm   A raised bank, usually of frozen snow, along a taxiway or runway.

Bjerknes, Vilhelm   Norwegian meteorologist who developed the polar front theory at the beginning of the twentieth century.

bleed air   Small extraction of hot air from turbine engine compressor.

blocking high   An upper-level area of high pressure that blocks approaching weather systems. See *omega block.*

boundaries   Zones in the lower atmosphere characterized by sharp gradients or discontinuities of temperature, pressure, or moisture and often accompanied by convergence in the wind field. Examples include surface fronts, dry lines, and outflow boundaries. Outflow boundaries are produced by a surge of rain-cooled air flowing outward near the surface from the originating area of convection. In an unstable air mass, thunderstorms tend to develop along these zones and especially at intersections of two or more boundaries.

bridging   A formation of ice over the deicing boot that is not cracked by boot inflation.

BWER/WER/LEWP   Bounded weak echo region/weak echo region/line echo wave pattern. All of these weather radar terms are indicators of strong thunderstorms and the development of severe weather.

calorie   A unit of heat required to raise the temperature of 1 gram of water 1°C.

carburetor ice   Ice formed in the throat of a carburetor due to the effects of lowered temperature by decreased air pressure and fuel vaporization.

celestial dome   A portion of the sky that would be visible provided, due to the absence of artificial structures, that there was an unobstructed view of the horizon in all directions.

Celsius   A temperature scale in which 0 is the melting point of ice and 100 is the boiling point of water.

Chinook   A Native American name given to a foehn wind on the eastern side of the Rockies.

chop   Reported with turbulence, a slight, rapid, and somewhat rhythmic bumpiness without appreciable changes in altitude or attitude.

clear air turbulence (CAT)   Nonconvective, wind shear turbulence occurring at or above 15,000 ft; in practice, the term usually refers to turbulence above 25,000 ft.

clear ice   Glossy, clear, or translucent ice formed by relatively slow freezing of large supercooled droplets. The large droplets spread out over the airfoil prior to complete freezing, forming a sheet of clear ice.

closed low   An area of low pressure aloft completely surrounded by a contour.

closed-cell stratocumulus   Common over water, satellite-viewed oceanic stratocumulus associated with an inversion. These clouds are associated with high pressure.

cloud band   A nearly continuous cloud formation with a distinct long axis, a length-to-width ratio of at least four to one, and a width greater than 1° of latitude (60 nautical miles).

cloud element   The smallest cloud form that can be resolved on satellite imagery from a given satellite system.

cloud line   A narrow cloud band in which individual elements are connected. The cloud line is less than 1° of latitude in width. Indicates strong winds, often 30 kn or greater over water.

cloud shield   A broad cloud formation that is not more than four times as long as it is wide. Often it is formed by cirrus clouds associated with a ridge or the jet stream.

cloud streets   A series of aligned cloud elements that are not connected. Several cloud streets usually line up parallel to each other, and each street is not more than 10 mi wide.

coalescence   The merging of two water drops into a single larger drop.

coefficient of friction   The ratio of the tangential force needed to maintain uniform relative motion between two contacting surfaces (aircraft tires to the pavement surface) to the perpendicular force holding them in contact (distributed aircraft weight to the aircraft tire area). The coefficient is often denoted by the Greek letter μ. It is a simple means used to quantify the relative slipperiness of pavement surfaces. Friction values range from 0 to 100 where zero is the lowest friction value and 100 is the maximum friction value obtainable.

cold-core low   A low-pressure area that intensifies aloft. When this type of low contains closed contours at the 200-mb level, its movements tend to be slow and erratic.

cold pool   Generally refers to an area at 500 mb in which the air temperature is colder than adjacent areas. Other atmospheric conditions being equal, thermodynamic instability is greater beneath cold pools, thus making thunderstorm development more likely.

comma cloud system   A cloud system that resembles the comma punctuation mark. The shape results from differential rotation of the cloud border and upward and downward moving air.

comma head   The rounded portion of the comma cloud system. This region often produces most of the steady precipitation.

comma tail   The portion of the comma cloud that lies to the right of, and often nearly parallel to, the axis of maximum winds.

conditional instability   A condition of the atmosphere in which a parcel will spontaneously rise as a result of its becoming saturated when forced upward.

conduction   The process of transferring energy by means of physical contact.

confluence   A region where streamlines converge. The speed of the horizontal flow will often increase where there is confluence. It is the upper-level equivalent of surface convergence.

Coordinated Universal Time (UTC)   Formerly Greenwich Mean Time, also known as Z or ZULU time, Coordinated Universal Time is the international time standard. UTC is used in all aviation time references.

convergence   Air flowing together near the surface being forced upward due to convergence. It is a vertical motion producer that tends to destabilize the atmosphere near the surface.

convergence zone   An area in which surface convergence is occurring, often associated with a large body of water.

Coriolis force   An apparent force that causes winds to blow across isobars at the surface.

Coriolis, Gaspard de   A French mathematician, who in 1835 developed the theory of an apparent deflection force produced by angular rotation.

COR   A METAR/SPECI report indicating that the original report was transmitted with an error that has now been corrected.

cumuliform Clouds that are characterized by vertical development in the form of rising mounds, domes, or towers, and an unstable air mass.

cutoff low See *closed low*.

cyclongenesis The development or strengthening of a cyclone.

cyclonic Having a counterclockwise rotation in the northern hemisphere. Associated with the circulation around a cyclone or low-pressure area.

deformation zone An area within the atmospheric circulation where air parcels contract in one direction and elongate in the perpendicular direction. The narrow zone along the axis of elongation is called the *deformation zone*. Deformation is a primary factor in frontogenesis and frontolysis.

dendritic pattern A branchy, sawtooth, pattern that identifies areas of snow cover. Mountain ridges above the tree line are essentially barren and snow is visible; in the tree-filled valleys, most of the snow is hidden beneath the trees.

density The weight of air per unit volume.

density altitude Pressure altitude corrected for nonstandard temperature.

dew point The temperature to which air must be cooled, water vapor remaining constant, to become saturated.

dew point front See *dry line*.

diabatic A process that involves the exchange of heat with an external source, or nonadiabatic. The loss may occur through radiation, resulting in fog or low clouds, or conduction through contact with a cold surface.

dig or digging A trough with a strong southerly component of motion. These troughs contain considerable strength and are difficult to forecast accurately.

difluence The spreading apart of adjacent streamlines. The speed of horizontal flow often decreases with a difluent zone. It is the upper-air equivalent of surface divergence and activates or perpetuates thunderstorm development.

direct user access terminal (DUAT) A computer terminal through which pilots can directly access meteorological and aeronautical information, plus file a flight plan without the assistance of an FSS.

divergence The spreading of subsiding air at the surface. Divergence is a downward motion producer that tends to stabilize the atmosphere near the surface.

drainage wind A wind directed down the slope of an incline and caused by greater air density near the slope than at the same levels some distance horizontally from the slope.

dry adiabatic lapse rate The rate at which unsaturated air cools or warms when forced upward or downward (3°C per 1000 ft).

dry snow Snow that has insufficient free water to cause cohesion between individual particles; generally occurs at temperatures well below 0°C.

dry line An area within an air mass that has little temperature gradient but significant differences in moisture. The boundary between the dry and moist air produces a lifting mechanism. Although not a true front, it has the potential to produce hazardous weather. It is also known as a *dew point front*.

dry slot A satellite meteorology term used to describe a cloud feature associated with an upper-level short wave trough. Generally speaking, the cloud system is shaped like a large comma. As the system develops, sinking air beneath the jet stream causes an intrusion of dry, relatively cloud free air on the upwind side of the comma cloud. The air of the intrusion is the dry slot. It is commonly the location where lines of thunderstorms subsequently develop.

embedded thunderstorm A thunderstorm that occurs within nonconvective precipitation. A thunderstorm that is hidden in stratiform clouds.

enhanced infrared (IR) imagery A process by which infrared imagery is enhanced to provide increased contrast between features to simplify interpretation. This enhancement is done by assigning specific shades of gray to specific temperature ranges.

enhanced V   A cloud top signature sometimes seen in enhanced infrared imagery in which the coldest cloud top temperatures form a V shape. Storms that show this cloud top feature are often associated with severe weather.

eutectic temperature/composition   A deicing chemical that melts ice by lowering the freezing point.

eye   An area of clear skies that develops in the center of a tropical storm.

eye wall   The area of thunderstorms that surrounds the eye of a tropical storm.

Fahrenheit   A temperature scale in which 32° is the melting point of ice and 212° is the boiling point of water.

fall streaks   Ice crystals or snowflakes falling from high clouds into dry air where they sublimate directly from a solid to a gas.

fine line/thin line   Weather radar indication of dust or debris that appears as a fine or thin line caused by a dry front or gust front. It indicates the presence of low-level wind shear.

flight level   Pressure altitude given by an altimeter set to standard pressure of 29.92; altitude used in the United States above 17,999 ft.

foehn wind   A warm, dry wind on the lee side of a mountain range.

freezing level   As used in aviation forecasts, the level at which ice melts.

freezing point depressant   A fluid that combines with supercooled water droplets, forming a mixture with a freezing temperature below the ambient air temperature.

front   A boundary between air masses of different temperatures, moistures, and wind.

frontal zone   See *front.*

frontogenesis   The process by which frontal systems are formed.

frontolysis   The process of frontal system dissipation.

general circulation   See *planetary scale.*

Geostationary Operational Environmental Satellite   See *GOES.*

glaciation   The transformation of liquid cloud particles to ice crystals.

global circulation   See *planetary scale.*

Global Positioning System (GPS)   A network of Earth satellites that provides highly accurate position, velocity, and time information to ground-based or airborne receivers.

GOES   Geostationary Operational Environmental Satellites, normally located about 22,000 nautical miles above the equator at 75° W and 135° W. The satellites provide half-hourly visible and infrared imagery.

gravity wind   See *drainage wind.*

ground clutter   Interference of the radar beam due to objects on the ground.

ground icing   Structural icing that occurs on an aircraft on the ground, usually produced by snow or frost.

Great Basin   The area between the Rockies and Sierra Nevada mountains, consisting of southeastern Oregon, southern Idaho, western Utah, and Nevada.

gust front   A low-level windshift line created by the downdrafts associated with thunderstorms and other convective clouds. Acting like a front, these features might produce strong gustiness, pressure rises, and low-level wind shear.

Hadley Cell   A circulation theory describing how low pressure at the equator rises, moves toward the pole, and sinks in high pressure at the pole, then moves toward the equator.

Hadley, George   In 1735, the first scientist to propose a direct thermally driven and zonally symmetric circulation as an explanation for the trade winds.

hail shaft   A shaft of hail detected on weather radar.

heat   The total energy of the motion of molecules with the ability to do work.

heat burst   A rapid, but brief, temperature jump associated with a thunderstorm.

heat capacity   The amount of heat required to raise the temperature of air, or the amount of heat lost when air is cooled.

hectopascal   The international unit of atmospheric pressure is the hectopascal (hPa), equivalent to the millibar (mb).

high   Area of high pressure completely surrounded by lower pressure.

hook echo   Indicates the existence of a mesolow associated with a large thunderstorm cell. Such mesolows are often associated with severe thunderstorms and tornadoes. Hook echoes are not seen on ATC radars.

horizontal extent   The horizontal distance of an icing encounter.

hydroplaning   A condition in which a thin layer of water between the wheel and runway causes the tires to lose contact with the runway.

ice   The solid form of water consisting of a characteristic hexagonal symmetry of water molecules.

ice protection equipment   Equipment required for an aircraft to be certified for flight into known icing conditions.

icephobic liquid   A spray that reduces the adhesion of ice to the deicing boot surface, improving deicing.

impulse   A weak, mid- to upper-level and fast-moving short wave feature that can kick off thunderstorms.

inches of mercury   For aviation purposes, the relationship of atmospheric pressure to inches of mercury (in Hg)—altimeter setting.

indicated airspeed   The airspeed read from the airspeed indicator uncorrected for temperature and pressure.

indicated altitude   The altitude read from the altimeter uncorrected for temperature and pressure.

inflight visibility   See *air-to-air visibility.*

infrared   Satellite imagery that measures the relative temperature of clouds or the Earth's surface.

instrument flight rules (IFR)   Federal regulations that govern flight in instrument meteorological conditions—flight by reference to aircraft instruments.

intensity level   Used to describe radar precipitation intensity on RAREPs (SD) and the radar summary chart.

intermountain region   The area of the western United States, west of the Rocky Mountains and east of the Sierra Nevada mountains, which includes Idaho and Arizona.

international standard atmosphere (ISA)   A hypothetical vertical distribution of atmospheric properties (temperature, pressure, and density). At the surface, the ISA has a temperature of 15°C (59°F), pressure of 1013.2 mb (29.92 in), and a lapse rate of approximately 2°C in the troposphere.

intertropical convergence zone (ITCZ)   The dividing line between the southeast trade winds and the northeast trade winds of the southern and northern hemispheres, respectively.

inversion   A lapse rate at which temperature increases with altitude.

isobars   Lines connecting equal values of surface.

isohumes   Lines of equal relative humidity.

isopleths   Lines of equal number or quantity.

isotachs   Used on charts and graphs, lines of equal wind speed.

isothermal   A constant lapse rate.

isotherms   Used on charts and graphs, lines of equal temperature.

jet streaks   See *jet stream.*

jet stream   A segmented band of strong winds that occur in breaks in the tropopause.

jetlets   See *jet stream.*

katabatic wind   Any wind blowing down an incline.

Kollsman, Paul   A German-born aeronautical engineer who invented the method to correct the altimeter for nonstandard pressure in 1928.

lapse rate   The decrease of an atmospheric variable, usually temperature, with height.

lee-side low   Low-pressure area that develops east of mountain barriers.

latent heat   The amount of heat exchanged (absorbed or released) during the processes of condensation, fusion, vaporization, or sublimation.

liquid water content (LWC)   The total mass of water contained in all the liquid cloud droplets within a unit volume of cloud. LWC is usually expressed in units of grams of water per cubic meter of air (g/m^3).

level of free convection (LFC)   The level at which a parcel of air lifted adiabatically until saturation becomes warmer than its surrounding air, and becomes unstable.

LEWP   See *BWER/WER/LEWP.*

lifted index   A measure of atmospheric instability that is computed on a thermodynamic chart by lifting a parcel of air near the surface to the 500-mb level and subtracting the temperature of the parcel from the temperature of the environment. A negative index means the lifted parcel is buoyant at 500 mb and will continue to rise, which is an unstable condition.

lifted condensation level (LCL)   The level at which a parcel of air lifted adiabatically cools and becomes saturated. The level at which clouds would form.

location identifier   Consisting of three to five alphanumeric characters, location identifiers are contractions used to identify geographical locations, navigational aids, and intersections.

long wave   See *Rossby wave.*

low   An area of low pressure completely surrounded by higher pressure.

low-level wind shear (LLWS)   See *wind shear.*

manual observation   A human-made observation of weather conditions.

mean effective diameter (MED)   The droplet diameter that divides the total water volume present in the droplet distribution in half. Half the water volume will be in larger drops, and half the volume in smaller drops.

median volumetric diameter (MVD)   The droplet diameter that divides in half the total water volume present in the droplet distribution. Half the water volume will be in larger drops, and half the volume, in smaller drops.

mean wind vector   The direction and magnitude of the mean winds from 5000 ft AGL to the tropopause. It can be used to estimate cell movement.

melting level   The temperature at which ice melts; often referred to as the *freezing level.*

mesolow   Also known as *mesocyclone*; a small area of low pressure within a severe thunderstorm. Tornadoes can develop within the vortex.

mesoscale   Small-scale meteorological phenomena that can range in size from that of a single thunderstorm to an area the size of the state of Oklahoma.

mesoscale convective complex (MCC)   A large homogeneous convective weather system on the order of 100,000 m^2. MCCs tend to form during the morning hours.

METAR   Meteorological Aviation Routine surface weather report.

micron (μm)   One-millionth of a meter.

microscale   See *subsynoptic scale.*

millibar (mb)   For aviation purposes, atmospheric pressure related to inches of mercury.

mixed icing conditions   An atmospheric environment in which supercooled liquid water and ice crystals coexist.

mixing ratio   The ratio of water vapor to dry air; expressed in grams of water vapor per kilogram of dry air.

microburst   A small-scale, severe, storm downburst less than two and a half miles across. Reaching the ground, the burst continues as an expanding outflow, producing severe wind shear.

moisture convergence   An objective analysis field combining wind flow convergence and moisture advection. Under certain circumstances, this field is useful for forecasting areas of thunderstorm development.

mixing layer   The layer of the atmosphere, usually within several thousand feet of the surface, in which wind speed and direction are affected by frictional forces with the earth's surface. Also a layer of atmosphere in which the air is thoroughly mixed by convection.

negative tilt   A trough in which the axis in the horizontal plane is tilting from northwest to southeast. These systems tend to cause more weather in California because they bring in warm, moist air.

negative vorticity advection (NVA)   Area of low values of vorticity producing downward vertical motion.

neutral stability   An atmospheric condition in which after a parcel is displaced it remains at rest—even when the displacing force ceases.

NEXRAD   Next-generation Doppler weather radar system.

nucleation   In meteorology, the initiation of either of the phase changes from water vapor to liquid water, or from liquid water to ice.

omega block   A blocking high that on weather charts resembles the Greek letter omega Ω.

open cell   On satellite imagery, a pattern of clear air surrounding individual convective cells.

orographic   A term used to describe the effects caused by terrain, especially mountains, on the weather.

outflow boundary   A surface boundary left by the horizontal spreading of thunderstorm-cooled air. The boundary is often the lifting mechanism needed to generate new thunderstorms.

overrunning   A condition in which air flow from one air mass is moving over another air mass of greater density. The term usually applies to warmer air flowing over cooler air as in a warm frontal situation. It implies a lifting mechanism that can trigger convection in unstable air.

parcel   A small volume of air arbitrarily selected for study. It retains its composition and does not mix with the surrounding air.

Pascal, Blaise   Blaise Pascal (1623–1662) was a French philosopher and mathematician for whom the international unit of atmospheric pressure, the hectopascal (hPa), is named.

patchy conditions   Areas of bare pavement showing through snow- and/or ice-covered pavements. Patches normally show up first along the centerline in the central portion of the runway in the touchdown areas.

partial obscuration   In METAR/SPECI reports, a condition in which between one-eighth and seven-eighths of the sky is hidden by a surface-based obscuring phenomenon.

planetary boundary layer   The frictional layer between the surface and the atmosphere.

planetary scale   Also called *global* or *general circulation*; consists of the jet stream, subtropical high, polar front, and intertropical convergence zone.

polar front   A semipermanent, semicontinuous front separating air masses of tropical and polar origin.

polar jet   The jet stream located at the break between the polar tropopause and subtropical tropopause.

positive tilt   Troughs in which the axis in the horizontal plane is tilting from northeast to southwest.

positive vorticity advection (PVA)   Usually applies to the 500-mb level and refers to areas where the wind flow implies advection from higher values of absolute vorticity to lower values. These areas are presumed to mark zones where upward vertical motion will be supported or enhanced. A vertical motion producer.

precipitation   Any or all of the forms of water particles, whether liquid or solid, that fall from the atmosphere and reach the ground.

pressure   Force per unit area.

pressure altitude   The altitude above the mean sea level constant pressure surface; indicated altitude with the altimeter set to 29.92 in Hg.

prevailing visibility   Visibility reported in manual observations—the greatest distance that can be seen throughout at least half the horizon circle, which need not be continuous.

QFE   Altimeter setting used so that the altimeter will read zero when the aircraft is on the ground.

QNE   Altimeter setting used to obtain pressure altitude.

QNH   Altimeter setting used to obtain indicated altitude.

radiation   The process of transferring energy through space without the aid of a material medium.

rapid update cycle (RUC)   A high-speed computer model that updates every 3 h, designed for short-term forecasting.

RAREP (SD)   Automated, textual radar weather report.

relative humidity   The ratio, expressed as a percentage, of water vapor present in the air compared to the maximum amount the air could hold at its present temperature.

ridge   An elongated area of high pressure.

rime   A white or milky and opaque granular deposit of ice.

rime ice   A rough, milky, opaque ice formed by the instantaneous freezing of small, supercooled droplets as they strike the aircraft.

Rossby waves   Also known as *major waves, planetary waves,* or *long waves;* waves characterized by their large length and significant amplitude. They tend to be slow moving.

runback   Icing that occurs when local heating of accumulated ice melts, causing water to run back to unheated areas and refreeze.

Santa Ana   The local name given to a foehn wind that occurs in late fall, winter, and early summer in southern California.

saturated adiabatic lapse rate   The rate at which saturated air cools or warms when forced upward or downward.

scud   Shreds of small detached clouds moving rapidly below a solid deck of higher clouds.

severe thunderstorm   A thunderstorm that produces winds of 50 kn or more or hail of 3/4-in diameter or greater.

severe wind shear   See *wind shear.*

shear axis   An axis indicating maximum lateral change in wind direction, as in an elongated circulation. This lateral change or shear might be either cyclonic or anticyclonic.

short waves   Wavelengths shorter than long waves. These waves tend to move rapidly through the long wave circulation. They can intensify or dampen weather systems.

showers   Rain showers are characterized by the suddenness with which they start and stop and the rapid changes in their intensity.

SIGMET A significant meteorological advisory that warns of phenomena that affect all aircraft.

slant range visibility The visibility between an aircraft in the air and objects on the ground.

slush Snow that has a water content exceeding its freely drained condition such that it takes on fluid properties (e.g., flowing and splashing).

Sonora storms Local name given to storms that approach southern California from the southeast.

snow A porous, permeable aggregate of ice grains that can be predominately single crystals or close groupings of several crystals.

SPECI A special surface aviation weather report.

squall line An organized line of thunderstorms.

stability The property of an air mass to remain in equilibrium—that is, its ability to resist displacement from its initial position.

stagnation point The point on a surface where the local free-stream velocity is zero; the point of maximum collection efficiency for a symmetric body at zero degrees angle of attack.

standard atmosphere See *international standard atmosphere.*

storm detection equipment Real-time weather radar or lightning detection equipment that can be accessed by airborne planes.

stratiform Clouds of extensive horizontal development, which indicate a stable air mass.

stratosphere The atmospheric layer above the tropopause. It is characterized by a slight increase in temperature with height and the near absence of water vapor. Occasionally severe thunderstorms will break through the tropopause into the stratosphere.

streamline A line that represents the wind flow pattern and that is parallel to the instantaneous velocity. Streamlines indicate the trajectory of the flow.

subsynoptic scale Small, microscale events that often are not within the detection system currently available—for example, tornadoes and microbursts.

sublimation The process by which ice changes directly to water vapor or water vapor directly to ice. The sublimation of ice to vapor is a cooling process and from water vapor to ice is a warming process.

subsidence Downward vertical motion of the air.

subtropical jet The jet around 20° to 30° N latitude at approximately 39,000 ft, located in the break between the midlatitude and tropical tropopause.

supercell A thunderstorm in which updrafts and downdrafts coexist, prolonging the life of the cell, often producing severe thunderstorms and tornadoes.

supercooled Liquid water or water vapor that exists at temperatures below freezing.

surface visibility The visibility at and along the surface of the Earth.

synoptic scale Large-scale weather patterns the size of the migratory high- and low-pressure systems of the lower troposphere with wavelengths on the order of 1000 mi.

TAF Terminal aerodrome forecast.

temperature A measurement of the average speed of molecules.

thermal high An area of high atmospheric pressure caused by the cooling of air by a cold surface. Thermal highs remain relatively stationary over the cold ground.

thermal low An area of low atmospheric pressure caused by intense surface heating. Thermal lows are common to the continental subtropics in summer and remain stationary, and cyclonic circulation is generally weak and diffuse.

thin line See *fine line/thin line.*

Torricelli, Evangelista The Italian inventor of the mercurial barometer in 1643.

total air temperature   The result of ambient air temperature and aerodynamic heating.

towering cumulus   Growing cumulus that resembles a cauliflower but with tops that have not yet reached the cirrus level.

transverse cirrus banding   Irregularly spaced, bandlike cirrus clouds that form nearly perpendicular to a jet stream axis. They indicate turbulence associated with the jet.

tropical cyclone   A general term applied to any low-pressure area that originates over tropical oceans.

tropopause   The boundary between the troposphere and the stratosphere. The tropopause consists of several discrete overlapping leaves, rather than a single continuous surface, and it acts as a lid trapping almost all water vapor in the troposphere. It is marked by a decrease in wind speed and constant temperature with an increase in height.

troposphere   The lower layer of the atmosphere, extending from the surface to an average height of 7 mi. Temperature normally decreases with height, and winds increase with height. It is the layer of the atmosphere where almost all weather occurs.

trough   An elongated area of low pressure.

true altitude   Indicated airspeed corrected for temperature and pressure.

upslope   The orographic effect of air moving up a slope, which tends to cool adiabatically.

vertical motion   Upward or downward motion in the atmosphere.

video integrator and processor (VIP)   Equipment formerly used to indicate precipitation intensity, on a scale of 1 to 6, on the radar summary chart. VIP was used prior to the introduction of NEXRAD.

virga   Rain that evaporates before reaching the surface.

visible moisture   Moisture in the form of clouds or precipitation.

visual flight rules (VFR)   Federal regulations that govern flight in visual meteorological conditions. Flight by reference to the natural horizon and surface.

vortex   In the most general sense, any flow possessing vorticity; more specifically, a flow with closed streamlines.

vorticity (vort) lobe   Usually applied to the 500-mb level; identifies an area of relatively higher values of vorticity. It is synonymous with *short wave trough* or *upper-level impulse*. Generally speaking, there is rising air ahead of the vort lobe and sinking air behind it.

vorticity maximum (vort max)   Usually applied to the 500-mb level; identifies a point along a vorticity lobe at which the absolute vorticity reaches a maximum value.

vorticity   A circulation or rotation within the atmosphere.

wall cloud   See *eye wall.*

warm air advection   A condition in the atmosphere characterized by air flowing from a relatively warmer area to a cooler area. This condition is often accompanied by upward vertical motion that, in the presence of sufficient instability, leads to thunderstorm development.

warm-core low   An area of low pressure that is warmer at its center than at its periphery. Thermal lows and tropical cyclones are examples.

warm nose   A prominent northward bulge of relatively warm air.

water vapor   The invisible water molecules suspended in the air.

wave   A pattern of ridges and troughs in the horizontal flow as depicted on upper-level charts. At the surface, a wave is characterized by a break along a frontal boundary. A center of low pressure is frequently located at the apex of the wave.

WER   See *BWER/WER/LEWP.*

wet ice   An ice surface covered with a thin film of moisture caused by melting, insufficient to cause hydroplaning.

wet snow   Snow that has grains coated with liquid water that bonds the mass together but has no excess water in the pore spaces.

whiteout   An atmospheric optical phenomenon in which the pilot appears to be engulfed in a uniformly white glow. Neither shadows nor horizon nor clouds are discernible; sense of depth and orientation is lost.

wind chill factor   The cooling effect of temperature and wind.

wind field   Winds plotted at a specified level in the atmosphere (surface, 5000 ft, 300 mb, etc.). Wind fields show areas of convergence, divergence, and advection, and thus they provide meteorologists with a valuable forecast tool.

wind shear   Any rapid change in wind direction or velocity. Low-level wind shear (LLWS) is generally shear that occurs within about 2000 ft of the surface. LLWS is classified as severe when a rapid change in wind direction or velocity causes an airspeed change greater than 15 kn or vertical speed change greater than 500 ft/min.

zonal flow   A wind flow that is generally in a west-to-east direction.

INDEX

Pages shown in **boldface** have illustrations on them; with an *italic t*, tables.

ABOUT THE AUTHOR

Terry T. Lankford was a Weather Specialist with the Federal Aviation Administration for nearly 30 years. Now a flight instructor, he is the author of *Weather Reports, Forecasting, and Flight Planning*, now in its Third Edition; *Cockpit Weather Decisions; Understanding Aeronautical Charts*, now in its Second Edition; and *Aircraft Icing*, all published by McGraw-Hill. He lives in Pleasanton, California.